WILEY SERIES IN MICROWAVE AND OPTICAL ENGINEERING

KAI CHANG, Editor
Texas A&M University

FIBER-OPTIC COMMUNICATION SYSTEMS
Govind P. Agrawal

COHERENT OPTICAL COMMUNICATIONS SYSTEMS
Silvello Betti, Giancarlo De Marchis and Eugenio Iannone

HIGH-FREQUENCY ELECTROMAGNETIC TECHNIQUES: RECENT
ADVANCES AND APPLICATIONS
soke K. Bhattacharyya

OMPUTATIONAL METHODS FOR ELECTROMAGNETICS AND MICROWAVES
chard C. Booton, Jr.

CROWAVE SOLID-STATE CIRCUITS AND APPLICATIONS
Chang

LTICONDUCTOR TRANSMISSION-LINE STRUCTURES: MODAL ANALYSIS
HNIQUES
Brandão Faria

OSTRIP CIRCUITS
ardiol

-SPEED VLSI INTERCONNECTIONS: MODELING, ANALYSIS,
IMULATION
oel

REQUENCY ANALOG INTEGRATED CIRCUIT DESIGN
r Goyal (ed.)

COMPUTING: AN INTRODUCTION
ad A. Karim and Abdul Abad S. Awwal

AVE DEVICES, CIRCUITS AND THEIR INTERACTION
Lee and G. Conrad Dalman

S FOR RADAR AND COMMUNICATIONS: A POLARIMETRIC
H

S AND THEIR APPLICATIONS
ain (ed.)

MULTICONDUCTOR TRANSMISSION LINES
ul

ON TO ELECTROMAGNETIC COMPATIBILITY
l

AL PROCESSING, COMPUTING AND NEURAL NETWORKS
and Suganda Jutamulia

Solar Cells and Their Applications

Solar Cells and Their Applications

Edited by

LARRY D. PARTAIN
Edward L. Ginzton Research Center
Palo Alto, California

A WILEY-INTERSCIENCE PUBLICATION

JOHN WILEY & SONS, INC.

New York / Chichester / Brisbane / Toronto / Singapore

This text is printed on acid-free paper.

Library of Congress Cataloging in Publication Data:

Solar cells and their applications / edited by Larry D. Partain.
p. cm. -- (Wiley series in microwave and optical engineering)
Includes index.
ISBN 0-471-57420-1
1. Solar cells. I. Partain, L. D. II. Series.
TK2960.S652 1995
621.31'244--dc20 94-29987

Printed in the United States of America

10 9 8 7 6 5 4 3 2 1

To my wife Deborah
and daughters Lauren and Catherine

Contents

PART FOUR OTHER ASPECTS

PART FIVE SUMMARY

Contributors

A. CATALANO
National Renewable Energy Laboratory, Golden, CO 80401

E.A. DeMEO
Electric Power Research Institute, Palo Alto, CA 94303

F.J. DOSTALEK
Electric Power Research Institute, Palo Alto, CA 94303

DENNIS J. FLOOD
NASA Lewis Research Center, Cleveland, OH 44135

LEWIS M. FRAAS
JX Crystals, Inc., Issaquah, WA 98027

F.R. GOODMAN, JR.
Electric Power Research Institute, Palo Alto, CA 94303

MARTIN A. GREEN
Centre for Photovoltaic Devices and Systems, University of New South Wales, Kensington, NSW 2033, Australia

STEPHEN J. HOGAN
Spire Corporation, Bedford, MA 01730

ROBERT HILL
Newcastle Photovoltaics Application Centre, University of Northumbria, Newcastle-upon-Tyne NE1 8ST, England

P.A. ILES
Applied Solar Energy Corporation, Industry, CA 91749

CHRISTINA JENNINGS
Department of Research and Development, Pacific Gas & Electric Company, San Ramon, CA 94583

MARTIN E. KLAUSMEIER-BROWN
Edward L. Ginzton Research Center, Varian Associates, Palo Alto, CA 94304

YUKINORI KUWANO
Sanyo Electric Company, Ltd., Harikata, Osaka 573, Japan

KIM MITCHELL
Siemens Solar, Camarillo, CA 93011

PAUL D. MOSKOWITZ
Brookhaven National Laboratory, Upton, NY 11973

ROBERT E. NEFF
Space Systems/Loral, Palo Alto, CA 94304

MARK J. O'NEILL
Entech, Inc., DFW Airport, TX 75261

LARRY D. PARTAIN
Edward L. Ginzton Research Center, Varian Associates, Palo Alto, CA 94304

CAROL RIORDAN
National Renewable Energy Laboratory, Golden, CO 80401

S. RUSTY SAILORS
Power Systems, Spectrum Astro, Inc., Gilbert, AZ 85234

J.C. SCHAEFER
Electric Power Research Institute, Palo Alto, CA 94303

R.A. SINTON
Sinton Consulting, San Jose, CA 95129

WILLIAM WALLACE
National Renewable Energy Laboratory, Golden, CO 80401

STUART R. WENHAM
Centre for Photovoltaic Devices and Systems, University of New South Wales, Kensington, NSW 2033, Australia

Y.C.M. YEH
Applied Solar Energy Corporation, Industry, CA 91749

Foreword

The photovoltaic effect was discovered more than 150 years ago, but since the demonstration of the crystalline silicon, p/n junction solar cell in 1954, photovoltaics and microelectronics have been closely linked. As Si microelectronics developed and expanded, there were similarly high hopes for terrestrial power generation by solar cells. By 1956, two companies were already manufacturing solar cells for terrestrial applications, but solar cells have struggled to displace existing technologies for large-scale, terrestrial power generation. During the past 40 years, the technical performance of solar cells has made remarkable strides, but the development of a new commercial technology involves much more than the demonstration of high performance in the laboratory. This volume, written as photovoltaics is moving from the laboratory to the terrestrial marketplace, surveys the status of the technology, as well as the systems, market, and related issues that photovoltaics now faces.

Both microelectronics and photovoltaics were fortunate to find market niches where they could establish a base for growth. While it was difficult at first for transistors to compete with vacuum tubes in high-performance markets, they were ideal for personal, portable products. As the markets grew, the profits funded the continued development of the technology, which then expanded into new markets. Similarly, a few innovative engineers realized that solar cells were the ideal technology for space power, where weight and endurance were key factors. There was, however, a key difference between microelectronics and photovoltaics. In the words of K. Kurokawa of Fujitsu, microelectronics enjoyed a path for the "graceful growth of technology." The technology could expand bit by bit without directly competing with established technologies. As new markets were developed, the profits were plowed back into the further development of the technology. The resulting positive feedback fueled the microelectronics revolution. Photovoltaics found a niche in the space power market and quickly dominated it, but the next step—to terrestrial photovoltaics—was a very large one. A small terrestrial market for remote power applications did develop, but the profits were not sufficient to fund the kind of development effort necessary for large-scale applications. As a result, the development of photovoltaic technology has been largely funded by governments, motivated by their awareness that fossil fuels are being depleted and by an increasing sensitivity to the environmental consequences of conventional power generation technologies.

Solar cells and microelectronics face very similar technical issues. Both need high quality materials, junctions, and contacts. However, for solar cells the requirements are exceptionally demanding. Bulk and surface recombination must be suppressed to extremely low levels, junctions must be ideal, and contact resistances low. Since 1954, the saturation current density of silicon solar cells—a measure of the recombination losses—has been reduced by a factor of ~10,000. The optical requirements are also demanding. Reflection of the incident light must be minimized, and virtually all of the photons with energy above the band gap must be absorbed. Modern cells meet these optical requirements with the use of sophisticated light-trapping techniques and back-surface reflectors. Since the first 6% efficient crystalline Si solar cells, one sun conversion efficiencies have risen to ~25% for both Si and GaAs cells, a value that is quite close to the thermodynamic limit of ~30%.

For the semiconductor researcher, photovoltaics has provided a rich source of intellectually challenging problems. Because the devices are so simple, and because of the need to extract the maximum performance, deep issues concerning the fundamental performance limits, device physics, and material imperfections had to be addressed in order to guide the development of high-efficiency, low-cost cells. The field has been marked by an especially strong interplay between science and technology, and the knowledge gained on topics such as band gap narrowing, Auger and surface recombination, minority-carrier transport, and the electronic properties of defects benefited the larger electronic device community as well. Photovoltaic researchers have also played leading roles in examining new semiconductor material systems and heterostructures. One of the earliest applications of semiconductor heterostructures and liquid phase epitaxy was for passivating the surfaces of GaAs solar cells and, as III–V single and tandem junction cells were developed for space and terrestrial concentrator applications, photovoltaic researchers made important contributions to high-lifetime epitaxial growth techniques and to the growth of lattice-matched and lattice-mismatched films. While photovoltaics has benefited greatly by riding the learning curve of Si microelectronic technology, photovoltaics research itself has also produced new knowledge that benefited the microelectronics industry.

A crucial difference between microelectronic and terrestrial photovoltaic technology is the overriding importance of cost in photovoltaics. The achievement of solar cell efficiencies very near the thermodynamic limit is a remarkable technical accomplishment, but it does not provide the cost reduction necessary to make photovoltaic power generation competitive for large-scale terrestrial applications. To lower costs, two fundamentally different approaches have been explored. The first makes use of optical concentrating systems to replace expensive solar cell areas with less expensive optics, and the second makes use of low-cost, noncrystalline semiconductor films. Both approaches led to the examination of a very wide variety of semiconductor material systems. High-performance crystalline Si and GaAs solar cells have now been developed for concentrator applications, and a wide variety of III–V semiconductors has been developed for tandem cells (which use a series of solar cells with different band gaps to extract maximum efficiency). For low-cost, thin-film cells, even more materials have been explored. Multicrystalline Si cells provide high efficiency at reduced cost, and low-cost hydrogenated amorphous Si cells are now commercially available. Other materials such as CdTe and $CuInSe_2$ show promise for producing high efficiencies at low cost.

Forty years of research and development have brought photovoltaic technology to a high level of performance. Terrestrial markets continue to expand, but very significant challenges remain. Laboratory achievements must be transferred to the production line, and substantial cost reductions are still necessary. This book reviews the current status of solar cell technology as well as the systems and market issues. Eight chapters review the status of key solar cell technologies, crystalline and amorphous, silicon and compound semiconductors. Progress in device technology has set the stage for the wider application of photovoltaics where the important issues are low-cost manufacturing processes and the performance of solar cell modules and systems technologies. The nine chapters on module and systems technologies address these important issues. As the focus shifts from the underlying cell technology to large-scale applications, other issues such as environmental concerns, the solar resource itself, and projections for market growth must also be addressed, and the reader will find these topics examined as well. All in all, the editor has struck a good balance between surveying current cell technology and addressing the systems and market-related concerns that are now the central issues for those involved in commercializing photovoltaics. A realist must be cognizant of the significant challenges that remain, but the past 40 years of progress in photovoltaic science and technology are cause for optimism that clever minds will continue to find ways to meet the challenges.

MARK LUNDSTROM

Purdue University

Preface

This book is a survey of the underlying fundamentals and the status and prospects for solar cells. It considers their applications to a wide variety of energy consumption markets based on a unique ability to directly convert light into electricity without traditional fuels. The first eight chapters and the two appendices review the basic principles, the fundamental limits, and the actual demonstrated performance for a wide range of laboratory cells, some of remarkably high efficiency. These chapters are suitable for graduate level and advanced undergraduate level courses on photovoltaic devices. Chapters 9–14 on module technology discuss the interconnection of enough modules for hundred-watt level outputs. They explore the problems and tradeoffs of applying and deploying solar cell devices. These chapters should be of interest to engineers developing solar energy power sources. Chapters 15 through 17 summarize the experience and success of applying the technology to 20 kilowatt, 100 kilowatt, and even multimegawatt output terrestrial and space systems. The latter should help planners assess the potential of solar cells for large-scale production of electric power.

Chapter 12 contains a summary for the design of low power systems for consumer products. Chapter 15 and its references describes the design of higher-power, off-grid solar cell systems. Chapter 16 outlines important elements in megawatt-level system development. Many aspects affect the role solar cells will ultimately play in the world's energy market. Chapter 18 considers the important safety issues in manufacturing and disposal that require attention as the market grows. Chapters 19 and 23 assess the present size and composition of the solar cell market and make projections on where and how it is most likely to grow. Chapter 20 describes the magnitude, distribution, and properties of sunlight worldwide that systems designers need to supply given energy demands at specific geographical locations. Chapters 21 and 22 discuss how government programs have provided key support that has allowed this technology to develop, and why these programs will have important influence on the future progress of the industry. The latter topics are relevant to policy and strategic planning in business and government.

The initial work on solar cells started more than 150 years ago in 1839 with Becquerel's discovery of a photovoltage when he directed light onto an electrode in an electrolyte (Chapter 1). More than 100 years ago in 1877, Adams and Day reported similar behavior in solid selenium. Selenium and copper oxide solar cells achieved one percent efficiency levels by the early 1900s. Understanding of these effects and relatively high performance awaited two developments. One was the quantum theory of light introduced around the turn of the century. The other was the progress of quantum mechanics and solid state physics from the early to middle 1900s. These culminated in the 1940s and 1950s with the development of solid state, p/n junction diodes and transistors and relatively high-performance solar cells. These could convert incident sunlight directly into DC electricity with efficiencies of around 6%. Their first major commercial application by the late 1950s was to power orbiting satellites (Chapter 17). They are now the primary satellite power source (Chapters 4, 13, 17, and 21). During the energy crises of the 1970s, a major push was initiated to make solar cells practical for terrestrial solar energy conversion. By the early 1990s this had resulted in tens of megawatts per year production levels and hundreds of millions of dollars per year sales levels worldwide, primarily for off-grid industrial and residential uses (Chapters 15, 19, 22, and 23).

The dominant technology by the early 1990s was based on single and multicrystalline silicon cells (Chapters 2, 4, 9, 13, 15, and 17). However, hydrogenated amorphous silicon (a-Si:H) also provided a significant fraction of the total solar cell market, mainly for consumer applications like handheld calculators (Chapters 7, 12, and 19). Innovators have developed promising systems based on a variety of other materials, including GaAs (Chapters 4–6, 11, 13, 14, and 17), InP (Chapter 4), and $CuInSe_2$ and CdTe (Chapter 8). Solar array panels for space satellites sold at a volume of ~0.1 MW for \$300–1800/W in 1992, depending on the type of cells used, the mission requirements, and the rocket and satellite components assigned to array costs (Chapters 13 and 23). These arrays are a highly-leveraged enabling technology that allows the billions-of-dollars-per-year space launch and communications industry to exist in its present form. Terrestrial solar cell power modules profitably sold for ~\$6.50/W in 1992 and accounted for total worldwide sales on the order of \$0.4 billion (Chapters 9 and 23). The a-Si:H panels for calculators sold for ~\$20–50/W in 1992 (Chapter 19) and accounted for sales on the order of \$0.1 billion (Chapter 23).

Projections indicate that the potentially large, on-grid market requires solar cell modules selling for \$2/W or less with efficiencies of 10% or higher (Chapters 16 and 23). Large terrestrial demonstration projects of 20 to 100 kW sizes gave efficiencies of 8%–12% with crystalline silicon cells and 2%–4% with a-Si:H cells through 1993 (Chapters 1 and 16). Learning-curve analysis (Chapter 22) indicates that terrestrial module costs should fall below \$2/W in

(1992 dollars) when cumulative module production (in watts) increases by one to two orders-of-magnitude beyond 1992 levels unless some technical breakthrough introduces a steeper learning curve. If the 17% average annual growth rate of the late 1980s and early 1990s (Chapter 22) continues into the twenty-first century, an order of magnitude production increase to 650 MW/yr should occur by ~2007. A second order of magnitude growth to 6500 MW/yr would then follow by ~2021. The latter assumes a power-law growth in market size with time, while an "S"-shaped growth pattern (Chapter 19) is more typical.

Solar cell modules are intermediate between single cells of watt-level power and large kW to MW systems. Typical modules have power outputs in the 50 to 100 W range for terrestrial uses. Both 1000X GaAs concentrator modules and single crystal silicon, one sun modules have produced over 20% terrestrial efficiencies (Chapters 5, 11, and 23). Laboratory demonstration cells have made extraordinary progress over the past decade using sunlight concentration and multijunctions with differing band gaps. Concentrator Si and GaAs cells have given 26%–28% terrestrial efficiency performance (Chapters 1, 3, and 5). Two junction $GaInP_2$/GaAs has given over 29% terrestrial efficiency at one sun (Chapter 23). Two junction GaAs/GaSb has provided 33%–35% terrestrial efficiency under concentrated sunlight (Chapters 1 and 6). Unfortunately, large terrestrial system efficiencies are only a half or less of these best laboratory performance levels (Chapters 1 and 16). A primary reason is that cost-effective ways have not been developed to include advanced technology into commercial modules. A barrier to such inclusion is the billions of dollars spent on the learning curve of traditional crystalline silicon terrestrial modules (Chapter 23). New advances must typically start on their own learning curves and become competitive with much lower total investment levels unless they can begin in protected niche markets. Well-directed development and acceleration programs over the next 30 to 50 years should provide large terrestrial systems with field-measured efficiencies of 30%–50% at competitive prices based on ideal Shockley p/n diode solar cell and learning curve analyses (Chapters 1, 22, and 23).

Most of the higher-efficiency, laboratory solar cell parameters approach the limits of ideal Shockley p/n junction diode devices. Chapter 1 uses these Shockley properties to project the fundamental limits of solar cell performance. Most solar cells used in commercially competitive modules and systems have parameter values that fall substantially below the Shockley ideal for reasons that are not completely understood or controlled (Chapter 23). The best near-term opportunities to increase commercial system performance are to close the gap between laboratory and module device performance and to understand and control the factors that make cells in commercial systems perform well below the Shockley ideal.

The editor is deeply grateful to all the distinguished chapter authors for the time and effort they devoted developing the text based on their individual areas of expertise. Most of this was at the considerable expense of personal time. All had the full-time duties of their jobs largely undiminished by their writing commitments. The editor particularly appreciates Lewis Fraas' critical evaluation of early versions of the book. He and his students used it for a graduate level, university course on photovoltaics. Fraas' recommendations were particularly helpful in making the large cuts required to make Chapter 1 appropriate for this book. Mark O'Neill is also due special thanks for pointing out the need to correct published efficiency values for the Sandia Laboratory's simulator recalibration in January 1991.

LARRY D. PARTAIN

Palo Alto, California

CHAPTER ONE

Solar Cell Fundamentals

LARRY D. PARTAIN, Edward L. Ginzton Research Center,
Varian Associates, Palo Alto, CA 94304-1025

1. INTRODUCTION

The origin of solar cells can be traced back to 1839 and Becquerel's discovery of a photovoltage when light was shown on an electrode in an electrolyte solution (Fahrenbruch and Bube, 1983). A similar effect was reported in solid selenium material in 1877 by Adams and Day. Subsequent work in selenium and copper oxide led to the development of the selenium solar cell and its wide use in photographic exposure meters. By 1914 selenium solar cells had reached about 1% efficiency in directly converting sunlight into DC electricity. The modern semiconducting era of solar cells began in 1954. Chapin et al. (1954) reported a 6% conversion efficiency for a single-crystal silicon cell, and Reynolds et al. (1954) described a 6% efficiency from what was later understood to be a copper sulfide/cadmium sulfide cell.

Since the mid-1980s there have been dramatic increases in the efficiencies of laboratory demonstration cells. These came from improvements in current, significant increases in voltage, and splitting the sunlight among solar cells of differing band gaps that waste less of the sunlight input power. The higher voltages directly resulted from increases in the density of minority carriers generated by absorbed light, as discussed below. Three key ways for doing this are (1) reducing the minority-carrier recombination rate, (2) trapping light in active layers, and (3) increasing the intensity of light with concentrating optics. In single-crystal Si and GaAs laboratory cells, respectively, one or both of the first two has led to 23% and 24% efficient devices (see Chapters 1, 2, and 6), and inclusion of the third has given 26% and 28% devices (see Chapters 1, 3, and 6). Without the third key, concentrated light, two band-gap laboratory

Solar Cells and Their Applications, Edited by Larry D. Partain.
ISBN 0-471-57420-1

cells have given 27.6% and 29.5% efficiencies, respectively, in single-crystal AlGaAs/GaAs (Chung et al., 1989) and $GaInP_2$/GaAs (Bertness et al., 1993). With concentrated light, two band-gap single-crystal GaAs/GaSb laboratory cells gave 33%–35% efficiencies (Fraas et al., 1990; Benner, 1991; Green and Emery, 1993) (see Chapter 6). Hydrogenated amorphous silicon (a-Si:H) alloy materials have led to 14% laboratory cell efficiencies using multiple band gaps and light trapping without light concentration (see Chapter 7). The basic relationships determining and limiting these performance levels are developed below. The efficiencies of commercial power systems in the field remain in the 3%–12% range (Candelario et al., 1991) despite the above advances due to a number of barriers and considerations that are explored throughout this book.

2. PRINCIPLES OF SOLAR CELL OPERATION

A typical configuration for a solar cell is shown schematically in Figure 1.1. This is a standard n/p junction, rectifying diode with contact metalization partially covering its emitter to allow light entrance. An n on p device is shown with an electrostatically charged depletion region penetrating a depth x_E into the emitter of thickness t_E and also penetrating a depth x_B into the base of thickness t_B. Light absorption generates equal numbers of nonequilibrium electron-hole pairs in concentrations much higher than equilibrium minority-carrier levels but typically less than the equilibrium majority-carrier concentrations. It is these nonequilibrium minority carriers and their large potential energy changes that transform the absorbed photon energy into a DC voltage to drive DC current through the metal contacts for power delivery to an external electrical load. Outside the depletion region there is no net charge, so any potential energy change, for the mobile electrons and holes there, cannot be due to an accumulation of net electrostatic charge. However, potential energy is affected by carrier concentration. The challenge is to convert the nonequilibrium minority carriers into majority carriers with as little loss of their added potential energy as possible. It is the majority carriers that ultimately flow through the two metal contacts to an external circuit. However, the majority-carrier transport is simply described by the linear, Ohm's Law, current–voltage relationships, described by a single series resistance. The more important physics is contained in the nonlinear relationships that describe the minority-carrier transport, where the effects of concentration gradients are particularly important.

Consider a mobile particle moving through a concentration gradient in a host material. Take the simplest one-dimensional case of conduction electrons of concentration n(x) with a random walk scattering model. Consider two points x_1 and x_2 separated by the mean free path between scattering events $\Lambda = x_2 - x_1$ and a volume $V_\Lambda = \Lambda A$, where A is the cross-sectional area. Since velocity is random, only half these electrons at x_1 move toward x_2 at an average velocity v_x. The other half move away from the $x_2 - x_1$ interval. The

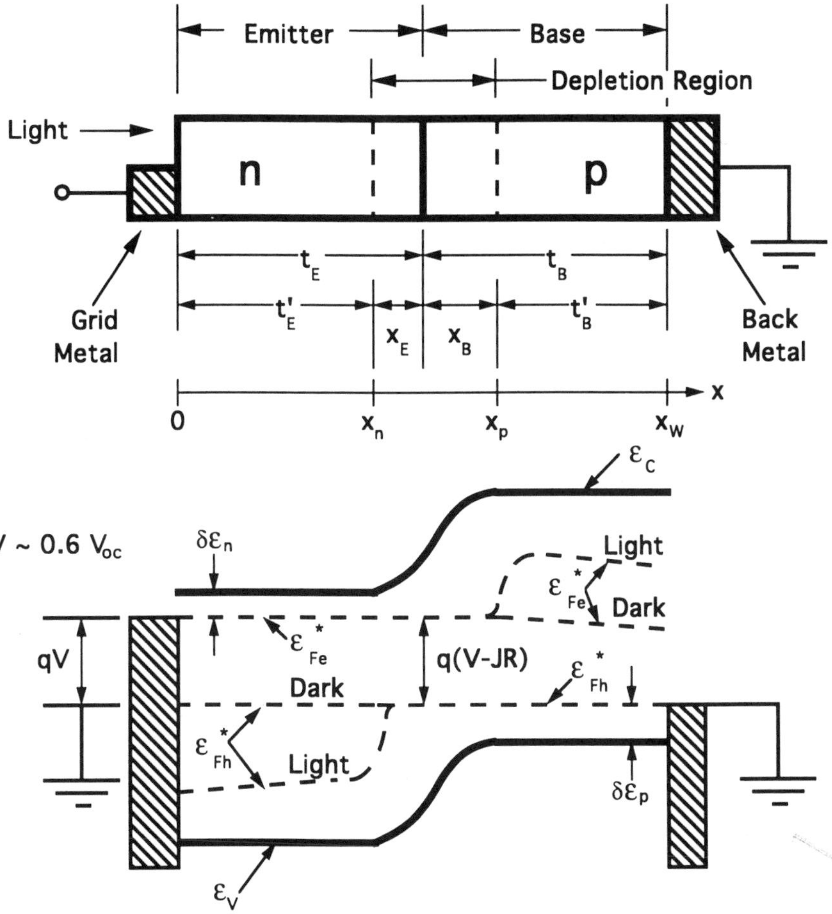

FIGURE 1.1 Schematic diagram for a typical n/p solar cell configuration. The energy band diagram shows the change in the quasi-Fermi energy levels between dark and light with a constant voltage bias of $\sim 0.6\ V_{oc}$.

force that these electrons (moving between x_1 and x_2) exert on the host lattice equals the number involved ($nV_\Lambda/2$), their change in momentum upon scattering ($2p_x$), and the rate at which they are scattered ($1/\tau'$), where τ' is the mean free time between scattering events. This time rate of change in momentum p and force is

$$\left.\frac{\Delta p}{\Delta t}\right|_L = \frac{n(x_2)}{2}\, 2p_x(x_2)\, \frac{V_\Lambda}{\tau'(x_2)} - \frac{n(x_1)}{2}\, 2p_x(x_1)\, \frac{V_\Lambda}{\tau'(x_1)}$$

$$= n(x_2)p_x(x_2)\,\frac{v_x(x_2)}{\Lambda}\, V_\Lambda - n(x_1)p_x(x_1)\,\frac{v_x(x_1)}{\Lambda}\, V_\Lambda \qquad (1)$$

since $\tau' = \Lambda/v_x$, where Δt is the differential time interval and the L subscript identifies force on the lattice. The average velocity squared v^2 equals three times the average velocity squared in any single direction (i.e., $v^2 = 3v_x^2$), and p_x equals $m_{de}v_x$, where m_{de} is the effective mass for conduction electrons. Also the mean kinetic energy $m_{de}v^2/2$ equals 3kT/2, where k is Boltzmann's constant (Feynman et al., 1963). The latter combined give $p_x v_x = kT$ and simplify Eq. (1) to

$$\left.\frac{\Delta p}{\Delta t}\right|_L = k\left[\frac{n(x_2)T(x_2) - n(x_1)T(x_1)}{x_2 - x_1}\right] V_\Lambda \tag{2}$$

Taking the negative of this and dividing by the number of electrons, nV_Λ, then gives the average force on a electron as $-(kT/n)\,d/dx[n(x)]$ in the limit of a small $x_2 - x_1$ interval and a uniform temperature T. Integrating this concentration gradient force over distance gives a change in potential energy that is just as real as that of integrating electric field force over distance. In solar cells Fermi energy ε_F is most consistently interpreted as potential energy (Partain and Sheldon, 1980), and it will be treated as such in this chapter. A gradient in potential energy (or Fermi energy) gives a force that can be multiplied times a mobility μ to give a transport velocity for mobile particles. When this is combined with the well-known relations between electron n and hole p concentrations and Fermi energy (Kittel, 1969; Sze, 1981),

$$n = N_c e^{-(\varepsilon_C - \varepsilon_F)/kT} \tag{3}$$

and

$$p = N_v e^{-(\varepsilon_F - \varepsilon_V)/kT} \tag{4}$$

the current transport equations are obtained for charged electrons and holes (Sze, 1981; Shockley, 1976) with an electric field force term plus the above concentration gradient force term. Here ε_C and ε_V are the conduction and valence band edges illustrated in Figure 1.1 and $N_c = 2M_c(2\pi m_{de}kT/h^2)^{3/2}$ and $N_v = 2(2\pi m_{dh}kT/h^2)^{3/2}$. The m_{de} and m_{dh} are the density-of-states effective masses for electrons and holes, respectively, h is Planck's constant, and M_c is the number of equivalent minima at the conduction band edge.

The energy levels plotted at the bottom of Figure 1.1 are negatively charged electron energies that increase when the levels move up. In contrast, the energies of positively charged holes increase as the levels move down. This opposite behavior is expected for oppositely charged particles in an electrostatic potential field. However, Eq. (3) shows that ε_F also increases as the electron concentration n increases. If the electron concentration were assembled by bringing in each electron one at a time, there would be an increasing concentration gradient force opposing each additional electon at increasing concentrations so that their average potential energy ε_F would be higher for higher concentrations. For consistency with the hole electrostatic energy, the increasing hole potential energy increases by moving down with increasing hole concentration as specified by Eq. (4). The potential energies of electrons and holes are in fact different. However, potential energies are always uncertain by a constant of integration. It is the convention to choose the difference between these constants of integration so that the electron and hole values align at equilibrium in Figure 1.1–type energy band diagrams. This defines a very useful common reference point that distinguishes the Fermi level potential energy from just any potential energy. Thus when there are equal concentrations of electrons and holes at equilibrium in an undoped "intrinsic" semiconductor, the Fermi level is almost halfway between ε_C and ε_V, deviating only slightly from the midpoint because of the differences in electron and hole effective masses as specified by Eqs. (3) and (4). With a constant Fermi energy at equilibrium there is no impetus for carriers to have net movement or for them to have concentration changes from net recombination of electrons and holes. In nonequilibrium the differences between electron and hole potential energies become evident, and they must each be described by separate quasi-Fermi levels ε_{Fe}^* and ε_{Fh}^*, respectively. Since they are referenced to a common equilibrium reference point ε_F, their split is actually a measure of the thermodynamic potential energy difference that drives the nonequilibrium electrons and holes to recombine if they are in the same location or move toward each other if in separate locations. The resulting transport and/or recombination allows the material to return to equilibrium once a disturbing external driving force is removed.

With the massive relative increases in the minority-carrier concentrations with light absorption, the minority-carrier hole quasi-Fermi level ε_{Fh}^* in the emitter (shown in Fig. 1.1) splits away from the majority electron one ε_{Fe}^* there. The latter's energy spacing $\delta\varepsilon_n = \varepsilon_C - \varepsilon_{Fe}^*$ from the conduction band edge is scarcely changed from its equilibrium value (in the dark), since the majority-carrier concentration is virtually unchanged. Comparable splits occur as shown for the quasi-Fermi levels in the base with a similar, almost constant, $\delta\varepsilon_p = \varepsilon_V - \varepsilon_{Fh}^*$ spacing from the valence band edge in the light and dark. A crucial boundary condition in Shockley's original treatment of the transistor (Shockley, 1976; Sze, 1981) is that the split between the quasi-Fermi levels at each side of the depletion region is the measurable voltage at the contacts (minus any ohmic voltage drop from majority-carrier transport in the neutral regions of the emitter and base). This is logical since the quasi-Fermi levels separately describe the differing potential energies of the nonequilibrium electrons and holes. Another crucial aspect of Shockley's original treatment is that there is negligible change in the quasi-Fermi energies across the depletion region compared with larger minority-carrier changes in the neutral regions. The latter implies that any minority carrier that enters this region and exits the other side loses relatively little of the potential energy it had on entrance, as indicated in Figure 1.1. Thus, if a minority carrier crosses, it exits as a majority carrier with almost unchanged potential energy.

Combining the transport equations with the one-dimensional continuity equations and the standard boundary conditions, and the assumption of current dominated by minority-carrier diffusion current at each edge of the depletion region (Hovel, 1975; Shockley and Queisser, 1961), one obtains the standard Shockley diode equations for current density J versus voltage V in an n/p junction solar cell as

$$J = J_0\{\exp[q(V - J\mathcal{R})/kT] - 1\} - J_L \tag{5}$$

or solving for V

$$V = (kT/q)\ln[(J/J_0) + (J_L/J_0) + 1] + J\mathcal{R} \tag{6}$$

where $J_0 = (J_{oh} + J_{oe})$ and $J_{oh} = q(n_i^2/N_D)v_{dh}f_h$ and $J_{oe} = q(n_i^2/N_A)v_{de}f_e$ and the light-generated current J_L is

$$J_L = \int_0^\infty qF(\lambda)[1 - r(\lambda)]QE(\lambda)\,d\lambda \tag{7}$$

Here

$$f_h = \frac{(S_h/v_{dh})\cosh(t'_E/L_h) + \sinh(t'_E/L_h)}{(S_h/v_{dh})\sinh(t'_E/L_h) + \cosh(t'_E/L_h)} \tag{8}$$

$$f_e = \frac{(S_e/v_{de})\cosh(t'_B/L_e) + \sinh(t'_B//L_e)}{(S_e/v_{de})\sinh(t'_B/L_e) + \cosh(t'_B/L_e)} \tag{9}$$

The $\mathcal{R} = RA$ is the specific series resistance, where R is the actual series resistance in ohms describing majority-carrier transport and A is the cross-sectional area of the device. For convenience, $\mathcal{R}$ typically also contains any additional series resistance contributed in the contact metals themselves plus their two metal contact resistances to the solar cell. The n_i is the intrinsic carrier concentration, N_D and N_A are the respective donor and acceptor doping concentrations of the emitter and base, the $v_d = D/L$ are the minority-carrier diffusion velocities, determined by minority-carrier diffusion coefficient D divided by minority-carrier diffusion length L, and the S are the minority-carrier recombination velocities at metal contact surfaces of the emitter and base. For these parameters, subscripts e and h, respectively, identify which describe minority-carrier electrons or holes. The diffusion lengths are related to D and the minority-carrier lifetimes τ by the relation $L^2 = D\tau$. The $F(\lambda)$ is the photon flux density of light of wavelength λ, $r(\lambda)$ is the light reflectivity of the emitter surface at λ, and $QE(\lambda)$ is the quantum efficiency of the solar cell at the indicated wavelength. The t'_E and t'_B are the undepleted widths of the emitter and base, respectively, shown in Figure 1.1. These nonlinear, non-ohmic, exponential relations essentially describe the controlling minority-carrier transport arising largely from the Eqs. (3) and (4) expressions through minority-carrier quasi-Fermi levels and through the continuity and transport equations and their boundary conditions.

Serious computer round-off errors are eliminated with light exposure (i.e., when $J_L > 0$) by defining $J = \Delta J - J_L$ and rewriting Eq. (6) as

$$V(\Delta J) = (kT/q)\ln[(\Delta J/J_0) + 1] + (\Delta J - J_L)R \qquad (10)$$

This was used to calculate the dark and light J–V curves plotted in Figure 1.2

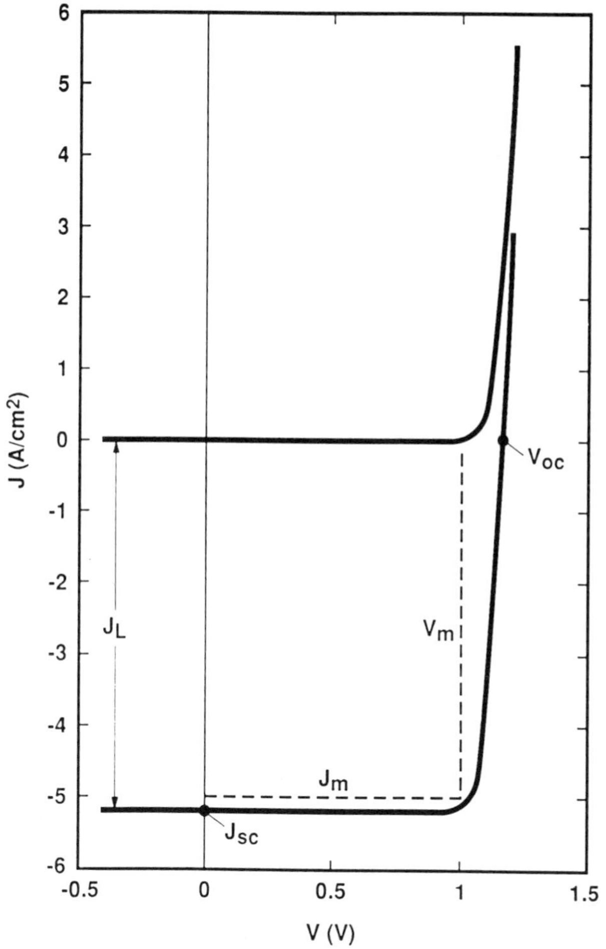

FIGURE 1.2 The dark and light J–V characteristics calculated with Eq. (10) for $J_0 = 9.23 \times 10^{-20}$ A/cm², $J_L = 5.17$ A/cm², and $R = 0.0091$ Ω-cm². The open circuit voltage V_{oc}, short circuit current density J_{sc}, maximum power voltage V_m, and current density J_m are shown.

for a J_0 of 9.23×10^{-20} A/cm^2 and a light-generated current J_L of 5.17 A/cm^2 using a specific series resistance of 0.0091 Ω-cm^2.

The open circuit voltage V_{oc} comes from setting $J = 0$ in Eq. (6) to give

$$V_{oc} = (kT/q)\ln[(J_L/J_0) + 1] \tag{11}$$

and the short circuit current density, J_{sc} (where $V = 0$), is given from Eq. (5) by the transcendental expression

$$J_{sc} - J_0[\exp(-qRJ_{sc}/kT) - 1] = -J_L \tag{12}$$

If R is known, the light J–V characteristics are uniquely determined when any two of the three coefficients $J_L(\sim -J_{sc})$, J_0, or V_{oc} are known. Equation (11) gives the third when the other two are specified. The power density is the JV product $(=J\{[kT/q]\ln[(J/J_0) + (J/J_L) + 1] + JR\})$ from Eq. [6] that is maximum when the $d(JV)/dJ = 0$, at a maximum power current density J_m and voltage V_m. This J_m is negative, with magnitude slightly less than J_L (as shown in Fig. 1.2), and it is specified by the transcendental expression

$$\ln[(J_L + J_m + J_0)/J_0] = -J_m/(J_L + J_m + J_0) - 2qRJ_m/kT \tag{13}$$

Substituting this J_m back into Eq. (6) then gives V_m.

The efficiency η of the solar cell is given by

$$\begin{aligned}\eta &= -V_m J_m A/P_{in}\\ &= -V_{oc}J_{sc}FF/(P_{in}/A)\end{aligned} \tag{14}$$

where P_{in} is the optical input power to the solar cell and A is the cell's cross-sectional area. The negative signs come from J_m and J_{sc} being negative. The fill factor FF is conveniently defined as the ratio

$$\begin{aligned}FF &= V_m J_m/V_{oc}J_{sc}\\ &= (J_m/J_L)(V_m/V_{oc})(J_L/J_{sc})\end{aligned} \tag{15}$$

that specifies the "squareness" of the light J–V (see Fig. 1.2), and it has a maximum value of one. When sunlight is concentrated through some optical means, a concentration ratio X can be defined as the concentrated optical input power P_{in}^{c} divided by the unconcentrated one sun value P_{in}^{1x}. If the spectral distribution is not changed too much by the concentrating optics, the light-generated J_L^c is just X times its one sun value J_L^{1X} since Eq. (7) is linear in

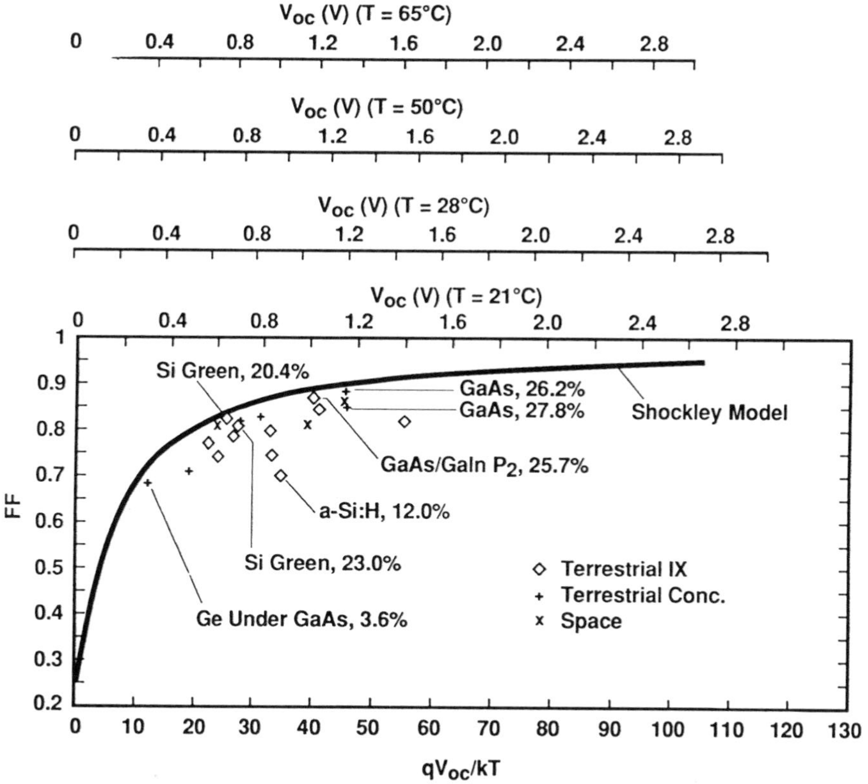

FIGURE 1.3 The variation of fill factor FF with qV_{oc}/kT. The solid line is the theoretical values for an ideal Shockley diode solar cell. The data points are all the experimental values listed in Table 1.5. The upper horizontal scales give actual V_{oc} values for four typical cell temperatures.

photon flux intensity $F(\lambda)$. Hence concentrator efficiency η_c is given as

$$\eta_c = -V_{oc} J_{sc}^{1X} FF/(P_{in}^{1X}/A) \tag{16}$$

In high-performance devices with a low R so that $J_{sc}/J_L = -1$ and with $J_0/J_L \ll 1$ so that $\ln(J_0/J_L)$ is well approximated in Eq. (11) by qV_{oc}/kT, the FF of Eq. (14) is totally specified by the factor qV_{oc}/kT independent of the value of J_L (Shockley and Queisser, 1961; Hovel, 1975), as can be shown by manipulating Eqs. (6), (12), and (13). The resulting relationship is plotted as the solid line theoretical curve in Figure 1.3 for qV_{oc}/kT values between 0 and 105. For qV_{oc}/kT between 50 and 105 the FF rises from 0.90 to 0.95. The corresponding V_{oc} values are shown along the top horizontal axis for typical T values of 21°, 28°, 50°, and 65°C. This theoretical FF asymptotically approaches 1 for $V_{oc} \gg kT/q$ and 0.25 for $V_{oc} \ll kT/q$ (Shockley and Queisser, 1961).

3. ANALYSIS OF 27.8% EFFICIENT GaAs SOLAR CELL

High-performance solar cells are beginning to demonstrate properties near those predicted by the ideal Shockley diode theory developed above. Analysis of the 27.8% efficient GaAs concentrator cell illustrates this agreement and the relation of quasi-Fermi level splitting to fundamental performance limits. The correlations of this section justify the predictions of solar cell performance limits developed in the following section.

The 27.8% efficiency is a corrected value calculated by multiplying the originally published 29.2% (Kaminar et al., 1988) by a correction factor of 0.952. In January 1991, Sandia National Laboratories implemented a recalibration of their primary reference cell (King and Hansen, 1991). They introduced a new procedure that is traceable to the National Institute of Standards and Technology in the United States. The result was agreement in short circuit current within $\pm 2\%$ and in open circuit voltage and FF to better than 1% for several different solar cells also measured by the National Renewable Energy Laboratory and the National Aeronautics and Space Administration in the United States and the Fraunhofer Institute in Germany. However, Sandia concluded that its prior measurements gave short circuit current densities and efficiencies that were 4.8% too high, and they specified the correction factor for both at 0.952. Using this, the corrected Sandia efficiency value for the point contact, crystalline silicon concentrator cell listed in Table 3.1 is 26.7% versus the uncorrected 28% value (Green and Emery, 1993).

Absolute light intensity and spectral distribution measurements are difficult to make with accuracy greater than 5%. Even with the considerable efforts and care taken by Sandia and its sister laboratories, uncertainties persist on the order of $\pm 2\%$.

The $J_0 = 9.23 \times 10^{-20}$ A/cm^2 for the Figure 1.2 plots was calculated from Eq. (11) using the V_{oc} of 1.155 V and the $J_{sc} \sim -J_L$ of -5.17 A/cm^2 of the 27.8% efficient GaAs cell with 0.242 cm^2 illuminated area at 21°C under 240× concentrated sunlight (Kaminar et al., 1988; Green and Emery, 1993). Here a series resistance R of 0.0091 Ω-cm^2 was used to match the observed FF of 0.857. The effects of various series resistance values on these characteristics are plotted in Figure 1.4 for R between 0 and 0.3 Ω-cm^2 using Eq. (10). The corresponding J_{sc}/J_L, J_m/J_L, V_m/V_{oc}, FF, and $J_L R/V_{oc}$ values are listed in Table 1.1. The theoretical maximum FF is 0.896 for an ideal device with these V_{oc} and J_{sc} values if R were equal to 0. With R small enough that $J_L R/V_{oc}$ is less than about 0.7, J_{sc}/J_L equals -1 to within three significant figures. When $J_L R$ is larger than 70% of V_{oc}, J_{sc}/J_L begins to differ substantially from the -1 value as shown in Figure 1.4. Table 1.2 lists the same ratios for even smaller R values between 0.005 and 0.0001 Ω-cm^2. For the effect on FF to be at most reduction of 0.001, $J_L R$ needs to be less than about 0.04% of V_{oc}.

The transport equation solutions that give Eqs. (5) and (6) are formulated in terms of the nonequilibrium changes, induced by light and/or voltage, in the

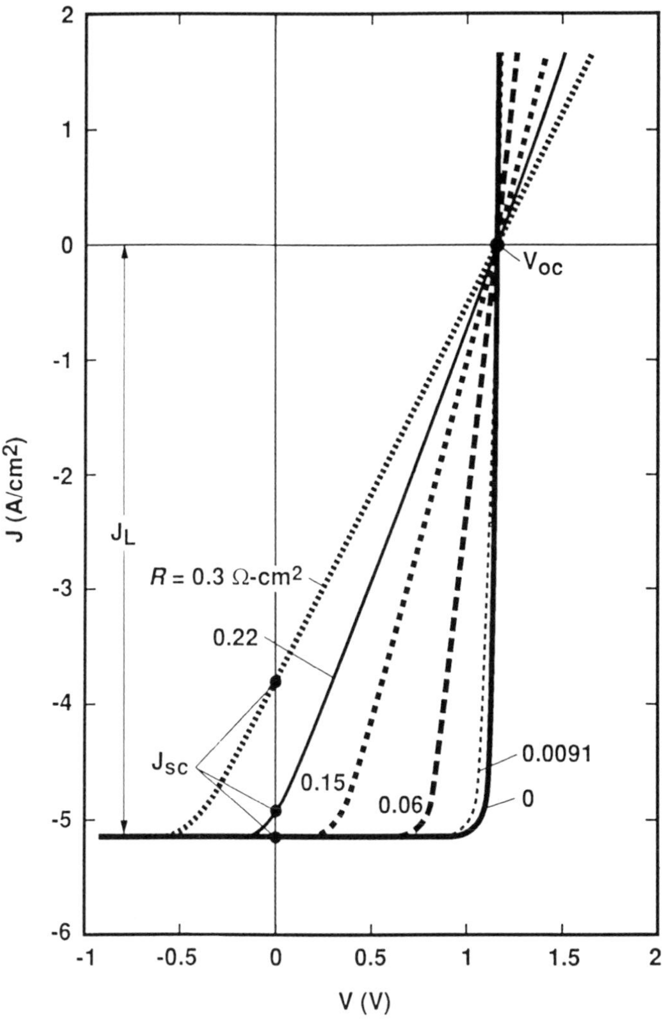

FIGURE 1.4 The effects of various series resistance values R on the light J–V curves and the short circuit current density calculated with Eq. (10) for $J_0 = 9.23 \times 10^{-20}$ A/cm^2 and $J_L = 5.17$ A/cm^2.

minority carrier hole $\Delta p(x)$ and electron $\Delta n(x)$ concentrations as a function of position x in the emitter and base. These can be combined with their equilibrium values p_0 and n_0 to give the positions of the quasi-Fermi levels as a function of x using Eqs. (3) and (4). These parameters are plotted in Figure 1.5 for $V = 0$, in Figure 1.6 for $V = V_{oc}$, and in Figure 1.7 for $V = V_m$ for the parameters that fit the 27.8% efficient GaAs concentrator cell data. The materials and transport parameter values that give this fit are listed in

TABLE 1.1 The Dependence of Calculated Ideal Solar Cell Properties on the Series Resistance *R*, Corresponding to the Curves Plotted in Figure 1.4 and to the J_0 and Other Values Determined From the J_L and V_{oc} of the 27.8% Efficient GaAs Concentrator Cell

R (Ω-cm^2)	J_{sc}/J_L	J_m/J_L	V_m/V_{oc}	FF	$J_L R/V_{oc}$
0	−1	−0.977	0.918	0.896	0
0.0091	−1	−0.975	0.880	0.857	0.0407
0.06	−1	−0.951	0.679	0.645	0.269
0.15	−1	−0.689	0.512	0.353	0.671
0.22	−0.949	−0.490	0.503	0.260	0.985
0.3	−0.724	−0.364	0.501	0.252	1.34

Table 1.3. The minority carrier lifetimes τ_h and τ_e values shown allowed the observed J_0 and J_{sc} to be matched. These τ are a factor of 20 below the radiative recombination limit specified by the listed radiative recombination coefficient B at the specified doping levels. The optical parameter approximations used for the fit are listed in Table 1.4. They give the measured J_{sc} of $-5.17\ \text{A/cm}^2$ using Eq. (7) for the AM1.5D spectra of Appendix A appropriate for terrestrial concentrator solar cells.

Figure 1.5 shows the short circuit current case where 0 volts V is applied to the terminal as diagrammed in Figure 1.1. Figure 1.5a, b shows that the peak changes in minority-carrier concentration due to light exposure are $1.82 \times 10^{13}\ \text{cm}^{-3}$ for Δp_{nL} in the n-type emitter and $9.74 \times 10^{12}\ \text{cm}^{-3}$ for Δn_{pL} in the p-type base. Figure 1.5c, d shows the corresponding energy band diagram and quasi-Fermi energy profiles. Since $\delta\varepsilon_n$ is negative at the $10^{18}\ \text{cm}^{-3}$ emitter doping level, ε_{Fe}^* penetrates into the conduction band. The maximum splits of the quasi-Fermi energies are 1.103 eV in the emitter and 1.069 eV in the base. Because R is not zero, the bias voltage across the depletion region is $-J_{sc}R$, which is the positive value 0.047 V since J_{sc} is negative. The penetrations of the depletion region into the emitter and base, x_E and x_B, are relatively

TABLE 1.2 The Same Properties and Parameters as Table 1.2 But Calculated for Smaller Series Resistance Values *R*

R (Ω-cm^2)	J_{sc}/J_L	J_m/J_L	V_m/V_{oc}	FF	$J_L R/V_{oc}$
0.005	−1	−0.976	0.897	0.875	0.0224
0.002	−1	−0.976	0.909	0.888	0.00895
0.001	−1	−0.976	0.913	0.892	0.00448
0.0001	−1	−0.977	0.917	0.896	0.000448

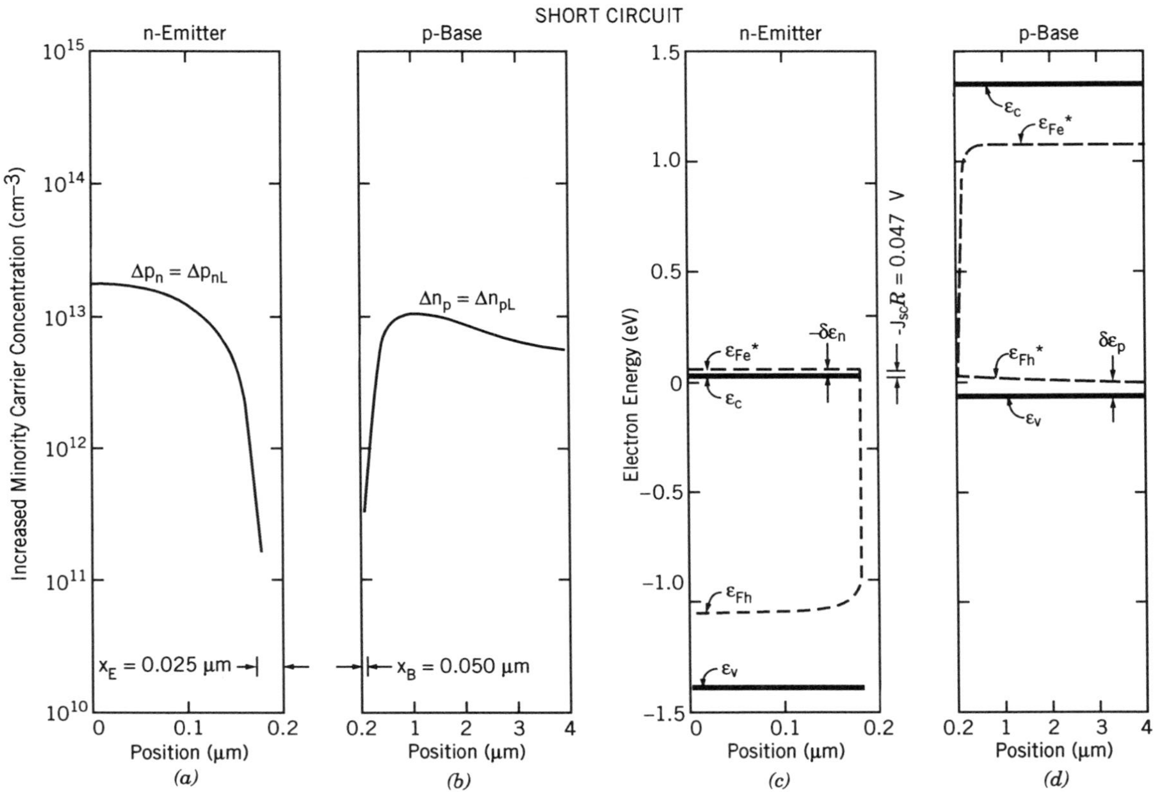

FIGURE 1.5 The increased minority-carrier concentration profiles and the electron energy band diagram for the 27.8% efficient n/p GaAs terrestrial concentrator cell modeled with the ideal Shockley diode theory using the parameters listed in Tables 1.3 and 1.4 for the specific case of short circuit.

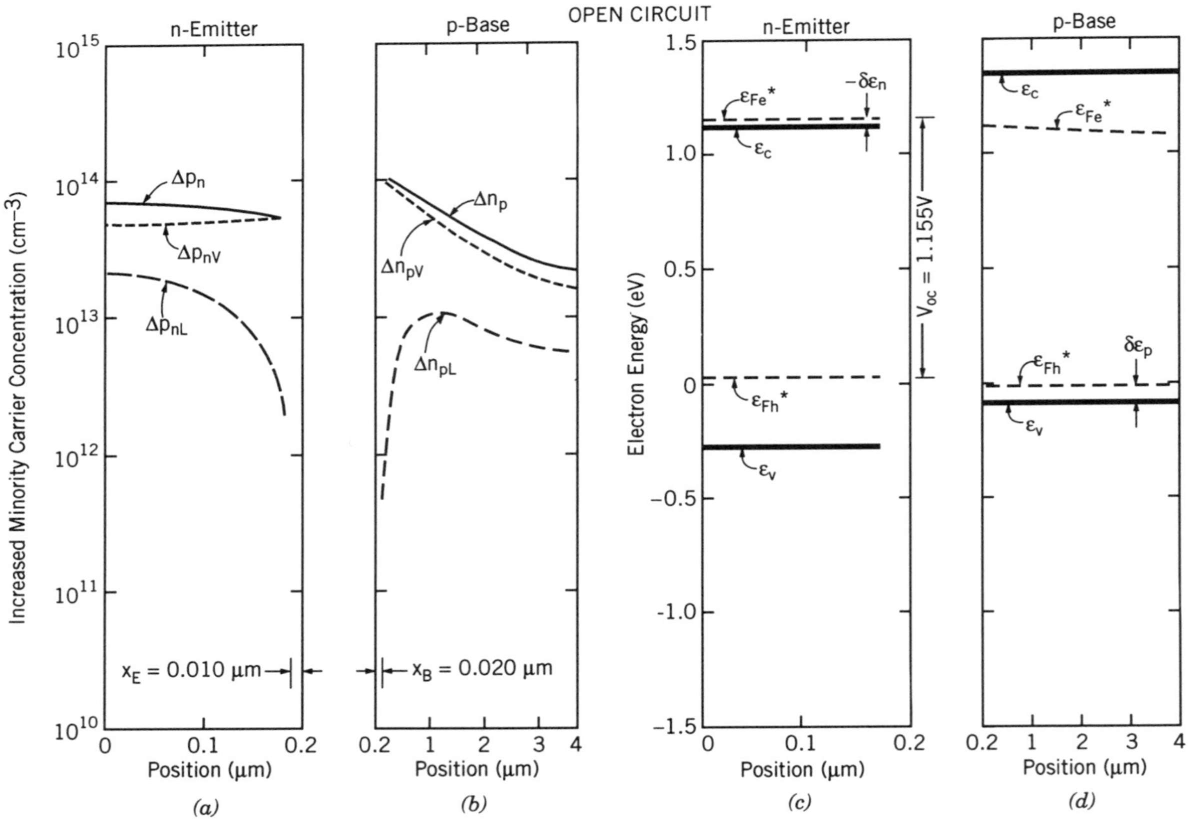

FIGURE 1.6 The same diagrams as in Figure 1.5 except the solar cell is biased at open circuit.

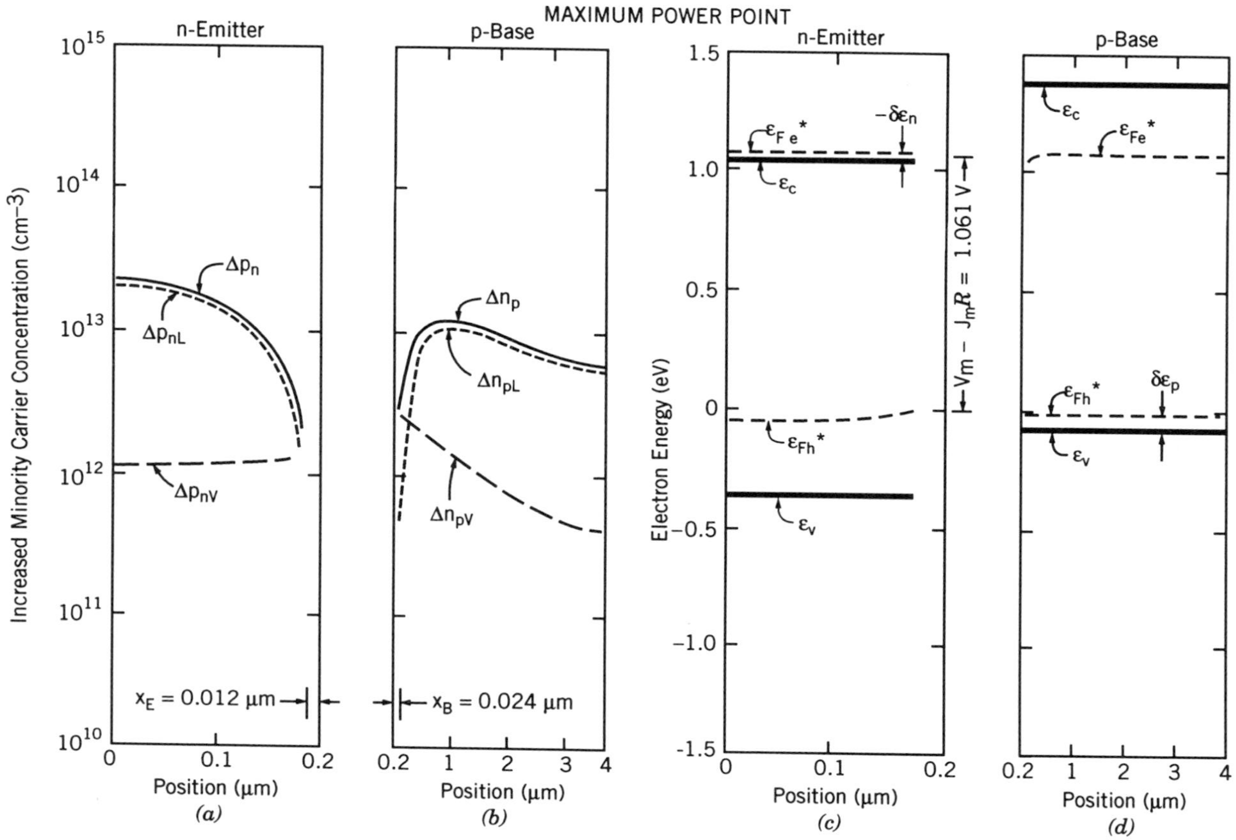

FIGURE 1.7 The same diagrams as in Figure 1.5 except the solar cell is biased at its maximum power point.

TABLE 1.3 The Materials and Transport Parameter Values Used to Model the 27.8% Efficient GaAs Concentrator Cell

Parameter	Value	Units	Reference
t_E	0.2	μm	MacMillan et al. (1988)
t_B	3.8	μm	MacMillan et al. (1988)
N_D	1.0×10^{18}	cm^{-3}	MacMillan et al. (1988)
N_A	5.0×10^{17}	cm^{-3}	MacMillan et al. (1988)
x_E	0.025	μm	Partain et al. (1990)
x_B	0.05	μm	Partain et al. (1990)
μ_h	295	$cm^2/V\,s$	Lovejoy et al. (1992)
μ_e	1,800	$cm^2/V\,s$	Klausmeier-Brown et al. (1989)
D_h	7.5	cm^2/s	Lovejoy et al. (1992)
D_e	745.7	cm^2/s	Klausmeier-Brown et al. (1989)
S_h	5,000	cm/s	Partain et al. (1987b, 1990)
S_e	5,000	cm/s	Partain et al. (1987b, 1990)
m_{dh}/m_0	0.5	—	Sze (1981)
m_{de}/m_0	0.063	—	Sze (1981)
M_c	1	—	Sze (1981)
n_i^2	8.6×10^{11}	cm^{-6}	Sze (1981)
T	21	C	
B	2.0×10^{-10}	cm^3/s	Partain et al. (1987b) Nelson and Sobers (1978)
R	9.1×10^{-3}	$\Omega\text{-cm}^2$	This chapter
τ_h	2.5×10^{-10}	s	$=(1/B\,N_D)/20$
τ_e	5.0×10^{-10}	s	$=(1/B\,N_A)/20$
L_h	0.432	μm	Calculated from above values
L_e	1.51	μm	Calculated from above values
v_{dh}	1.73×10^5	cm/s	Calculated from above values
v_{de}	3.02×10^5	cm/s	Calculated from above values
$\delta\varepsilon_n$	-0.024	eV	Calculated from above values
$\delta\varepsilon_p$	0.072	eV	Calculated from above values
J_0	9.23×10^{-20}	A/cm^2	Calculated from above values
J_{oe}/J_{oh}	8.47	—	Calculated from above values

small as plotted in Figure 1.5a, b. In all of the derived equations, a positive Vcorresponds to forward bias of an n/p junction. For the Figure 1.1 configuration with a grounded p-type base terminal, forward bias is a negative voltage applied to the n-type emitter terminal. Since the electron charge is negative, the shift in the electron energy levels is positive for forward bias with magnitude qV as shown in the bottom of Figure 1.1.

The Shockley approximations assume no change in the quasi-Fermi levels (and their associated carrier potential energies) across the depletion region of width $x_E + x_B$ and current transport only by minority-carrier diffusion at the depletion region edges. The former causes $\Delta p_n(x_n) = 0 = \Delta n_p(x_p)$ so that the

TABLE 1.4 Optical Parameters Used to Model the 27.8% GaAs Concentrator Cell[a]

Parameter	Value	Units	Reference
λ_1	0.61	μm	
λ_2	0.84	μm	
λ_3	0.905	μm	
λ_4	1.04	μm	
$\varepsilon(\lambda_1)$	2.03	eV	
$\varepsilon(\lambda_2)$	1.48	eV	
$\varepsilon(\lambda_3)$	1.37	eV	
$\varepsilon(\lambda_4)$	1.19	eV	
$\alpha(\lambda_1)$	5.03×10^4	cm^{-1}	Sell and Casey (1974)
$\alpha(\lambda_2)$	1.2×10^4	cm^{-1}	Sell and Casey (1974)
$r(\lambda_1)$	0.02	—	MacMillan et al. (1988)
$r(\lambda_2)$	0.01	—	MacMillan et al. (1988)
$QE(\lambda_1)$	0.919	—	Partain et al. (1990)
$QE(\lambda_2)$	0.721	—	Partain et al. (1990)
$F(\lambda_1)$	1.99×10^{19}	$cm^{-2} s^{-1}$	Appendix A
$F(\lambda_2)$	2.01×10^{19}	$cm^{-2} s^{-1}$	Appendix A
$\Phi(\lambda_2)/\Phi(\lambda_1)$	0.735	—	Appendix A
$\Phi(\lambda_3)/\Phi(\lambda_1)$	0.579	—	Appendix A
$\Phi(\lambda_4)/\Phi(\lambda_1)$	0.535	—	Appendix A

[a]The solar spectrum is approximated with four photon wavelengths, λ_i, of energy $\varepsilon(\lambda_i)$ (two of which are greater than the GaAs band gap of 1.42 eV); the associated optical absorption and reflections coefficients $\alpha(\lambda_i)$ and $r(\lambda_i)$, respectively; photon fluxes $F(\lambda_i)$; spectral irradiances $\Phi(\lambda_i)$; and quantum efficiencies $QE(\lambda_i)$. These combine to give the J_{sc} of -5.17 A/cm^2 from an optical input power density P_{in}/A of 18.4 W/cm^2, which is the 240 concentration ratio times the 76.7 mW/cm optical power density in the integrated AM1.5D spectra appropriate for terrestrial concentrator cells given in Appendix A.

latter requires an infinite gradient (with zero nonequilibrium carriers) at x_n and x_p to carry the finite J_{sc} value since no current flows at equilibrium. Thus these approximations produce an artifact of a vertical slope in the quasi-Fermi levels at the depletion region edges as shown in Figure 1.5c, d. In reality, the change in quasi-Fermi levels are not zero across the depletion region but are "negligible" (as illustrated in Fig. 1.1) in the sense that their depletion region changes are much smaller than in the low-field, minority-carrier regions outside the depletion region (Fahrenbruch and Bube, 1983). Hence the electron and hole potential energy changes in the depletion region are in reality very small but not absolutely zero. However, these important Shockley simplifications produce little error in terms of observed J–V characteristics of the best n/p junctions.

Figure 1.6 shows the effects of increasing the terminal voltage V to the open circuit value 1.155 V. It is the same as the junction voltage since J = 0 here.

Because the differential equations are linear, the voltage-induced changes in minority-carrier concentrations Δp_{nV} and Δn_{pV} just superimpose on the light-induced changes Δp_{nL} and Δn_{pL}. The latter two are essentially the same as their Figure 1.5a, b short circuit values—differing slightly from the decreased depletion layer width with forward bias. Their sums Δp_n and Δn_p give minority-carrier concentration gradients at the depletion region edges that make the net diffusion current here zero. Their peak total concentrations are 6.79×10^{13} and $1.03 \times 10^{14}\,cm^{-3}$, respectively. Stated another way, the voltage-induced "dark" current opposes the light-induced short circuit current due to Shockley's linearizing assumptions. The result is superposition of "dark-" and "light-induced" currents that exactly cancel at the open circuit voltage.

The maximum splits of the quasi-Fermi level are 1.161 eV at the front surface of the emitter and the open circuit voltage 1.155 eV value at the depletion region edge of the base. Note that the open circuit value is only about one-fourth of a kT (0.006 eV) below the peak splitting that occurs in the emitter. At 21°C, kT equals 0.0254 eV. These maximum splits are larger than in the short circuit case of Figure 1.5 because of the minority-carrier injection with forward voltage, specified by Δp_{nV} and Δn_{pV}. This causes the light-generated carriers to be created in a region with higher minority-carrier concentration and thus higher potential energy than when V equals zero. Since there is no net charge outside the depletion region, there is no electrostatic contribution to potential energy change, only the concentration component.

Maximum power is achieved at the terminal voltage of 1.015 V, where the maximum power current density ratio $J_m/J_{sc} = 0.975$ (see Table 1.1). Since $J_{sc} = -5.17\ A/cm^2$ and $R = 0.0091\ \Omega\text{-}cm^2$, the maximum power voltage at the junction is 0.046 V higher or 1.061 V, as shown in Figure 1.7. The latter is a 0.094 V reduction from the junction voltage at open circuit. This is a 3.7 kT reduction that decreases the minority-carrier concentrations at the depletion region edges by a factor of 40 ($= e^{3.7}$, see Eqs. [3] and [4]) to 2.80×10^{12} and $3.35 \times 10^{12}\,cm^{-3}$, respectively, in the emitter and base. The minority carriers are collected by diffusion only. This reduction at the edges produces a large enough gradient to collect 97.5% of the light-generated current with only a slight (0.094 eV) reduction in their usable potential energy. The remainder is the potential energy after they are converted from minority to majority carriers in traversing the depletion region. Once they are majority carriers, further potential energy loss is described by Ohm's law through the series resistance R value. Stated another way, the 3.7 kT reduction in junction voltage reduces the "dark current" by 97.5%. Decreasing the terminal voltage by another couple of kT (~ 0.05 V) reduces the "dark current" by another factor of 7 to a negligible value. Then the collected current is essentially the total light-generated value J_L (see the J–V plot in Fig. 1.2).

This exponential dependence on a given $\Delta V/kT$ to reduce the injected minority-carrier concentration at the depletion region edge by a given fraction (from open circuit to the maximum power point) and thus the relative magnitude of the "dark current" by the same fraction is independent of the open circuit voltage. Thus, for a given T, the $\Delta V/V_{oc}$ becomes a smaller fraction as V_{oc} increases. Hence the FF increases with V_{oc} as shown in Figure 1.3. As T decreases, the magnitude of the ΔV needed for the same exponential change decreases. Thus FF inversely varies with T as also shown in Figure 1.3.

The potential performance of this n/p GaAs concentrator cell can be assessed by substituting in reported low values of surface recombination velocity $S_h = 100\,\text{cm/s} = S_e$ (Dawson and Woodbridge, 1984) and the limiting values of minority-carrier lifetimes set by radiative recombination. For the specified doping values the latter are $\tau_h = 5 \times 10^{-9}\,\text{s}$ and $\tau_e = 1 \times 10^{-8}\,\text{s}$. These specify a J_0 of 1×10^{-20} A/cm^2, a J_{sc} of -5.74 mA/cm^2, and a V_{oc} of 1.214 V from Eq. (11) assuming the same sunlight intensity. The corresponding FF value calculated from Eqs. (12), (13), and (15) is 0.863 for the R value of 0.0091 Ω-cm^2, and the efficiency is 30.6%. If R were negligibly small, the FF could increase to 0.900 and the efficiency to 32.0%. Hence improvement in the present configuration cell to limiting materials properties should result in efficiency increases from the existing 27.8% to the 30%–32% range.

Under these materials-limiting conditions, the peak minority-carrier concentrations generated by light Δp_L and Δn_L have maximum values of 1.9×10^{13} and $2.1 \times 10^{13}\,\text{cm}^{-3}$ in the emitter and base, respectively. These saturated values are only slightly higher than the light-generated ones shown in Figure 1.6a, b. They are controlled by the term $\alpha\tau/(\alpha^2L^2 - 1)$ in the Δn and Δp expressions (see Eqs. [14] and [18] in Hovel, 1975) that can be expressed as one over a characteristic velocity $1/v_c$. In the limit of short τ values where $\alpha^2L^2 \ll 1$, $v_c = 1/\alpha\tau$ and the light-generated concentrations increase linearly with τ. When $\alpha^2L^2 \gg 1$ for the limiting case above, v_c saturates to equal αD, independent of τ, since $L^2 = D\tau$. The major increase in open circuit voltage is instead produced by the voltage-induced components Δp_V and Δn_V. At open circuit (and the above materials limits), the maximum values of the total Δp and Δn are 5.9×10^{14} and $1.15 \times 10^{15}\,\text{cm}^{-3}$, respectively, and the corresponding maximum quasi-Fermi level splittings, in the emitter and base, respectively, are 1.217 and 1.214 eV. The voltage that produces a given "dark" current is the excess minority-carrier concentration times either the diffusion velocity or the surface recombination velocity or the appropriate average of the two specified by Eq. (8) or (9). Thus the voltages (for a given "dark" current) continue to increase with larger τ or smaller S values.

As minority-carrier concentrations approach majority-carrier values, the near-equilibrium, Boltzmann transport, perturbation theory definition for mobility μ begins to break down (Ziman, 1965; Partain and Lakshminarayana, 1976). Also the minority-carrier lifetimes τ and their derivative minority-carrier diffusion lengths L are no longer well defined (Sze, 1981). Hence the Shockley diode solar cell theory of this chapter reaches its fundamental limits as minority carriers approach majority-carrier concentrations.

The above maximum minority-carrier concentrations are still well below the $(0.5 - 1.0) \times 10^{18}\,\text{cm}^{-3}$ levels of the majority carriers. Hence ultimate efficiencies for an ideal Shockley diode solar cell with a GaAs band gap can be obtained by letting excess minority concentrations approach the majority-carrier levels — to within an order of magnitude. The maximum V_{oc} is then the 1.42 eV band gap ε_G minus the maximum $\delta\varepsilon_n + \delta\varepsilon_p$ for the majority and the minority carriers in either the emitter or base. This gives $V_{oc} = 1.30$ V for the present case. The maximum FF with negligible series resistance R is then 0.905 (Fig. 1.3) at 21°C, and the maximum J_{sc} is 24.27 mA/cm^2 for the AM1.5D spectra of total one sun energy of 76.72 mW/cm^2 from Appendix A. These combine for a maximum efficiency (Eq. [14]) of 37.2%. The maximum $J_{sc} = -J_L$ is calculated in Appendix A from the spectral irradiance and Eq. (7), assuming unity quantum efficiency and zero reflectance as discussed in more detail in the following section.

With peak quantum yields over 90% (see Table 1.4), major improvements in efficiency need the increased V_{oc}, toward the indicated value, with the accompanying increases in FF. From Eq. (11), V_{oc} can be raised in two ways — by either increasing J_L or decreasing J_0. Either way the maximum splitting of the quasi-Fermi levels is increased at the maximum power point. As shown above, reducing J_0 by decreasing $S_e = S_h$ to 100 cm/s and raising τ_e and τ_h to the radiative recombination limits only increases V_{oc} to 1.214 V and the efficiency to at most 32%. Further decreases in J_0 require reducing the base layer thickness t_B to be much less than the base diffusion length L_e so that J_0 is controlled by the front and back surface recombination velocities. For the J_L of 5.17 A/cm^2 of the GaAs cell modeled above, a V_{oc} of 1.3 V requires a J_0 of 3.06×10^{-22} A/cm^2. This J_0 is only realized when t_E and t_B are decreased to 0.053 and 0.06 μm, respectively, at the radiative recombination limit and the 100 cm/s recombination velocities. With thinning, the light-generated current can only be maintained by light trapping where the light is reflected from the back of the cell and its morphology textured to increase total internal reflection of the photons. At present there is no known way to achieve this level of light trapping in such ultrathin layers of GaAs, although the trapping techniques are becoming quite sophisticated (see Chapters 2 and 3).

The second way of increasing V_{oc} is to increase J_L by concentrating the light. The latter term rises linearly with concentration ratio X. The theoretical increase in efficiency with X for the parameters in Tables 1.3 and 1.4 for n/p GaAs are shown by the solid line curve in Figure 1.8 for the case of zero series resistance. This line terminates at the minority-carrier concentration of 10^{17} cm^{-3}, where the predicted efficiency is 34%. The dashed curves show the effects of different nonzero series resistance values from 0.9 to 9×10^{-6} Ω-cm^2. The square data points are the measured concentration × dependence of the 27.8% efficient cell (Kaminar et al., 1988, after correction for the Sandia recalibration). For $R = 0.0091$ Ω-cm^2, the theoretical curve passes directly through the 27.8% efficiency at 240× concentration. This R value is actually an overestimate of series resistance (Liu and Sites, 1994). The width of the depletion region actually decreases with forward bias, as shown by the x_E and x_B values plotted in the a and b parts of Figures 1.5 through 1.7. These were calculated from the appropriate expressions given by Partain et al. (1990). This

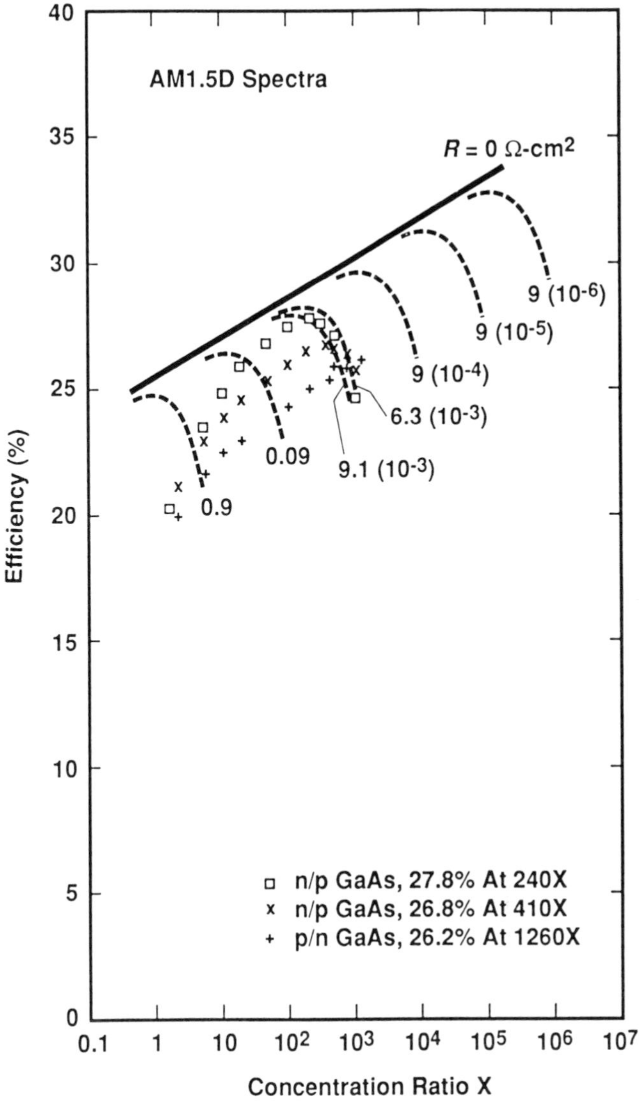

FIGURE 1.8 The theoretical curves show the variations in the ideal Shockley diode solar cell efficiencies with terrestrial sunlight concentration up to 1 million × for series resistance values R between 0 and 0.9 Ω-cm^2 for the Tables 1.3 and 1.4 parameter values used to fit the 27.8% efficient GaAs concentrator cell. The three data point sets are described in the text.

shrinking width of a region assumed to have unity quantum efficiency adds a small V dependence to J_L that has not been included in the ideal Shockley diode theory developed above. When included, the J_L decreases by 0.3% as the voltage is raised from 0 to its maximum power value. The R value that then matches the observed FF of 0.857 at 240× reduces by 30% to 0.0063 Ω-cm². When the latter is plotted in Figure 1.8, its curve slightly exceeds the 27.8% point because the theory used there does not include the V dependence of J_L.

Apparently, this 27.8% device does not begin to approach ideal Shockley diode behavior until 200× levels of concentration. However, two lower-peak-efficiency GaAs concentrator cells have the predicted ideal slope in efficiency versus concentration (with negligible *R*) for × values above approximately 100× to as high as 1,000×, as shown by the cross and plus data points in Figure 1.8 (MacMillan et al., 1988, after correction for the Sandia recalibration). If near-term improvements could reduce *R* by an order of magnitude, Figure 1.8 suggests that the efficiency would increase to the 29%–30% level due as much to decreased resistive losses (at 240× concentration) as to operation at increased concentration. Thus present technology limits GaAs concentrating cells to maximum efficiencies around 30%, although the fundamental Shockley diode limit is around 34%–37%. The latter would require some as yet undefined technology advances. This is consistent with the analysis in Chapter 5.

The modeling in this section was essentially a parameter fitting exercise based on the 27.8% GaAs concentrator cell. Its observed V_{oc} and J_{sc} were used to calculate J_0 from Eq. (11), and the measured FF was used to determine the series resistance *R*. GaAs is one of the best-characterized semiconductor materials as evidenced by its lists of relevant parameters in Tables 1.3 and 1.4. Yet this fitting essentially had to choose surface recombination values from a plausible range and then determine minority-carrier lifetimes (reduced from their theoretical maxima) to fit both the measured J_{sc} and J_0 values. This process is too cumbersome for convenient use in assessing the wide range of solar cell materials whose parameter values are not well specified, particularly for cell geometries that do not match the model's simple one-dimensional symmetry.

4. PREDICTION OF LIMITING SOLAR CELL PERFORMANCE

Several papers were written in the middle to late 1950s on the potential efficiency of p/n junction solar cells as a function of their band gap (Prince, 1955; Loferski, 1956; Rappaport, 1959; Wolf, 1960). These have provided helpful guidance for improving the performance of silicon cells and for assessing the potential benefits of using materials with other band gaps. Soon afterwards, Shockley and Queisser (1961) indicated that these predictions were based more on empirical constants than on fundamental limits. They specified potential efficiency levels much higher than earlier claims based on detailed balance arguments. Indeed recently demonstrated efficiencies are beginning to exceed the pre-Shockley–Queisser predictions for silicon cells as developed by Green et al. (see Chapter 2) and by Sinton et al. (see Chapter 3) and for GaAs developed by Varian, Spire, and others (see Chapter 5). Here these limit concepts are reviewed.

The maximum short circuit current of a Shockley n/p junction diode cell as a function of band gap can be calculated from Eq. (7) assuming unity quantum efficiency for photon energies greater than the band gap and zero quantum efficiency for energies less than the band gap and zero surface reflectivity.The photon flux densities are specified by the tabulated solar spectra that are coming to be widely accepted (see Chapter 20). Three of the most widely used ones are the AM0 of Appendix B and the AM1.5G and AM1.5D of Appendix A.

The AM0 indicates the air-mass zero sunlight intensity in space just above the earth's atmosphere at the mean distance of the earth from the sun (see Appendix B). An air-mass 1.0 indicates the spectra obtained when sunlight penetrates to the surface of the earth through a "standard" atmosphere at a sun angle 90° to the earth's surface. The air masses of 1.5 indicate a shallower sun angle that increases the atmospheric path length by 50% from the 90° case. This is intended to simulate the "average" spectra with the sun moving across the sky on a "standard" day. A final initial "G" specifies the global spectra of direct sunlight from the solar disk plus the scattered sunlight from the surrounding "blue" sky. A final "D" identifies only the direct spectra from the solar disk since scattered sunlight can usually not be focused by the optical elements of concentrator systems.

The three resulting sets of one sun maximum J_{sc}^{max}s are tabulated in Appendices A and B and plotted as three solid line curves in Figure 1.9. The one sun, total, input power densities are 136.7, 96.25, and 76.72 mW/cm^2, respectively, for these AM0, AM1.5G, and AM1.5D spectra. For comparison to measured data below, GaAs can be stacked on cells of lower band gap. Then the latter's maximum AM1.5D J_{sc}^{1X} is shifted down by the maximum 24.29 mA/cm^2 of GaAs, as indicated by the short broken line segment at the bottom left in Figure 1.9.

For reference, the measured photovoltaic parameters of a range of high-performance solar cells of differing band gaps are listed in Table 1.5. The listed one sun J_{sc}^{1X} values are normalized ones obtained by solving Eq. (14) or (16) with η (or η_c), V_{oc}, FF, and P_{in}/A specified by their measured values (as listed in Table 1.5). Normalization is necessary since various spectra other than the three listed above have been used in the past to determine these cells' performance. The resulting experimental values are plotted in Figure 1.9 as plusses for terrestrial concentrator AM1.5D cells, as diamonds for terrestrial one sun AM1.5G cells, and as "x's" for space AM0 cells. Note that three silicon cells (Nos. 7–9, Table 1.5) have experimental short-circuit-current densities of 90.6%–93.4% of their theoretical maximum value as listed as a performance ratio column in this table. A GaAs cell (No. 18) has a short-circuit-current density of 88.7% of its maximum. These closely approach their theoretical maxima, but most of the data are substantially below their fundamental current limit values.

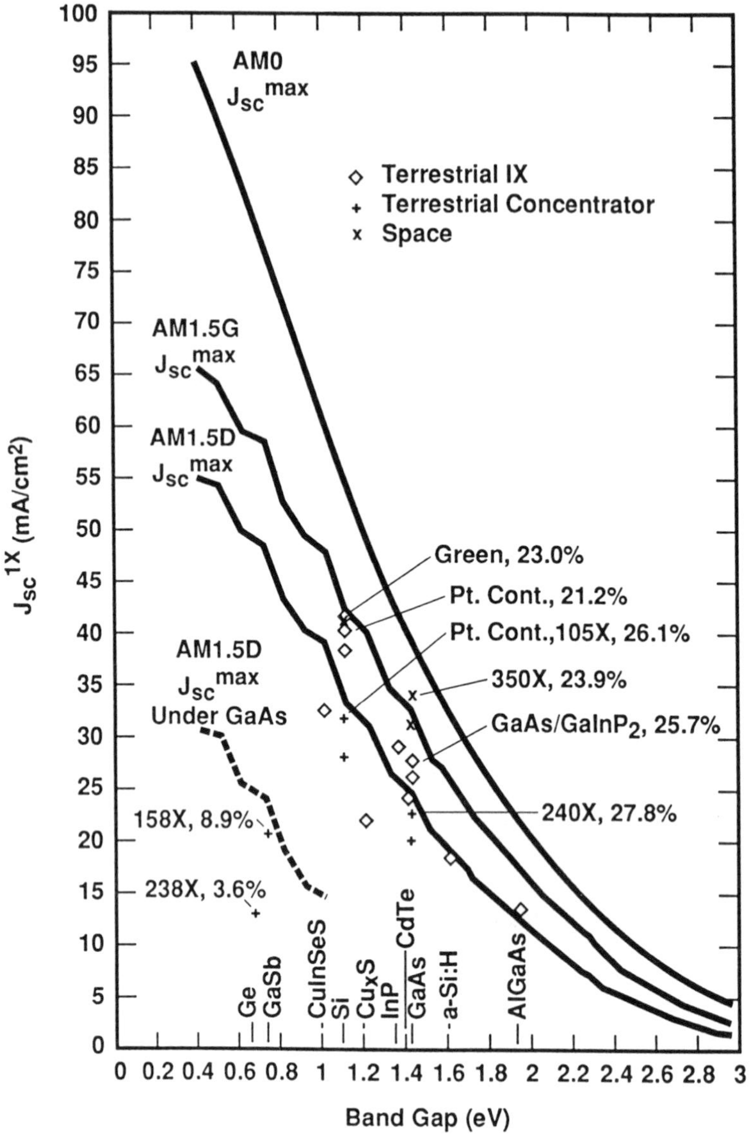

FIGURE 1.9 The theoretical maximum short circuit current densities at one sun for a solar cell with zero front surface reflection and unity quantum efficiency for photon energies greater than its band gap. These curves are given as a function of band gap for the standard space input light spectra (AM0) of Appendix B and the standard terrestrial spectra for unconcentrated (global AM1.5G) and concentrated (direct normal AM1.5D) sunlight given in Appendix A, all normalized to one sun input intensities. The lower, broken line curve is for a cell placed under a 1.42 eV band gap GaAs solar cell assumed to have unity transmission for photon energies less the GaAs band gap for concentrator spectra sunlight. The data points are from the experimental cell measurements listed in Table 1.5.

TABLE 1.5 Summary List of Performance Parameters of a Range of High-Efficiency Solar Cells[a]

Band Gap ε_G (eV)	Cell Type	Conc (×)	Measured Values: η (%)	V_{oc} (V)	FF	$J_{sc}^{1\times}$ (mA/cm²)	T (°C)	Performance Ratios: J_{sc}/J_{sc}^{max}	qV_{oc}/ε_G	V_{oc}/V_{oc}^{max}	FF/FF^{max}: $V_{oc}=V_{oc}^{meas}$	FF/FF^{max}: $V_{oc}=V_{oc}^{max}$	η/η^{max}
1. 0.67	Ge under GaAs	238	3.6	0.306	0.69	13.2	21	0.542	0.46	0.610	0.947	0.855	0.283
2. 0.73	GaSb under GaAs	158	8.9[c]	0.48	0.713	19.8[c]	21[d]	0.839	0.66	0.854	0.891	0.867	0.621
3. 1.0	CuInSeS/Cds	1	15.2[e]	0.613	0.74	32.2[e]	25	0.671	0.61	0.739	0.890	0.855	0.424
4. 1.1	Si[f]	1	15.0	0.615	0.809	41.2	28	0.760	0.56	0.664	0.975	0.925	0.467
5. 1.1	Si green	1	20.4[c]	0.654	0.826	36.2[c]	25	0.861	0.59	0.704	0.985	0.943	0.572
6. 1.1	Si green	21	20.4[c]	0.711	0.818	26.8[c]	25	0.805	0.65	0.765	0.965	0.934	0.575
7. 1.1	Si Pt. cont.	1	21.2[c]	0.68	0.787	38.1[c]	25	0.906	0.62	0.732	0.932	0.898	0.596
8. 1.1	Si green	1	23.0[c]	0.696	0.81	39.3[c]	24	0.934	0.63	0.749	0.957	0.925	0.647
9. 1.1	Si Pt. cont.	105	26.1[c]	0.80	0.83	30.2[c]	24	0.907	0.73	0.861	0.963	0.947	0.740
10. 1.2	Cu_xS/Cd(ZnS)	1	10.2	0.578	0.773	22.0	28.5	0.551	0.48	0.563	0.940	0.874	0.271
11. 1.35	InP	1	20.0[e]	0.83	0.80	29.0[e]	21[d]	0.862	0.61	0.702	0.924	0.891	0.539
12. 1.4	CdTe/CdS	1	15.8[e]	0.843	0.745	24.2[e]	25	0.734	0.77	0.686	0.860	0.828	0.417
13. 1.42	GaAs/Ge Subst[f]	1	18.5	1.01	0.808	31.0	28	0.792	0.71	0.811	0.914	0.897	0.576
14. 1.42	GaAs	1	24.0	1.05	0.845	26.0	25	0.817	0.74	0.841	0.953	0.921	0.633
15. 1.42	GaAs[f]	350	23.9[c]	1.18	0.862	32.1[c]	28	0.821	0.83	0.945	0.962	0.958	0.744
16. 1.42	$GaAs/GaInP_2$	1	25.7	1.04	0.868	27.4	28	0.860	0.73	0.835	0.981	0.964	0.692
17. 1.42	GaAs	1050	26.2[c]	1.18	0.89	19.1[c]	28	0.787	0.83	0.947	0.993	0.989	0.738
18. 1.42	GaAs	240	27.8[c]	1.16	0.857	21.5[c]	21	0.887	0.82	0.927	0.956	0.950	0.782
19. 1.6	a-Si:H	1	12.0	0.891	0.701	18.4	25	0.720	0.56	0.624	0.804	0.769	0.345
20. 1.93	AlGaAs	1	16.0	1.41	0.82	13.3	25	0.778	0.73	0.778	0.901	0.886	0.536

[a]For consistency, the one sun short circuit current densities ($J_{sc}^{1\times}$) were calculated with Eqs. (14) and (16) from the measured values of efficiency (η), V_{oc}, and FF and from the tabulated sunlight input power densities (Appendices A and B) of 96.25 mW/cm² for the terrestrial "flat-plate" AM1.5G spectra of 76.72 mW/cm² for terrestrial concentrator AM1.5D spectra and of 136.7 mW/cm² for the space AM0 spectra. The superscript "f" identifies the three space cells in the "Cell Type" column. All others are terrestrial cells. The "×" is the sunlight concentration ratio.

[b]References: 1. Partain et al. (1987). 2. Fraas et al. (1990). 3. Chapter 8. 4. Chapter 4. 5. Wenham et al. (1990). 6. Wenham et al. (1990). 7. Sinton et al (1986). 8. Zhao et al. (1990). 9. Sinton et al. (1986). 10. Hall et al. (1981). 11. Sugo et al. (1987). 12. Chapter 8. 13. Chapter 4. 14. Bertness et al. (1988). 15. Hamaker et al. (1988). 16. Kurtz et al. (1990). 17. MacMillan et al. (1988). 18. Kaminar et al. (1988). 19. Catalano et al. (1987). 20. Chung et al. (1989).

[c]Sandia recalibration factor 0.952 applied.

[d]Assumed T values.

[e]Active area values.

[f]Space cells.

The V_{oc} is the split of the quasi-Fermi levels at the depletion region boundaries when J = 0. This is specified by

$$V_{oc} = (\varepsilon_G - \delta\varepsilon_{MJ} - \delta\varepsilon_{MN})/q \tag{17}$$

where ε_G is the energy band gap and $\delta\varepsilon_{MJ}$ and $\delta\varepsilon_{MN}$ are the spacings, respectively, of the majority, and minority-carrier quasi-Fermi levels from the band edges at the depletion region boundaries as specified by Eqs. (3) and (4). The Shockley diode model is only valid when the minority-carrier concentrations are much less than those of the majority carrier ones. These spacings $\delta\varepsilon$ are plotted versus carrier concentrations in Figure 1.10 for various density-of-states effective mass values m_d normalized by m_0 the free electron mass. These effective masses are $0.063\,m_0$ and $0.5\,m_0$ for electrons and holes, respectively, in GaAs and $0.328\,m_0$ and $0.549\,m_0$, respectively, in Si (Sze, 1981).

The above V_{oc} expression is a fundamental result arising from the realization that the quasi-Fermi level splitting specifies real potential energy differences. A

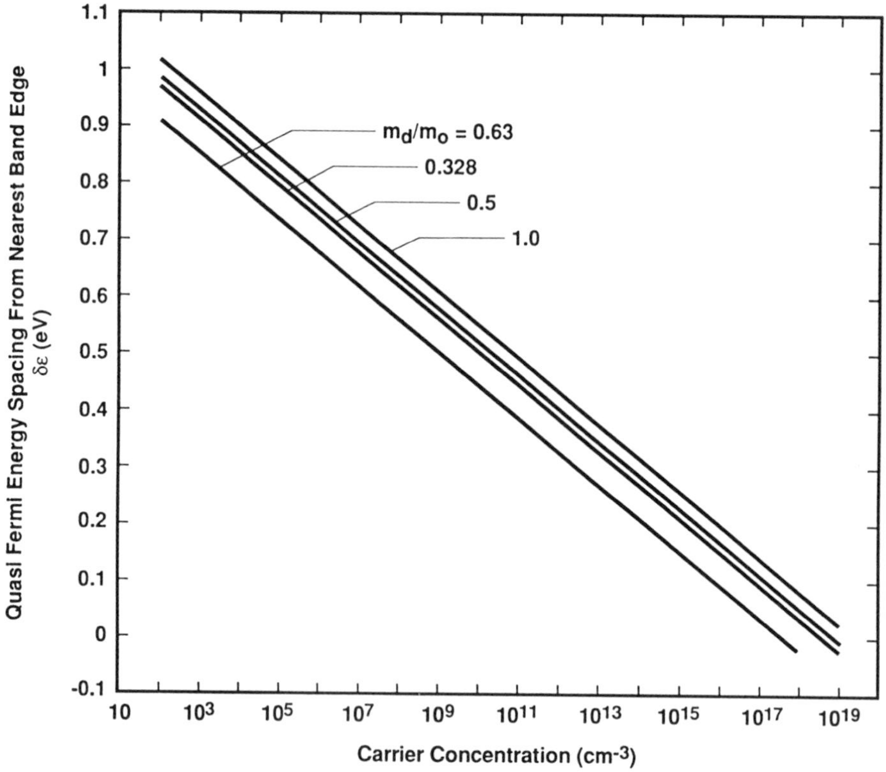

FIGURE 1.10 The spacing $\delta\varepsilon$ in eV of the quasi-Fermi energy from the nearest band edge for carrier concentrations between 100 and 10^{19} cm^{-3} for various density-of-states effective mass values.

majority-carrier doping maxima for Shockley diode solar cells is empirically estimated at $\sim 10^{18}$ cm^{-3} in standard GaAs due to the inverse dependence of radiative recombination lifetimes and the accompanying decrease in minority-carrier diffusion lengths at higher doping levels that decrease J_L (Partain et al., 1987b). Similar effects appear in Si due to Auger recombination (see Chapter 3). This majority-carrier doping level and a "typical" $m_d/m_0 = 0.5$ are used in Figure 1.11 to calculate V_{oc} and qV_{oc}/ε_G versus band gap as a function of the nonequilibrium minority-carrier concentration at the depletion layer edge. In high-performance devices this edge value is very near the peak minority-carrier concentrations. An empirical value for "much less" minority carriers (also termed "low injection") is taken as an order of magnitude or 10^{17} cm^{-3} to stay well below the 10^{18} cm^{-3} empirical majority-carrier limit. Above this, a different theory for solar cell J–V properties and fill factor would be needed. At minority-carrier concentrations $\geqslant 10^{14}$ cm^{-3} and band gaps $\geqslant 1.1$ eV, the V_{oc} value is at least two-thirds of the band gap according to Figure 1.11b for its assumed m_d/m_0 value. At lower concentrations and band gaps, the V_{oc} is less than two-thirds the band gap because the constant spacing of the quasi-Fermi levels from the band edges is then too large a fraction of the band gap.

For reference, all the measured values for the cells in Table 1.5 are plotted in Figure 1.11. Again, all the terrestrial values are plotted as diamonds for one sun devices and as plusses for concentrator devices. The one sun and concentrator space cells are shown as "x's." Note that most of the data lie below the 10^{14} cm^{-3} carrier density line. Only 4 of the 20 devices lie above it, and these are thus identified as approaching Shockley diode open circuit voltage limits. These four are one Si, two GaAs, and one GaSb cells. The 27.8% GaAs cell actually equalled 10^{14} cm^{-3} as shown in Figure 1.8. Figure 1.11 overestimate is due to the approximation of m_d/m_0 as 0.5.

The empirical upper 10^{17} cm^{-3} line appears to be a reasonable limit, at least for the current best-performing cells. The cells with the lowest indicated minority-carrier densities, $Cu_xS/Cd(ZnS)$ and a-Si:H, are both down near the 10^8 cm^{-3} density line—six orders of magnitude lower and ~ 0.2 V below the quasi-Fermi level splitting limit for "low injection." The voltage performance ratio column in Table 1.5 was computed from the highest 10^{17} cm^{-3} minority-carrier density line. Once V_{oc} is known, the ideal FF is determined by the Figure 1.3 relation scaled by the device temperature T. The data points shown in Figure 1.3 are all the measured values listed in Table 1.5. Again, all the terrestrial values are plotted as diamonds for one sun devices and as plusses for concentrator devices. The one sun and concentrator space cells are shown as "x's." Two FF performance ratios are listed in Table 1.5. One applies to the ideal FF for the measured V_{oc}^{meas} and the other applies to each cell's maximum V_{oc}^{max} determined by the upper line in Figure 1.11. Note that all data lie below the solid line theoretical maximum curve in Fig. 1.3. Four Si cells and four GaAs cells have measured FF values that are within 90% of the theoretical values. The latter are listed in the FF performance ratio columns in Table 1.5.

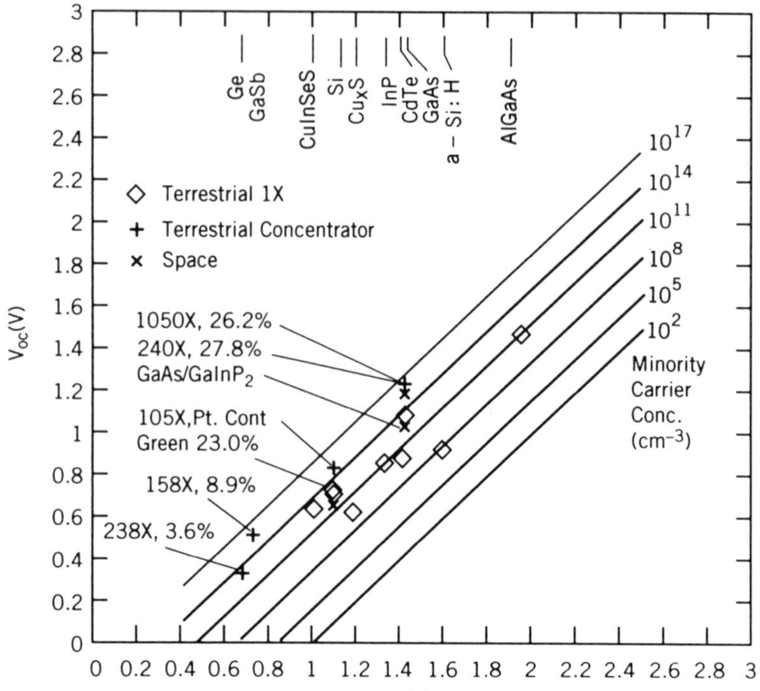

(a)

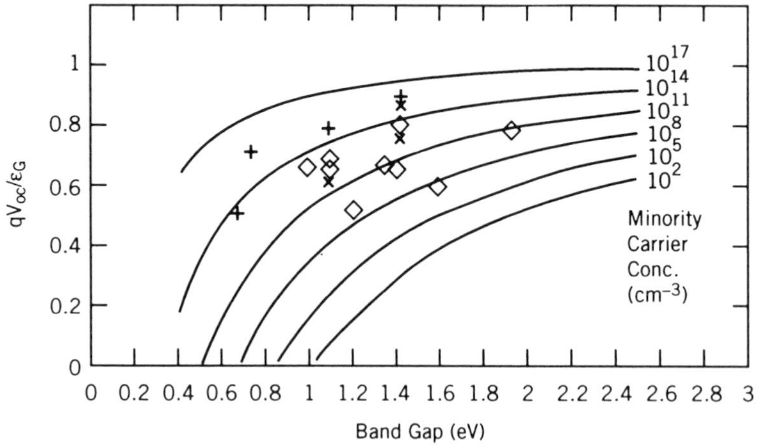

(b)

FIGURE 1.11 Open circuit voltages as a function of band gap. The solid lines are theoretical values for ideal Shockley diodes with various minority-carrier concentrations assuming 10^{18} cm^{-3} majority-carrier concentrations and the density-of-states effective mass ratio of $m_d/m_0 = 0.5$. In **(a)** the actual voltages are given that are normalized by the band gap in **(b)**. The data points are from Table 1.5.

All of these theoretical limit values can now be combined to calculate efficiencies. The results for the terrestrial AM1.5G sunlight spectra are given in Figure 1.12, for the AM1.5D spectra in Figure 1.13, and for the AM0 spectra in Figure 1.14. As before, the Table 1.5 measured data are plotted as diamonds and plusses, respectively, for the terrestrial one sun (Fig. 1.12) and concentrator (Fig. 1.13) cases and as "x's" for the space cell (Fig. 1.14) cases. The ratio of the measured-to-maximum efficiencies are listed as the last performance ratio column in Table 1.5. The best GaAs cell (No. 18, AM1.5D) has an efficiency of 78% of its calculated maximum. The best Si cell (No. 9, AM1.5D) has a comparable measured efficiency, but it is a lower 74% of its indicated maximum. The reason is illustrated in the figures. At the lower minority-carrier concentrations, the maximum efficiencies occur at higher 1.7–1.8 eV band gaps. For medium minority-carrier concentrations, the peaks are nearer 1.4 eV. At the highest limiting minority-carrier concentration, the maxima are around 1.1–1.2 eV but not strongly decreased at 1.4 eV. This is due to the vanishingly small quasi-Fermi level splitting at low, photo-generated, minority-carrier concentrations in low band gap materials. When the light-generated minority concentrations do not exceed their equilibrium values (which are much higher at low band gaps), essentially zero open circuit voltage is available, as indicated in Figure 1.11. This disadvantage at lower band gaps decreases as the nonequilibrium, minority-carrier concentration increases. At the 10^8 cm^{-3} minority-carrier concentration indicated by the V_{oc} (Table 1.5) of the highest efficiency a-Si:H cell plotted in Figure 1.11, the maximum AM1.5G efficiency should occur at an $\sim$1.5 eV band gap, near a-Si:H's 1.6 eV value. The V_{oc} values of the high-performance one sun GaAs cells from Table 1.5 specify minority-carrier concentrations of almost 10^{14} cm^{-3} in Figure 1.11. This concentration level corresponds to maximum AM1.5G efficiencies at a band gap of 1.4 eV from Figure 1.12, very near GaAs's 1.42 eV value.

The peak predicted efficiencies are 37.9%, 37.3%, and 33.0%, respectively, for the AM1.5G, AM1.5D, and AM0 spectra, all at a 1.2 eV band gap. Such high but as yet unachieved efficiencies have been predicted before (see Fig. 7, Chapter 14 in Sze, 1981). However, now the performance below theory can be separated into deficits in current, voltage, and FF. The first of these is due to imperfect quantum efficiency and nonzero surface reflection; the second, to low minority-carrier concentrations; and the third, to excess "leakage currents," which is discussed below. These three deficit factors are listed as performance ratios at the right side of Table 1.5, and their product specifies the ratio of measured-to-maximum efficiencies. For the one sun cells with efficiencies of 23% or greater, the largest deficit is in V_{oc} due to low minority-carrier concentrations. The latter deficit is reduced in the concentrator cells with efficiencies greater than 26%, particularly in GaAs, where the 27.8% efficient device has nearly equivalent deficits in all three factors.

Figures 1.12 and 1.13 are similar to Shockley and Queisser's Figure 5 (1961) except that these workers allowed the equivalent of 10^{18} cm^{-3} and higher minority-carrier densities (where $V_{oc} \sim \varepsilon_G$) and they considered only non

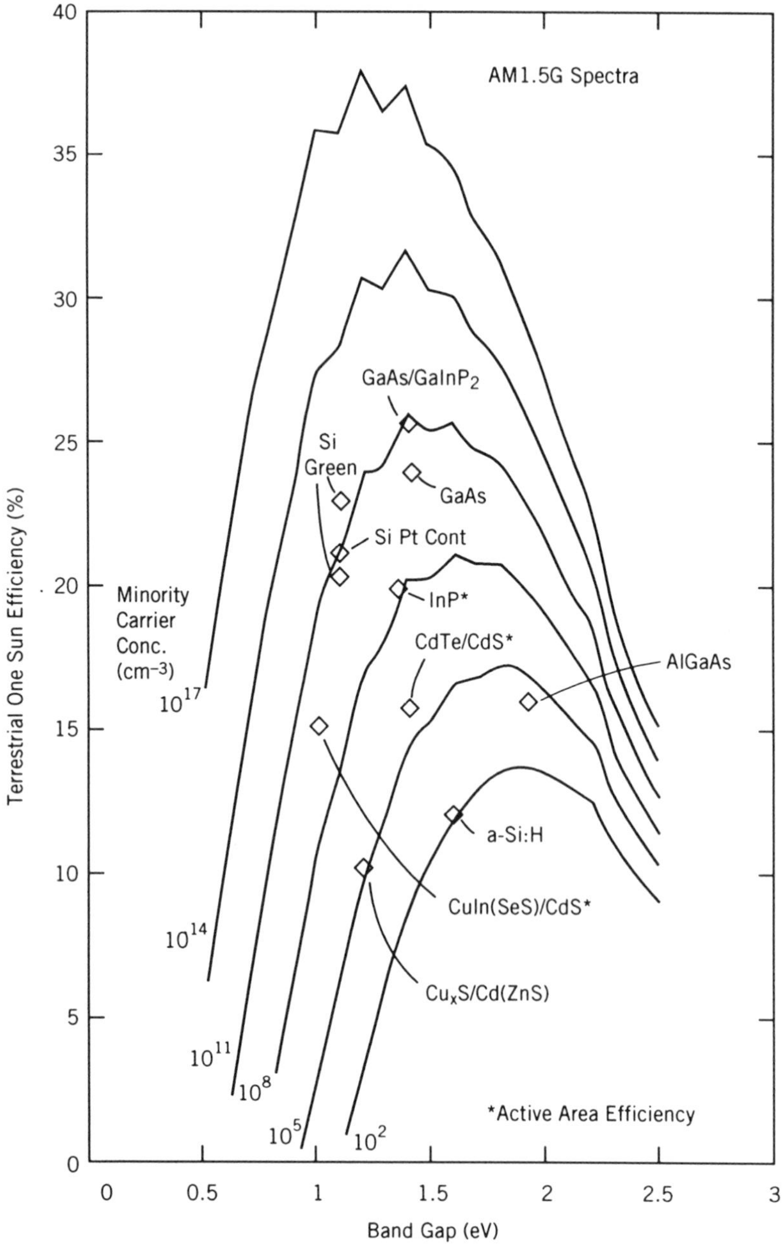

FIGURE 1.12 Terrestrial one sun solar cell efficiencies as a function of band gap. The curves are ideal Shockley diode values for the Figure 1.11 conditions and the maximum short circuit currents in Figure 1.9. The data points are from Table 1.5.

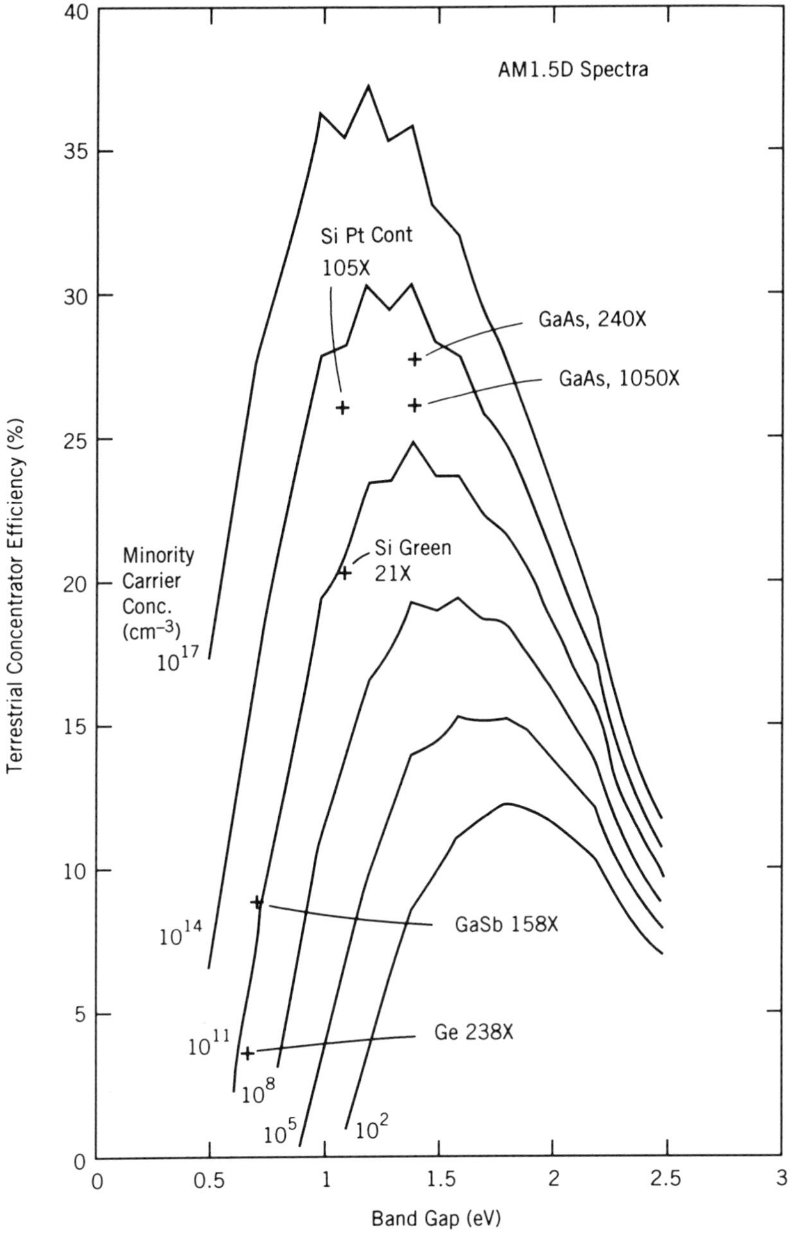

FIGURE 1.13 Terrestrial concentrator solar cell efficiencies as a function of band gap. The curves are ideal Shockley diode values for the Figure 1.11 conditions and the maximum short circuit currents in Figure 1.9. The data points are from Table 1.5.

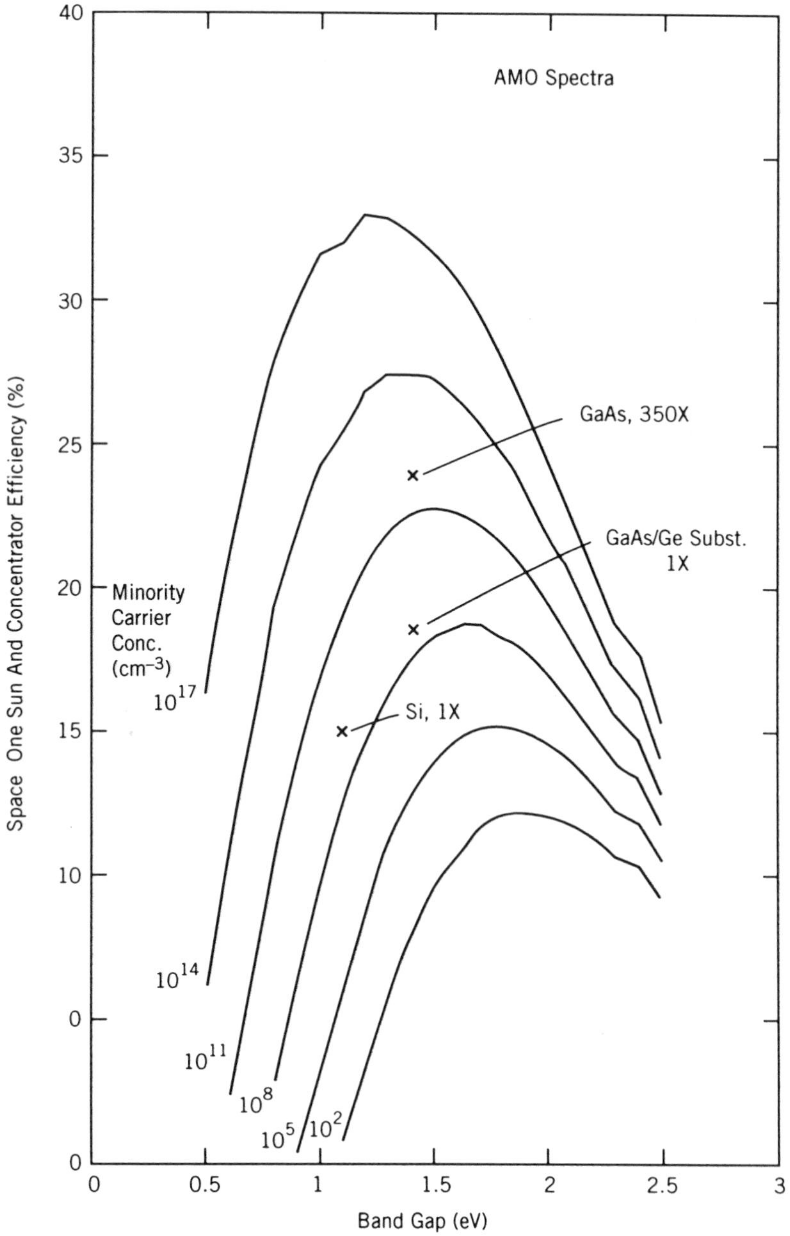

FIGURE 1.14 Space solar cell efficiencies as a function of band gap. The curves are ideal Shockley diode values for the Figure 1.11 conditions and the maximum short circuit currents in Figure 1.9. The data points are one sun and concentrator values from Table 1.5.

radiative recombination lowering of minority-carrier lifetime (and its resultant density) and a smooth black body spectral irradiance at 6,000°K. The figures in the present chapter apply to any mechanism that lowers minority-carrier concentration (and lifetime), both radiative and nonradiative, and they are based on the tabulated spectral irradiances in Appendices A and B.

Theoretically, any energy of an absorbed photon greater than the splitting of the quasi-Fermi levels is wasted because it cannot be converted into a usable, majority-carrier potential energy difference or voltage that can be transported to an external electrical load. To be absorbed and produce an electron–hole pair, the photon energy must exceed the band gap ε_G. According to Eq. (17), the open circuit voltage then increases linearly with band gap for given majority- and minority-carrier concentrations. Since the available current decreases with band gap (Fig. 1.9), there is a maximum efficiency versus band gap that varies with the assumed majority- and minority-carrier concentrations (Figs. 1.12 to 1.14) for a single band gap solar cell.

The loss of absorbed photon energy, down to the solar cell output voltage, can be divided into three parts. The first component is the photon energy excess greater than the band gap. It is reduced by multiple band gaps that allow photon absorption in sections where the band gap is only slightly less than the photon energy. The second component is the loss in converting light-generated minority carriers into majority ones. The latter equals the band bending across the depletion region at the maximum power point (refer to Figs. 1.1 and 1.7c, d). It is reduced by splitting the quasi-Fermi levels to as near the band gap (or flat band) value as possible. Low injection Shockley theory limits this to slightly less than the band gap as indicated in Figure 1.11b. Fundamentally flat band operation is possible but a new "high injection" theory would then be needed to specify the FF. This second component is reduced by concentrating the light to increase the minority-carrier concentration. It is also reduced by decreasing minority-carrier recombination rates with longer lifetimes and less surface recombination and by thinning the cell with light trapping for higher minority carrier generation rates per unit volume with a constant light input flux. The third component is the spacing of the majority-carrier quasi-Fermi levels from their nearest band edge (for this, compare Figs. 1.1 and 1.7c, d). This last loss can be reduced to zero with the degenerate majority carrier doping illustrated in Figure 1.7c.

Figure 1.15 shows the predicted efficiencies for a two band gap cell with the higher band gap device stacked on top of the lower band gap one for the concentrated AM1.5D terrestrial sunlight spectra. It is assumed here that the two terminals of each device are accessible for independent operation as a four terminal cell and that all photons with energies below the top cell band gap are transmitted without loss to the lower device. Here the particular case of a 1.4 eV band gap top cell is plotted versus the energy gap difference with the lower band gap bottom cell. With two band gaps, the peak predicted efficiency is increased to 49.3% for a band gap difference of 0.7 eV at the limiting 10^{17} cm^{-3} minority-carrier concentration. This corresponds to a 13.4% efficiency

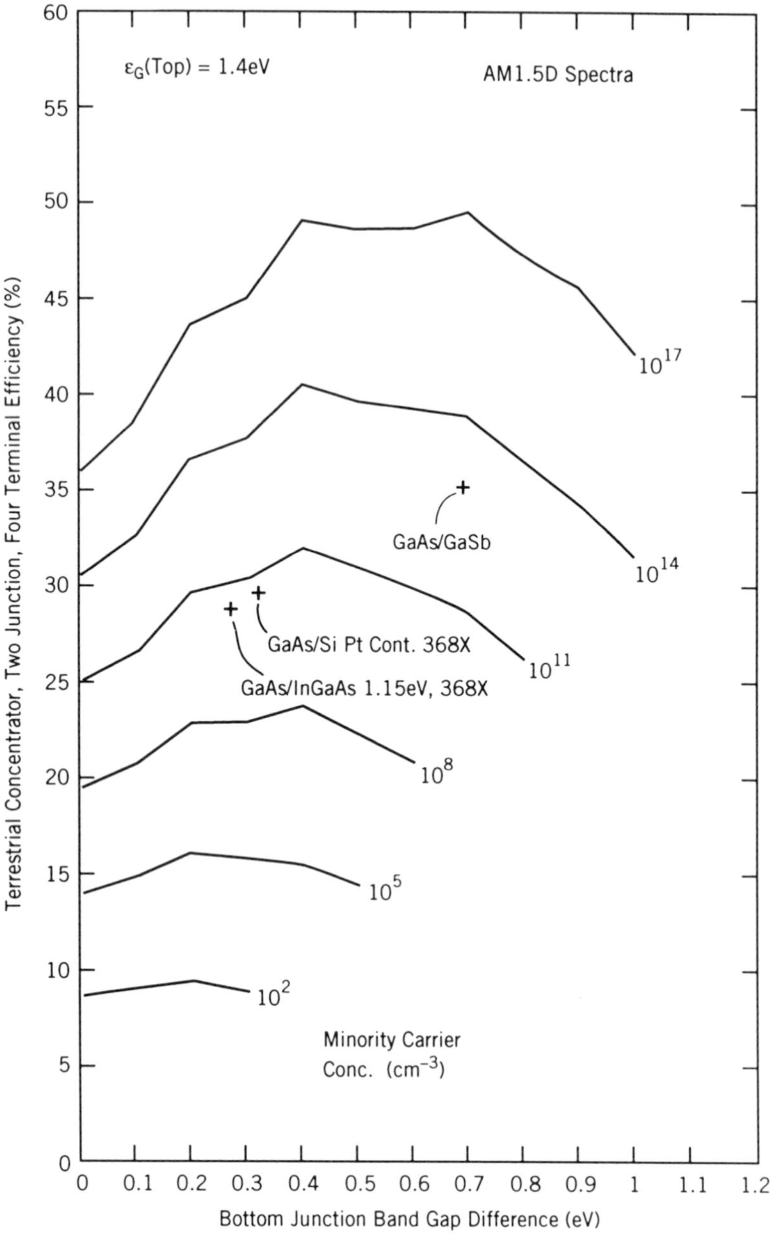

FIGURE 1.15 Two junction, terrestrial concentrator solar cell efficiencies as a function of bottom junction band gap difference, from the top cell, for four terminal devices. The curves are ideal Shockley diode values for the Figure 1.11 conditions and the maximum short circuit currents in Figure 1.9, when the top junction band gap is 1.4 eV.

contribution from the bottom device as seen by subtracting the 35.9% contribution from the top device indicated at the zero band gap difference. At lower minority-carrier concentrations, the peak efficiency occurs at lower band gap differences. The resulting larger band gap of the bottom device gives lower equilibrium minority-carrier concentrations for greater quasi-Fermi level splitting when nonequilibrium minority-carrier concentrations are low. Below the $10^5\,cm^{-3}$ concentrations, little meaningful contribution comes from the bottom device. All the lower curves are truncated at the point where there is zero voltage and efficiency contribution from the bottom device.

The experimental data point at 35% efficiency is for a GaAs concentrator cell, mechanically stacked on a GaSb one, for four terminal operation (Fraas et al., 1990, after correction for Sandia recalibration). The 8.9% contributed from the bottom device (Table 1.5) corresponds to 84% of its maximum current density (Fig. 1.9), 87% of its maximum FF (Fig. 1.3), and 85% of its maximum V_{oc}. The latter V_{oc} specifies a minority-carrier concentration approaching the $10^{17}\,cm^{-3}$ limit line shown in Figure 1.11. These illustrate how the analysis in this chapter can evaluate how close a device's operation approaches its theoretical maximum to judge how much improvement can be expected from additional development (or from increased sunlight concentration ratios to increase minority-carrier concentration and V_{oc}). The two data points at 29.5% and 28.7% are also four terminal values for GaAs concentrator cells mechanically stacked onto an Si point contact cell and an InGaAs cell, respectively (Gee and Virshup, 1988, after correction for Sandia recalibration).

Figure 1.16 gives a two junction, four terminal, AM1.5D efficiency plot similar to that in Figure 1.15 except for a 1.8 eV band gap for the top device. The peak efficiency of 51.3% is similar but for a larger 0.8 eV band gap difference for the bottom device. At higher top junction band gaps, there is a smaller relative contribution to total efficiency by the top device.

Two junction devices can be constrained to be current matched. Then the two can be connected in series for two terminal operation. The predicted peak AM1.5D efficiency of 50.1%, as shown in Figure 1.17, is similar to the four terminal cases. However, at the specified 1.6 eV top junction band gap for maximum performance, there is a single 1.01 eV band gap lower junction band gap that current matches it. For reference, the 27.6% and 29.5% AM1.5G efficiency, two junction, two terminal data points are shown for the one sun 1.93 eV AlGaAs/GaAs cell (Chung et al., 1989) and the 1.9 eV $GaInP_2$/GaAs cell (Bertness et al., 1993). AM1.5G and AM1.5D efficiencies are similar as illustrated in Figures 1.12 and 1.13. Such one sun, two junction devices are in their infancy with much room for potential improvement. Currently both AlGaAs and GaAs cells perform well below their potential at one sun, as illustrated in Table 1.5 and Figures 1.9 and 1.11. They underperform in both open circuit voltage and short circuit current. The voltage deficit can be reduced substantially by light concentration as illustrated by the concentrator values in Table 1.5 and by the 35% GaAs/GaSb datum point (Fraas et al.,

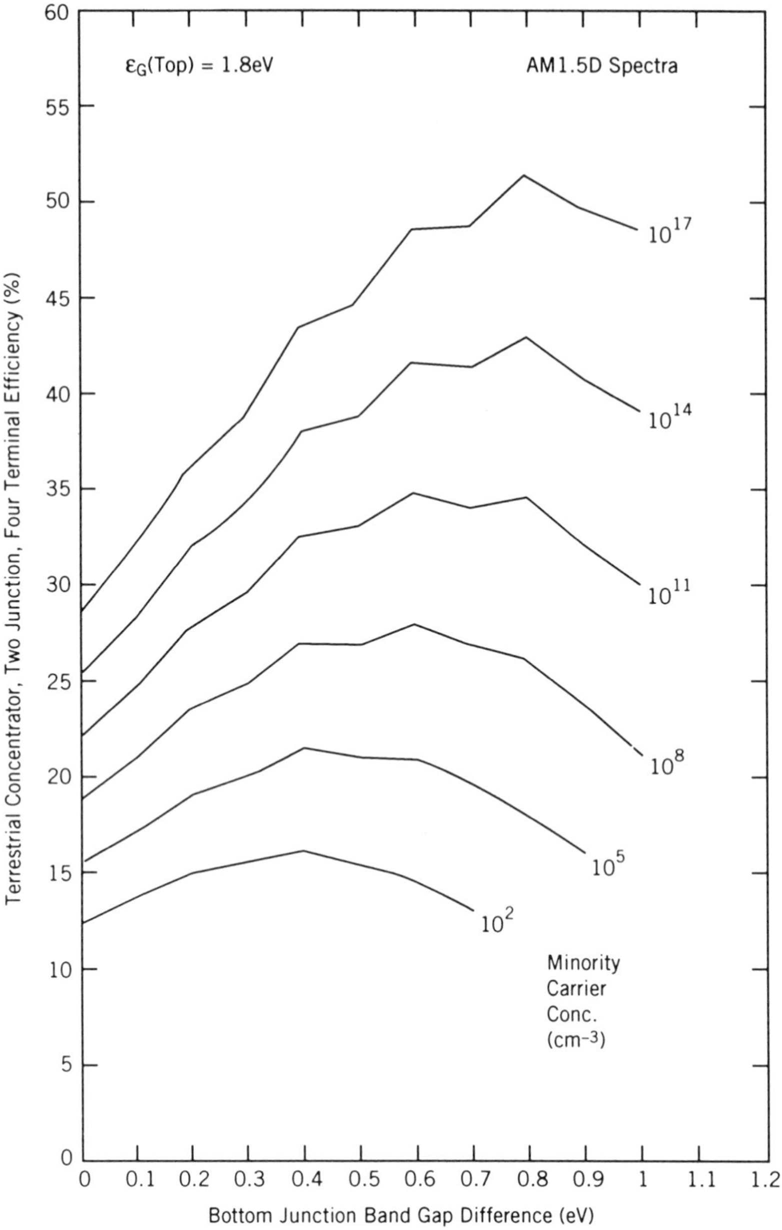

FIGURE 1.16 Two junction, terrestrial concentrator solar cell efficiencies as a function of bottom junction band gap difference, from the top cell for four terminal devices. The curves are ideal Shockley diode values for the Figure 1.11 conditions when the top junction band gap is 1.8 eV.

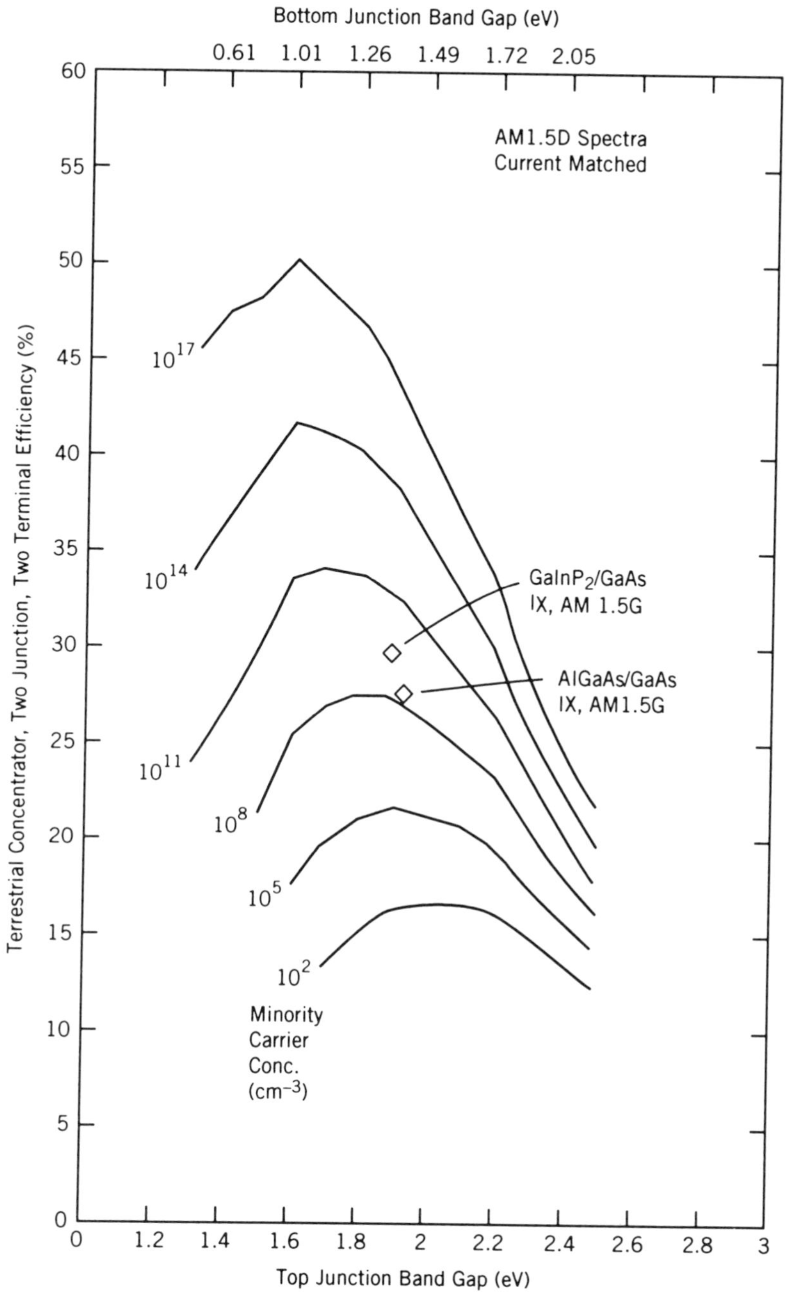

FIGURE 1.17 Two junction, terrestrial concentrator solar cell efficiencies as a function of top junction band gap for two terminal devices. The curves are ideal Shockley diode values for the Figure 1.11 conditions when the top junction band gap is 1.4 eV. The bottom junction band gap required for two terminal current matching is shown on the top horizontal axis. The data points are unconcentrated one sun values.

1990, after correction for Sandia recalibration) in Figure 1.15 that is much nearer its predicted maximum performance.

At modest, nonequilibrium, minority-carrier concentrations, only modest increases in efficiency come from adding extra junctions at lower band gaps whose higher equilibrium concentrations give lower voltage from less splitting of their quasi-Fermi levels (Fig. 1.11). For example, such 10^5–10^8 cm^{-3} concentrations and a second junction would provide 1–3 absolute percent extra concentrator (AM1.5D) efficiency under a 1.4 eV top device band gap (Fig. 1.15) and 4–6 absolute percent under a 1.8 eV top device band gap (Fig. 1.16). This is consistent with the 14% one sun AM1.5G efficiency (~AM1.5D efficiency) reported for multiple (triple!) junction a-Si:H–based cells (see Chapter 8). This is 2% higher than the best 12% single junction efficiency cell listed in Table 1.5 whose V_{oc} and 1.6 eV band gap specify a 10^8 cm^{-3} minority-carrier concentration (in Fig. 1.11). Theoretically additional junctions add extra efficiency for increasingly better matches between quasi-Fermi level splitting and photon energy when the former concentrations are high. Such predicted multiple junction efficiencies are as large as 72% for 36 different band gaps under concentrated AM1.5D sunlight (Sze, 1981). Indeed, a triple junction, two terminal, one sun AM0 space cell of AlGaAs/GaAs/InGaAsP has an efficiency of 25.2% (Chung et al., 1992). This exceeds the best two junction efficiency of 23.0% for AlGaAs/GaAs under the same conditions (Chung et al., 1989), which in turn exceeds the highest reported one junction efficiency of about 22% for GaAs under the same conditions (Bertness et al., 1988). Clearly such one sun multijunctions have great potential for improved performance, particularly through increasing their minority-carrier concentrations. For example, the best AlGaAs top cell listed in Table 1.5 has a V_{oc} and a 1.93 eV band gap that correspond to only a 10^{11} cm^{-3} minority-carrier concentration (Fig. 1.11).

5. EFFICIENCY IMPROVEMENT PROSPECTS

The technology solutions to closer approaches to fundamental Shockley diode performance boundaries are limited only by one's imagination. The highest one sun efficiency listed in Table 1.5 (No. 16) is 25.7% for the GaAs/$GaInP_2$ cell (Kurtz et al., 1990). The paper of Kurtz et al. (1990) only describes heteroface geometry with a vague reference to "variations on this structure." The record performance device was actually a heterojunction device with a $GaInP_2$ emitter and a GaAs base (J.M. Olson, personal communication). Its performance advantages compared with the one sun homojunction GaAs cell in Table 1.5 are higher short circuit current and FF. The higher current may be due to better emitter surface passivation for improved blue response, and the higher FF may be due to reduced "leakage currents." If one starts at the open circuit voltage and traces down to the maximum power point (Fig. 1.2), ideally the current decrease is described by the exponential diode term in Eq. (5) that varies with qV/kT for negligible series resistance in the Shockley diode model.

Real data are invariably displaced up from this "ideal" to a higher diode current at maximum power in a manner not well described by series resistance. This decreases the measured FFs below the ideal. These differences between "ideal" and "measured" diode currents, at voltages around and below the maximum power point, are often referred to as excess "leakage currents". Apparently the $GaAs/GaInP_2$ heterojunction suppresses such "leakage currents" to levels significantly exceeded by light-generated current at one sun rather than at the concentrations required for similar performance levels in traditional GaAs cells.

Fundamentally, lower leakage means that a higher concentration of minority carriers must be injected, across the junction, to provide a given "diode dark current" to oppose the light-generated current for any specific net current value. The results are a larger quasi-Fermi level splitting and higher voltage at this net current level. As "leakage current" increases, it first lowers FF values and then, at higher values, decreases V_{oc}. The above heterojunction success suggests trying similar experiments with AlGaAs/GaAs heterojunctions for better one sun performance. In addition, replacing the p-type side of a GaAs cell with AlGaAs of increased band gap could inhibit minority-carrier electron injection and decrease the "ideal" dark current J_0 by up to a factor of 8 (see $J_{oe}/J_{oh} = 8.47$ in Table 1.3) and thus further improve both the V_{oc} and FF after "leakage current" levels have been exceeded by the light-generated current.

Another obvious way to improve efficiency toward fundamental Shockley diode limits (through increases in minority-carrier concentrations and thus voltage) is to concentrate the sunlight. This has been done up to $\sim 1{,}000\times$ in GaAs cells as illustrated in Figure 1.8. The controlling problem here is series resistance. The prismatic coverslide described in Chapter 10 can allow up to 50% top grid metal coverage with almost no loss of light and thus reduce resistance for high current densities. The cover refracts light around the metal grid fingers. This capability has not yet been fully exploited. However, the record concentrator efficiencies reported thus far (27.8% GaAs $240\times$ and 33%–35% GaAs/GaSb $\sim 100\times$, AM1.5D) used the prismatic cover. Series resistance is particularly troublesome in silicon with lower carrier mobility than in GaAs. The 21X concentration listed for silicon cell No. 6 in Table 1.5 increased the open circuit voltage compared with the unconcentrated cell No. 5. However, the FF decreased, apparently due to series resistance (see Fig. 1.4).

An innovative way around this series resistance problem is the point contact cell covered in Chapter 3. It is a major departure from standard n/p junction solar cell geometry modeled in the present chapter (Fig. 1.1). An undoped, long minority-carrier lifetime, silicon wafer of high quality (typically float zone) has one side heavily doped ($\sim 10^{19}\,cm^{-3}$), by diffusion through an oxide mask, in a polka-dot pattern of alternating p- and n-type dots of small diameter. Light enters the opposite surface and immediately generates nonequilibrium carrier concentrations well in excess of the equilibrium majority-carrier value. This immediately violates the "low injection" assumption of the one-dimensional, Shockley diode n/p junction model and rigorously requires a more involved treatment, described in Chapter 3 and its references (particularly for the three-dimensional structure). However, this type of cell behavior does not yet exceed the limit values defined in this chapter (see cell Nos. 7 and 9 in Table 1.5 and Figs. 1.3, 1.9, and 1.11 to 1.13). These suggest that an approximate treatment with the present, one-dimensional, Shockley model may be useful. In this vein, the light-generated current may be approximated as that collected when the polka dots impose zero, nonequilibrium, carrier concentration boundary conditions at the back surface. Nonequilibrium carrier collection is then by diffusion just as in the standard model, except that "minority" carrier loses its meaning here since both electrons and holes greatly exceed their equilibrium values and both are collected at the same back "surface." Very little of the charge transport distances are traversed as majority carriers, described by a series resistance *R*, and then only through thin, heavily doped polka dots that contribute low series resistance values. These dots directly contact metal conductors that remove the current to the external circuit with little added series resistance.

Such a p/i/n geometry often results in increased "leakage current" that lowers V_{oc} and FF, particularly in defect-filled materials such as an a-Si : H cell (see Table 1.5 and Chapter 7) or a GaInAs cell with lattice mismatch to a GaAs substrate (Schultz et al., 1993) or in a high-defect-selected GaAs cell (Partain and Liu, 1989). However, such increased "leakage" may not be a necessary restriction in very high quality, defect-free material. High "emitter" and "base" polka-dot doping should raise the dark current J_0 through increased Auger recombination in indirect band gap materials like silicon (and by radiative recombination in direct band gap materials like GaAs), but this is offset by the polka dots covering only a small fraction of the back surface. High 10^{19} cm^{-3} "emitter" and "base" doping allows minority carriers to reach 10^{18} cm^{-3} levels before "low injection" is exceeded in the "emitter" and "base" regions. Hence, for point contact cells, even higher 10^{18} cm^{-3} limit curves should be added to Figures 1.11 to 1.14. The net result is predicted V_{oc} values (Eq. [17]) increased by $2 \times (kT/q) \ln 10$ ($=0.117$ V at 21°C), with corresponding increases in FF from Figure 1.3.

The silicon band gap of 1.1 eV is closer to the efficiency maximum when minority carriers in the dot emitters and bases approach 10^{17} cm^{-3} and higher concentrations. Low injection conditions are always violated in the junction depletion region of standard cells, but this region's thickness is usually so thin (see Figs. 1.5 to 1.7) that its effects are considered negligible in the Shockley model. The point contact geometry could allow GaAs concentrator cells to approach their 36% (and higher) efficiency potential, indicated in Figure 1.8, with less restrictions from standard geometry series resistance. Although there is no immediately obvious way to generate the alternating small polka dots in GaAs with the metal-organic-chemical-vapor-deposition process typically used, ion implantation might offer a viable approach for point contact GaAs fabrication. The CLEFT process could provide a thin device (Gale et al., 1987) and surface texturing, the light trapping.

The other way to increase minority-carrier concentration and voltage is by light trapping. If the surfaces are well passivated and the volumes of the emitter and base are decreased, the number of recombination sites should decrease, allowing higher minority-carrier concentration as long as the same number of photons per second is absorbed. Light trapping was used to produce the first thin-film polycrystalline cell with a reported efficiency exceeding 10% (see cell No. 10 in Table 1.5) (Rothwarf et al., 1975; Bragagnolo, 1978). This lattice mismatched heterojunction had a 0.2 μm thick emitter of Cu_xS ($\varepsilon_G = 1.2\,eV$) on a 30 μm thick base of "transparent" Cd(ZnS) ($\varepsilon_G > 2.4\,eV$) whose textured geometry and reflecting back contact provided light trapping. A III–V analog could have a thin GaAs emitter layer lattice matched to an AlGaAs base that is textured with a reflecting back surface. Chapters 2 and 3 and their references describe how recent approaches to light trapping have become quite sophisticated. Very thin cells may require external structures for efficient light trapping (Landis, 1990).

6. FIELD EFFICIENCY PERFORMANCE

The efficiencies measured in 15–18 kW demonstration systems (Candelerio et al., 1991) are listed in the first three rows in Table 1.6 and compared with the best laboratory cell efficiencies. The net de-rating factors are given for converting "best-ever" laboratory cell measurements to system demonstration values.

TABLE 1.6 Comparison of Measured Demonstration System Efficiencies to Efficiencies of Best Individual Laboratory Cells and the Net De-Rating Factor to Convert the Latter Into the Former

	Demonstration System			Best Laboratory Cell		
Cell Type	Efficiency (%)	Size (kW)	Concentration (×)	Efficiency (%)	Concentration (×)	Net De-rating Factor
Tandem junction amorphous silicon	2.5–4.0	15.7–17.3	1	13.7	1	0.18–0.29
Single crystal silicon	10–12	18.7	1	23.0	1	0.43–0.52
Single crystal silicon	11–12	16.5	22	26.1	100	0.42–0.46
GaAs	19–20[a]	0.04[b]	942	27.8	240	0.68–0.72 (0.59–0.63)[c]

[a]Efficiency for direct normal sunlight not adjusted for "lost" diffuse sunlight offset by two axis tracking.
[b]Single module, not a multi-kW system.
[c]Corrected de-rating accounting for median/best-efficiency ratio of 0.87.

These de-ratings vary from 0.18–0.29 for tandem junction a-Si:H to 0.4–0.5 for single crystal silicon. The tandem junction, a-Si:H systems are installed in Davis, California, and on the island of Maui in Hawaii, and their "best-cell" comparison values come from Chapter 7. The single crystal silicon 1 × and concentrator systems are installed in Davis, and their "best-cell" values come from Table 1.5. Thus far, the highest performance GaAs concentrator cells have only been tested as a single 40 W module installed in Tempe, Arizona, and its "best-cell" comparison comes from Table 1.5. The "corrected" de-rating factor-for GaAs is between 0.59 and 0.63 as described below. As also discussed below, a larger GaAs system could fall into the same 0.4–0.5 de-rating range as the single crystal silicon. The factors that determine these de-ratings are developed below, and rough estimates are given for their individual values. More precise values should be a topic for future study.

Individual solar cells typically provide watt-level or less of electrical power each on exposure to normal unconcentrated sunlight. Hence significant power delivery requires them to be connected in series to provide desired voltage levels and in parallel for desired current levels. The starting point for interconnected cell behavior is the individual cell current–voltage characteristics such as in Figure 1.2. The zero current and voltage lines divide the characteristics into four quadrants, with I in the upper right hand corner and II through IV counted counterclockwise from there. In quadrants I and III the product of current and voltage is positive, and the cell consumes electrical power. In quadrant IV the product is negative, and the cell delivers power. When one cell is completely shaded in a series voltage string it becomes passive, operating in quadrant III. The voltage from the remaining cells tries to force the reverse current J_L through the shaded one and reverse bias it at the voltage sum of the other cells at open circuit. If the latter voltage is high enough, the shaded cell will go into reverse bias break down, heat up, and typically fail by becoming either a shorted or an open circuit. This is the first type of the "hot spot" failure mode. The second type is when there is partial shading of a single cell in a series string. The remaining cells will try to maintain the given total current by forcing greater than J_L through the unshaded part of the shaded cell, again operating it in quadrant III. This second type of "hot spot" failure mode also biases the cell to its reverse bias break down, but now at high current for even more severe heating and more rapid failure.

Shading is just the most drastic example of cell nonuniformity. Production runs typically provide cells with a distribution of values for short circuit currents, open circuit voltages, and FFs. For space panels, individual cells are typically sorted by their short circuit currents. Series strings are only assembled from those with nearly identical currents to avoid the quadrant III heating and failures. First connecting all the cells in parallel would make the assembly "average" all the currents and probably eliminate the need for sorting by individual cell current values. These parallel strings can then be connected in series for the required voltage. Unfortunately this "parallel first" interconnection requires about twice the wiring of "series first" interconnection schemes, and it is not widely used. Nevertheless the "parallel first" provides the greatest protection against "hot spot" failures, gives the least fractional power loss with shading or short circuit current non uniformity, and requires the fewest protective diodes (Partain and Sayed, 1974; Sayed and Partain, 1975). The latter are shunted across parallel strings to short out reverse bias voltages and eliminate "hot spot" failures. Thus far protective shunt diodes are too expensive for widespread use on large terrestrial arrays.

Parallel-first interconnection does short out cell output voltage variations. This should affect array performance less than short circuit current variations in series-first strings because open circuit voltage varies only logarithmically with J_L/J_0 in ideal Shockley diode solar cells (Eq. [11]). Similarly the FF and maximum power voltage are determined only by the open circuit voltage. Spatial J_0 variations are "averaged" over a single cell's area. Parallel-first connection has the same effect as increasing cell area. Individual cell-averaged J_0's should be similar in well-controlled manufacturing operations.

A major point here is that real manufacturing processes produce a distribution of different individual cell properties. Each cell cannot operate at its own maximum power point but at some group "average" influenced by the interconnection scheme. Thus there are unavoidable interconnection losses in efficiency. The exact loss depends on individual case particulars, including ohmic wiring losses, but can be a few percent. Hence an interconnect de-rating factor of say 0.95 can be used as a ball park figure for a well-designed system. This estimated value is listed in Table 1.7. Even if identical cells could be made, some shading (and mismatch) is inevitable, particularly for terrestrial systems, from clouds and birds and adjacent structures or trees or leaves or service personnel, and so forth. Shading losses can easily exceed a few percent and can produce temporary and permanent losses in system power (Partain and Sayed, 1974; Sayed and Partain, 1975).

TABLE 1.7 Estimated Values of De-Rating Factors to Account for the Lower Performance of Solar Cells in Large-Scale Systems Compared with the "Best-Ever" Efficiencies Reported for Carefully Constructed, Individual Laboratory Cells

De-Rating Condition	Estimated Factor Value
Interconnect losses	0.95
Parameter distribution versus best	0.5–0.8
Design differences	0.5–0.85
Surface fill factor	0.85–0.95
Operating temperature	0.94
Net estimated de-rating factor	0.19–0.58

Large-scale manufacturing cannot match the best-ever performance reported for carefully fabricated laboratory cells of each type. By definition, "best" is taken from the extreme end of a distribution with all the other devices located some performance distance below. Table 1.5 is such a list of "best laboratory cells" with one exception. This exception is the average value of the one sun Si space cells taken from Chapter 4. In systems of large numbers of cells, real performance can substantially differ from "laboratory best" by 20% and more. Hence a fabrication distribution de-rating of 0.5–0.8 can be taken as a ball park figure as listed in Table 1.7. The lower value applies to the least developed cells. The 19%–20% GaAs module of Table1.6 was assembled with 27% efficient cells matched and selected for their high efficiency (Kuryla et al., 1991, single cell efficiencies corrected for Sandia recalibration). The median efficiency for 750 such cells was 24.3%. The latter is 0.87 times the "best cell" value of Table 1.6 and is the factor used to correct the "net de-rating factor" listed there in parentheses. If the GaAs fabrication process were constrained to provide tens of kWs of power at the low fixed prices of the other Davis and Maui systems, this distribution de-rating factor could well fall into the 0.8 range listed in Table 1.7 and lower the Table 1.6 "net de-rating factor" for GaAs into the same 0.4–0.5 range as the single crystal silicon.

The cell designs and/or the materials quality used in large-scale manufacturing are those that are low enough in cost that the resulting product can be sold at a competitive price. Best-ever laboratory cell fabrication typically spares no expense to squeeze out the highest possible device performance. A de-rating factor to account for basic design and materials differences between commercial (or demonstration) and "best laboratory" cells can be substantial and vary widely between solar cell types. An estimated value of 0.5–0.85 is listed in Table1.7.

The Table 1.5 performance values are all for cell temperatures between 21° and 28°C. Real systems operate in a variety of ambient temperatures that are generally highest when the sun is the brightest, providing the most input power. In addition, the cell temperatures typically exceed the ambient by ~25°–30°C for one sun cells (Candelario et al., 1991) and ~30°–40°C for concentrator cells (Kuryla et al., 1991; O'Neill et al.,1991). Chapter 13 gives temperature de-rating factors that are similar for space Si and GaAs one sun cells. Here a "ball park" temperature de-rating factor of 0.94 is used for Table 1.7. It is a measured value for GaAs concentrator cells 30°–40°C above room temperature (Kuryla et al., 1991). The latter assume an "average" 21°C ambient temperature. The real de-rating factor depends on the detailed cell and module thermal design and on the average yearly ambient temperature of operation that is location and altitude dependent. An even stronger variable is average wind speed, because the resulting convective cooling substantially lowers the cell temperature rise above ambient.

The surface of solar cell arrays cannot avoid some inactive regions due to interconnect wiring and mechanical framing at the array perimeter. This is described by a surface fill factor (SFF) that is the fraction of active area to total array area. Estimated typical values of SFF are the 0.85 to 0.95 listed in Table 1.7. Dead regions occur in the spacing between cells and these also lower the SFF. This loss is larger in nearly square cells cut from circular substrates to minimize loss of expensive semiconductor material that leave missing diagonal edges.

The optical components of terrestrial concentrator systems typically cannot collect the diffuse scattered light from a "blue" or cloudy sky, but only that portion that comes directly from the unobscured solar disk. The system operator typically does not care about the nuances of diffuse and direct sunlight, but only about how much power he or she can expect from a given area of installed solar cell collectors in a given location. Thus a "fair" efficiency comparison divides all electrical outputs by a single solar input value (such as AM1.5G onto a fixed, optimal tilt surface) installed at a single location (Hester and Hoff, 1985). For the ideal AM1.5D and AM1.5G spectra of Appendix A, this would impose as diffuse light de-rating factor of $76.72/96.25 = 0.80$ on concentrator systems. However, this is offset by the two axis tracking of concentrating systems by an amount that varies from one geographical location to the next. This net ratio is listed in Table 1.8 for the direct normal energy with two axis tracking divided by the global energy striking a fixed, optimal tilt surface (Zweibel, 1990). In southwest regions like Las Vegas, one gains 6% more energy in tracking than is gained by collecting the diffuse light onto a fixed surface. In areas with more clouds, there is a net loss for concentrators with ratios of 0.76 and 0.89 for Miami and Oklahoma City, respectively. If one or two axis tracking is added to an $1\times$ nonconcentrating system, from 16% to 38% more energy can be made to strike the system surface compared with a fixed tilt system (Zweibel, 1990).

The optical collectors of concentrator systems typically transmit from 80% to 90% (Kuryla et al., 1991; O'Neill et al., 1991) of the incident sunlight onto the solar cell. The Table 1.6 GaAs module had an optics transmission of about 0.85. This can be used in place of the "design and materials difference factor" in Table 1.7, since these GaAs cells had essentially the same design and materials as the "best-ever" GaAs concentrator cell.

TABLE 1.8 Ratio of the Direct Normal Sunlight Energy Striking a Two Axis Tracking Surface to the Global Sunlight Energy Striking a Surface of the Same Area But Fixed at a Constant, Optimal Tilt Angle to the Earth's Surface

Location	Ratio
Las Vegas, NV	1.06
Miami, FL	0.76
Oklahoma City, OK	0.89

The individual de-rating factors in Table 1.7 are rough estimates whose products are in the range of the measured "net de-rating factors" in Table 1.6. Accurate measurement and modeling of these individual factors may lead to improved performance of demonstration and commercial systems. Temperature and interconnect losses do not appear to be dominant. The distribution and materials difference factors seem to be the largest. An additional example of the de-rating effects is the 25% efficiency measured at 50°C operating temperatures on GaAs mechanically stacked on GaSb (see Chapter 14) for three stacks interconnected for voltage matching and assembled under 56 × concentrating lenses into a minimodule. The latter is 71%–76% of the 33%–35% efficiency measured for "best-ever" laboratory cells at room temperature. In tens of kW level demonstration systems, fabrication distribution effects could lower it to the 60% levels listed in Table 1.6. Resulting 19%–21% conversion efficiencies would still be impressive.

7. SUMMARY

Sunlight is converted into DC electric power in band gap semiconductors through changes in minority-carrier concentration and their potential energy that light absorption and electron–hole pair generation induce. The equilibrium minority concentrations are so low in semiconductors with typical band gaps on the order of 1 eV that sunlight absorption can produce enormous relative increases in minority-carrier concentration with scarcely any measurable changes in majority-carrier concentration. Any concentration gradients that exist exert as real a force on mobile charged carriers as electrostatic electric fields. This gradient force changes charged carrier potential energies with distance and, by implication, with concentration (just as with electrostatic potential energy of charged particles, but now in regions of no net charge). Essentially, if the concentrations were assembled one particle at a time, subsequent particles would climb steeper gradients to add increasing potential energy increments until the final concentration and potential energy were reached. The relatively large increases in minority-carrier concentration directly translate into large relative increases in minority-carrier potential energy that can be a large fraction of the band gap. The latter occurs as minority-carrier densities approach majority-carrier values (within an order of magnitude). The increased potential energy can be efficiently used to do work, external to the semiconductor, when the minority carriers are converted into majority carriers with little loss in their potential energy gain. The majority carriers dominate the conditions of current transport through the metal contacts for power delivery to an external electrical load.

Low loss conversion of minority to majority carriers is provided by the depletion region of p/n junction diodes. At equilibrium, the electrostatic and concentration gradient forces exactly cancel at every point of the depletion region for both conduction electrons and holes. With the enormous relative increases in minority carrier concentration with light absorption, the concentration gradients and their related forces decrease in the depletion region. The electric field that opposes majority carrier diffusion then is unbalanced, and it sweeps minority carriers across the depletion region for conversion to majority carriers. A fundamental simplifying assumption of Shockley's n/p junction diode model (Sze,1981; Shockley, 1976) is that there is negligible change in the quasi-Fermi energies — and the potential energies of their related electrons and holes — across the depletion region. The charge transport expressions (following from Shockley's assumption) accurately describe the behavior of the highest performance n/p junction solar cells.

The dominant optical absorber and generator of minority carriers, under steady-state conditions, is band-to-band electron–hole pair generation for photon energies greater than the band gap. However, any photo-excited minority-carrier energy gain greater than the quasi-Fermi level splitting at the depletion region edges is not available as a potential energy gain following conversion from minority to majority carriers. Thus the highest cell potential energy differences, and thus voltage, occur for the largest quasi-Fermi level splitting. This is limited by the low-level injection assumptions so that the parameter definitions of Shockley diode transport remain valid, like minority-carrier mobility, lifetimes, and diffusion lengths. Formally the cell voltage is determined by the band gap minus the quasi-Fermi level spacings from the band edges at the depletion region edges. Minority-carrier lifetime limitations, like Auger and radiative recombination, practically limit the minimum majority-carrier spacing to levels corresponding to maxima of $10^{18}\,cm^{-3}$ concentrations in standard, one-dimensional symmetry, abrupt n/p junction solar cells. The prime determinant of cell voltage is thus the spacing of the minority-carrier quasi-Fermi levels from the band edges at the edges of the depletion region. In good cells this is very near this minimum spacing anywhere in the solar cell. Typically this peak minority-carrier concentration (and its spacing from the band edge) is so much lower than the majority-carrier concentration that the cell voltage is significantly less than the band gap. However, the cell voltage can approach the band gap value when the minority-carrier concentration is increased, up to the limit of about an order of magnitude less than the majority-carrier concentration. Solar cell current is fundamentally limited, by the quantum nature of light, to one collected electron–hole pair per incident photon of energy greater than the band gap. The best cell currents approach this perfect quantum efficiency limit.

The exponential qV/kT shape of the Shockley diode equations determines the maximum power points for solar cells in terms of a FF that multiplies open circuit voltage and short circuit current density J_{sc} to give conversion efficiency (after division by solar input power density). Hence efficiency plots can be generated as a function of minority-carrier concentration and band gap assuming a majority-carrier concentration. The band gap gives the maximum J_{sc}. It and the minority carrier concentration define the maximum V_{oc}, which then gives the FF. For $10^8\,cm^{-3}$ minority-carrier concentrations in low performance cells, the maximum single junction efficiencies of ~18%–20% occur at band gaps around 1.5 eV. This compares with reported 12% best efficiency for a single junction a-Si:H cell with a 1.6 eV band gap and a 10.2% $Cu_xS/Cd(ZnS)$ at a 1.2 eV band gap under one sun illumination listed in Table 1.5. The measured V_{oc} for these two cells correspond to $\sim 10^8\,cm^{-3}$ minority-carrier concentrations (see Fig. 1.11). For the $10^{14}\,cm^{-3}$ minority-carrier concentrations, the maximum efficiencies of ~28%–30% are realized at ~1.4 eV band gaps. For comparison 23.0% has been reported for 1.1 eV silicon cell and 24.0% and 25.7% have been reported for 1.4 eV GaAs andGaAs/$GaInP_2$, respectively, at one sun. The V_{oc} of the latter three correspond to $\sim 10^{14}\,cm^{-3}$ levels.

The measured V_{oc} in Figure 1.11 indicate that limiting $10^{17}\,cm^{-3}$ minority-carrier concentrations have only been approached thus far with concentrator cells like GaSb at ~150×, Si point contact devices at 100× and GaAs at 240× and 1050×. If high quasi-Fermi level splitting can be retained when the solar spectrum is split between two cells of differing band gap, even higher efficiencies around 50% are theoretically possible. This is due to the better match (between the maximum splitting and its changes with band gap) to the different solar photon energies. Thus far the best two junction efficiency reported is 33%–35% for GaSb mechanically stacked under GaAs at ~100× concentration. Here there is still much room for improvement.

Prospects for further improvement, particularly at one sun, appear to rest on suppression of excess leakage currents and decreased cell thicknesses with light trapping to achieve high quasi-Fermi level splitting. The series resistance losses at high concentration may be alleviated by better exploitation of prismatic covers and point contact geometries. Field efficiency measurements are 50% and less of the best laboratory cell performance levels with the lowest percentage for a-Si:H demonstration systems. Elevated temperature and interconnect losses are significant but small contributors to these differences. Basic design variations and differing materials qualities between laboratory and demonstration cells, and the distribution of properties between "best-ever" and "typical" cells, and surface fill factors appear to account for the bulk of these performance level differences.

REFERENCES

Benner, J.P. (1991), *22nd IEEE Photovoltaic Specialists Conf. Record,* IEEE, New York, pp. 7–11.

Bertness, K.A., Friedman, D.J., Kebbler, A.E., Kramer, C., Kurtz, S.R., and Olson, J.M. (1993), AIP Conf. Proc., 12th National Renewable Energy Laboratory Photovoltaic Program Review, Denver, CO.

Bertness, K.A., Ristow, M.L., and Hamaker, H.C. (1988), *20th IEEE Photovoltaic Specialists Conf. Record*, IEEE, New York, pp. 769–770.

Bragagnolo, J.A. (1978), *20th IEEE Photovoltaic Specialists Conf. Record,* IEEE, New York, pp. 769–770.

Candelario, T.R., Hester, S.L., Townsend, T.U., and Shipman, D.J. (1991), *22nd IEEE Photovoltaic Specialists Conf. Record,* IEEE, New York, pp. 493–500.

Catalano, A., Arya, R.R., Fortman, C., Morris, J. Newton, J., and O'Dowd, J.G. (1987), *19th IEEE Photovoltaic Specialists Conf. Record,* IEEE, New York, pp. 1506–1507.

Chapin, D.M., Fuller, C.S., and Pearson, G.L. (1954), *J. Appl. Physol.* **25**, 676–677.

Chung, B.-C., Virshup, G.F., Hikido, S., and Kaminar, N.R. (1989), *Appl. Phys. Lett.* **55**, 1741–1743.

Chung, B.-C., Virshup, G.F., Klausmeier-Brown, M., Ristow, M.L., and Wanlass, M.W. (1992), *Appl. Phys. Lett.* **60**, 1741–1743.

Dawson, P., and Woodbridge, K. (1984), *Appl. Phys. Lett.* **45**, 1227–1229.

Fahrenbruch, A.L., and Bube, R.H. (1983), *Fundamentals of Solar Cells,* Academic Press, New York, pp. 9, 75.

Feynman, R.P., Leighton, R.B., and Sands, M. (1963), *The Feynman Lectures on Physics,* Addison-Wesley, Reading, MA, Vol. I, p. 39-9, Vol. II, pp. 4–6.

Fraas, L.M., Avery, J.E., Martin, J., Sundaram, V.S., Girard, G., Dinh, V.T., Davenport, T.M., Yerkes, J.W., and O'Neill, M.J. (1990), *IEEE Trans. Electron Devices* **37**, 443–449.

Gale, R.P., Zavracky, P.M., McClelland, and Fan, J.C.C. (1987), *19th IEEE Photovoltaic Specialists Conf. Record,* IEEE, New York, pp. 63–66.

Gee, J.M., and Virshup, G.F. (1988), *20th IEEE Photovoltaic Specialists Conf. Record,* IEEE, New York, pp. 754–758.

Green, M.A., and Emery, K. (1993), *Progress in Photovoltaics Research and Applications,* John Wiley, New York, p. 227.

Hall, R.B., Birkmire, R.W., Phillips, J.E., and Meakin, J.D. (1981), *15th IEEE Photovoltaic Specialists Conf. Record,* IEEE, New York, pp. 198–202.

Hamaker, H.C., Grounner, M., Kaminar, N.R., Kuryla, M.S., Ladle, M.J., Liu, D.D., MacMillan, H.F., Partain, L.D., Virshup, G.F., Werthen, J.G., and Gee, J.M. (1988), Space Photovoltaic Research and Technology Conf. Proc., NASA Lewis Research Center, April 1988.

Hester, S. and Hoff, T. (1985), *22nd IEEE Photovoltaic Specialists Conf. Record,* IEEE, New York, pp. 777–779.

Hovel, J.J. (1975), *Solar Cells, Semiconductors And Semimetals,* Vol. 11, R.K. Willardson and A.C. Beer, Eds., Academic Press, New York, pp. 16, 61.

Hulstrom, R., Bird, R., and Riordan, C. (1985), *Solar Cells* **15**, 365–391.

Kaminar, N.R., Liu, D.D., MacMillan, H.F., Partain, L.D., Ristow, M.L., Virshup, G.F., and Gee, J.M. (1988), *20th IEEE Photovoltaic Specialists Conf. Record,* IEEE, New York, pp. 766–768.

King, D.L., and Hansen, B.R. (1991), *Validation of PDML and PTEL Primary Reference Cell Calibrations,* Sandia National Laboratories Internal Memo.

Kittel, C. (1969), *Thermal Physics,* Wiley, New York, pp. 46, 49, 61, 73, 81, 140, 151–156, 161.

Klausmeier-Brown, M.E., Lundstrom, M.S., and Melloch, M.R. (1989), *IEEE Trans. Electron Devices* **36**, 2146–2155.

Kurtz, S.R., Olson, J.M., and Kibbler, A. (1990), *19th IEEE Photovoltaic Specialists Conf. Record*, IEEE, New York, pp. 138–140.

Kuryla, M.S., Ristow, M.L., Partin, L.D., Bigger, J.E. (1991), *22nd IEEE Photovoltaic Specialists Conf. Record*, IEEE, New York, pp. 506–511.

Landis, G.A. (1990), *21st IEEE Photovoltaic Specialists Conf. Record*, IEEE, New York, pp. 1304–1307.

Liu, X.X., and Sites, J.R. (1994), *J. Appl. Phys.* **75**, 577–581.

Loferski, J.J. (1956), *J. Appl. Phys.* **27**, 777–784.

Lovejoy, M.L., Melloch, M., Lundstrom, M.S., and Ahrenkiel, R.K. (1992), *Appl. Phys. Lett.* **61**, 2683–2684.

Lush, G.B., MacMillan, H.F., Keyes, B.M., Levi, D.H., Melloch, M.R., Ahrenkiel, R.K., and Lundstrom, M.S. (1992), *J. Appl. Phys.* **72**, 1436–1442.

MacMillan, H.F., Hamaker, H.C., Kaminar, N.R., Kuryla, M.S., Ristow, M.L., Liu, D.D., Virshup, G.F., and Gee, J.M. (1988), *20th IEEE Photovoltaic Specialists Conf. Record*, IEEE, New York, pp. 462–468.

Nelson, R.J., and Sobers, R.G. (1978), *J. Appl. Phys.* **49**, 6103–6108.

O'Neill, M.J., McDanal, A.J., Walters, R.R., and Perry, J.L. (1991), *22nd IEEE Photovoltaic Specialists Conf. Record*, IEEE, New York, pp. 523–528.

Partain, L.D., Kuryla, M.S., Weiss, R.E., Ransom, R.A., McLeod, P.S., Fraas, L.M., and Cape, J.A. (1987a), *J. Appl. Phys.* **62**, 3010–3015.

Partin, L.D., Kuryla, M.S., Fraas, L.M., McLeod, P.S., and Cape, J.A. (1987b), *J. Appl. Phys.* **61**, 5150–5158.

Partain, L.D., and Liu, D.D. (1989), *Appl. Phys. Lett.* **54**, 928–930.

Partain, L.D., Liu, D.D., Kuryla, M.S., Ahrenkiel, R.K., and Asher, S.E. (1990), *Solar Cells* **28**, 223–232.

Partain, L. D., and Lakshminarayana, M. R. (1976), *J. Appl. Phys.* **47**, 1015–1022.

Partain, L. D., and Sayed, M. (1974), *9th Intersociety Energy Conversion Engineering Conf. Proc.*, American Society of Mechanical Engineers, New York, pp. 362–369.

Partain, L.D., and Sheldon, J.F. (1980), *Quasi-Fermi Level Interpretation as Potential Energy*, Lawrence Livermore Laboratory Report No. UCRL-83787, p. 16.

Prince, M.B. (1955), *J. Appl. Phys.* **26**, 534–540.

Rappaport, P. (1959), *R.C.A. Rev.* **20**, 373–397.

Reynolds, D.C., Leies, G., Antes, L.L., Marburger, R.E. (1954), *Phys. Rev.* **96**, 533–534.

Rothwarf, A., Burton, L.C., Hadley, H.C., and Storti, G.M. (1975), *11th IEEE Photovoltaic Specialists Conf. Record*, IEEE, New York, pp. 476–481.

Sayed, M., and Partain, L. (1975), *Energy Conversion* **14**, 61–71.

Schultz, J.C., Klausmeier-Brown, M.E., Ristow, M.L., Partain, L.D., Al-Jassim, M.M., and Jones, K.M. (1993), *J. Electronic Mtls* **22**, 755–761.

Sell, D.D., and Casey, H.C. (1974), *J. Appl. Phys.* **45**, 800–807.

Shockley, W. (1976), *Electrons and Holes in Semiconductors*, Kreiger, Huntington, New York, pp. 302, 308.

Shockley, W., and Queisser, J. (1961), *J. Appl. Phys.* **32**, 510–519.

Sinton, R.A., Kwark, Y., Gan, J.Y., and Swanson, R.M. (1986), *IEEE Electron Dev. Ltr.* EDL-7, pp. 567–569.

Sugo, M., Yamamoto, A., and Yamaguchi, M. (1987), *IEEE Trans. Electron Devices* **34**, 772–777.

Sze, S.M. (1981), *Physics of Semiconductor Devices*, 2nd Ed., Wiley-Interscience, New York, pp. 12, 17–19, 75, 84, 87, 250, 796, 798, 851.

Wenham, R., Zhang, C.M., and Green, M.A. (1990), *21st IEEE Photovoltaic Specialists Conf. Record*, IEEE, New York, pp. 323–326.

Wolf, M. (1960), *Proc. I.R.E.* **48**, 1246–1263.

Zhao, J., Wang, A., and Green, M.A. (1990), *21st IEEE Photovoltaic Specialists Conf. Record*, IEEE, New York, pp. 333–335.

Ziman, J.M. (1965), *Principles of the Theory of Solids*, Cambridge, London, p. 184.

Zweibel, K. (1990), *Harnessing Solar Power*, Plenum, New York, p. 230.

PART ONE

CELL TECHNOLOGIES

CHAPTER TWO

Silicon Cells: Single Junction, One Sun, Terrestrial, and Single and Multiple Crystalline

MARTIN A. GREEN and STUART R. WENHAM, Centre for Photovoltaic Devices and Systems, University of New South Wales, Kensington, NSW 2033, Australia

1. INTRODUCTION

Despite enthusiasm for selenium cells in the nineteenth century (Fritts, 1883) and low-cost cuprous oxide cells in the 1930s (Grondahl, 1933), the rapid evolution of silicon technology in the 1950s provided the first prospects for practical power generation using photovoltaics. Bulk silicon remains the "work-horse" for outdoor applications, withstanding challenges by thin-film cells based on cadmium sulphide in the 1970s and on amorphous silicon alloy in the 1980s. Improved cell designs and increasingly streamlined manufacturing, combined with excellent field reliability, ensure an on-going role for this technology into the future.

This chapter describes the evolution of bulk silicon cell technology and its present commercial status. Recent progress with laboratory cells is also discussed showing the potential for substantially improved commercial product. Present pilot-line work aimed at transferring recent improvements into production are also described. Finally, advanced substrates including ribbons and thin-film multicrystalline silicon plus recent topics such as spheral and implanted "bubble" cells, as well as the likely development of silicon technology over coming decades, are discussed.

Solar Cells and Their Applications, Edited by Larry D. Partain.
ISBN 0-471-57420-1

2. SILICON CELL EVOLUTION

Operational silicon cells were described by Russell Ohl of Bell Laboratories in the early 1940s (Ohl, 1941). He made cells from natural junctions in recrystallized melts as in Figure 2.1a. He designated the side of the junction becoming positive under illumination as *positive* or *p type* and the other *n type* (the role of acceptor and donor impurities was only subsequently clarified). This fortunate nomenclature has trivialized calculation of cell voltage polarity for later generations! Helium implantation into purer silicon was later reported as a more controllable way of forming cells (Kingsbury and Ohl, 1952). In the 1950s, rapid developments in silicon crystal growth and in the high-temperature diffusion of dopants to form junctions produced the first crystalline silicon cells, shown in Figure 2.1b (Chapin et al., 1954). Cell design then evolved rapidly to that shown in Figure 2.2a by the early 1960s.

This design remained standard until the early 1970s. It was then realized that sintered Al along the rear improved performance, probably by a combination of gettering and formation of a heavily doped rear interface known as a *back surface field*. By suppressing minority carriers, rear recombination is reduced, improving cell current and voltage. Additional improvements originated from COMSAT Laboratories. Using photolithography, finer, more closely spaced top contact fingers were possible than with earlier metal masking approaches. This reduced the phosphorus doping required along the top surface, eliminating an inactive layer near the surface and improving response to blue light (Lindmayer and Allison, 1973). Better antireflection coatings complemented this improvement. Subsequent improvement came from surface texturing using anisotropic etches (Haynos et al., 1974). Pyramids on the surface as in Figure 2.2b reduce reflection so that, after antireflection coating, cells look like black velvet. Pyramids also couple light obliquely into

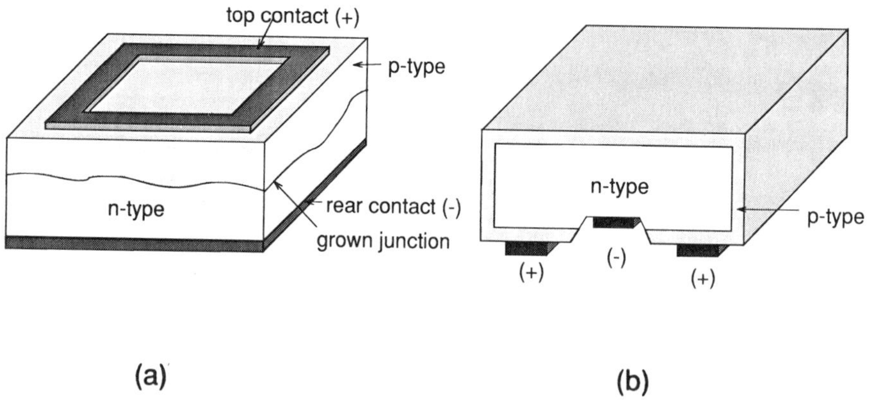

FIGURE 2.1 Early Bell Laboratory silicon cells: (**a**) Grown junction cell (1941). (**b**) Crystalline "wrap-around" cell (1954).

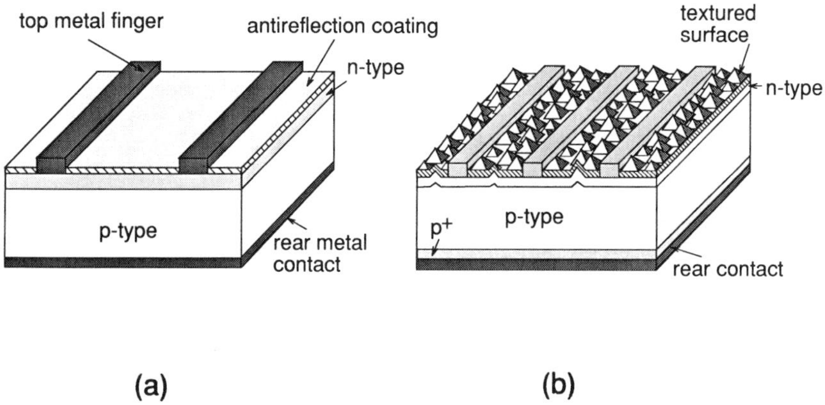

FIGURE 2.2 (**a**) Standard silicon space cell of the 1960s. (**b**) "Black" cell (1974).

such "black" cells, allowing absorption closer to the surface. By the mid-1970s, cell design had evolved to that shown in Figure 2.2b. The fine, closely spaced fingers (Ti/Pd/Ag multilayer) were defined photolithographically, allowing shallow top surface diffusion and increased blue response, with a rear Al alloyed region often incorporated. Terrestrial efficiencies were around 17% (Rittner and Arndt, 1976).

Concurrently, a Middle East embargo highlighted the western world's dependence on oil. The search for energy self-sufficiency generated interest in terrestrial cell applications. In an important program, the U.S. government stimulated the terrestrial industry by a series of block purchases of solar modules (Christensen, 1985). Increasingly stringent specifications for successive blocks guided the industry's development. The earliest modules used cells as shown in Figure 2.2a but with varied metallization approaches. Some combined vacuum evaporation with photolithography. Others used electrolessly plated nickel followed by solder dipping; others, screen-printed silver pastes, borrowing technology from thick-film microelectronics (Ralph, 1975). In the second block purchase in 1976, Spectrolab, Inc., combined screen-printing with a module design based on lamination as in automobile windscreens. This combination is now used by virtually all manufacturers.

In another important development, Wacker (Fischer and Pschunder, 1976) and Solarex (Lindmayer, 1976) developed "cast" multicrystalline silicon ingots. Cost potential was much lower than with the standard Czochralski (CZ) process, while giving acceptable quality. Multicrystalline silicon accounted for half the silicon used in commercial cells in 1992. In the late 1970s, the Jet Propulsion Laboratory (JPL), coordinated a comprehensive, mission-orientated program aimed at reducing costs of both cells and modules (Christensen, 1985). Issues addressed ranged from silicon feedstock purification to field testing of completed modules. In the early 1980s, with substantial private funding, several large ($\geqslant 1$ MW) photovoltaic power stations were installed. Most used modules incorporating the screen-printing/lamination technology earlier developed at Spectrolab, with increased production stimulating a 3 to 4 times decrease in cell costs over the decade.

Large increases in laboratory cell efficiency from the early 1980s to the present dominate recent developments. Efficiency has increased from 17% for "black" cells to above 23%, increasing expectations as to commercially feasible efficiencies. For example, in 1983, an expert panel convened by the Electric Power Research Institute (EPRI) suggested that 15% silicon module was the highest expected commercially (Witwer, 1983). In 1989, a similar EPRI-convened panel assigned high probability to 20% efficiency being demonstrated commercially by silicon modules (Steitz and Associates, 1989). Cell structures responsible for this increased expectation are discussed in Section 5.

3. PRESENT COMMERCIAL TECHNOLOGY

3.1. Ingot Growth

To make silicon cells of reasonable performance, large-grained multicrystalline or single crystalline substrates of high purity are required. The main technique for preparing crystalline silicon is the CZ method of Figure 2.3a. High-purity, fine-grained polysilicon is melted in a quartz crucible. A seed is inserted into this melt and slowly withdrawn. Oxygen from the crucible is incorporated as an impurity into the growing crystal. Carbon is also introduced from heating elements, with boron usually deliberately added as a p-type dopant. While oxygen is closely controlled for microelectronics, solar cells can tolerate lower

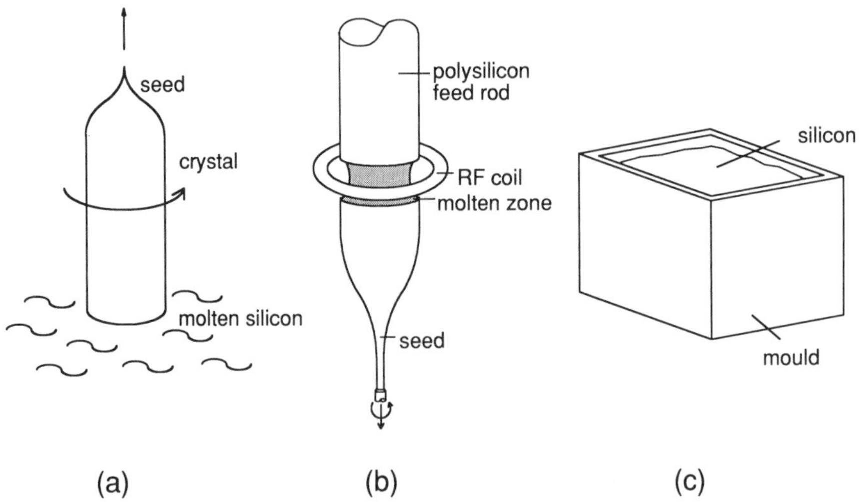

FIGURE 2.3 (**a**) Czochralski (CZ) growth of single crystal silicon. (**b**) Float-zone (FZ) process. (**c**) Multicrystalline ingot casting.

grade material, and oxygen content is ofter not specified. Similarly, the high-purity polysilicon source material need not be microelectronics quality. Cell manufacturers growing their own ingots often use polysilicon "off-specification" for microelectronics. Purchased wafers correspondingly are either "off-specification" or grown under relaxed quality control.

A recent refinement in microelectronics is the use of magnetic confinement (MCZ growth). By placing strong magnets around crucibles, melt convection can be controlled, improving growth conditions. MCZ silicon is suited for fabricating very high- efficiency cells, with better control possible over oxygen, carbon, and crystal defects. Although equipment costs are higher, higher growth rates are feasible under the better growth conditions. The float zone (FZ) process shown in Figure 2.3b also produces crystalline silicon. A rod of fine-grained polycrystalline is crystallized by passing a molten zone from a seeded end to the other. This produces cyrstals incorporating much lower oxygen and carbon than the CZ process and of better quality for cell fabrication. Reduced consumables such as crucibles offset higher equipment costs to some extent.

All of the above techniques were originally developed for microelectronics. Others producing multicrystalline ingots were developed specifically for photovoltaics. These are based on slowly solidifying molten silicon in crucibles, as shown in Figure 2.3c. Differences between these include the way crucibles are loaded, either by solid or molten material; crucible material, commonly graphite, quartz coated with silicon nitride, or silica ceramic; and the way the melt is cooled and growth nucleated, either from the bottom or crucible wall. Compared with crystalline approaches, equipment is simpler and less costly with less stringent feedstock requirements. Ingots are larger with more ideal square cross section. Disadvantages include poorer material quality and poorer uniformity both within an ingot and between different ingots.

3.2. Shaping and Wafering

After growth, ingots can be cut to more appropriate geometries, e.g., edges sawn off cylindrical ingots or large multicrystalline ingots cut into smaller blocks. Alternatively, cylindrical ingots are ground to uniform diameter for processing as round wafers.

After shaping, ingots are sawn into wafers, normally by inner diameter blade sawing, which uses a thin metal blade held rigid under tension. The blade has an internal hole with the perimeter diamond coated to provide the cutting edge. An alternative method, increasingly used in production, is continuous wire cutting. A thin wire is passed over rollers to form a parallel array to cut hundreds of wafers simultaneously by grinding an abrasive slurry through the ingot. Similar techniques include the multiblade slurry technique using fixed blades to guide the slurry and the FAST technique, using continuous wire impregnated with diamonds (Briglio et al., 1986).

3.3. Etching and Texturing

Frequently, NaOH etches are used to remove saw damage, to prepare surfaces for texturing, and to texture surfaces to produce the square-based pyramids shown in Figure 2.2b. The damage removal etch is typically 300 g/L NaOH at 80°C. Additives can improve the surface for subsequent texturing (Guelden and Wenham, 1992). After rinsing, wafers are texture etched, typically using 20 g/L NaOH at 90°C. The altered concentration increases etching rate anisotropy, with (111) crystallographic planes etched slowly. With wafers of (100) surface orientation, this etch exposes (111) equivalent planes that intersect to form the pyramids shown. An important texturing etch component is isopropanol.

Texturing is one of the more difficult processing steps. Common production problems include repeatability, pyramid nucleation toward the end of texturing, which destroys existing pyramids, lack of control over pyramid size, etching of pyramid peaks through excessive solution turbulence, and the presence of untextured regions between pyramid bases through low pyramid nucleation density and reflective pyramid bases where adjacent (111) planes intersect. Correct isopropanol concentrations and evaporation rates, solution surface area to volume ratio, temperature, NaOH concentration, length of texturing, and solution turbulence can solve these problems (Guelden and Wenham, 1992). Poor texturing quality from inferior parameter combinations may be partially compensated by modifying the surface finish from the prior damage removal etch or through use of appropriate additives in the texturing solution as partial substitutes for isopropanol. After texturing, wafers are prepared for juction diffusion by de-ionized water rinsing. For multicrystalline wafers, only a small fraction of grains are correctly oriented for texturing. Anisotropic etching also gives different etch rates for different grains producing steps at grain boundaries. This can complicate the subsequent screening of metal pastes. Often, isotropic etching is used for multicrystalline material to give flat surfaces, with texturing deliberately eliminated.

3.4. Diffusion and Edge Junction Isolation

Since wafers are usually boron doped, an n-type junction is generally diffused into the cell. Phosphorus is the usual n-type diffusant, with various phosphorus sources used commercially. In some cases, phosphine gas is passed down the diffusion tube. In other cases, a carrier gas is passed down the tube after bubbling through liquid sources such as $POCl_3$ or PBr_3. Alternatively, a solid source such as P_3O_5 is heated at the end of the furnace tube. In all these cases, oxygen is simultaneously passed down the tube with wafers forming phosphorus-doped surface oxides. At temperatures involved (850°–950°C), phosphorus diffuses from the oxide into the cell. Other techniques include the deposition of phosphorus-doped oxides on cell surfaces from liquid sources by spinning or spraying before loading into furnace tubes, the deposition from mists, use of interleaved solid source wafers of the same size as processed wafers, and implantation of phosphorus ions. Regardless of the source, sufficient phosphorus is introduced to give sheet resistivity of 25-50 Ω/square for screen-printed metallization sequences. After diffusion, processing can diverge depending on manufacturer. In some sequences, diffusion oxides are removed in HF-based etches. More simply, the diffusion oxide is left and metal contacts fired through it.

Phosphorus diffuses not only into the desired wafer surface but also into the side and the opposite surface, to some extent. This gives a shunting path between the cell front and rear. Removal of the path around the wafer edge, "edge junction isolation," is commonly effected by "coin stacking" the cells. Stacked cells are placed into a plasma etching chamber to remove exposed edges. Usually no attempt is made to remove the rear junction. Contact firing conditions are chosen to neutralize its effects, often only partially.

3.5. Contact Screen and Firing

Next, contacts are screened onto the wafer front and rear. For top contacts, the screened paste normally consists of Ag powder combined with frit (low melting point glass composites) and organic binders. Sometimes phosphorus compounds are added to dope underlying regions more heavily n type and improve contact resistance. For rear surface pastes, Al is often added to dope underlying regions p type. Immediately after screening each side, pastes are dried by heating to 350°–400°C.

Contacts must be fired at above 700°C to give reasonable metal resistivity, still three times worse than pure Ag. The final contact resistance to silicon can be very sensitive to firing conditions. Infrared lamps are often used for firing rather than normal furnace heating elements. Rear contact firing can also be critical to ensure that the rear junction is neutralized. Temperature gradient zone melting can be an important issue when attempting to form "back surface fields" by alloying of Al paste components (Chong, 1989).

3.6. Antireflection Coating

For textured crystalline cells, surface reflection is low without antireflection coating. Even in this case, about 4% performance boost after encapsulation can be obtained by antireflection coating at the end of processing. Since multicrystalline cells cannot be readily textured, antireflection coating is essential in present sequences, usually by chemically or spray-deposited TiO_2. Silicon nitride is also sometimes used. Steps are often taken to ensure that contact regions are free from this deposited layer for solderability.

3.7. Interconnection and Module Assembly

After testing and sorting, cells are connected into modules (see Chapter 9). Briefly, by soldering interconnects, cells are assembled into their final layout within the module. The interconnected cells are then placed in a stack consisting of the glass superstrate, a layer of ethylene vinyl acetate (EVA), the interconnected cells, another layer of EVA, and then a backing layer, usually Tedlar. Sometimes "scrim" or spacer layers are interleaved in the stack, which is then laminated together under vacuum at elevated temperature.

EVA was specifically developed for this use during the JPL program mentioned in Section 2. Polyvinyl buteral (PVB) had been used in a similar role, but EVA is cheaper and more easily stored. Although standard for many years in modules, EVA has recently been shown to discolor, in some situations, after several years in the field, particularly when modules reach high temperatures (Pern and Czanderna, 1992). Discoloring is associated with the deterioration of ultraviolet absorbers added to the EVA. Although not a major problem unless module temperature is enhanced, such as by reflectors, some performance loss can be expected after several years in the field until improved formulations are developed.

3.8. Multicrystalline Cells

Although the screen-printing sequence can be equally well applied to multicrystalline and single crystalline wafers with the minor differences already noted, more complicated sequences can give rewards in terms of increased multicrystalline cell performance. In particular, hydrogen exposure during processing can neutralize grain boundary activity. High-throughput equipment for hydrogen ion implantation near the end of processing has been described (Johnson et al., 1985).

Alternatively, the whole processing sequence can be built around hydrogen incorporation. One sequence involves silicon nitride deposition after top surface diffusion under conditions encouraging hydrogen incorporation (Kimura, 1984). During subsequent processing, the nitride prevents hydrogen egress. Other processing options effective for multicrystalline silicon are described in Section 5.2, although not yet used in production.

3.9. Manufacturing Costs

Recent studies suggest that manufacturing costs for modules lie between \$3.00 and \$3.50 per watt (1992 U.S. dollars), assuming annual production of about 10 MW. Both multicrystalline and single crystalline CZ substrates have similar module costs (Wohlgemuth et al., 1990; Hogan et al., 1991).

For crystalline substrates, about half this cost is due to the wafer. Processing of wafers into cells and encapsulation are roughly equal in cost and account for the remainder. Multicrystalline cells have slightly lower wafer costs but higher costs in other areas, due to lower cell efficiency. Wafer costs, therefore, dominate present module costs, with a large fraction of these due to the cost of cutting wafers from ingots.

4. COMMERCIAL CELL PERFORMANCE

The screen-printing process that has become the commercial standard provides a simple way of manufacturing moderate efficiency cells with the structure shown in Figure 2.4a. For independent measurements under the global air-mass 1.5 (AM1.5) spectrum at 25°C, corresponding efficiencies for crystalline cells would be 13%-15% and 10%-13% for multicrystalline cells. The best laboratory cells use different technology and have efficiencies above 23%, nearly twice these figures, for reasons addressed below.

One major difference is the top contact design. To minimize the sum of top contact losses, metallization should have a high aspect ratio (height divided by width), fine line width, high conductivity, and low contact resistance to the underlying silicon (Green, 1982). Screen-printing pays for its simplicity by being poor in all four areas. The normal screening process gives line widths of about 150 μm (although special screens can do better, these have not proved suitable for high throughput production). Upon baking and firing, the screen-printed metallization contracts in thickness by a factor of 2. Typically, the paste is printed to 20 μm thickness, reducing to 10 μm after firing. The aspect ratio is therefore very poor, only about 0.06. The conductivity of the fired paste is only about one-third that of pure silver, exaggerating the effect of poor aspect ratio. Additionally, the metallization has a high contact resistance to silicon, arising from the glass frit required for bonding to silicon. Contact resistance depends critically on firing conditions and surface doping.

Due to poor aspect ratios and low conductivity, screen-printed fingers can carry current over only short distances. An interconnect strap design has

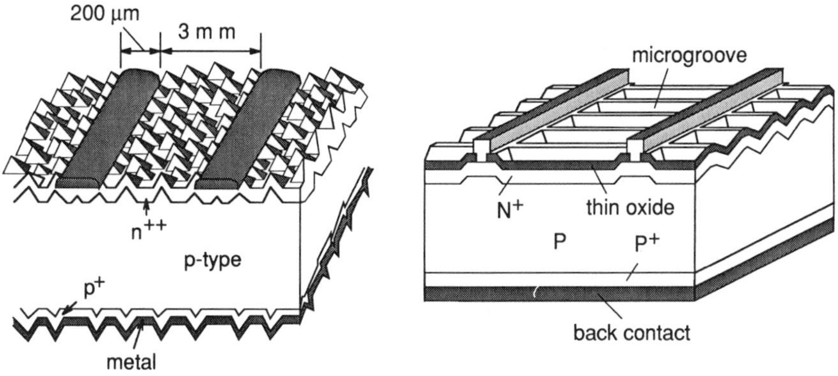

(a) (b)

FIGURE 2.4 (**a**) Commercial screen-printed silicon cell. (**b**) UNSW PESC cell (passivated emitter solar cell) (1985).

therefore proved popular for screen-printed cells. By having two interconnects running across the cell, effective finger length is reduced below 2.5 cm for a 10 cm square cell. To accommodate differential thermal expansion, these interconnects have to be thin and reasonably wide, typically shading 4%–5% of the cell. The contact resistance problem means the underlying silicon must be heavily doped, even with phosphorus incorporated into the paste, while the coarse line width forces screen-printed fingers to be reasonably far apart for low shading loss. This supplements the need for heavy junction diffusion, since carriers have to flow reasonable distances laterally along the top surface.

The performance penalty from the simple screen-printing process is therefore enormous. For the reasons mentioned, top surfaces of screen-printed cells are shaded by about 10%–15% metal, compared with 3% in a 23% efficient cell. Even with this increased shading, resistive losses are much higher. Combined resistive losses in a commercial cell give a further 10% loss compared with 1%–2% in the 23% efficient cell. The total resistive and shading losses in a screen-printed cell are therefore above 20% compared with below 5% in the 23% efficient cell. Additionally, the heavy diffusion, necessary for tolerable contact and top diffused layer resistance, reduces cell blue response since it creates an inactive surface layer. This reduces short circuit current density (J_{sc}) by a further 10%–20%.

There are additional penalties. The heavily diffused region contributes to the cell's dark saturation current, limiting open circuit voltage (V_{oc}) to below 630 mV regardless of substrate quality. The 23% cells mentioned have V_{oc} about 10% higher than this limit. Screen-printed metal is not an effective internal reflector for weakly absorbed light, accounting for a few percent loss for normal commercial thicknesses (0.3–0.5 mm). Other differences relate to material quality. The post processing quality of commercial CZ material is much poorer than that of the FZ material used in the 23% cells. Diffusion lengths are several millimeters in the latter, about 10 times those in commercial cells. This reduces J_{sc} by about another 5%–10%, while further reducing voltages below 630 mV by about 3%. Taking all these factors into account, a 23% efficient cell with V_{oc}, J_{sc}, and fill factor of 700 mV, 41 mA/cm^2, and 81%, respectively, becomes a 14% efficient cell with parameters of 610 mV, 30 mA/cm^2, and 75%.

Multicrystalline cells have additional disadvantages. Reflection is higher since cells are not textured, giving an additional 5% loss. Diffusion lengths are 2 or 3 times shorter than in CZ material, resulting in an additional few percent loss in voltage and 5%–10% in current, accounting for 10%–20% lower performance than in CZ cells.

To improve the efficiency from screen-printing sequences, finer line width and lower contact resistance, or a technique for simply aligning the reduced line width screen-printed finger to a locally heavily diffused area are required. No satisfactory solution has yet been found.

5. HIGH-EFFICIENCY LABORATORY CELLS

5.1. Crystalline Cells

As described in Section 2, the "black" cell in Figure 2.2b remained the highest efficiency cell throughout the 1970s. Substantially improved V_{OC} was demonstrated at the University of New South Wales (UNSW) toward the end of that decade, leading to efficiency increases to 18% and then 19%, with the first 20% efficient cell confirmed in 1985. This was a milestone in silicon cell development, since 20% efficiency had long been regarded as a practical limit.

The cell that first surpassed this milestone was the UNSW passivated emitter solar cell (PESC cell), shown in Figure 2.4b. The cell used an alloyed Al rear to form a conventional "back surface field." On the top surface, a thin oxide layer was used to reduce recombination. The improved surface quality allowed shallow phosphorus diffusion while maintaining high V_{oc}. Also to improve V_{oc}, top metal contact was made by narrow stripes through this oxide, with only 0.3% of the top surface contacted. Conventional metallization (vacuum evaporated Ti/Pd plus plated Ag) was used for this contact. This gives low contact resistance to the lightly diffused region and high finger conductivity due to the good conductivity of plated Ag. The metallization aspect ratio is reasonable. Oxide contact stripes were 4 μm wide, and Ag was plated to 8 μm thickness, giving aspect ratios of approximately 0.3. The cells also used crystallographic surface texturing. In this case, photolithography allowed microgrooves to be anisotropically etched into the surface. Grooves were chosen rather than randomly located pyramids primarily for processing convenience. Since the oxides for these cells were thin, about 100 Å, a double layer antireflection coating of MgF_2/ZnS deposited on the oxide surface gave very low overall reflection.

The improved top surface allowed advantage to be taken of high-quality FZ silicon. The normal commercial sequences could also, of course, be applied to FZ silcon. However, little performance improvement would be noticed, since cell structures are unable to capitalize on the superior FZ properties. Due to the good top surface of the PESC cell and its relatively poor rear, the best results were obtained when low resistivity FZ material was used (0.2 Ω-cm, boron doped), giving high V_{oc} and reasonable J_{sc}. Since corresponding diffusion lengths are only comparable to wafer thickness, the importance of the rear quality is de-emphasized.

An efficiency of 20.1% was measured for a PESC cell by the present National Renewable Energy Laboratory (NREL) in October 1985, with subsequent refinements at UNSW increasing this to 20.6% in 1986. These results were duplicated at Hitachi with 19.8% independently confirmed (global spectrum) for a nearly identical structure (Saitoh et al., 1987). JPL, with a similar structure but randomly textured, also fabricated cells in 1986, initially reported as 20% efficient (Bickler and Callaghan, 1987), although independently confirmed as 19.3% under current calibrations. More recently, PESC technology has been transferred to Applied Solar Energy Corporation with excellent results (Khemthong et al., 1991).

The next major improvement came with the Stanford University rear point contact cell shown in Figure 2.5a. Although developed for concentrator cells (see Chapter 3), this structure also performs well when designed for nonconcentrating conditions. In 1988, 22.3% aperture area efficiency was demonstrated for an 8.5 cm^2 cell (King et al., 1991). Subsequently, a larger 37.5 cm^2 cell demonstrated an efficiency converting to 21.6% under current standards (King et al., 1991). Contacting to small regions of the rear places far more severe demands on both the bulk silicon quality and the surface passivation than previous designs. Photogenerated carriers, most generated within microns of the cell surface, must diffuse to the rear while squeezing into small contact areas. Carrier diffusion lengths must be many times the wafer thickness for carriers to be collected with high probability. Similarly, top surface recombination has to be extremely low. This makes the approach unsuited for space cells where the top surface is quickly degraded by radiation damage. Modifications to the original design were also required to enable cells to withstand high energy ultraviolet photons in terrestrial sunlight, although most would be filtered by lenses, superstrates, or laminating materials in many applications.

The cells used high-resistivity n-type substrates and microelectronics-quality oxide growth for surface passivation. The original cell design required four photolithographic masks, with minimum mask features around 5 μm. A simplified design reduces this to two masks. One possible production difficulty is that the rear oxide has to isolate the rear metal from the substrate over nearly the total cell area. Similarly, the two rear contact metals are spaced only

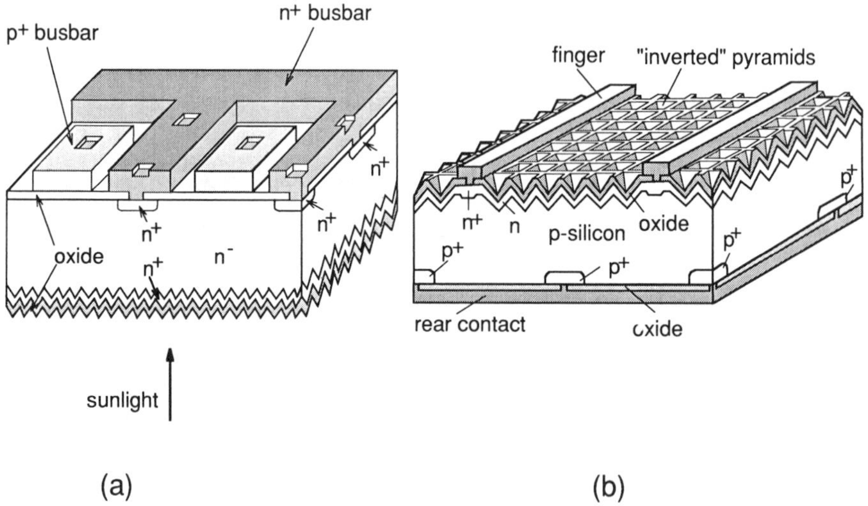

FIGURE 2.5 (**a**) Stanford rear contact cell. (**b**) UNSW PERL cell (passivated emitter, rear locally diffused).

5 μm apart over an enormous length, inviting shunts. Workers at the University of Louvain developed a similar structure in the mid-1980s with slightly lower performance (Verlinden et al., 1985). Moderate success in reproducing device performance has been reported by SERA Solar and Amonix. The SunPower Corporation, involving key members of the Stanford group, has recently commercialized this approach.

The next major improvement came at UNSW by combining the earlier PESC approach with features of the point contact cell to give the PERL cell (passivated emitter, rear locally diffused cell) shown in Figure 2.5b. Microelectronics-quality oxide enshrouds the cell surface. Chlorine-based sequences, both in cleaning tubes and during oxide growth, gave improved oxide quality and higher post processing lifetimes. With these changes, Al gettering at the rear to ensure high postprocessing lifetime could be eliminated.

The top surface of the PERL cell resembles that of the earlier PESC device shown in Figure 2.4b. However, major changes are apparent at the rear. Most is passivated by oxide, with p-type contact made to small regions diffused with boron, similarly to the approach for both contacts in the point contact cell. One important difference is that the rear metal is isolated from the cell by an oxide overlying material of only one polarity. Pin holes, while undesirable, are not disastrous. This rear treatment provides far superior surface passivation than the earlier PESC approach and allows a much wider choice of substrate resistivity. The separation of the rear metal from silicon by an intervening oxide also improves its reflectance (Campbell, 1989). For internal light incident perpendicularly, rear reflectance is about 95%. This decreases as the critical angle for total internal reflection at the Si/SiO_2 interface (24.7°) is approached, to about 90%. Once this angle is exceeded, the internal reflectance approaches 100%. The inverted pyramids along the top surface exploit this improved rear reflectance. For external light, reflection is reduced as for upright pyramids and microgrooves. However, weakly absorbed light is obliquely reflected from the rear. Upon reaching the top surface, light internally striking a pyramid face, oppositely orientated to the face that coupled it in, is coupled out of the cell. Light striking other faces is totally internally reflected and trapped into the cell. An additional refinement is use of separate doping levels in contacted and noncontacted areas of the top surface, allowing separate optimization for their conflicting requirements. These cells demonstrate efficiencies up to 23.5% constituted by V_{oc} around 705 mV, J_{sc} of 41 mA/cm^2, and fill factors around 81% (Zhao et al., 1994).

Each of these parameters can be improved. V_{oc} as high as 717 mV has been confirmed for recent test devices, reduced top surface reflection can improve J_{sc} since these cells use only a rudimentary oxide antireflection coating, improved light trapping can also improve current, and fill factor can be increased (by improving rear oxide quality!). Efficiency around 25% would appear feasible without changing the basic cell structure. Interestingly, recent performance improvements arise largely from improved designs and processing rather than material quality. For example, the 5.08 cm diameter FZ wafers used, until recently, in UNSW PERL cells are prepared with identical equipment as in the mid-1970s.

Recently, other groups have reproduced these results. The Fraunhofer Insitute, using similar structures but with rear p^+ regions formed by Al alloying, have had efficiencies of 21% confirmed at NREL (Kopp et al., 1992). Using a similar structure but with random texturing, Sharp has recently reported efficiencies of 22% ("in house" measurements) using 200 μm thick CZ substrates (Saitoh et al., 1993). V_{oc} is modest (628 mV), fill factor is good (81.5%), while J_{sc} is enormous (43 mA/cm^2). Given the reduced thickness and lower grade substrate, it would seem unlikely that such high J_{sc} would be measured under independent testing to present international standards, with estimated J_{sc} under these conditions about 5%–10% lower. The correspondingly reduced efficiency remains creditable for a CZ subrate.

The above discussion highlights the importance of independent confirmation of high-efficiency cell measurements. Reference cells are, almost inescapably, of lower performance than improved cells. This can produce errors from spectral mismatch between reference and test cells. Independent test centers such as JRC-ISPRA in Europe, the Fraunhofer-ISE and PTB in Germany, ENEA in Italy, JQA in Japan, RAE in the United Kingdom, and NREL and Sandia in the United States (Green and Emery, 1994) have experience in testing a wide variety of cells and are better able to adjust for such mismatch. Standardized techniques are also used for measuring other critical parameters such as cell area.

5.2. Multicrystalline Cells

"Cast" multicrystalline material is likely to remain lower quality than the best crystalline material. Different approaches may be needed to extract its full potential. As noted, some manufacturers use specialized processing geared for multicrystalline material. The best laboratory results with multicrystalline material to date were obtained using an earlier version of the PESC structure shown in Figure 2.4b. In that version, flat or planar cell surfaces were used, appropriate since multicrystalline material is not amenable to anisotropic texturing.

An additional modification was that phosphorus was diffused much more heavily than in the standard sequence. This enhances gettering by this diffusion, as well as allowing diffusion preferentially down grain boundaries and other crystallographic defects. This converts these regions from liabilities to assets in terms of current collection. The heavy diffusion is subsequently etched to the desired sheet resistivity. Using low oxygen multicrystalline wafers (Osaka Titanium), an independently confirmed efficiency above 17% has been reported with this sequence (Narayanan et al., 1990), a result reproduced at Georgia Tech (Rohatgi et al., 1992).

To improve multicrystalline cell performance further, surface texturing is extremely desirable. Not only is reflection reduced, but the oblique light coupling is particularly important in multicrystalline cells due to shorter collection distances. Several texturing techniques based on laser and mechanical grooving, as well as photolithography combined with chemical and plasma etching, have been reported. Efficiencies above 16% have been confirmed on intermediate grade multicrystalline material using a combination of laser texturing and the laser grooving approach described in Section 6 (Zolper et al., 1989). More recently, 17.1% efficiency for a large, 100 cm^2 cell was reported by Sharp using mechanical texturing (Saiton et al., 1993). A systematic study of mechanical texturing using mechanical dicing wheels was recently reported (Willeke et al., 1992).

Special structures can be geared to the shorter carrier collection distances of multicrystalline material, such as the double junction cell in Figure 2.6. By having collecting junctions on both surfaces, current is collected from throughout the cell even when diffusion lengths are only half the wafer thickness. V_{oc} can be shown always to be lower than for equal thickness devices with perfect "back surface fields." Ideally, the voltage difference is 18 mV at worst, decreasing as diffusion lengths increase. The structure has recently been implemented at Hitachi and at UNSW. In the former case, an efficiency of 16.8% was reported for a large area multicrystalline cell (Warabisako et al., 1992). In the latter case, the device has been implemented only in planar cells. The most interesting result has been the demonstration of J_{SC} almost identical to that of an FZ control with the same reflection loss. An 8% relative performance advantage was demonstrated over a single junction cell on the same material (Green et al., 1992).

Evolution in casting processes and implementation of customized cell structures are expected to increase multicrystalline cell performance further over coming years. Based on present quality, an efficiency of 17% would appear the limit for commercial production on multicrystalline material. With crystalline material, the corresponding efficiency would be 21%, a 20%–25% performance margin.

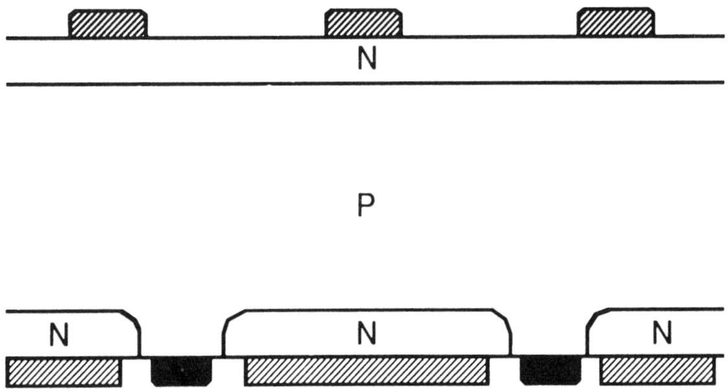

FIGURE 2.6 Double junction cell suited for multicrystalline and ribbon substrates. (Reproduced from Green, 1987, with permission of the publisher.)

6. BURIED CONTACT SOLAR CELLS

6.1. Cell Structure

Differences between present commercial and laboratory cells are discussed in Section 2. Most arose from limitations of the screen-printed contact process used commercially. Offsetting these limitations is the simplicity and availability of equipment for screen-printing and the relatively uncomplicated dry processing sequence. Much research has addressed the previous limitations without success. It appears that a major break from screen-printing is required if the efficiency improvements described in the previous section are to impact commercial production.

The UNSW buried contact cell shown in Figure 2.7a provides such a dramatic break. The device was developed as a low cost version of the PESC cell shown in Figure 2.4b. Recent results show that it can match PESC cell efficiency with much lower processing costs. The associated technology has been licensed to several major cell manufacturers with commercial modules based on this technology available since 1991, at prices not markedly different from standard product. Full scale production commenced in 1994.

6.2. Processing

Wafers are processed initially as in a normal screen-printing operation. After saw damage removal and texturing, the wafer surface is diffused, although more lightly than normally, and the wafer oxidized. The deep grooves apparent in Figure 2.7a are then formed in the top surface by either laser or mechanical grooving. This is followed by a groove etch and a second heavy diffusion confined to the grooves by the oxide covering nongrooved areas. Al is then applied to the cell rear, by either screen-printing or evaporation, and alloyed.

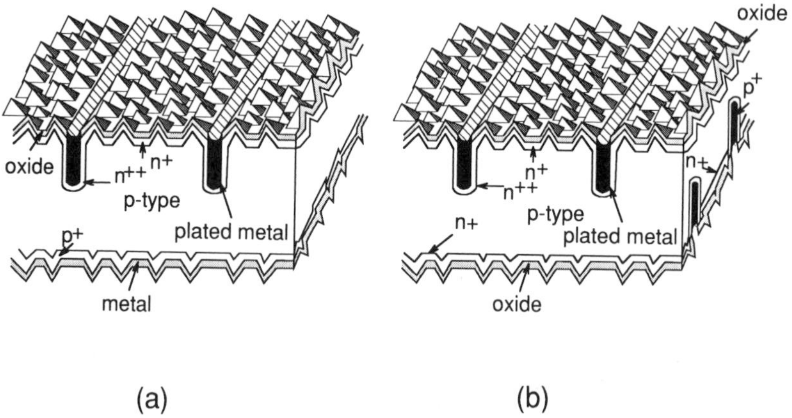

FIGURE 2.7 (**a**) UNSW buried contact solar cell. (**a**) Double-sided buried contact cell.

The cells are then plated by electroless nickel, sintered, and plated in electroless copper and silver solutions. These metals plate only to the conductive cell areas, namely, grooved areas on the top and the entire rear surface. Edge junctions are isolated by laser scribing of the cell at this stage or by normal "coin- stack" etching after the first junction diffusion.

An elegant feature is the way the top surface oxide is used in several different ways both during fabrication and in the final cell. It acts as a shroud protecting against impurity ingress during processing, a diffusion mask, a plating mask, a surface passivation layer, and a rudimentary antireflection coating. A sequence has also been developed using silcon nitride in place of the oxide, reducing reflection in the completed cells (Mason and Jordan, 1991). Compared with a screen-printing process, additional steps are laser grooving of the top surface, groove etching and diffusion, and contact plating. The top surface screen-printing process is not required, and the process eliminates expensive silver pastes. a survey of published results of the cost of this sequence suggests that it is comparable with that of the screen-printing process (Green et al., 1991). These costs were for laser grooving. Mechanically grooved sequences offer lower cost.

6.3. Cell Performance

The structure offers similar efficiency to PESC cells, the first 20% efficient silicon cells, but has inherent efficiency advantages. One arises from the high aspect ratio of the cell metallization. As opposed to aspect ratios of 0.06 for screen-printing and 0.3 for PESC metallization, the aspect ratio can be as high as 5. This is particularly important for large area cells. It allows much larger cells to be used than present 10 cm diameter or 10 cm square cells. A second advantage over the PESC sequence is that doping in contacted and noncontacted areas is separately optimized, increasing V_{oc} potential. Voltages as high as 693 mV have been confirmed for buried contact cells with upgraded rear passivation.

6.4. Production Experience

Buried contact solar cell technology has been licensed to several major cell manufacturers, with pilot production experience reported by some. A high-efficiency array was fabricated by Telefunken (Boller et al., 1989) for the Swiss car "Spirit of Biel", which convincingly won the 1990 World Solar Challenge, the solar car race from Darwin to Adelaide. Array efficiency was 17%, then the highest ever for silicon. The array gave 25% more power than that of the second placed car, which used the best available Japanese screen-printing technology (Kyle, 1991).

BP Solar has reported on both manufacturing yields and process economics (Mason and Jordan, 1991). Using the same "solar grade" CZ substrates as in their screen-printing process, BP reports substantial efficiency improvement for the technology (circa 30%) and cell efficiencies of 17.5%–18%. Economic analysis shows that the approach, as developed by BP with nitride antireflection coating, is well suited for polycrystalline material giving module costs below U.S. $3/peak Watt. Solarex has reported preliminary work using mechanical dicing wheels to form the grooves (Wohlgemuth and Narayanan, 1991). A small pilot production line operated by Unisearch Ltd. in conjunction with UNSW has also given good yields of cells with efficiencies of 19%–20% in production volumes of 10,000 cells per year (Wenham et al., 1992).

The consensus of these pilot studies is that the process, when transferred into production, can give efficiency margins of 25%–40% over screen-printing. Although more processing steps are involved than in the simplest screen-printing approach, expensive silver pastes are eliminated so that processing costs per unit area are not greatly different, with costs per watt of product likely to be lower (Bruton, 1994). Marketing experience has shown that higher selling prices are feasible for this product due to the lower balance of systems costs in installed systems and the perceived higher quality due to the superior performance.

6.5. Future Improvement

The double-sided grooved structure shown in Figure 2.7b holds the promise of further simplifying buried contact cell processing while increasing efficiencies to around 21% for large cells. A complementary improvement is the cutting of grooves by mechanical dicing wheels. Experiments with 35 ganged dicing blades have produced grooves of 2% depth uniformity over the wafer surface with a processing time of 3 s/cell. The attraction of this approach is lower equipment costs, although consumable costs are higher than with laser grooving. Both Al deposition and alloying steps and the need for edge junction isolation are eliminated, but an additional boron diffusion is introduced. The cell also responds to light from both directions. This allows use in modules with transparent rear backings with about 20% additional power expected in some applications.

7. ADVANCED APPROACHES

7.1. Ribbon and Sheet Growth

The cost of sawing ingots into wafers limits achievable costs by the previous ingot approaches. Approaches where silicon is grown directly as ribbons or sheet, eliminating wafering, clearly have lower cost potential. Various approaches have been explored and are reviewed elsewhere (Eyer et al., 1990). Two of the earliest and possibly the most developed are the edge-defined film-fed growth (EFG) and dendritic web approaches shown in Figure 2.8.

The EFG approach has been the focus of most commercial activity. Molten silicon moves upward between two faces of a graphite die by capillary action,

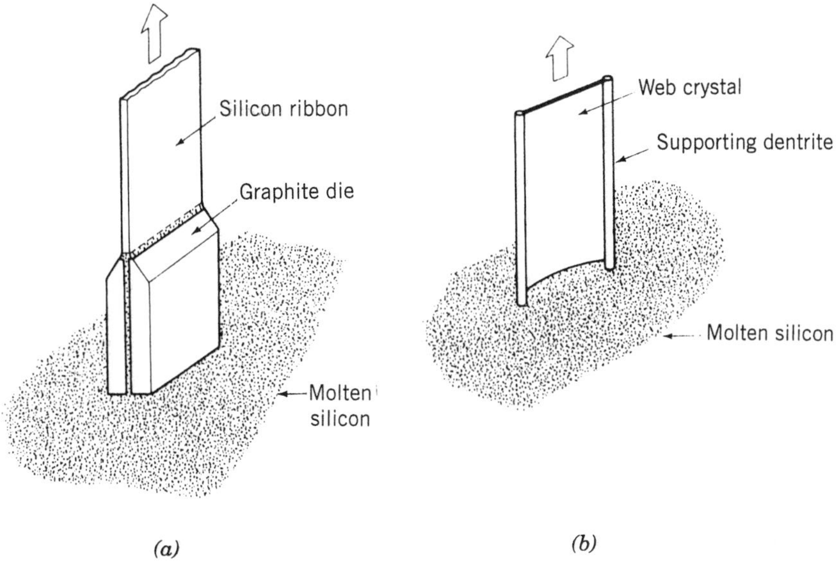

FIGURE 2.8 Ribbon growth approaches: (**a**) Edge-defined film-fed growth (EFG). (**b**) Dendritic web.

forming a narrow molten region along the die top. By appropriate seeding, thin silicon ribbons can be pulled from this molten region as shown in Figure 2.8a. A major issue has been the quality of the large-grained multicrystalline material produced, with impurities from the die reported as a problem. Carrier lifetimes in the material also reduce rapidly as the bulk dopant concentration increases, giving highest quality in higher resistivity substrates that are not well suited for conventional cell designs. The rough ribbon surface also restricts metallization options. Quality appears similar to that of cast multicrystalline wafers. As for the latter, normal texturing is not applicable.

The dendritic web approach in Figure 2.8b is a cleaner process, relying on close temperature control to define ribbon shape. Two silicon dendrites of controlled separation are grown. As these are pulled from the melt, they trap a thin layer of molten silicon that solidifies to form the ribbon. The potential material quality is better than with most ribbon approaches. However, a twinning plane half-way across the ribbon thickness nucleates impurities. Although this material has a reputation for producing high-quality devices (Rohatgi et al., 1985), this reputation may be based more on sophisticated cell processing than on material quality. As with other ribbon techniques, attracting sufficient investment to scale to production volumes where the lower fabrication costs would give a marketing advantage, has been difficult.

7.2. Spheral Cells

Surface tension allows silicon powders to conglomerate into small spheres when melted on a platform. During solidification, impurities are swept to the outer shell of the sphere. By crude processes such as grinding, the outer shell can be removed and the process repeated. Crude silicon grades can thereby produce relatively pure sperical balls The balls can be diffused, slotted into an aluminum mesh, and processed to convert each into a small cell, connected in parallel with neighbors. Inoperative balls are disconnected during processing. In this way, large cells 10 cm square have been fabricated with efficiencies up to 11% (Levine et al., 1991).

Attractions are that crude silicon grades can be used and wafering eliminated. Disadvantages include the processing complexity and the inherently low efficiency potential. Given the small cell size and consequent handling constraints, it is difficult to imagine "state-of-the-art" performance being obtained. This difficulty is compounded by relatively low packing densities of the spheres (about 70%), suggesting little scope for improvement beyond the 11% efficiency already demonstrated. The approach's material intensiveness is an additional long-term disadvantage.

7.3. Implanted Bubble Cell

Recently, extremely high efficiencies have been reported for cells using a layer of "bubbles" near the junction, formed by annealed hydrogen implants (Li et al., 1992). Cells were otherwise processed conventionally using boron junction diffusion. A J_{sc} of 69.1 mA/cm^2 was reported, about twice that expected from conventional cells of this structure. Spectral response measurements showed unusual features, including quantum efficiencies apparently well above 100%, but, perhaps more importantly, infrared response beyond silicon's band gap. Given the small cell size (2 mm square) and that the infrared tail would only marginally enhance J_{sc}, the most likely explanation for the enormous current reported is measurement error due to the very small device size (Godfrey and Green, 1977). A researcher involved states that yields were extremely low with only two cells of hundreds fabricated showing these characteristics. Both apparently have been damaged or rendered inoperative so that no samples remain for independent evaluation, as recommended for high-efficiency results (see Section 5.1, above). The work does, however, show the potential for improving cell infrared response, if able to be sustantiated.

7.4. Thin-Film Silicon Cells

Recent theoretical and experimental developments have reawakened interest in thin-film cells based on multicrystalline silicon. Initial interest dates to the early 1970s, although poor progress shifted emphasis to the ribbon and cast ingot approaches previously described. One difficulty is silicon's weak light absorption. This implied that silicon had to be thick to absorb a reasonable fraction of sunlight. Developments in the early 1980s with "light trapping" have changed this assessment. Theoretically, weakly absorbed light can be trapped for at least 50 passes across the cell, with over 30 passes demonstrated experimentally. This allows cell thickness less than 10 μm to be seriously considered. Reducing cell thickness also reduces allowable grain size in multicrystalline material, since this only has to be several times the material thickness for good performance.

Experimentally, good progress has been made with the growth of relatively thin silicon cells on ceramic substrates. Although process details are scanty, it appears that a seeding layer may be deposited onto an expansion-matched ceramic substrate and this layer either recrystallized (Barnett, 1992) or a layer grown from solution onto the seeds (Barnett, 1989). Efficiency of 15% has been independently confirmed for cells produced with this general approach.

Interest in thin films of silicon on foreign substrates also comes from the consumer electronics area. Some active-matrix liquid–crystal television displays currently use thin films of polycrystalline silicon, chemically deposited over large areas, to form transistors (Stix, 1989). Although these films are too thin for photovoltaics, only about 0.1 μm thick, this work suggests that technology for depositing good-quality silicon on foreign substrates will significantly advance over the coming decade. A thin-film, high-efficiency silicon technology may provide the path for evolution of photovoltaics from current dependence on thick, self-supporting substrates.

8. CONCLUSIONS

Crystalline and multicrystalline silicon has been the work-horse for outdoor photovoltaic applications since the development of the first crystalline silicon cells in the 1950s. Recent improvements in laboratory cell performance have greatly increased expectations as to what is feasible with this technology. The consensus appears to be that 20% efficient modules based on silicon technology are commercially feasible (Steitz and Associates, 1989).

The present commercial screen-printing technology for cell fabrication does not appear capable of such efficiencies. This is due to limitations of screened silver in terms of aspect ratio, contact resistance, line width, and conductivity. The buried contact cell approach satisfies all criteria for producing high efficiency and is suited for large-scale commercial manufacture, as supported by recent pilot-line studies. Crystalline silicon ingot technologies are the most compatible with reaching 20% module efficiency. Multicrystalline ingot technologies are capable of further improvement but are unlikely to produce the highest performance. The eventual limitation with ingot technologies may be the costs of sawing ingots into wafers. Ribbons and associated approaches including assembled sheets, such as with spheral cells, offer efficiencies comparable with cast multicrystalline cells while avoiding wafering costs.

It is argued that the long-term future of silicon photovoltaics lies with a thin-film approach. Theoretical developments over the last decade show that silicon films only a few microns thick can give high efficiency with present device concepts. Low film deposition temperature and multijunction cell concepts (Green et al., 1994) may be particularly relevant to this future.

ACKNOWLEDGMENTS

The Centre for Photovoltaic Devices and Systems is supported by the Australian Research Council under the Special Research Centres Scheme and by Pacific Power.

REFERENCES

Barnett, A. M. (1989), *Thin Silicon-Film Solar Cells on Ceramic Substrates*, Conf. Record, 4th Photovoltaic Science and Engineering Conference, Sydney, pp. 151–158.

Barnett, A. M. (1992), *Thin Polycrystalline Silicon Solar Cells on Low Cost Substrates*, Conf. Record, 6th International Photovoltaic Science and Engineering Conference, New Delhi, pp. 737–744.

Bickler, D. B., and Callaghan, W. T. (1987), *The Economic Payoff for a State-of-the-Art High-Efficiency Flat-Plate Crystalline Silicon Solar Cell Technology*, Conf. Record, 19th IEEE Photovoltaic Specialists Conference, New Orleans, pp. 1424–1429.

Boller, H.-W., and Ebner, W. (1989), *Transfer of the BCSC-Concepts Into an Industrial Production Line,* Conf. Record, 9th E.C. Photovoltaic Solar Energy Conference, Freiburg, September, pp. 411–413.

Briglio, A., Dumas, K., Leipold, M., and Morrison, A. (1986), *Flat Plate Solar Array Project: Final Report: Vol. 3: Silicon Sheet: Wafers and Ribbons.* Jet Propulsion Laboratory Publication 86-31, Pasadena, CA.

Campbell, P. R. (1989), *Light Trapping and Reflection Control in a Crystalline Silicon Solar Cell.* PhD Thesis, University of New South Wales.

Chapin, D. M., Fuller, C. S., and Pearson, G. L. (1954), *J. Appl. Phys.* **25**, 676–677.

Chong, C. M. (1989), *Buried Contact Solar Cells.* PhD Thesis, University of New South Wales.

Christensen, E., Ed. (1985), *Flat Plate Solar Array Project: Ten Years of Progress*, Jet Propulsion Laboratory Publication 400-279, Pasadena, CA.

Eyer, A., Rauber, A., and Goetzberger, A. (1990), *Optoelectronics* **5**, 239–257.

Fischer, H., and Pschunder, W. (1976), *Low Cost Solar Cells Based on Large Area Unconventional Silicon*, Conf. Record, 12th IEEE Photovoltaic Specialists Conference, pp. 86–92.

Fritts, C. E. (1883), *Proc. Am. Assoc. Advancement Sc.* **33**, 97.

Godfrey, R. B., and Green, M. A. (1977), *Appl. Phys. Lett.* **31**, 705–707.

Green, M. A. (1982), *SOLAR CELLS: Operating Principles, Technology and System Applications*, Prentice-Hall, New Jersey.

Green, M. A. (1987), *High Efficiency Silicon Solar Cells*, Trans. Tech. Publications, Aedermannsdorf.

Green, M. A., and Emery, K. (1994), *Prog. Photovoltaics* **2**, 27–34.

Green, M. A., Wenham, S. R., Zhao, J., Bowden, S., Milne, A. M., Taouk, M., and Zhang, F. (1991), *Present Status of Buried Contact Solar Cells*, Conf. Record, 22nd IEEE Photovoltaic Specialists Conference, Las Vegas, pp. 46–53.

Green, M. A., Wenham, S. R., Zhao, J., Wang, A., Yun, F., and Campbell, P. (1992), *High Efficiency Silicon Solar Cells*, Conf. Record, 6th Photovoltaic Science and Engineering Conference, New Delhi, pp. 863–868.

Green, M. A., Wenham, S. R., Zhao, J., Shi, Z., and Honsberg, C. B. (1994), *23.5% Efficiency and Other Recent Improvements in Silicon Solar Cell Efficiency*, Conf. Record, 12th E.U. Photovoltaic Solar Energy Conference, Amsterdam, in press.

Grondahl, L. O. (1933), *Rev. Modern Phys.* **5**, 141.

Guelden, M., and Wenham, S. R. (1992), *Internal Report 9202*, University of New South Wales Centre for Photovoltaic Devices & Systems.

Haynos, J., Allison, J., Arndt, R., and Meulenberg, A. (1974), *The COMSAT Non-Reflective Silicon Solar Cell: A Second Generation Improved Cell*, Conf. Record, Int. Conference on Photovoltaic Power Generation, Hamburg, p. 487.

Hogan, S., Darkazalli, G., and Wolfson, R. (1991), *An Analysis of High Efficiency Si Processing*, Conf. Record, 10th E.C. Photovoltaic Solar Energy Conference, Lisbon, pp. 276–279.

Johnson, J. E., Hanoka, J. I., and Gregory, J. A. (1985), *Continuous Mode Hydrogen Passivation*, Conf. Record, 18th IEEE Photovoltaic Specialists Conference, Las Vegas, pp. 1112–1115.

Khemthong, S., Cabbanis, S., Zhao, J., and Wang, A. (1991), *Low Cost Silicon Solar Cells with High Efficiency at High Concentrations*, Conf. Record, 22nd IEEE Photovoltaic Specialists Conference, Las Vegas, pp. 268–272.

Kimura, K. (1984), *Recent Developments in Polycrystalline Silicon Solar Cell*, Conf. Record, 1st International Photovoltaic Science and Engineering Conference, Kobe, pp. 37–42.

King, R. R., Sinton, R. A., and Swanson, R. M. (1991), *One-Sun, Single-Crystalline Silicon Solar Cell Research Report SAND91-S003* Sandia National Laboratories, Albuquerque, NM.

Kingbury, E. F., and Ohl, R. S. (1952), *Bell Syst. Tech. J.* **31**, 802–815.

Kopp, J., Knobloch, J., and Wettling, W. (1992), *Influence of Emitter Oxide Thickness on the Passivation of High Efficiency Silicon Solar Cells*, 11th E.C. Photovoltaic Solar Energy Conference, Montreux, pp. 49–52.

Kyle, C. (1991), *Racing with the Sun: The 1990 World Solar Challenge,* Engineering Society for Advancing Mobility: Land, Sea, Air and Space, SAE Order No. R-111.

Levine, J., Hotchkiss, G. B., and Hammerbacher, M. D. (1991), *Basic Properties of the Spheral Solar Cell*, Conf. Record, 22nd IEEE Photovoltaic Specialists Conference, Las Vegas, pp. 1045–1048.

Li, J., Chong, M., Zhu, J., Li, Y., Xu, J., Wang, P., Zhang, Z., Yang, Z., Zhu, R., and Cao, X. (1992), *Appl. Phys. Lett.* **60**, 2240.

Lindmayer, J. (1976), *Semi-Crystalline Silicon Solar Cells*, Conf. Record, 12th IEEE Photovoltaic Specialists Conference, Baton Rouge, pp. 82–85.

Lindmayer, J., and Allison, J. (1973), *COMSAT Tech. Rev.* **3**, 1–22.

Mason, N. B., and Jordan, D. (1991), *A High Efficiency Silicon Solar Cell Production Technology*, Conf. Record, 10th E.C. Photovoltaic Solar Energy Conference, Lisbon, pp. 280–283.

Narayanan, S., Wenham, S. R., and Green, M. A. (1990), *IEEE Trans. Electron Devices* **37**, 382–384.

Ohl, R. S. (1941), *Light-Sensitive Electric Device*, U.S. Patent No. 2,402,622; *Light-sensitive Device Including Silicon*, U.S. Patent No. 2,443,542.

Pern, F. J., and Czanderna, A. W. (1992), *Solar Energy Materials Solar Cells*, **25**, 3–23.

Ralph, E. L. (1975), *Recent Advancements in Low Cost Solar Cell Processing*, Conf. Record, 11th IEEE Photovoltaic Specialists Conference, Scottsdale, pp. 315–316.

Rittner, E. S., and Arndt, R. A. (1976), *J. Appl. Phys.* **47**, 2999–3002.

Rohatgi, A., Meier, D. L., O'Keefe, T. W., and Rai-Choudhury, P. (1985), *High-Efficiency Solar Cells on Low-Resistivity Dendritic Web Silicon Ribbon*, Conf. Record, 18th IEEE Photovoltaic Specialists Conference, Las Vegas, pp. 50–54.

Rohatgi, A., Sana, P., and Salami, J. (1992), *Record High Efficiency Solar Cells on Cast Polycrystalline Silicon*, 11th E.C. Photovoltaic Solar Energy Conference, Montreux, pp. 159–163.

Saitoh, T., Uematsu, T., Kida, Y., Matsukuma, K., and Morita, K. (1987), *Design and Fabrication of 20% Efficiency, Medium-Resistivity Silicon Solar Cells*, Conf. Record, 19th IEEE Photovoltaic Specialists Conference, New Orleans, pp. 1518–1519.

Saitoh, T., Shimokawa, R., and Hayashi, Y. (1993), *Prog. Photovoltaics* **1**, 11–24.

Steitz and Associates (1989), *EPRI Workshop on Photovoltaic Cell and Module Technology Status*, Summary Memorandum, Electric Power Research Institute, Palo Alto.

Stix, G. (1989), *IEEE Spectrum*, **26**, 36–40.

Verlinden, P., Van de Wiele, F., Stehelin, G., and David, J. P. (1985), *Optimized Interdigitated Back Contact (IBC) Solar Cell for High Concentrated Sunlight*, Conf. Record, 18th IEEE Photovoltaic Specialists Conference, Las Vegas, pp. 55–58.

Warabisako, T., Matsukuma, K., Kokunai, S., Kieda, Y., Uematsu, T., and Yagi, H. (1992), *A 16.8% Efficient, 100 cm^2 Polycrystalline Silicon Solar Cell with Triode Structure*, 11th E.C. Photovoltaic Solar Energy Conference, Montreux, pp. 172–175.

Wenham, S. R., Wu, Y., Xiao, R. D., Taouk, M., Guelden, M., Green, M. A., and Hogg, D. (1992), *Pilot Line Production of Laser Grooved Silicon Solar Cells*, 11th E.C. Photovoltaic Solar Energy Conference, Montreux, Switzerland, pp. 416–422.

Willeke, G., Nussbaumer, H., Bender, H., and Bucher, E. (1992), *Solar Energy Materials Solar Cells* **26**, 345–356.

Witwer, J. G. (1983), *Photovoltaic Power Systems Research Evaluation*, Report AP-3351, Electric Power Research Institute, Palo Alto.

Wohlgemuth, J., and Narayanan, S. (1991), *Buried Contact Concentrator Solar Cells*, Conf. Record, 22nd IEEE Photovoltaic Specialists Conference, Las Vegas, pp. 273–277.

Wohlgemuth, J. H., Narayanan, S., and Brenneman, R. (1990), *Cost Effectiveness of High Efficiency Cell Processes as Applied to Cast Polycrystalline Silicon*, Conf. Record, 21st IEEE Photovoltaic Specialists Conference, Kissimimee, pp. 221–226.

Zhao, J., Wang, A., and Green, M. A. (1994), *Prog. Photovoltaics* **2**, 227–230.

Zolper, J. C., Narayanan, S., Wenham, S. R., and Green, M. A. (1989), *Appl. Phys. Lett.* **55**, 2363–2365.

CHAPTER THREE

Terrestrial Silicon Concentrator Solar Cells

R.A. SINTON, Sinton Consulting, San Jose, CA 95129-1454

1. INTRODUCTION

The primary motivation behind using silicon solar cells in concentrator applications is to reduce the cost of the energy produced. High-efficiency flat-plate solar cells are fabricated on high-quality crystalline or semicrystalline silicon. The manufacturing cost of flat-plate silicon modules is \$3.00–\$3.50 per watt, with half of this due to the wafer alone (see Chapter 2). This low cost is achieved only with the use of very inexpensive cell fabrication techniques. Typically, the desired processing cost of a solar cell is one to two orders of magnitude less than the cost of processing the same wafer into integrated circuits.

The use of concentration can directly reduce the effects of the solar cell costs on the system generated power cost. If the cost of concentrating the light is significantly less expensive per unit area than the cost of flat-plate modules, then concentration allows a lower price. This concentration is usually achieved by the use of Fresnel lenses or reflective optics. Since the cell cost in concentrator systems is a smaller fraction of the total system cost, the efficiency of the cell is more important. Improvements in cell efficiency are worthwhile until such a point as the cost of such improvements increases the entire system cost proportionately to the efficiency gain.

Concentrator cells can also be more efficient than one-sun cells. More sophisticated designs and fabrication techniques including those typically used for integrated circuits can be used in fabrication. Higher quality substrates can also be used. The operation of solar cells at higher concentration itself gives

Solar Cells and Their Applications, Edited by Larry D. Partain.
ISBN 0-471-57420-1

higher efficiency, since the voltage output of the solar cell increases logarithmically with the concentration.

A plot of the demonstrated laboratory efficiencies for silicon one sun and concentrator cells is given in Figure 3.1 for the last decade of progress. Concentrator cells have a 4%–4.5% advantage in absolute efficiency over one sun cells. However, in a typical situation, 85% of the power in the sunlight is direct beam (light available for concentration), the concentrator cell temperature is 5°C hotter than the cells in a one sun module, and the Fresnel lens in the module is 85% transmissive to sunlight. Hence, the power output from state-of-the-art flat-plate and concentrator silicon module efficiencies will be similar.

2. DESCRIPTIVE DEVICE PHYSICS

The key requirements for obtaining high efficiency in a concentrator solar cell are largely the same as for a one sun cell. The primary difference is that for a concentrator cell the fabrication cost is not so tightly restrained. A wider range of options are available for addressing the various problems, since in the final energy cost the cell cost is not as critical.

The types of silicon concentrator solar cells fall into two groups. The first are cells quite similar to flat-plate cells. These are fabricated on doped material, usually p type, and operate in a regime where the density of photogenerated electron–hole pairs in the silicon never approaches the dopant density. These cells are said to operate under low-level-injection conditions. The second group of solar cell designs for concentrator applications uses nominally undoped

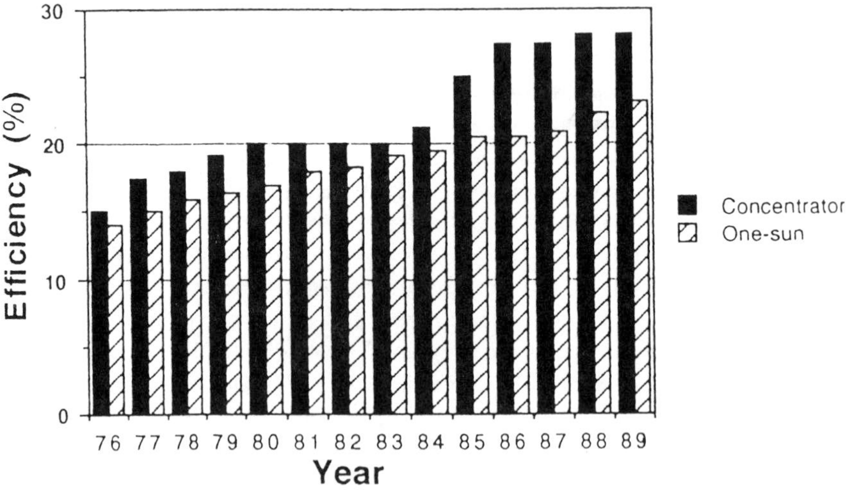

FIGURE 3.1 The efficiency versus time for a decade of progress in concentrator and one sun silicon solar cells.

substrates. In this case, the photogenerated carriers are necessary not only to generate voltage but also to lower the internal resistance drops within the cell. The device physics describing the two cases is quite different. These issues have been reviewed in detail in recent monographs (Green, 1987; Boer, 1990).

2.1. General Features of High-Efficiency Solar Cells

Many of the main features for high-efficiency silicon solar cells are common to both high- and low-level-injection solar cell designs. These features are summarized here.

Some of the primary factors present in high-efficiency designs are

- Well-passivated surfaces
- High minority-carrier lifetimes in the substrate material
- Good light trapping
- Thin substrates
- High quantum efficiency in the front doped region
- Low series resistance
- Low grid shadowing

2.2. Surface Recombination

Recombination at the cell surface presents a major loss to cell efficiency. Detailed studies of the optimization of dopant diffusions that serve as surface passivation indicate that a complete optimization includes the following factors (King et al., 1990):

- A very high-quality Si–SiO_2 interface
- Minimal area of metal–silicon contacts
- High-surface-concentration, deep-dopant diffusions under metal contacts
- Undoped or lightly doped surfaces except under metal contacts

The last item, the use of undoped or lightly doped surfaces, is of particular current interest. To date, undoped Si–SiO_2 surfaces have the lowest recombination rates at operating conditions typical for high-concentration applications. However, a degradation of this passivation quality has been observed upon exposure of the oxide to high-intensity ultraviolet light (Gruenbaum et al., 1988). Detailed studies of this effect showed that sunlight injects hot electrons into the oxide. This injection is primarily for photons exceeding 2.9 eV, and it increases the surface interface-state density, causing an increase in the surface recombination velocity (Gruenbaum et al., 1989). High-concentration solar cells must be designed to withstand exposure to such photons.

The rate of surface recombination velocity degradation after UV exposure is dependent on precise details of the wafer surface preparation and oxide growth schedules. These physical effects seem to be quite similar to the hot electron effects that plague CMOS integrated circuit devices.

One solution has been to dope the front surface of the cell n type, with a surface concentration sufficient to create a built-in voltage that passivates the surface (Gruenbaum et al., 1990). This has been shown to be effective, although with a certain loss in performance at the higher concentrations (Sinton et al., 1990; Ruby and Schubert, 1991). This is discussed later in the context of various cell designs.

2.3. Minority-Carrier Lifetime

The minority-carrier lifetime in the substrate material is a critical factor in determining the efficiency of concentrator cells. Concentrator cells designed for moderate to high concentration can make use of the highest quality substrates without significantly impacting the final cell cost. The substrates that have shown the highest minority-carrier lifetimes are wafers cut from float-zone-grown boules. More recently, magnetically grown Czochralski wafers have shown characteristics very similar to float-zone wafers. Normal Czochralski-grown silicon, the standard for the integrated circuit industry, generally has a lower lifetime after processing than either the float zone or the magnetic Czochralski silicon and results in lower solar cell efficiencies.

The use of high-quality substrates does not guarantee higher efficiencies for the solar cells. The final minority-carrier lifetime in the bulk of the solar cell can depend critically on the detailed processing schedule. Care must be taken to optimize the process and maintain the original lifetime in the material. Key elements of this processing include very careful avoidance of chemical contamination of the wafers and HCl gettering of the high-temperature furnace tubes (Sinton et al., 1985; Zhao et al., 1990). In the case of ideal material with state-of-the-art processing that maintains the material quality, fundamental limits to the bulk lifetime will eventually limit the achievable efficiency. This lifetime limit is determined by the Auger and radiative recombination rates intrinsic to silicon (Tiedje et al., 1984; Green, 1987; Sinton and Swanson, 1987a).

2.4. Light Trapping

Much of the improvement in the efficiencies of silicon solar cells during the decade of the 1980s was due to light trapping. Solar cells with high substrate lifetime and well-passivated surfaces have a collection efficiency approaching 100% of the photogenerated carriers, independent of where in the solar cell these carriers are generated. A significant number of photons in the solar spectrum have enough energy to create electron–hole pairs in the silicon, but are very weakly absorbed in the solar cell. *Light trapping* refers to schemes utilized to maximize the absorption of the weakly absorbed light by maximizing the path length traversed by the photon in the solar cell. One of the simplest implementations of this concept is to optimize the backside metalization for high reflectance, ensuring that the weakly absorbed photons will cross the cell twice. Another simple scheme is to structure the front cell surface to refract the photons into an angle with a larger path across the cell. More complex schemes have been investigated that use structured front and back surfaces as well as mirrored surfaces to increase the effective path length of the photons by, in principle, up to 50 times the wafer thickness (Green, 1987; Yablonovitch and Cody, 1982; Wenham et al., 1991). This can increase the current from the solar cell by up to 15% for a 100 μm thick solar cell. Even higher degrees of light trapping are made possible by the use of a very high concentration with a limited acceptance angle (Green, 1987; Sinton and Swanson, 1987c; Luque et al., 1991).

These features of light trapping are important for any silicon solar cell, one sun or concentrator. They are especially critical for concentrator cells. Concentrator solar cells have very high current densities that cause transport losses due to bulk series resistance and carrier-diffusion gradients. A minimization of these effects demands thin substrates. In addition, the best available concentrator cells are operating very near the Auger recombination limit to performance. This is a bulk recombination effect that is minimized by using thin substrates. Light trapping is the means by which a very thin, optimized concentrator solar cell can be engineered to absorb most of the available light.

2.5. Front Junction Optimization

In a conventional solar cell with an n^+ diffusion on the cell frontside, there are some electron–hole pairs generated by uv light very near the cell surface that recombine in the heavily doped surface layer before they can contribute to the cell output current. Much work has been done to address this problem both experimentally and with modeling using the properties of heavily doped regions. This work has identified the key design parameters for this front dopant-diffused region that allow virtual elimination of this loss. These studies indicate that a phosphorous diffusion with a sheet resistance greater than 50 Ohms/square will limit this loss to about 1% of the current independent of the precise surface concentration and junction depth (King et al., 1990). This limit to the doping of the front layer is particularly crucial for concentrator cells. It opposes the use of highly conductive front dopant diffusions to reduce the series resistance losses due to the high current densities of concentrator solar cell operation.

The substrate thickness is an important parameter in two different ways. First, a thin substrate will suffer less bulk minority-carrier recombination. Second, the transport of carriers over large distances results in voltage losses due to series resistance effects. The transport of minority carriers is driven by diffusion gradients that represent voltage losses as well. An accurate modeling of these effects is generally done with computer models (Sinton and Swanson, 1987b; Basore et al., 1988).

2.6. Series Resistance and Grid Shadowing

The last two factors that need careful optimization to achieve high efficiency in silicon concentrator cells are the series resistance and the grid shadowing. There are several components of this series resistance.

- Spreading resistance in the dopant-diffused layers
- Resistance in the bulk substrate
- Resistance in the metal fingers
- Contact resistance of the metal–silicon interface

Because the current density is higher for concentrator cells than for one sun cells, the grid lines must be much closer. This demands much finer fingers in order to maintain low losses from grid-line shadowing. To keep the same spreading resistance losses in the dopant-diffused layers, these grids must be spaced closer than for one sun cells by the square root of the concentration ratio if the sheet resistance of the doped layer is held constant. The actual area grid coverage fraction for concentrator cells need not be different than for one sun cells as long as the total current remains the same for the two cases (i.e., the area-concentration product is constant) and the metalization sheet resistance is the same.

Contact resistance is a critical parameter for concentrator cells. For the same contact coverage fraction and absolute current, the current density through each individual contact scales up with concentration. This generally requires that special attention be given to the minimization of the contact resistance of concentrator cells. This is addressed in one of several ways.

- By increasing the contact coverage fraction
- By increasing the surface doping concentration to reduce the specific contact resistance
- By choosing inherently low-contact-resistance metals

An increase in the contact coverage fraction increases recombination within the cell, since the metal–silicon contact is a sink for minority carriers. Also, the potential for reducing the contact resistance is only linear with the increase in contact area. On the other hand, the contact resistance drops several orders of magnitude for each factor of 10 in doping level, often without as much effect on the recombination.

The most common metalization systems for concentrator solar cells are Ti/Pd/Ag, Al-Si, and Ni. An example of the extreme doping dependence of this contact resistance is the specific contact resistance of Ti to n-type layers. It has been found to decrease by 3 orders of magnitude for a factor of 10 increase in doping density for n-type doping surface concentrations around $10^{19}\,cm^{-3}$ (Swanson and Swirhun, 1987).

The grid shadowing is a simple problem with a very large effect. In a conventional concentrator solar cell, the combination of the grid shadow and the grid series resistance can cause a 10%–20% loss. Less grid area decreases the shadowing but at the expense of series resistance. This is covered in more detail in a discussion on current cell designs. The various and creative ways that the grid shadow versus series resistance compromise are addressed is the largest single feature distinguishing various cell designs.

2.7. Demonstrated Versus Theoretical Solar Cell Efficiencies

Figure 3.2 illustrates for perspective the relative importance of these effects on the cell performance. The intrinsic efficiency limit for a 100× concentrator solar cell is generally believed to be determined by the Auger recombination in the bulk of the cell. For the case of perfect light trapping, very thin cells can be used to minimize this bulk Auger recombination. By balancing the current generated through the photon absorption in silicon with this Auger recombination, a limit of 36%–37% is calculated (Green, 1987: 83–109).

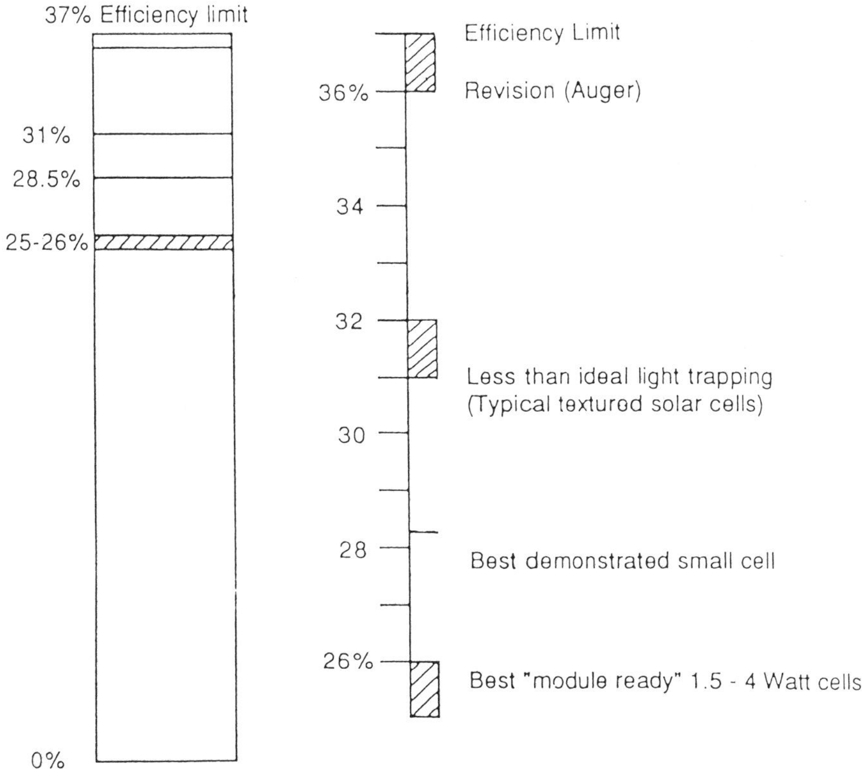

FIGURE 3.2 A perspective of the performance of concentrator cells relative to the Auger recombination limit at 100× (10 W/cm^2).

The most common light-trapping scheme, a textured front surface with a perfect backside reflector, is far from the ideal limit for light trapping. The limit imposed by the photogeneration calculated for this geometry is 31%–32%.

The highest efficiency ever demonstrated for a silicon concentrator cell at 100× is 28.2% (Sinton et al., 1986; Sinton and Swanson, 1987b). This was for a backside point contact solar cell that has all of the metalization grids for both positive and negative contacts on the cell backside in order not to have grid shadowing losses (Fig. 3.3). The efficiency versus concentration curve for this cell is shown in Figure 3.4. At 1.0 W/cm^2 (10 suns), the cell was 26% efficient. It attained 28% for the range between 3 and 15 W/cm^2 before dropping off to 26% at 45 W/cm^2. The cell incorporated all of the high-efficiency features discussed above. The surfaces were entirely undoped except for very heavily doped regions under the metal–silicon contacts. The substrate material was undoped float-zone material, with minority-carrier lifetimes in the finished cell of over 3 ms. A front surface texture combined with a back surface mirror (SiO_2 and Al) provided state-of-the-art light trapping. Since the front surface was undoped and the backside doping was restricted to small areas around each metal contact, there were virtually no losses associated with photogeneration in a heavily doped layer. The series resistance was negligible at 100× because the grids were on the cell backside, with nearly 50% coverage of each polarity. Also, the cell was quite small, 0.15 cm^2. The losses in this cell that reduced its efficiency from the possible limit were

- Practical limits on the device thickness (120 μm)
- Recombination in the dopant diffusions
- Surface recombination
- Front surface reflection, about 4%
- Transport losses, finite carrier mobilities

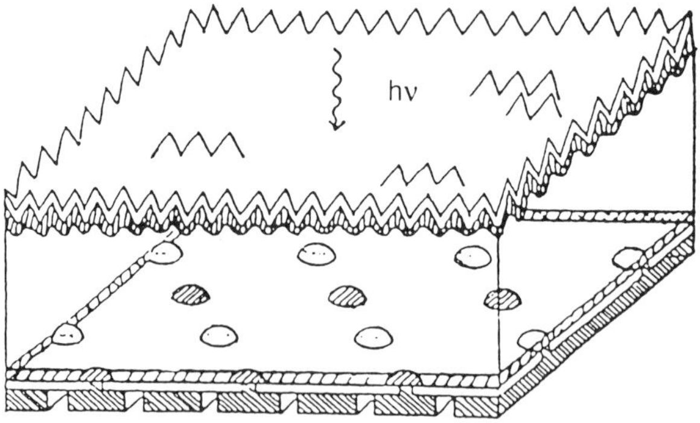

FIGURE 3.3 A cross section of a backside point contact solar cell. Both the p$^+$ and n$^+$ dopant diffusions are on the back of the solar cell so that no grid obscures the incident sunlight (hν).

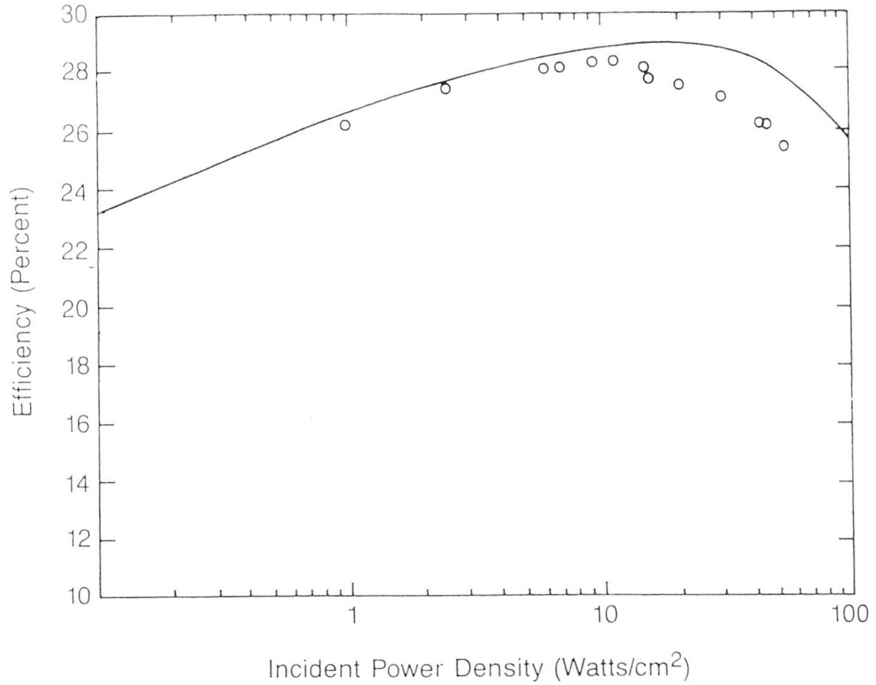

FIGURE 3.4 The efficiency versus concentration for a 0.15 cm^2 backside point contact solar cell as measured by Sandia National Laboratories. The solid curve is a modeling result based on the cell geometry.

- Series resistance
- Imperfect light trapping, absorption in backside mirror

A comparison of the parameters measured at Sandia National Laboratories with the calculated parameters for a 120 μm thick cell operating at the Auger recombination limit is shown in Table 3.1. The calculation considers the theoretical current from photogeneration in an isotropically textured solar cell with a perfect backside mirror.

TABLE 3.1 Real Versus Theoretical-Limit Cells (100 ×)

	Auger Limit	Point Contact Cell	Percent of Limit
Short circuit current density (A/cm^2)	4.33	4.17	96
Open circuit voltage (mV)	825	815	99
Fill factor	0.90	0.83	92
Efficiency (%)	32	28	88

The open circuit voltage of the cell is 99% of the theoretical limit. The short circuit current is also very nearly achieved, about 96% of that predicted. The primary difference is in the fill factor. A detailed three-dimensional analysis for this solar cell design indicates that the fill factor losses are due to recombination in the dopant-diffused regions and bulk series resistance in the areas near the point contacts (Sinton and Swanson, 1987b).

The best module-ready cells are in the 25%–26% range (Sinton et al., 1990; Green et al., 1989). These cells are less efficient for an additional set of reasons. They are designed for actual concentrator modules and are larger with outputs of 1.5–4 W. They have significant series resistance and numerous compromises in efficiency made in the interest of manufacturability. These types of cells are discussed in more detail below.

3. CURRENT STATUS OF DEVICE PERFORMANCE

High-efficiency features have been implemented in several full-sized, module-ready cell designs. The 28% efficient cell described above circumvented the heat-sinking and series-resistance problems with small size and low-duty-cycle testing. Cells intended for actual use in modules must address these problems. A survey of high-performance cells demonstrates the varieties of solutions that have been demonstrated.

3.1. Low-Concentration Solar Cells

Some of the most successful concentrator modules fielded to date use linear Fresnel lenses at a geometric concentration of 22 × (O'Neill et al., 1990). A key idea behind this module design is the advantage that slightly modified one sun cells that are already in mass production may be used. This ensures a ready supply of inexpensive cells. The applicability of one sun cells for this concentrator was enhanced by an innovation, the prismatic cell cover (Fig. 3.5), that

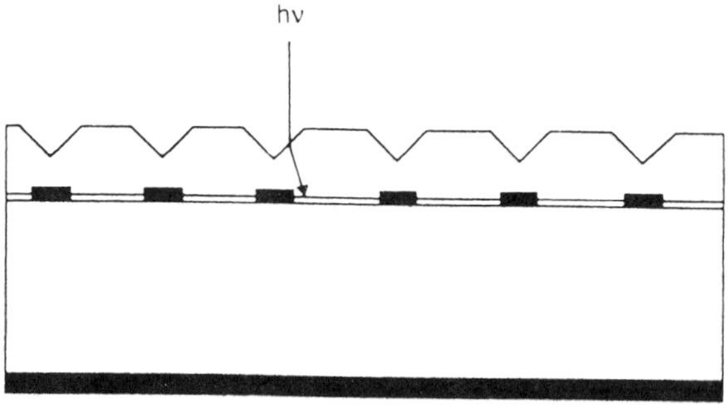

FIGURE 3.5 The prismatic cell cover refracts incident sunlight (hν) to miss the grid lines. This allows much larger grid areas without the grid obscuration that would usually result.

allows a much higher grid coverage fraction without the expected grid obscuration. This approach has only been partially successful, due to contact resistance and line resistance limits of the screen-print technology in widest use for production of one sun cells (O'Neill, 1985). Recent advances in this area have allowed the demonstration of cells with 17% efficiency at 19 suns with prismatic covers. A design optimized for use at 6 suns was 16.7% efficient without covers (King et al., 1993).

A future leading cell contender for low-concentration systems is the buried contact solar cell (see Fig. 2.7a, Chapter 2). This cell, originally designed for one sun applications, has an inherently low metalization series resistance and incorporates a surprising number of high-efficiency features (Green, 1987; et al., 1991). A full discussion of one sun Si buried contact cells is given in Chapter 2.

The application of this design to low concentration is very natural. The 22 × concentrator cell is 5 cm wide, with busbars on both sides and a grid of parallel lines running across the cell. This reduction in cell width from a typical 10 cm wide one sun cell means that the longest metal line length to a bus is 2.5 cm. This factor of 4 reduction in line length results in a metalization series resistance that is already optimized for 16 × ! The contact resistance is also naturally addressed by the nickel-sintered contacts and the highly doped contact regions with relatively large (three walls of a groove) contact area. These cells incorporate most of the high-efficiency features discussed above. A clever fabrication sequence results in a process with cost projections ($/watt) similar to screen-printed cells, where the additional costs of processing are offset by higher efficiencies.

The high-efficiency features evident in Figure 2.7a include the front surface texture, low grid resistance and shadowing losses from the high aspect ratio lines, and an advanced doping schedule that results in a lightly phosphorus-doped surface despite the heavily doped regions under the metal contacts. The only key feature missing is optimization of the reflectance of the cell backside. For these cells the back surface is simply a sintered Al layer, with about a 50% measured reflectance for infrared light. This light-trapping combination has been measured to have an effective cell-width enhancement (for purposes of light absorption) of 5 compared with a best-demonstrated case of 23 (Green et al. 1992).

These cells have demonstrated efficiencies of 22% at 2 W/cm^2, 20 suns, when used with the prism covers. Further optimization of the groove geometries has resulted in cells that have achieved 22% at 2 W/cm^2 without the prism covers (Green et al., 1992).

This design in a one-sun configuration has been installed into a pilot production facility. High-efficiency results with one sun cells (18%) and optimization work on low-concentration designs (16.5% at 13 suns) at this facility promise production of such high-efficiency low-concentration cells in the near future (Mason et al., 1991).

3.2. Conventional High-Concentration Cells

Conventional one sun solar cells generally have an n-type front emitter, a front grid, a p-type substrate, and a backside contact. One very successful thrust into concentrator solar cells has essentially involved a generalization of the best available one sun cell results into the concentrator regime.

Originally, this involved cells with Al-sintered full-area backside contacts. These cells were fabricated in moderate quantities with 24% efficiencies (Green, et al., 1989). More recently, the design has evolved into that shown in Figure 3.6. The inverted pyramids have been shown to have a superior light-trapping performance, slightly edging out random or regular upright pyramids (Zhao et al., 1990). The p^+ point contacts on the cell backside enhance the performance in two ways. First, the majority of the surface is an optimized reflector, with SiO_2 and Al. Second, the doping over the metal contact provides a lower recombination contact than the metal contact alone. This doping also reduces the contact resistance.

When coupled with a prismatic cell cover to reduce the grid shadowing losses, a cell with an active area of $1.56\,cm^2$ has a demonstrated performance of 26% at 20 suns, dropping to 21% at 140 suns. This design is being implemented into an industrial environment, with demonstrated efficiencies of over 21% for a range of power densities from 3 to $20\,W/cm^2$ (30–200 suns) (Khemthong et al., 1991).

Another cell design with front grids has also achieved very high efficiency (Fig. 3.7) (Cuevas et al., 1990). This cell differs in two important respects. First, it has an undoped substrate. Second, rather than using a prism cover, the grid shadow is minimized by using a structured surface so that grid reflection strikes the cell.

Under conditions of concentrated light, the undoped substrate is high-level

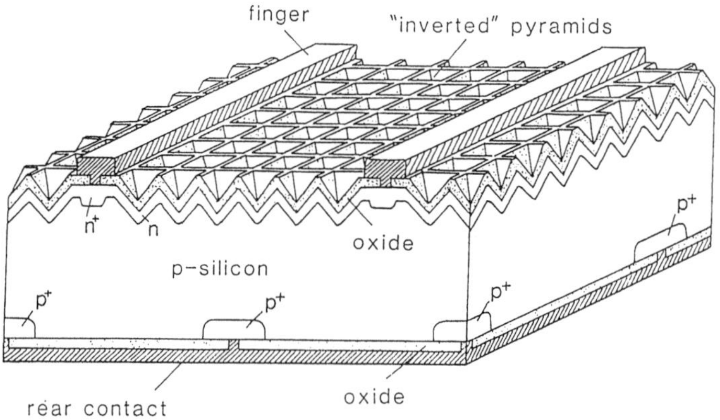

FIGURE 3.6 The PERL cell combines attributes of the point contact cell with those of a highly optimized front-gridded solar cell. (Reproduced from Zhao et al., 1990, with permission of the publisher.)

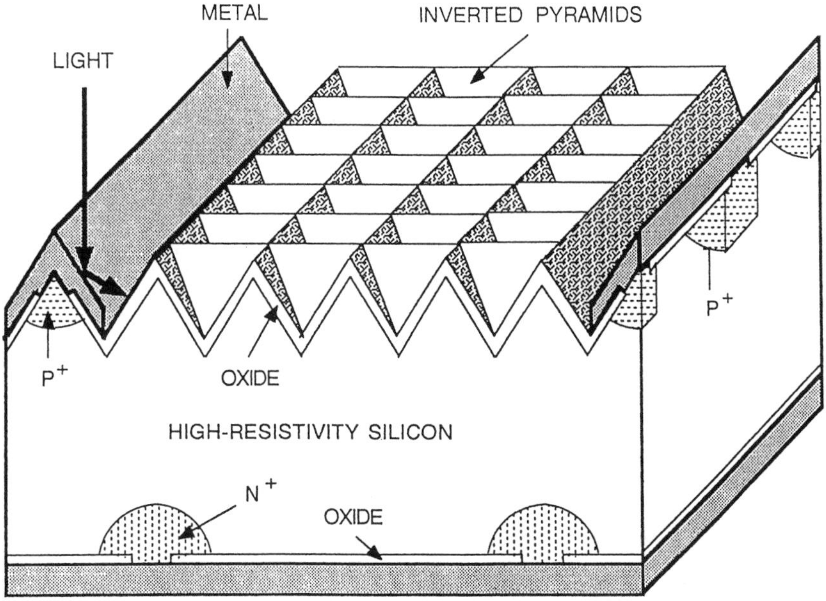

FIGURE 3.7 A concentrator cell where the silicon surface is shaped with selective etchants to form ridges for the grid line metalization. These ridges are designed so that reflected light strikes the cell. (Reproduced from Cuevas et al., 1990, with permission of the publisher.)

injected. These photogenerated carriers are crucial in providing a low series resistance within the cell as well as a high terminal voltage. The p^+ point contacts are placed at the front of the cell due to the mobility difference between electrons and holes. The preferred configuration to minimize transport losses for a high-injection cell is to minimize the distance between the average point of photogeneration and the collection for holes. The contact and grid geometry was optimized using similar modeling to the backside point contact solar cell in order to minimize the losses due to the strongly three-dimensional effects near the point contacts. This cell had a measured efficiency of 26% between 5 and 9 W/cm^2, dropping to 25% at 15 W/cm^2.

3.3. Backside Contact Solar Cells

Another approach to the design for concentrator cells has been backside contact solar cells. These cells have evolved specifically for concentrator applications. Early examples were interdigitated backside contact solar cells for concentrator applications and cells designed for thermophotovoltaic systems (Schwartz, 1982). The 28% backside point contact solar cell is the prime example of this type of cell (Sinton and Swanson, 1987b).

Recent results for backside-contact cells have focused on the problems of manufacturability. Simplified versions have been demonstrated that have performance exceeding 25% at high concentration and a relatively simple fabrication schedule that has allowed pilot production lots of cells for demonstration modules (Sinton et al., 1990). The cell shown in Figure 3.8 is essentially an interdigitated backside contact design. Some attractive features with respect to manufacturability are that the p^+ and n^+ dopant diffusions are allowed to overlap. In this way, the geometry for both diffusions is determined by a single photolithographic mask. The metalization is aligned to the diffusion such that pin holes will not short the device. The cell is soldered directly to the heat sink/electrode assembly with a solder-bump technique that has proven to have high yield compared with a double-level metalization. This cell achieved efficiencies approaching 26% at $10\,W/cm^2$. A concentrator module constructed with these cells demonstrated a 20% efficiency at 25°C and an operating temperature efficiency of 19% (Maish et al., 1991).

These cells have a frontside n^+ diffusion in order to be stable to UV light. The backside contact solar cell has been found to be especially susceptible to UV degradation. There are two reasons for this. First, the original backside contact solar cells had an undoped front surface as an optimum passivation layer for concentration solar cells. The undoped $Si-SiO_2$ interface was subsequently found to be unstable to UV light. The second reason for the sensitivity of backside contact cells to frontside passivation properties is that the minority-carrier current is collected at the back of the cell. The gradients in minority-carrier density necessary to transport the minority carriers from the front to the back of the solar cell always ensures that the minority-carrier density is highest at the cell front surface. Even under short circuit conditions there is recombination at the front surface of a backside contact solar cell. This makes the front surface passivation extremely critical to this design.

An advantage of the backside contact solar cell is that it is essentially fabricated entirely on one side of the wafer. This opens the interesting possibility of wafer-scale integration of modules. Until recently, the best results

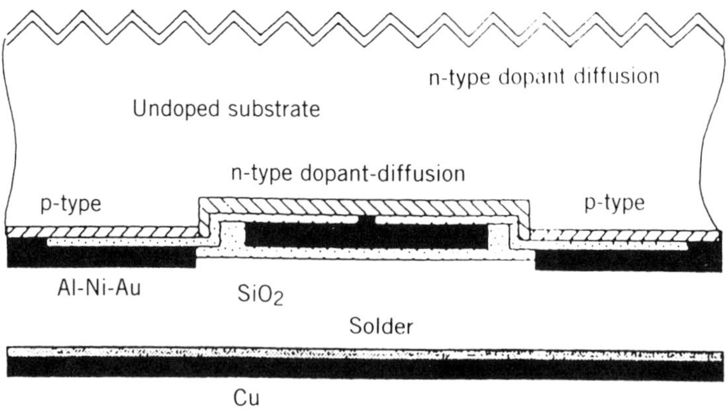

FIGURE 3.8 A simplified backside contact solar cell. The dopant diffusions are defined by a single photolithography and are allowed to abut directly. The cell is attached to an electrode using a solder-bump technique.

for such series interconnected cells was 12% (Chappel, 1978). The primary problem in series connecting the cells on a silicon substrate is in the electrical isolation of adjacent cells.

Figure 3.9 shows results from a recent demonstration of a 36 cm^2 module consisting of five series-connected cells fabricated on a single 4 inch wafer (Verlinden et al., 1991). The current–voltage curves clearly indicate the voltage addition of the adjacent cells without shunting effects. This module was 22% efficient for an incident power density of 6 W/cm^2, dropping to 19.7% at 20 W/cm^2. Mounted on a water-cooled plate, this module achieved 19.8% (17 W/cm^2) efficiency at operating temperature behind a parabolic dish concentrator. The heat transfer to ambient was accomplished through a simple closed-loop heat exchanger.

4. PROSPECTS FOR FUTURE IMPROVEMENTS

A recent study compared the prospects for backside contact vs. front-gridded solar cells (Cuevas et al., 1991). Technologically proven parameters were

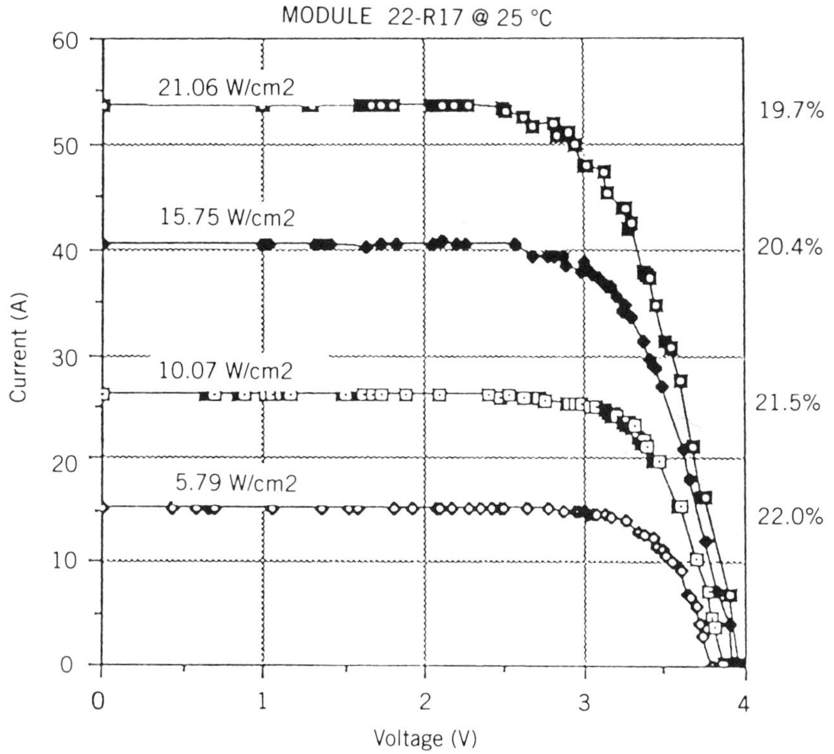

FIGURE 3.9 The current–voltage characteristics for a 36 cm^2 single-wafer integrated module at several different incident power densities. The total area efficiency, including the spaces between the five series-connected cells on the wafer, is shown for each curve.

assumed for all parameters, including bulk recombination, surface recombination, and dopant-diffused-region recombination. Cell dimensions were chosen that are technologically achievable. Surface passivations were chosen that have tested to be stable under accelerated UV light exposure.

The modeling was done using both PC-1D (Basore et al., 1988) and a quasi-three-dimensional code developed for modeling the backside point contact solar cells (Sinton and Swanson, 1987b). Key results from this study are shown in Figure 3.10. The comparison is between a backside contact solar cell, n/i/p solar cells, $n^+/p/p^+$ cells, and $p^+/i/n^+$ cells. The last three of these have front grids. All of these cells are of designs similar to those mentioned in the previous sections, except that the p/i/n cell modeled here has a front n^+ layer in between the p^+ point contacts to achieve optimum passivation qualities with UV stability. All of the front-gridded cells have locally highly doped point contacts around the grid regions. The backside cell considered here is the simplified version discussed in the previous section rather than the backside point contact solar cell.

The first point to note is the similarity in performance between such disparate designs. The designs never differ in efficiency by more than 10% (2% in absolute efficiency) over a range of operation from 1 to 100 W/cm^2 (10 to 1,000 suns). Essentially, this reflects the fact that optimum cells have long diffusion lengths. The minority carriers sample the properties of the front and the back of the cells. In this case, the properties of each of the dopant diffusions is important irrespective of precise geometric placement.

The backside contact cell is 9% (relative) more efficient at low concentra-

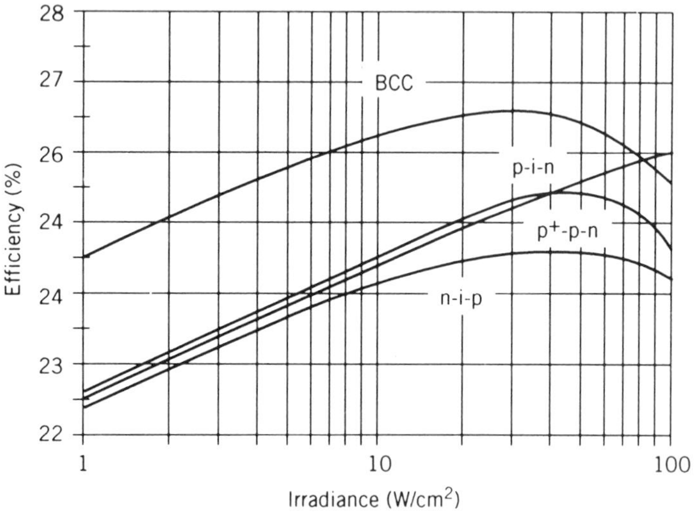

FIGURE 3.10 A modeled comparison for four distinct varieties of silicon solar cells. The top curve, the backside contact solar cell, is compared with a p/i/n cell, a $p^+/p/n$ cell (0.2 Ω-cm substrate), and an n/i/p cell.

tions. This result is entirely due to a modeling assumption that the backside contact solar cell could achieve a 2% total series-resistance shadowing loss while manufacturable designs for front-gridded cells could only achieve a minimum loss of 10%. The use of prism covers or structured and mirrored grid lines as discussed in the last section can reduce these losses for the front-gridded cells, boosting them up toward the performance of the backside contact solar cell. This comes at a price in cost and complexity, and these optical solutions may reduce the acceptance angle for the cells, limiting their functionality behind some lenses or other concentrators.

The only other obvious feature in this curve is the superiority of the p/i/n cell at the highest concentration levels. This is due to the differences in mobility between electrons and holes. In devices operating at very high current density, it is preferred that the holes be collected as near the point of photogeneration as possible. In the case of the p/i/n cell, this configuration minimizes the minority-carrier density near the cell front surface, reducing the front surface recombination.

5. CONCLUSION

An improved understanding of the device physics and fabrication techniques for silicon concentrator solar cells has allowed rapid progress to be made. Efficiencies climbed from about 20% in 1980 to 28% by 1987. These improvements are now becoming available in pilot production cells.

Two approaches have proven effective for bringing high efficiency into the low-concentration regime. The first innovation was the prismatic cell cover that allowed common one sun cell technology to perform well under low concentration by making possible very large grid lines without the usual shadowing losses. A second approach for low concentration cells, the buried contact cell design, has achieved 22% at 20×. Pilot production facilities have successfully demonstrated industrial fabrication of buried contact one sun cells and are beginning to develop concentrator versions.

In the high-concentration area, module-ready cells have been made with efficiencies approaching 25%–26%. Demonstrations of laboratory devices with 28% efficiencies promise future improvements. If full use is made of the optimization of the system optics for light trapping, theory indicates that efficiencies up to 36%–37% are possible. These factors will guide module-ready cells toward 30%. New concepts in high-concentration solar cells include the wafer-scale integration of concentrator modules. Twenty-two percent efficiencies for 36 cm^2 modules have been demonstrated for such a module operating at 5 W/cm^2. This module maintains efficiencies over 20% at power densities up to 15 W/cm^2 where the power output is 110 W.

All of these improvements at the cell level promise that efficiencies of production concentrator systems will improve very rapidly during the next few years.

REFERENCES

Basore, P.A., Rover, D.T., and Smith, A.W. (1988), *20th IEEE Photovoltaic Specialists Conf. Record,* IEEE, New York, pp. 583–585.

Boer, K., Ed. (1990), *Advances in Solar Energy*, Plenum Press, New York, Vol. 6, pp. 422–484.

Chappel, T.I. (1978), *13th IEEE Photovoltaic Specialists Conf. Record,* IEEE, New York, pp. 791–796.

Cuevas, A., Sinton, R.A., and King, R.R. (1991), *10th Eur. Commun. Photovoltaic Solar Energy Conf.,* Kluwer Academic Publishers, Dordrecht, pp. 23–26.

Cuevas, A., Sinton, R.A., Midkiff, N., and Swanson, R.M. (1990), *Electronic Device Lett.* **EDL-11**, 6–8.

Green, M.A. (1987), *High Efficiency Silicon Solar Cells*, Trans Tech Publications, Switzerland.

Green, M.A., Blakers, A.W., Zhao, J., Wang, A., Milne, A.M., Dai, X., and Chong, C.M. (1989), *Sandia Contractor Report Sand89-7041*, Sandia National Laboratories, Albuquerque, NM, pp. 3–23.

Green, M.A., Wenham, S.R., Zhao, J., Bowden, S., Milne, A.M. Taouk, M., and Zhang, F. (1991), *22nd IEEE Photovoltaic Specialists Conf. Record*, IEEE, New York, pp. 46–53.

Green, M.A., Wenham, S.R., Zhang, F., Zhao, J., and Wang, A. (1992), *Sandia Contractor Report Sand91-7016*, Sandia National Laboratories, Albuquerque, NM, pp. 10–29.

Gruenbaum, P.E., Gan, J.Y., King, R.R., and Swanson, R.M. (1990), *21st IEEE Photovoltaic Specialists Conf. Record,* IEEE, New York, pp. 317–322.

Gruenbaum, P.E., King, R.R., and Swanson, R.M. (1989), *J. Appl. Phys.* **66**, 6110–6114.

Gruenbaum, P.E., Sinton, R.A., and Swanson, R.M. (1988), *Appl. Phys. Lett.* **52**, 1407–1409.

Khemthong, S., Cabaniss, S., Zhao, J., and Wang, A. (1991), *22nd IEEE Photovoltaic Specialists Conf. Record,* IEEE, New York, pp. 268–272.

King, R.R., Sinton, R.A., and Swanson, R.M. (1990), *IEEE Trans. Electronic Device* **37**, 365–371.

King, R. R., Mitchell, K. W., Walle, J. R., Aldrich, D. L., Gee, J. M., O'Neill, M., Kaminar, N., McEntee, J., and Start, P. (1993), *23rd IEEE Photovoltaic Specialists Conf., Record,* IEEE, New York, pp. 167–171.

Luque, A., Minano, J.C., Davies, P., Terron, M.J., Tobias, I., Sala, G., Alonso, J., and Olivan, J. (1991), *22nd IEEE Photovoltaic Specialists Conf. Record,* IEEE, New York, pp. 99–104.

Maish, A.B., Hund, T.D., Quintana, M.A., and Chiang, C.J. (1991), *Proc. 10th Eur. Photovoltaic Solar Energy Conf.,* Kluwer Academic Publishers, Dordrecht, pp. 988–991.

Mason, N.B., Jordan, D., Bruton, T.M., Summers, J. G., Hughes, A.E. Whitfied, G.R., and Bentley, R.W. (1991), *22nd IEEE Photovoltaic Specialists Conf. Record,* IEEE, New York, pp. 124–127.

O'Neill, M.J. (1985), *Sandia Contractor Report Sand85-0791-1*, Sandia National Laboratories, Albuquerque, NM, pp. 128–138.

O'Neill, M.J., McDanal, A.J., Walters, R.R., and Perry, J.L. (1990), *21st IEEE Photovoltaic Specialists Conf. Record,* IEEE, New York, pp. 523–528.

Ruby, D.S., and Schubert, W.K. (1991), *22nd IEEE Photovoltaic Specialists Conf. Record,* IEEE, New York, pp. 111–117.

Schwartz, R.J. (1982), *Solar Cells,* **6**, 17–38.

Sinton, R.A., Crane, R.A., Beckwith, S., Cuevas, A., Gruenbaum, P.E., Kane, D.E., Midkiff, N., and Swanson, R.M. (1990), *21st IEEE Photovoltaic Specialists Conf. Record,* IEEE, New York, pp. 302–306.

Sinton, R.A., Kwark, Y.H., Swirhun, S.E., and Swanson, R.M. (1985), *IEEE Electronic Device Lett.* **EDL-6**, 405–407.

Sinton, R.A., Kwark, Y.H., Gan, J.Y., and Swanson, R.M. (1986), *IEEE Electronic Device Lett.* **EDL-7**, 567–569.

Sinton, R.A., and Swanson, R.M. (1987a), *IEEE Trans. Electronic Device* **ED-34**, 1380–1389.

Sinton, R.A., and Swanson, R.M. (1987b), *IEEE Trans. Electronic Device* **ED-34**, 2116–2122.

Sinton, R.A., and Swanson, R.M. (1987c), *IEEE Electronic Device Lett.* **EDL-8**, 547–549.

Swanson, R.M., and Swirhun, S.E. (1987), *Contractor Report Sand87-7019,* Sandia National Laboratories, Albuquerque, NM.

Tiedje, T., Yablonovitch, E., Cody, G.D., and Brooks, B.G. (1984), *IEEE Trans. Electronic Device,* **ED-31**, 711–716.

Verlinden, P., Sinton, R.A., Swanson, R.M., and Crane, R.A. (1991), *22nd IEEE Photovoltaic Specialists Conf. Record,* IEEE, New York, pp. 124–127.

Wenham, S.R., Campbell, P., Bowden, S., and Green, M.A. (1991), *22nd IEEE Photovoltaic Specialists Conf. Record,* IEEE, New York, pp. 105–110.

Yablonovitch, E., and Cody, G.D. (1982), *IEEE Trans. Electronic Device* **ED-29**, 300–305.

Zhao, J., Wang, A., and Green, M.A. (1990), *21st IEEE Photovoltaic Specialists Conf. Record,* IEEE, New York, pp. 333–335.

CHAPTER FOUR

Silicon, Gallium Arsenide, and Indium Phosphide Cells: Single Junction, One Sun Space

P.A. ILES and Y.C.M. YEH, Applied Solar Energy Corporation, Industry, CA 91749

1. BACKGROUND

We describe the technology and performance of single junction solar cells designed to operate under one sun illumination in the space environment. Operating conditions in space necessitate different cell designs; these differences are discussed, along with the similarity with cells used on earth.

For over 30 years, solar cells have provided electric power for a variety of extraterrestrial spacecraft and missions. Space power is an extreme case of a remote application, an area where terrestrial photovoltaic (PV) cells have been used widely.

Using the available sunlight in space removes the need to carry in-board fuels into orbit and also increases the active life of storage batteries, which can be recharged by the solar cell array. The technology developed for space cells has contributed to, and also drawn from, advances in the overall PV field. Because real estate in space is expensive and limited, space cells must have high efficiency. However, many spacecraft are exposed to charged particle radiation, requiring that high conversion efficiency be traded-off against the degradation experienced from the radiation.

Each mission imposes a different set of operating conditions. Spacecraft power designers select the cell designs that best meet the specific mission requirements. These selections lead to very detailed specifications that are

Solar Cells and Their Applications, Edited by Larry D. Partain.
ISBN 0-471-57420-1

closely monitored to ensure that the cell manufacturing processes are consistent and that the mission requirements will be met with high reliability. After the cells have performed successfully in orbit, the conservative nature of the cell users tends to lock in proven designs for succeeding missions. As a result solar cells have an established niche market, comprising a succession of "customized" production runs. Considerable testing and validation is required to justify changing to a different design, and some specific designs have persisted for many years. Despite this conservative background, there has been continuing evolution of many different cell materials, cell designs, and processing techniques. Chronologically, the p/n silicon space cells, based on the original Bell Laboratories cells of the 1950s, were replaced in the early 1960s by n/p silicon solar cells. In the 1970s, n/p silicon cells were further improved, and development of GaAs cells began. In the 1980s, most of the space cells manufactured were still advanced n/p silicon cells, but high-efficiency GaAs cells were made in production numbers and also used. While not rivaling the worldwide technology developments associated with microwave devices, integrated circuits, or other devices, U.S. GaAs solar cell production processed comparable areas of GaAs substrates and consumed comparable quantities of gas sources.

In the 1990s, GaAs cells are gaining wider acceptance, although Si cells are most used. Consideration is being given to moderate production numbers of InP cells.

Over this period, there has been considerable interaction between space cell technology and the general PV field. The many attempts to improve efficiency, to reduce degradation under irradiation, and to meet all the detailed specifications developed by users have provided a good testing ground for many basic concepts of cell operation and of many different cell designs.

The heightened interest in PV technology during the 1970s included significant effort on DOE-supported programs, and there was considerable two-way assessment to check if space cell technology could be useful for terrestrial applications and vice versa. In the past decade, renewed efforts have raised silicon cell efficiencies to higher levels. The groups most responsible for these impressive advances, at the University of New South Wales (UNSW) (see Chapter 2) and Stanford University (see Chapter 3), adapted and improved some of the space cell technology, in addition to adding considerable theoretical understanding and developing innovative cell designs and processing methods. In the early 1990s, these reciprocal tests continue to evaluate some of the high-efficiency silicon cell technology for selected space missions or to adapt GaAs space cell manufacturing experience for terrestrial applications.

2. OPERATING CONDITIONS IN SPACE

The operating conditions in space determine the selection of cell designs that can perform to meet all of the mission requirements. In this section are discussed the major operating conditions (solar radiation, charged particle radiation, and temperature control), and other conditions specific to space operation are listed.

2.1. Major Space Operating Conditions

2.1.1. Solar Radiation. The solar spectrum outside the earth's atmosphere is the radiation associated with a 6,000°K black body and has the intensity/wavelength distribution shown in Figure 4.1. The spectral distribution near the earth, but outside its gaseous atmosphere, is called *air-mass zero* (AM0), and the total intensity for the wavelength range from 0.25 to 2.5 μm is 135.3 mW/cm^2. Absorption in the earth's atmosphere results in the AM1.5 spectrum also shown in Figure 4.1. The same cell tested under the two spectra will deliver more power under AM0 illumination, but the AM0 efficiency is 0.85–0.9 of the AM1.5 efficiency.

For a wider range of space applications, the intensity of solar radiation falls off with the inverse square of the distance from the sun. Table 4.1 shows the solar intensity for planets at various distances from the sun. Also indicated are estimated temperatures for oriented solar arrays operating near the planets (see Section 2.1.3, below).

2.1.2. Charged Particle Radiation in Space. A most striking difference between operating conditions in space and those on the earth's surface is the possibility of exposure to charged particle radiation (electrons and protons). Although there is a general background of radiation from cosmic rays, most of

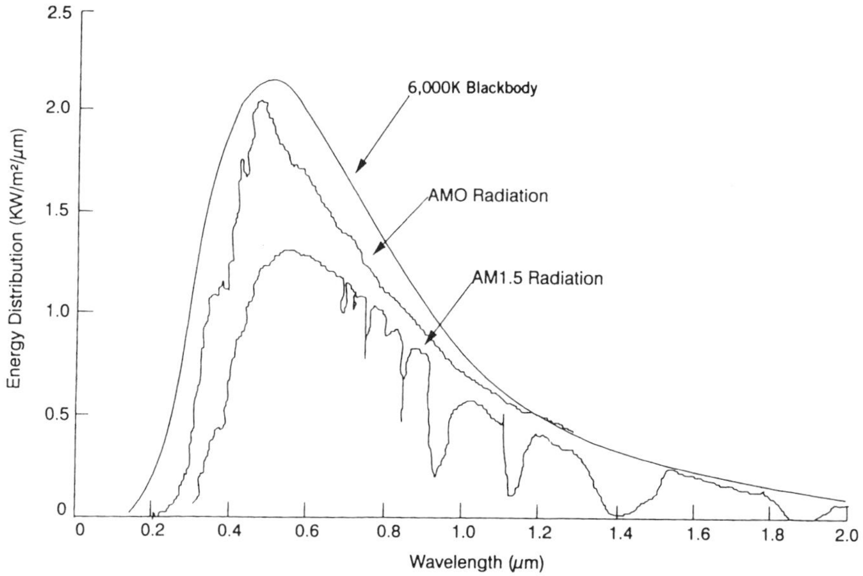

FIGURE 4.1 Spectral distribution of sunlight (AM0, AM1.5) and the radiation from a black body at 6,000K.

TABLE 4.1 Solar Intensity Around Planets

Planet	Distance From Sun (au)[a]	Solar Intensity (mW/cm^2)	Approximate Array Temperature (°C)
Mercury	0.387	890	190 to 210
Venus	0.723	260	70 to 100
Earth	1.0	135.3	20 to 40
Mars	1.524	59	−20 to −40
Jupiter	5.203	5.0	−135 to −145
Saturn	9.539	1.5	−170 to −180

[a] 1 au = 9.3×10^7 miles, the distance of the earth from the sun.

the charged particles are emitted from the sun. The emission of particulate radiation is intense during solar flares, and these relatively infrequent events, lasting only a few days, can cause up to 5% loss in spacecraft power.

The steady-state particle radiation from the sun is not intensive, but the charged particles emitted are trapped in a magnetic field, such as those fields surrounding the earth and Jupiter. Both planets have belts of trapped charged particles, which can provide severe radiation exposure to spacecraft orbiting in the belts. The radiation belts near the earth are called the *Van Allen belts* (after their discoverer), and they are shown diagrammatically in Figure 4.2. The radiation belts contain energetic electrons (around 1 MeV energy) and protons, with a wide range of energies from very low values (50 KeV) to high values (to 50 MeV).

Solar cells are usually placed on the outer surfaces of satellites, and they are directly exposed to the radiation. Transparent shields (coverglasses) that are used to reduce cell temperature can minimize degradation from the low-energy particles, but this shielding is less effective for energetic particles. Several computer programs are available to predict the time rate of exposure for different orbits (Anspaugh et al., 1982).

2.1.3. Operating Temperatures. Heat is generated from the excess solar energy that is not converted to electrical energy. Photons with wavelengths above the cut-off wavelengths ($\approx 1.15\,\mu m$ for Si, $0.9\,\mu m$ for GaAs) are not converted to charge carriers, and if these long wavelengths are directed out of the cell the operating temperature can be reduced (see Section 3.3, below).

In the vacuum of space, heat cannot be transferred by conduction or convection, leaving radiation as the only means for heat loss from spacecraft. The equilibrium temperature of a spacecraft is determined by the balance between the absorbed solar energy and the re-radiation (emission) of the heat generated.

Dielectric materials have high emissivity so that heat can be re-radiated through the glass (or quartz) covers bonded to the front of the cells. If the rear

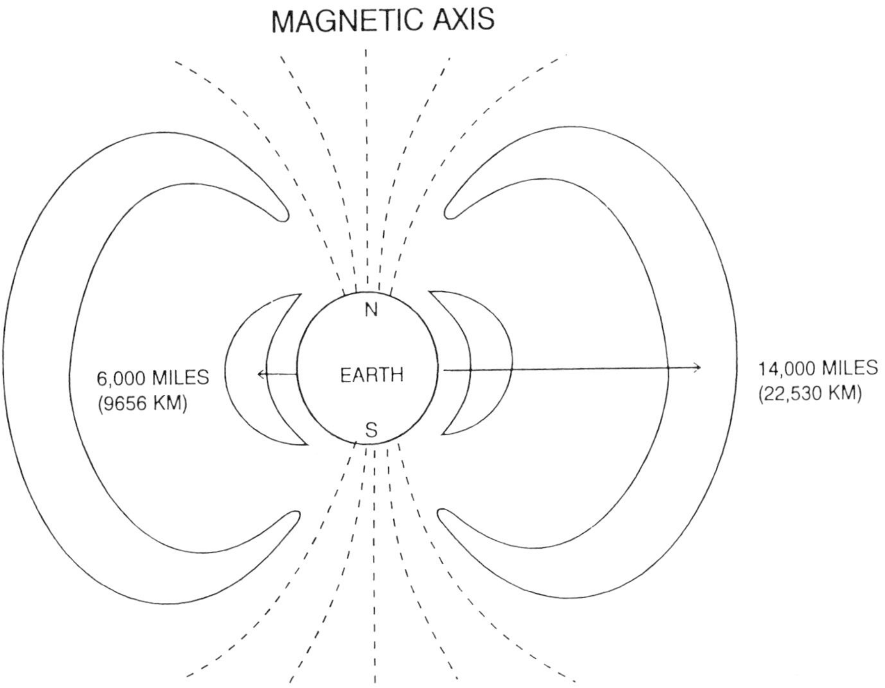

FIGURE 4.2. Van Allen belts of charge particles in crescent-shaped areas. Fluxes (particles $cm^{-2}s^{-1}$) of energetic electrons ($>10^4$) and protons ($>10^6$).

surface of the solar array re-radiates into space, high emissivity layers are used on the rear surface.

2.2. Other Space Operating Conditions

The space environment adds several factors not found in terrestrial applications:

(1) High vacuum $\approx 10^{-10}$ torr
(2) Residual gas atmosphere in near-earth orbits
(3) Intense ultraviolet radiation (from the sun)
(4) Solid particles from spacecraft debris or micrometeorites
(5) Electrostatic charge build-up
(6) Temperature cycling, for spinning satellites or as satellites move in and out of the earth's shadow
(7) Possible shadowing of parts of the solar cell array

Combination of factors 2 and 5 can lead to bombardment by atomic oxygen that is very reactive. Precautions are needed to minimize degradation of metal contacts or substrate materials, especially kapton.

Before launching, cells may have to survive storage in warm humid atmospheres, and they are subjected to severe vibration during launching.

3. DESIGN FACTORS FOR SPACE CELLS

The major requirement for satellite applications is to provide sufficient electric power, usually for extended periods. The power supply must occupy minimum area, without much weight, so that currently available launch vehicles and propulsion systems can be used. Good operating stability and high reliability are essential. An extensive set of specifications, with many in-line and qualification tests, serves to ensure that cells perform consistently with high reliability. The cell material, processing, inspection, and documentation procedures are expensive, leading to cell costs well above terrestrial levels. However, during the past three decades, space cell costs, when corrected for inflation, have continued to decrease steadily.

Production runs for space cells are well below the numbers needed for terrestrial applications. The cell designs that meet the necessary requirements must be manufactured under tightly controlled conditions, with throughput meeting the users' delivery schedules.

3.1. Cell Efficiency

To date, Si and GaAs have been mostly used for space cells (see fuller discussion in Chapters 1–3). To obtain high efficiency, good-quality, single crystal substrates are used, and the active layers are grown or formed in these substrates. The cell design is based on solutions to the equations describing charge generation and collection, leading to structures that increase the current and maximize the cell voltage (usually at a p/n junction). External factors such as contacts or coatings are also optimized.

The processing steps are optimized to retain high carrier lifetime (diffusion length) while following the cell design. Features such as surface passivation, light trapping, and controlled doping levels are included, using methods appropriate for specific semiconductors. As indicated in the following section, high cell efficiency must often be traded off against the radiation resistance.

It is also important to select the starting material and the sequence of process steps to give a tightly controlled manufacturing sequence. For all production designs, the cell properties, particularly for the ohmic contacts, must provide the required stability for prolonged operation in space. In later sections, the evolution of space cells is discussed to show how technology improvements leading to higher efficiency have been developed or followed.

3.2. Cell Radiation Resistance

3.2.1. Radiation Effects on Cell Design. The damage resulting from particle impact is generally dislodgement of semiconductor atoms from their normal lattice position. This gives rise to vacancies (single or multiple), and the defects created can trap minority carriers enroute to the voltage barrier, reducing the collected current. The increased recombination currents also decrease the voltage generated.

High-energy particles (>500 KeV) create defects relatively deep in the semiconductor so that, if the absorption of light is relatively low (the case for Si), energetic particles reduce the response to longer wavelengths that are absorbed deep in the silicon. For Si, as the bulk minority-carrier diffusion lengths are increased (to improve efficiency), the output decreases sharply at moderately low fluence levels, requiring that efficiency and radiation resistance be traded off in the selection of the cell design. For a wider range of missions where radiation exposure is lower, some advanced Si cell designs with high carrier diffusion lengths may be usable. For other materials, including most III–V compounds, that do not have such high dependence on carrier collection in the base of the cell, significant improvements in efficiency can be obtained along with good radiation resistance.

Low-energy particles (principally protons below 0.5 MeV) lose their energy close to the surface. This results in damage near the surface, which reduces the current collected from the emitter layer and decreases junction voltage. For semiconductors with high optical absorption coefficients at shallow depths (the case for GaAs and InP), these low-energy particles cause more relative damage than in silicon. In orbit, some protection against low-energy particle damage is provided by transparent covers.

3.2.2. Orbital Estimates of Radiation. The charged particle environment near the earth is complex, comprising different particles, mostly electrons and protons, with a wide range of energies, ranging from tens of KeV to tens of MeV. The particle concentrations and energies vary with orbit altitude and inclination and generally impinge on the spacecraft from all angles (omnidirectional). The rate of arrival is described by the flux, the number of particles per area per time; the accumulated flux is the fluence, the number of particles per area.

The design sequence used to predict solar cell degradation for the projected orbits is to

- Combine flight measurements and geomagnetic models to establish the spacial and temporal particle parameters
- Transform the effects of omnidirectional incidence and array shielding to that of monoenergetic particles (usually 1 MeV electrons), impinging at normal incidence
- Apply this transformation to the projected orbital exposure to derive an annual equivalent fluence of normal incidence, 1 MeV electrons/cm^2
- Carry out ground tests on candidate cells, using a Van de Graaf accelerator to produce well-controlled 1 MeV electron beams (The flux

used is much higher than those experienced in orbit so that long-term effects can be stimulated in a few hours. Preliminary tests confirm that the damage is not flux dependent so that these accelerated tests can predict degradation expected in extended operation.)

- Analyze the degradation measured on the production (or development) cells to determine the interaction of the cell design and the semiconductor properties with the radiation conditions (Supplementary tests using protons with a range of energies help to clarify the degradation mechanisms. These ground tests provide "damage coefficients" for the different cell types.)
- Perform iterative radiation tests on upgraded cell designs to confirm the corresponding performance under irradiation

This design sequence provides a good basis for selecting cells meeting specific mission goals. Full details are given in the JPL *Solar Cell Radiation Handbook* (Anspaugh et al., 1982). Iterative feedback from flight tests have confirmed these design methods.

Figure 4.3 shows the decrease in efficiency for a range of cells versus the fluence of 1 MeV electrons plotted on a logarithmic scale. The Si and GaAs

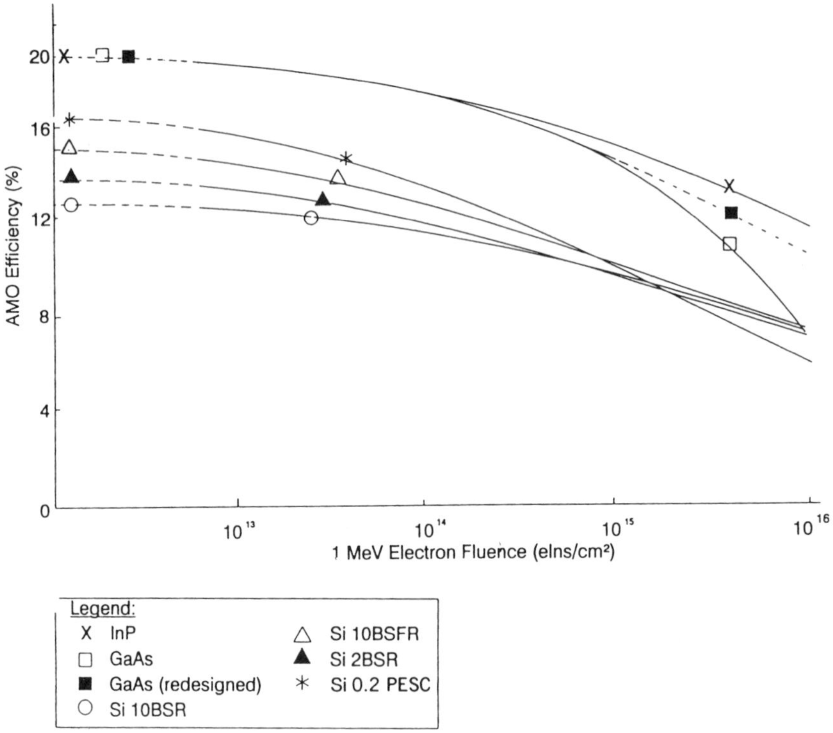

FIGURE 4.3 Cell efficiency (AM0) versus 1 MeV electron fluence for several cell designs.

cells are typical production cells, although the GaAs cell design can be modified to decrease the fall-off for fluences above 10^{15} electrons/cm^2. The other cells are development cells, including InP cells, and an Si cell (0.2 Ω-cm resistivity), resembling the UNSW-developed passivated emitter cell.

3.3. Cell Orbital Temperature

For all cells, there is a range of wavelengths that is absorbed in the semiconductor material comprising the cell, and these wavelengths generate the electric charge carriers that provide the cell current. In this intrinsic absorption process, pairs of charge carriers are generated by photons with energy at or above the band gap of the semiconductor. For wavelengths below the intrinsic threshold wavelength, the excess energy beyond the band gap energy is converted to heat (phonons) in the semiconductor crystal. For best cell efficiency, maximum absorption in the useful spectral range is preferred so that any diversion of these "useful" wavelengths to cool the cell is usually counter-productive in obtaining best efficiency.

For longer wavelengths beyond the intrinsic band edge wavelength, lattice absorption is very low, and free carrier absorption does not lead to useful current generation. Therefore, re-radiating these beyond-gap wavelengths (which comprise about 30% of the total energy in the AM0 spectrum) out of the cell will reduce the heating resulting from their absorption (usually at the back surface of the cell). There are three methods that have been successfully used to reduce solar absorptance (and thereby decrease the orbital cell temperature):

(1) Increase the reflectance of the back surface (BSR)
(2) Increase the transmittance of the back surface (BST)
(3) Use optical coatings to reject wavelengths that do not contribute to useful output

3.3.1. Back Surface Reflectance. Figure 4.4 shows the reflectance versus wavelength plot for a range of Si cells with various BSRs (Iles and Khemthong, 1978). The solar absorptance (αs) is calculated by summing the average reflectance in 25 wavelength increments, each containing around 4% of the energy in the AM0 spectrum. The calculated αs values are shown on the curves.

To increase BSR, the surface must be polished and then coated with a highly reflecting mirror. In manufacture, Si wafers are generally chemically polished on both surfaces, before formation of the p/n junction near the front surface and application of contacts. The polished surface finish resulting from etching can provide high reflectance in the near infrared.

Highly conducting metals (Al, Au, Ag, and Cu) all have high infrared reflectance. Al has been used most for convenience. If the metal is deposited directly on the semiconductor back surface, to reduce roughening, which

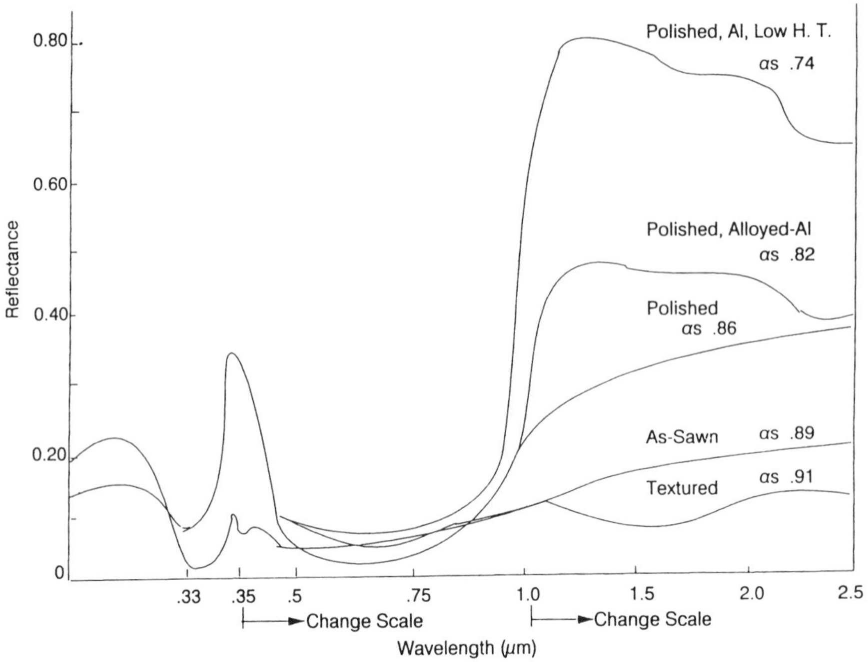

FIGURE 4.4. Reflectance versus wavelength for silicon cells with various back surface conditions (αs values shown on curves).

reduces reflectance, contact heating should not approach alloying temperatures. For some production runs, use of Si with (111) orientation led to relatively flat platelets after alloying, but usually (100) orientation is preferred. For Al-Si, temperatures below 450°C are used; for Au-Si, temperatures below 350°C are required.

Vacuum deposition of the metals is preferred to the use of metallic pastes. In some cases, high BSR is required on slices with a back surface field (BSF). The latter can be obtained for p-Si by alloying Al, but the alloy temperature reduces the BSR. If the BSF is incorporated by boron diffusion, the back surface retains high polish, giving high BSR after the mirror metal is deposited.

In some Si space cells, particularly thin cells $<100\,\mu m$ thick, discontinuous (usually gridded) back contacts are used to minimize bowing of the cell, and these contacts can be applied through slots formed in a thin insulating SiO_2 layer. The BSR Al layer can be applied over the grids and the SiO_2 and heated to high temperatures, giving high reflectance without interaction with the Si back surface. Very low αs values (<0.70) are possible.

Around 1980, analysis of thin Si cells with both BSR and BSF features, and often with textured front surfaces, led to the conclusion that significant internal trapping of the wavelengths near and beyond the Si band edge was occurring.

In later advanced Si cell designs, like those developed at UNSW, the back surface is passivated with a thin SiO_2 layer. An Al contact deposited over this

SiO_2 layer can provide effective light trapping when combined with the slant photon pattern resulting from front surface texturizing (Haynos et al., 1974).

For space cells, this internal trapping does not reduce the operating temperature and gives αs values around 0.90. The resultant power loss resulting from the higher temperature for cells with randomly textured surfaces offsets the increased power from the slant paths, and, as a result, randomly textured cells have not been widely used, except for spinning satellites, where cells have lower temperature because they are not exposed to full sunlight. The regular textured surfaces (microgrooves or regular pyramids) on advanced Si cells can give slight increase in BSR, giving αs values around 0.80.

Because GaAs cuts off at a shorter wavelength, at 0.9 μm rather than at 1.15 μm for Si, a good BSR on GaAs can lead to low αs values, around 0.60–0.65. However, the temperature coefficient of GaAs cells is less than one-half that of Si cells, so that the de-rating resulting from elevated temperatures is less serious. For this reason, designs for GaAs cells with high BSR have not yet been specified, although slightly higher orbital output would be obtained. For very thin GaAs cells, separated from the substrate, back surface reflectors are required to enhance cell output, and high BSR values are obtained.

When GaAs cells are grown on Ge substrates, the wavelengths beyond the cut-off wavelength for GaAs are absorbed in the Ge, and only those wavelengths beyond the Ge cut-off wavelength ($\sim$1.8 μm) can be directed out of the cell. Some increase in BSR has been obtained by using Al or Cu mirrors on the Ge back surface, but αs values from 0.85 to 0.92 are typical.

3.3.2. Back Surface Transmittance. If the back contact is mostly transparent (gridded or dot contacts), a polished back surface with suitable AR coating can transmit most of the long wavelength radiation out of the cell, giving low αs values. If the array material is also transparent to infrared radiation (kapton, for example), the long wavelengths are radiated into space.

Values of αs around 0.65 (corresponding to orbital operating temperatures of 30°–40°C) have been obtained. Figure 4.5 shows both sides of an Si cell, 8×8 cm, with BST design and wrapthrough contacts, designed for possible use on Lockheed Missiles and Space Company flexible arrays on Space Station Freedom.

3.3.3. Rejection Filters. The optical filters on the coverglass usually reflect all UV wavelengths below 350 nm to minimize darkening of the coverglass adhesive, and this increased reflectance decreases αs by $\sim$0.02.

In the 1960s, OCLI developed a multilayer blue-red (BR) optical filter for Si cells, with R-λ characteristics as shown in Figure 4.6. The filter rejected unwanted wavelengths, reduced cell temperature, and thereby reduced the temperature fall-off of output power. However, the low reflectance band was difficult to extend so as not to restrict the useful wavelength band. Therefore, these filters were discontinued.

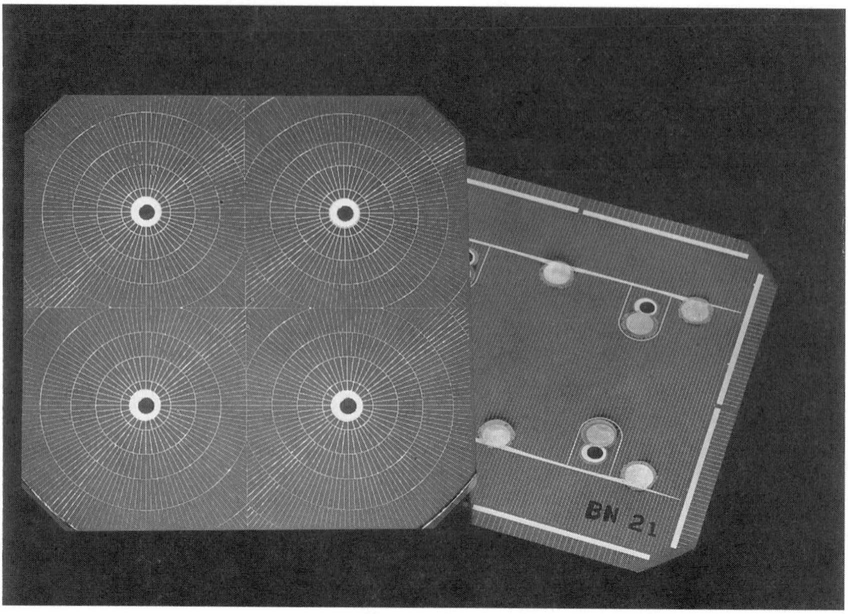

FIGURE 4.5 Front and back views of 8 × 8 cm² Si Cell, BSF, BST.

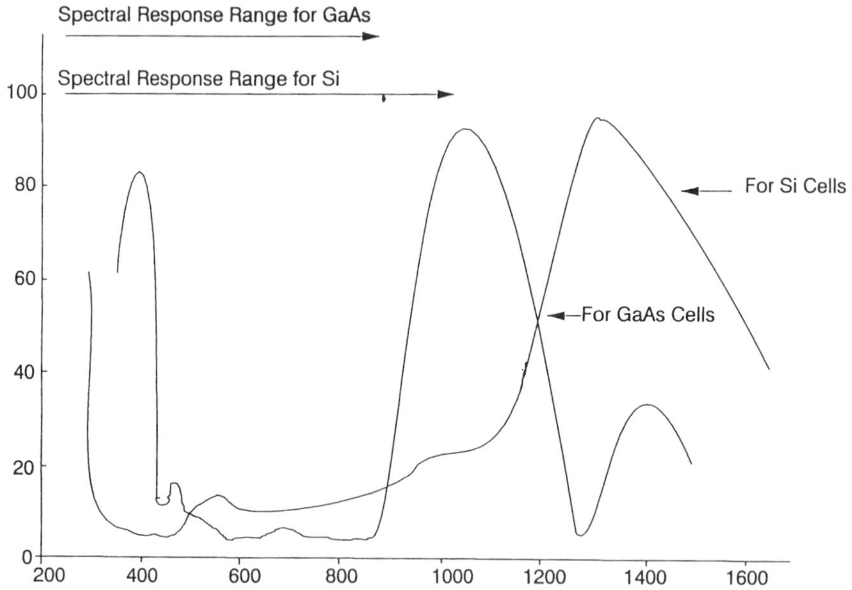

FIGURE 4.6 Reflectance versus wavelength for BR filters.

The band width of the spectral response of GaAs cells is slightly less than that of Si. With advanced coating designs now available, cover suppliers have developed BR filters that reduce αs for GaAs/Ge cells typically from 0.87 to 0.74 by replacing internal wavelength control with these external coatings. The R-λ characteristics of these new BR filters are also shown in Figure 4.6.

3.4. Adjustment to Other Operating Conditions

The coverglasses and cell coatings are designed to minimize surface effects on cells operating in the space environment. Much work has been directed to the design and testing of cell contacts to withstand the effects of temperature cycling and vibration.

4. DEVELOPMENT OF SPACE SOLAR CELL TECHNOLOGY

In this section, the development and current performance of the three main solar cell materials (Si, GaAs, and InP) are summarized.

4.1. Evolution of Si Cell Technology

Early Bell Laboratories Si cells (Chapin et al., 1954) had the p/n configuration from boron diffused into n-type Si, with plated contacts. For Telstar satellites, Bell Labs used n/p cells that had improved radiation resistance (Smith et al., 1963). Phosphorus was diffused into polished p-type Si, and evaporated Ti Ag contacts were used. For the past 30 years, n/p Si cells have been used.

To increase cell efficiency, improvements in Si crystal quality and in processing procedures have led to increased efficiency in large manufacturing numbers. Other tests intended to increase efficiency included

- Use of back surface fields. These were an outgrowth of the Li-diffusion work, where a shallow Li layer was diffused into the back surface of p/n cells (Iles, 1970). n/p cells used Al diffused/alloyed into the back surface of n/p cells (Fischer et al., 1970).
- COMSAT's violet cell, which used Si with resistivity around 2 Ω-cm, a very shallow p/n junction (0.3 μm), a fine-spaced grid pattern formed by photolithographic lift-off, and an improved AR coating of Ta_2O_5, with a violet color (Lindmayer and Allison, 1972).
- COMSAT'S surface texturing, which reduced reflectance and led to slant paths for the incident light, giving absorption in the base nearer the p/n junction and the possibility of enhanced light trapping.
- Maximizing the lifetime of minority carriers in the base region, use of high-quality starting Si, use of reduced heating/cooling schedules during cell processing, and use of gettering processes. When Si with high resistivity is used, the higher lifetime increases the effectiveness of BSFs.

- Use of surface passivation. This effort begin in COMSAT cells and has been considerably extended by the UNSW and Stanford groups.

Recent advances in terrestrial Si cells have improved passivation of the front and back surfaces and at the contacts. Steadily increased open circuit voltage (V_{oc}) values ($\sim 720\,mV$ at one sun illumination) have been achieved, along with increased fill factor (FF) and short circuit current density (J_{sc}). Light trapping has also been improved. AM0 efficiencies as high as 20% have been obtained with these terrestrial Si cell designs.Some of those improvements have been incorporated into space cells, giving efficiencies approaching 18%. Studies are progressing to determine the trade off between increased efficiency and radiation resistance. These studies will provide data to select Si cell designs with the best overall performance for specific missions covering a wide range of radiation exposure.

To improve radiation resistance, the following methods have been evaluated:

- Use of bulk drift fields, produced by graded acceptor doping in the p-Si base. It was difficult to maintain high fields across the base layer, and the high doping levels increased the damage rate.
- Use of Li (a fast diffusing donor), which was found to offset the effect of radiation-produced defects (especially for cluster damage caused by protons or neutrons) (Berman, 1972). In most of these tests Li was diffused into the n-base of p/n cells. Later tests at NASA Lewis Research Center evaluated counter-doping with Li in the p-base of n/p cells. Several thousand Li-doped p/n Si cells were flown successfully, but the overall advantages were not sufficient to warrant continued use of Li.
- Use of shallower n layers, which increased short wavelength response, which degraded less under irradiation. n layers around $0.2\,\mu m$ thick have been used in Si cells since the early 1970s.
- Vertical multijunction cells, where the added complexity of processing was not offset by the increased radiation resistance.
- Use of thin Si cells ($< 100\,\mu m$).

In the 1970s, thin Si cells (2–3 mils), including both BSF and BSR features, demonstrated high radiation resistance and high power-to-weight ratios. In some cases, the back surface was gridded, leading to effective BST operation with very low αs. Also, these thin, gridded-back cells were bifacial, and the chance of increased output by using albedo radiation from the earth's surface was demonstrated. These bifacial cells could also provide supplementary power and battery charging if exposed to laser power beamed from the earth.

Si cell efficiencies have increased to 17% for passivated emitters and to over 20% for the best UNSW cells (Wang et al., 1990). These cells are being evaluated for orbits where low radiation exposure is experienced.

Si space cell technology is fairly mature, and a few different designs have accomodated most specific mission needs. All designs include well-proven shallow diffused p/n junctions and robust, stable contacts that can be soldered or welded. Special contact configurations (wraparound or wrapthrough) can be included. Contacts withstanding high temperatures (to 600°C) can be produced when required. Three main Si cell designs are used most:

(1) n/p, 2 Ω-cm, BSR: for medium output under moderate radiation
(2) n/p, 10 Ω-cm, BSFR: for high output under relatively low radiation
(3) n/p, 10 Ω-cm, BSR: for slightly lower output and reduced degradation under high radiation

As mentioned above, for the past three decades the cost per watt for Si space cells has decreased steadily, offsetting the rate of inflation. In the past few years, strong competition to provide cells for commercial satellites, and from GaAs cells, has led to significant cost reduction.

4.2. Gallium Arsenide Solar Cells

GaAs has the optimum band gap for high efficiency. GaAs cells were made in small numbers in the 1960s. In 1972, IBM made a significant advance in using liquid phase epitaxy (LPE) to form a heteroface p/n GaAs cell with a passivating p-type AlGaAs window layer (Hovel and Woodall, 1972). This cell structure was further developed giving cell areas up to 4 cm^2 and AM0 efficiencies as high as 19% (Kamath, 1983; Yoshida et al., 1979). This work demonstrated good radiation resistance and significant annealing of radiation damage at 100°–200°C. Parallel work at several locations used Metal Organic Chemical Vapor Deposition (MOCVD) to make efficient GaAs cells (to 22% AM0) in small sizes and number (Bozler and Fan, 1977; Hamaker, 1985; Bertness et al., 1988; Tobin et al., 1988a; Gale et al., 1984; Kato et al., 1985).

In the early 1980s, an Air Force manufacturing technology (MANTECH) contract at Applied Solar Energy led to demonstration of acceptable producibility (1,000 per week) of high efficiency (16.5%), 4 cm^2 area, space-qualified GaAs cells (Wright Patterson Air Force Base,f 1984). The heteroface cell structure is shown in Figure 4.7.

This work validated the continuous use of large MOCVD reactors ($\sim$950 cm^2 of GaAs substrates per run) with increased capacity gas sources. Sources of rectangular GaAs substrates with low dislocation density were developed. All safety and toxic requirements were met. In the mid-1980s, around 250,000 GaAs cells, 8 cm^2 area with efficiency around 17% (AM0), passed array manufacturer's tests and were delivered.

In the same period, several groups showed that Ge single crystal substrates were suitable substrates for CVD-grown GaAs cells since they were matched in thermal and lattice properties (Gale et al., 1981; Stirn et al., 1981; Tobin et

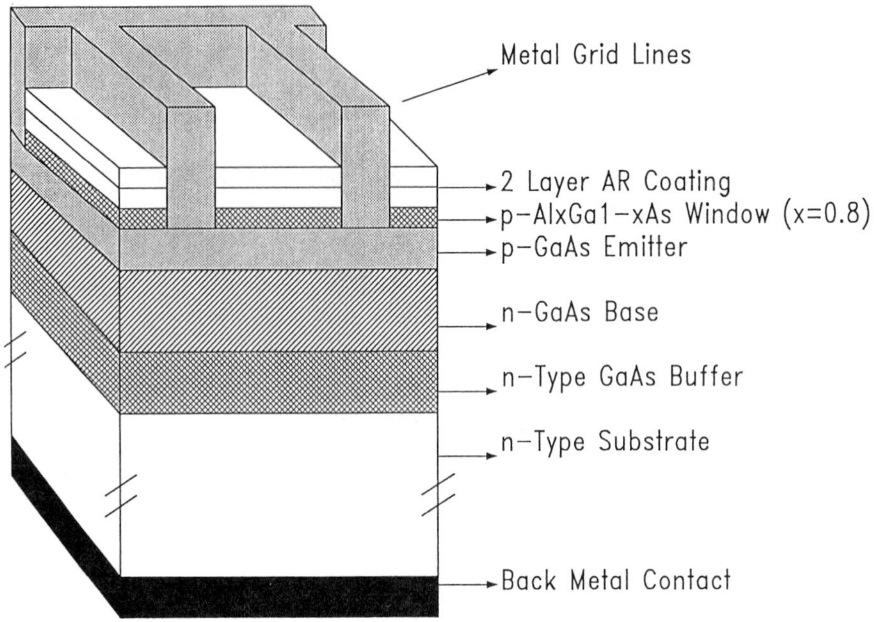

FIGURE 4.7 Section of GaAs heteroface cell.

al., 1985; Chang et al., 1987; Kato et al., 1985). The higher mechanical strength of Ge allowed larger and thinner GaAs cells to be made. Cell efficiencies over 17% were obtained for cells as large as 16 cm^2 area and 3.5 mils (90 μm) thick.

Often increased photovoltage was obtained from the GaAs/Ge interface, although this voltage decreased rapidly at higher operating temperatures (Chang et al., 1987; Tobin et al., 1988b). Further studies showed some disadvantages from this active GaAs/Ge interface, including the need for good Ge crystal quality to obtain matched currents. If the GaAs/Ge current was mismatched, the I–V curves were kinked, with reduced output under true AM0 simulating (requiring two separate light sources). GaAs/Ge cells have performed well at high concentrations (Wojtczuk et al., 1990).

GaAs growth conditions were developed to form GaAs cells on Ge substrates, with an inactive interface (Iles et al., 1990a; Chen et al., 1992). Efficiencies approaching 21% were obtained for cells 4–16 cm^2 (Iles et al., 1990a).

A second MANTECH program demonstrated the producibility of GaAs cells on inactive Ge substrates, providing cells above 18% average efficiency, with areas to 36 cm^2 and as thin as 90 μm (Wright Patterson Air Force Base, 1988). These large thin cells, alone or interconnected and bonded to lightweight panels, successfully passed space qualification tests. Contacts withstanding 600°C were developed. High-capacity Ge substrate suppliers were also developed. Scaled-up continuous MOCVD reactor operation was demonstrated, and increasingly severe safety and environmental regulations were complied with.

Beginning in 1989, a parallel production line was operated at Applied Solar, producing GaAs/Ge cells, 8 cm^2 area, 18% efficiency, and 200 μm thick. By 1992, over 200,000 of these cells were delivered, passing all user tests and requirements. This production experience has been well summarized (Datum and Billets, 1991; Cheng et al., 1991). Significant cost reductions were achieved by using Ge substrates, both from lower material costs and from the reduced need for high electronic quality.

Previously, GaAs/GaAs cells had shown the possibility of catastrophic failure when the cell were reverse biased, resulting from shadowing of the cells. The GaAs/Ge structure led to a serendipitous advantage of the heteroepitaxial structure that reduced the possibility of failure under reverse bias (Iles et al., 1990b). Growth of ionic GaAs on covalent Ge led to large numbers of growth features (possibly antiphase domains) that were incorporated throughout the growth layers, including the GaAs p/n junction. These growth features caused slight shunting in the forward diode direction, without significant reduction in FF. However, under reverse bias, these built-in leakage sites functioned as bypass paths, reducing the chance of catastrophic hot spots. For many applications, GaAs/Ge cells remove the need for separate bypass diodes on the array.

Array manufacturers have specified in-line reverse voltage stress tests to confirm that all cells delivered can withstand shadowing without degradation. The yield of GaAs/Ge cells after this screening is significantly higher than that of GaAs/GaAs cells.

Contacts for GaAs cells have been improved. The alloyed-Au systems used extensively are being replaced by nonalloyed metalization, with less costly metals and simplified processing procedures. High temperature contacts (525°C or 600°C) are available, if required.

Average production AM0 efficiencies for GaAs/Ge cells are increasing toward 19%. The increased use of commercial satellites and small satellites has led to trade studies of the overall satellite costs. Cells with higher efficiency reduce the area and weight of the power system, and, even though GaAs/Ge cells cost more than Si cells, their use can result in significant reduction in overall system costs. The recognition of the effect of cell efficiency on reducing system costs has increased the use of GaAs/Ge cells and has also accelerated demand for cascade cells with even higher efficiency.

Some recent developments in GaAs cell technology based on improved analysis and modeling (Lundstrom, 1988) or improved surface passivation are being evaluated on production cells. Tests continue of growing GaAs directly on Si substrates (Vernon et al., 1984, 1988; Okamota et al., 1988; Ohmachi et al., 1990). The severe mismatch between GaAs and Si generally results in high dislocation densities (or cracks) in the GaAs layers, with reduced output and doubts about space worthiness. Tests also continue using superlattices, intermediate compound layers, or extensive annealing procedures to improve the performance of GaAs cells grown on Si. Very thin GaAs cells (obtained by the CLEFT process [McClelland et al., 1980], epitaxial lift-off [Yablonovitch et al.,

1987], or removal of the substrate) with reflecting back surface layers are being investigated for increased output (resulting from photon recycling) as well as increased power-to-weight factors.

The demonstration of large-scale epitaxial growth, including the heteroepitaxial growth of GaAs on Ge, has led to increased interest in the possibility of manufacturing cascade solar cells with higher efficiency (see Chapter 6). Work on cascade cells has intensified, and production samples may be available in 1995.

4.3. Indium Phosphide Solar Cells

Indium phosphide (InP) has an energy gap of 1.35 eV, near the optimum value for high efficiency. Since the early 1960s, continuing evaluation of InP cells with many different processing technologies has continued. The resultant efficiencies were not high ($\sim$10% AM0), and, because of the perceived scarcity of In, the substrate costs were even higher than GaAs costs, and there appeared to be low chance of use in terrestrial or space applications.

However, in 1984, workers at Nippon Telegraph and Telephone (NTT) reported that they had observed minority-carrier injection annealing of InP cells after 1 MeV electron irradiation (Yamaguchi et al., 1984a). The carriers were injected by electrical bias or by exposure to photons (from simulated sunlight). They also reported that after gamma irradiation InP cells degraded less than Si or GaAs cells and that heavily damaged InP cells could be completely annealed around 100°C (Yamaguchi et al., 1984b). The cells were of n^+/p configuration, formed by diffusion of S or Se into p-type InP, and had an efficiency arund 14% AM0.

This large increase in radiation resistance was of great interest for space use, particularly with the prospect of continuous annealing of radiation damage by solar illumination. Intensive development of InP cells has continued, principally at NTT and in work sponsored by the NASA Lewis Research Center and by the Naval Research Laboratory.

4.3.1. Fabrication of InP Cells. Many methods have been used, but the most successful to date all being with single crystal InP substrates (much information in this section was obtained from three survey papers: Coutts and Yamaguchi [1988], Yamaguchi [1988], and Weinberg [1990]). The main designs and typical efficiencies are

- diffused n^+/p (or $p^+/i/n$), 15.5% AM0 (Yamamoto et al., 1984)
- ITO/InP, 15.5% AM0 (Li et al., 1980)
- MOCVD $n^+/p/p^+$, 18% AM0 (Sugo et al., 1987; Keavney et al., 1990)

In some cases, p-InP was grown on p^+ InP substrates, and p/n junctions were formed by ion implantation (S or Se) or by sputtered ITO layers.

Workable contacts and coatings have been developed, and preliminary space qualification tests have been carried out. Prgamatic estimates project efficiencies above 20% AM0. Cells sizes have increased to $4\,cm^2$.

4.3.2. Radiation Performance. Most InP cells degrade much slower than GaAs or Si cells when exposed to high fluences of energetic electrons or protons. Typical radiation behavior of all three cell materials was given in Section 3.3 (Fig. 4.4). This radiation resistance is partly the result of low-defect creation, and the ready annealing results from the increased removal of the defects. The radiation also removes majority carriers, perhaps reducing the effects of the defects, and as a result the radiation resistance also depends on the doping concentration. If the cell has a lightly doped base layer, carrier removal widens the depletion layer around the p/n junction. Since the cell performance is very sensitive to radiation-generated defects in this layer, InP cells show an opposite trend to other cells, namely, that radiation resistance is higher for highly doped base layers.

Most of the early InP cells that had high radiation resistance and were readily annealed also had relatively low efficiency ($\leqslant 15\%$), and this raised doubts whether cells with increased efficiency would have comparably high radiation resistance. However, recent cells with 18% efficiency showed low degradation after irradiation and also demonstrated annealing from photoinjected carriers (Walters et al., 1991). After bombardment by 1 MeV electrons (fluences up to 10^{16} electrons/cm^2), these cells had efficiencies over 11%. For plausible combinations of design parameters, efficiencies approaching 14% are projected after these high fluence levels. The detailed kinetics and energy levels of the various defects are well understood, and effective models of pre and postirradiation behavior, including annealing mechanisms, have been developed.

Most ground tests of radiation exposure require fairly high particle fluxes to simulate orbital exposure in reasonable times. For most orbits, the flux levels are low, and ground tests indicate that InP cells may experience continual annealing. This would further improve the current estimates for InP cell performance after exposure to typical orbital fluences.

4.3.3. Flight Experience. InP cells have been flown on LIPS-III (fairly high radiation levels) and on the Japanese scientific satellite MUSES-A (low radiation in lunar orbit). On both these flights, and on other small satellites, very little degradation was observed. It should be noted that some GaAs and copper indium selenide cells on LIPS-III also showed small degradation.

4.3.4. Status of InP Cells. The high radiation resistance makes InP cells most promising for use in orbits where high radiation fluences are experienced. These orbits include the regions in the Van Allen belts with high particle densities and also orbit transfer missions, where satellites move slowly through the Van Allen belts. However, this superior radiation performance must be available from production-level cells, which have been validated to meet all the other space operational requirements. Also the high cost of InP cells must be reduced.

While InP cells are being further developed, other cells may also become more competitive in radiation resistance and output. Homojunction GaAs cells are more radiation resistant than heteroface GaAs cells because of their very shallow emitters. The emitter of the heteroface cells can be further reduced to improve radiation resistance at high fluences. AlGaAs cells with low Al content have also been found to have high efficiency combined with high radiation resistance. Ultrathin GaAs cells should also have high radiation resistance, and the technology required to produce ultrathin high efficiency cells in large quantities is available and may result in cell costs below those of InP.

For missions with low or moderate radiation, there is also the possibility of doubling the active life of an irradiated array by increasing the area by about 20%. For missions with no area limitations, this may be a viable option. Nevertheless, the performance of InP cells has demonstrated that it is possible to retain high power output, even at fluences that were previously considered prohibitive. In the last few years, InP cells have changed attitudes concerning space cell operation at high fluence levels. The radiation advantages of InP are useful for cascade cells, either those using InP as top cells or for top cells using ternary alloys like $GaInP_2$ or quaternary alloys of Ga, In, As and P.

4.4. Other Candidate Space Cells

Most of the alternate cells are candidates for special missions. They are mostly thin-film cells, such as hydrogenated amorphous silicon (α-Si:H), copper indium selenide (CIS), or cadmium telluride (CdTe). Strictly speaking, most current α-Si:H cells are multijunction cells, but we can include these cells in this discussion.

The main advantage of thin-film cells is their high power-to-weight ratio; generally their low-to-medium efficiencies (5%–12%) give high radiation resistance. One likely application is to provide electric power for ion engines, which are used to propel spacecraft, especially for orbital transfer near the earth. The propulsion velocities are low, requiring long duration exposure to the Van Allen belts. The flexibility of these thin arrays is attractive for stowing in spacecraft traveling to the moon (or nearby planets). On arrival, the arrays can be unrolled and deployed to provide power at a manned base.

In the future, all the cells described in this section will continue to be evaluated, along with multijunction cells, to determine the best cell for specific missions.

5. FUTURE TRENDS

Work will proceed on the cell designs described above, with near-term emphasis on Si and GaAs cells. The future of single junction space cells will be determined by the demands of future space missions. Some single junction cells, especially those made from Si, GaAs, InP, and CIS, have promise for use in mechanically stacked cascade cells.

As mentioned earlier, some of the advanced terrestrial Si cell design features will be used for space cells when the radiation conditions can be met. The development of both GaAs and InP cells has accelerated the move to monolithic cascade cells, with further increase in efficiency. Small additional-growth time is required to add a tunnel diode and another cell structure. The increased cost is relatively small, and production versions of space-qualified cascade cells could be available as early as 1995.

5.1. Future Missions

Most military, government, and commercial satellites will operate near the earth. There are many mission requirements in this region, and radiation is a continuing factor. Detailed analysis of the specific mission conditions and requirements indicates which cell design is best suited to each mission.

The current trend is to use larger numbers of small satellites rather than launching large complex satellites. The power needs of these small satellites are around 1 kW, but since satellite (panel) area is limited, high-efficiency cells are needed. GaAs/Ge cells are being used more, although for some missions Si cells will be sufficient. InP and other radiation-resistant cells will be considered for orbital transfer missions and for orbits with high radiation exposure.

NASA will continue to fly satellites near the earth, mostly for monitoring environmental conditions, including the weather. NASA-manned missions (such as Space Station Freedom) will fly in Low Earth Orbit, requiring arrays with high efficiency and good reliability. Moderate-efficiency Si cells are now specified for Space Station Freedom, but several alternate cell types, including GaAs/Ge cells, are being evaluated.

NASA also has plans for continued solar system exploration. Analysis shows that PV cells can meet most of NASA's power needs, at least out to Saturn (Iles, 1992). For near-sun missions, GaAs cells are preferred, although, in the future, cascade cells will be competitive.

For far-sun missions, Si cells are competitive. With improved p/n junction passivation or contact improvements, Si cells can now operate efficiently at low intensities and temperatures out to Saturn.

For lander missions on Mars, Si cells, GaAs/Ge cells, or thin film cell arrays are promising. For the lunar stations, providing power during the long lunar night is a problem, and viable options combining PV and other power sources are being sought.

5.2. Nonsolar Applications

PV cells are also finding applications where sources other than sunlight are available (Iles, 1990b). For most of the nonsolar applications, space cells can be used.

For narrow-band sources (sometimes monochromatic), PV cells have higher efficiencies than under the broad band solar spectrum. Some of the applications to date include the following:

(1) *Thermophotovoltaic* converters, wherein infrared heat is converted directly to electrical energy. Black body sources operating at available temperatures can be used, in some cases with selective filters that reject the wavelengths beyond the band edge of the applicable PV cell material. Selective emitters have been developed that, when heated, emit with an output sharply peaked in the near infrared. By using different rare earth oxides, peak output can be around 0.9 μm (matched to the response of a high-efficiency Si cell, with conversion efficiency at 30%–40%) or at wavelengths out to 2.5 μm, where small band gap semiconductors can give efficiencies of 20%–30%. Most of the promising narrow gap semiconductors, usually compounds of III–V elements, can be used for infrared conversion, either alone or in cascade arrangements. For terrestrial uses, the heating can be by gas flames. For space use, heat generated by radioactive isotopes, or warmed lunar rocks, is being evaluated as a possible source for these cells.

(2) *Laser-beaming of power.* Laser beaming is being considered for use in space to transfer power from satellite to satellite, from satellites to the moon (or even Mars), from the earth to the moon, or to orbiting satellites. Solar power satellites have considered use of power-beaming to earth, either by microwaves or lasers. Single junction space cells have been used successfully to convert laser power, and the cell material is selected to match the laser wavelengths. For diode lasers based on GaAs or AlGaAs compounds, GaAs cells have demonstrated efficiencies over 50%. For Nd-YAG lasers, Si cells have shown efficiencies above 40%. For chemical lasers with output around 1.3 μm, cells made from GaSb or ternary III–V compounds can have efficiencies around 30%. There are also other applications where laser beams are directed down fibers to actuate equipment, often on aircraft. This is a growing field of use (fly-by-light), and the converters are small area solar cells, often series-stacked to give high voltage; conversion efficiencies over 50% have been demonstrated.

(3) *Other narrow band sources*, such as fluorescent compounds, sometimes used in a glass matrix for solar energy conversion, can be converted efficiently by space cells, made from either GaAs or Si.

5.3. Conclusions

Single junction space cells have served the space market successfully while contribution to progress in the overall PV field. There has been a steady development in the technology and performance, and it appears that these cells will continue to be used for a wide variety of space missions.

REFERENCES

Anspaugh, B.E., Downing, R.G., Tada, H.Y., and Carter, J.R. (1982), *Solar Cell Radiation Handbook*, JPL Publication, California, pp. 82–69. Addendum, 1989.

Berman, PlA. (1972), *Proc. 9th IEEE Photovoltaic Specialists Conf. Record*, IEEE, NY, pp. 281–291.

Bertness, K.A., Ladle Ristow, M., and Hamaker, H.C. (1988), *Proc. 21st IEEE Photovoltaic Specialists Conf. Record*, IEEE, NY, pp. 769–770.

Bozler, C.O., and Fan, J.C.C. (1977), *Appl. Phys. Lett.* **31**, 629–631.

Chang, K.I., Yeh, Y.C.M., Iles, P.A., Tracy, J., and Morris, R.K. (1987), *Proc. 19th IEEE Phot5ovoltaic Specialists Conf. Record*, IEEE, NY, pp. 273–279.

Chapin, D.M., Fuller, C.S. and Pearson, G.L. (1954), *J. Appl. Phys.* **25**, 676–677.

Chen, J.C., Ladle Ristow, M., Cubbage, J.I., and Werthen, J.G. (1992), *J. Electronic Materials* **21**, 347–353.

Cheng, C.H., Yeh, Y.C.M., Chu, C.L., and Ou, T. (1991), *Proc. 22nd IEEE Photovoltaic-Specialists Conf. Record*, IEEE, NY, pp. 393–398.

Coutts, T.J., and Yamaguchi, M. (1988), in *Current Topics in Photovoltaics*, T.J. Coutts and J.D. Measkin, Eds., Academic Press, New York, pp. 80ff.

Datum, G.C., and Billets, S.A. (1991), *Proc. 22nd IEEE Photovoltaic Specialists Conf. Record*, IEEE, NY, pp. 1422–1428.

Fischer, H., Link, E., and Pschunder, W. (1970), *Proc. 8th IEEE Photovoltaic Specialists Conf. Record*, IEEE, NY, pp. 70–77.

Gale, R.P., Fan, J.C.C., Turner, G.W., and Chapman, R.L. (1984), *Proc. 17th IEEE Photovoltaic Specialists Conf. Record*, IEEE, NY, pp. 1422–1425.

Gale, R.P., Tsaur, B.Y., Fan, J.C.C., Davis, F.M., and Turner, G.W. (1981), *Proc. 15th IEEE Photovoltaic Specialists Conf. Record*, IEEE, NY, pp. 1051–1055.

Hamaker, H.C. (1985), *J. Appl. Phys.* **58**, 2344–2351.

Haynos, J., Allison, J.F., Arndt, R., and Meulenberg, A. (1974), *Proc. Int. Conf. Photovoltaic Power Generation*, DGLR, Köln, Germany, pp. 487–500.

Hovel, H.J., and Woodall, J.M. (1972), *Appl. Phys. Lett.* **21**, 379–381.

Iles, P.A. (1970), *Proc. 8th IEEE Photovoltaic Specialists Conf. Record*, IEEE, NY, pp. 345–352.

Iles, P.A. (1990a), *Solar Cells* **29**, 205–223.

Iles, P.A. (1990b), *Proc. 21st IEEE Photovoltaic Specialists Conf. Record*, IEEE, NY, pp. 420–425.

Iles, P.A. (1992), *Proc. ASME/JSES/KSES International Solar Energy Conference*, pp. 853–860.

Iles, P.A., and Khemthong, S. (1978), *Proc. 13th IEEE Photovoltaic Specialists Conf. Record*, IEEE, NY, pp. 327–332.

Iles, P.A., Yeh, Y.C.M., Ho, F.H., Chu, C.L., and Cheng, C. (1990a), *IEEE Electron Device Lett.* **11** 140–142.

Iles, P.A., Yoo, H., Chu, C.L., Krogen, J., and Chang, K.I. (1990b), *Proc. 21st IEEE Photovoltaic Specialists Conf. Record*, IEEE, NY, pp. 448–454.

Kamath, S. (1983), *Proc. 18th IEEE Photovoltaic Specialists Conf. Record*, IEEE, NY, pp. 1224–1228.

Kato, M., Mitsui, K., Mizuguchi, K., Hayafuji, N., Ochi, S., Yukimoto, Y., Murotani, T., and Fujikawa, K. (1985), *Proc. 18th IEEE Photovoltaic Specialists Conf. Record*, IEEE, NY, pp. 14–19.

Keavney, C.J., Haven, V.E., and Vernon, S.M. (1990), *Proc. 2nd Int. Conf. InP Rel. Metals*, pp. 435–438.

Li, X., Wanlass, M.W., Gessert, T.A., Emery, K.A., and Coutts, T.J. (1989), *Appl. Phys. Lett.* **54**, 2674–2676.

Ldindmayer, J., and Allison, J.F. (1972), *Proc. 9th IEEE Photovoltaic Specialists Conf. Record*, IEEE, NY, pp. 83–84.

Lundstrom, M.S. (1988), *Solar Cells* **24**, 91–102.

McClelland, R.W., Bozler, C.O., and Fan, J.C.C. (1980), *Appl. Phys. Lett.* **37**, 560–562.

Ohmachi, Y., Ohara, T., and Kadota, Y. (1990), *Proc. 21st IEEE Photovoltaic Specialists Conf. Record*, IEEE, NY, pp. 89–94.

Okamota, H., Kadota, Y., Watanabe, Y., Fukuda, Y., O'hara T., and Ohmachi, Y. (1988), *Proc. 20th IEEE Photovoltaic Specialists Conf. Record*, IEEE, NY, pp. 475–480.

Smith, K.D., Gummel, H.K., Bode, I.D., Cuttriss, D.B., Nielsen, R.J., and Rosenweig, W. (1963), *Bell Sys. Tech.* **XLII**, 1765–1817.

Stirn, R.J., Wang, K.L., and Yeh, Y.C.M. (1981), *Proc. 15th IEEE Photovoltaic Specialists Conf. Record*, IEEE, NY, pp. 1045–1050.

Sugo, M. Yamamoto, A. and Yamaguchi, M. (1987), *IEEE Trans Electron Devices* **ED-34**, 772–776.

Tobin, S.P., Vernon, S.M., Bajgar, C., Haven, V.E., and Davis, S.E. (1985), *Proc. 18th IEEE Photovoltaic Specialists Conf. Record*, IEEE, NY, pp. 134–139.

Tobin, S.P., Vernon, S.M., Bajgar, C., Geoffroy, L.M., Keavney, C.J., Sanfacon, M.M., and Haven, V.E. (1988a), *Solar Cells* **24**, 103–115.

Tobin, S.P., Vernon, S.M., Bajgar, C., Haven, V.E., Geoffroy, L.M., Lillington, D.R. (1988b), *IEEE Electron Device Lett.* **9**, 256–258.

Vernon, S.M., Spitzer, M.B., Tobin, S.P., and Wolfson, R.G. (1984), *Proc. 17th IEEE Photovoltaic Specialists Conf. Record*, IEEE, NY, pp. 434–439.

Vernon, S.M., Tobin, S.P., Haven, V.E., Bajgar, C., Dixon, T.M., Al-Jassim, M.M., Ahrenkiel, R.K., and Emery, K.A. (1988), *Proc. 20th IEEE Photovoltaic Specialists Conf. Record*, IEEE, NY, pp. 481–485.

Walters, R.J., Messenger, S.R., Summers, G.P., Burke, E.A., and Keavney, C.J., (1991), *Proc. 22nd IEEE Photovoltaic Specialists Conf. Record*, IEEE, NY, pp. 1560–1565.

Wang, A., Zhao, J. and Green, M.A. (1990), *Appl. Phys. Lett.* **f57**, 602–604.

Weinberg, I. (1990), *Solar Cells* **29**, 225–244.

Wojtczuk, S.J., Tobin, S.P., Keavney, C.J., Bajgar, C., Sanfacon, M.M., Geoffrey, L.M., Dixon, T.M., Vernon, S.M., Scofield, J.D., and Ruby, D.S. (1990), *IEEE Trans. Electron Devices* **37**, 455–462.

Wright Patterson Air Force Base (1984), Contract No. F33615-84-C-2403. Final Report.

Wright Patterson Air Force Base (1988), Contract No. F33615-88-C-5415. Final Report.

Yamaguchi, M. (1988), *Proc. 20th IEEE Photovoltaic Specialists Conf. Record*, IEEE, NY, pp. 880–885.

Yamaguchi, M., Ando, K., Yamamoto, A., and Uemura, C. (1984a), *Appl. Phys. Lett.* **44**, 432–434.

Yamaguchi, M., Uemura, C., and Yamamoto, A. (1984b), *J. Appl. Phys.* **55**, 1429–1436.

Yamamoto, A., Yamaguchi, M., and Uemura, C. (1984), *Appl. Phys. Lett.* **44**, 611–613.

Yoshida, S., Mitsui, K., Oda, T., Sogo, T., Shirahata, K. (1979), *Jpn. J. Appl. Phys.* **19**, 563–566.

CHAPTER FIVE

Status, Prospects, and Economics of Terrestrial, Single Junction GaAs Concentrator Cells

MARTIN E. KLAUSMEIER-BROWN, Edward L. Ginzton Research Center, Varian Associates, Palo Alto, CA 94304-1025

1. INTRODUCTION

Thin-film solar cell technologies, such as those based on $CuInSe_2$, hydrogenated amorphous Si, or CdTe, rely on very low cell cost per unit area to deliver power at an acceptably low price. In contrast, concentrator system designs use fairly expensive cells. Inexpensive optics, which concentrate normally incident light by a factor of 1,000 or more, are used to reduce the needed cell material, thus maintaining low overall system cost. The allowed cell cost is approximately \$5–\$10/cm^2 in high-efficiency concentrator systems (for an installed system cost of \$1/annual kW-hr; Arvizu, 1985). Concentration not only makes GaAs technology potentially cost effective, but it actually increases the efficiency of GaAs cells as well.

The performance of GaAs concentrator cells and modules has increased steadily over the last several years. The AM1.5D efficiency of single cells has reached 28% (Chapter 1; Kaminar et al., 1988; Wojtczuk et al., 1990). A passively cooled, 12 cell module has demonstrated a peak efficiency over 22% at 1,000×, under real operating conditions (Kuryla et al., 1991). Cost projections

Solar Cells and Their Applications, Edited by Larry D. Partain.
ISBN 0-471-57420-1

for high-volume production are difficult to make; however, these performance levels are nearing those required for cost effectiveness.

This chapter begins with a survey of the development of high-efficiency GaAs cells, followed by an examination of the physics of GaAs concentrator cell operation. Next, the influence of the concentrator *system* on cell design and economics is presented, and the prospects for improved cell efficiencies are considered. Finally, issues of manufacturability, safety, and cost are addressed in the last two sections.

2. SURVEY OF BEST CELL RESULTS

To gauge the progress in the development of GaAs solar cells, it is helpful to know what is the highest achievable efficiency. It is possible to calculate a limiting efficiency that is approximately independent of a particular cell design, and therefore characteristic of the material itself, either by using a detailed balance argument or by assuming radiative-limit lifetimes. The limiting efficiency for GaAs under one sun, AM1.5D conditions is about 30%; for 1,000 × concentration, the limit is about 36% (Shockley and Queisser, 1961; Mathers, 1977; Henry, 1980; Gray et al., 1987).

Early high-efficiency GaAs cells were grown by the liquid-phase epitaxy technique. James and Moon (1975) reported a conversion efficiency of 19% at very high concentration (1,735 ×). An $(Al_xGa_{1-x})As$ layer was employed to passivate the front surface of the cells. The efficiency of GaAs cells first exceeded 20% in the late 1970s, when Woodall and Hovel (1977) fabricated GaAs heterostructure cells with a one sun efficiency of 22%. Fan et al. (1978) also demonstrated 20% efficient cells, using an n^+/p homojunction design. High efficiency was realized in these unpassivated cells by using anodic oxidation to thin the emitter to less than 500 Å.

A significant effort was progressing at Varian Associates for the development of terrestrial concentrator cells, modules, and arrays. Vander Plas et al. (1978) reported the fabrication of a large number of 1,000 × cells, some with efficiencies over 23%. Metal organic vapor phase epitaxy (MOVPE) was applied to the growth of GaAs concentrator cells at Varian in 1979, and a 400 × , 17% concentrator module was demonstrated (Owen and Borden, 1981). The introduction of MOVPE made it possible to grow much thinner $(Al_xGa_{1-x})As$ window layers.

Subsequently, most development efforts were focused on the "heteroface" cell design, which features a thin $(Al_xGa_{1-x})As$ window layer to passivate the surface of a relatively thin, heavily doped emitter. With continued refinements in vapor phase growth techniques, and improvements in the source gases available for such growth, reported cell efficiencies increased slowly throughout the 1980s. In 1985, Varian and Sandia researchers reported 26% efficient (AM1.5D, 753 ×) heteroface cells (Hamaker et al., 1985). More recently, groups at both Varian and Spire produced concentrator cells with efficiencies of 28% (Chapter 1; Kaminar et al., 1988; Wojtczuk et al., 1990).

While many of the highest efficiencies have been achieved with MOVPE-grown $(Al_xGa_{1-x})As/GaAs$ heteroface cells, significant contributions have been made using slightly different processes. For example, molecular beam epitaxy (MBE) has been used to produce cells comparable to the best MOVPE-grown devices (Melloch et al., 1990). At Kopin, researchers developed a process called CLEFT for separating thin, MOVPE-grown films from GaAs substrates. This process promises re-use of GaAs substrates and has yielded cells with efficiencies as high as 22.4% (one-sun, AM1.5D; Gale et al., 1988). Another alternative process used 1.9 eV (Ga, In)P for the window layer in place of $(Al_xGa_{1-x})As$, resulting in a record one sun, AM1.5G efficiency of 25.7% (Kurtz et al., 1991). Finally, interest in lighter weight GaAs cells for use in space has motivated investigation of growth on Ge substrates. Processes that result in both active (photon-producing) Ge substrates (Tobin et al., 1988b) and inactive substrates (Iles et al., 1990) have produced high-efficiency cells.

3. PHYSICS AND DESIGN OF GaAs CONCENTRATOR CELLS

The best GaAs concentrator cells have reached a very high level of development, comparable to that of the best Si cells. Cells made from both materials have potential (limiting) efficiencies of about 36% for high concentration, terrestrial operation (Shockley and Queisser, 1961; Mathers, 1977; Henry, 1980); the efficiency of concentrator cells for both materials has reached 26%–28% (Chapter 1; Sinton and Swanson, 1987; Kaminar et al., 1988). The operation of the "heteroface" cell, the "standard" design used to achieve such high efficiency in GaAs, is examined in this chapter.

3.1. Standard GaAs Concentrator Cell Design

Figure 5.1 shows a schematic diagram of the commonly used heteroface design; typical thicknesses and dopant concentrations are given in Table 5.1. Designs

TABLE 5.1 Typical Layer Thickness and Doping for a Standard Heteroface Concentrator Cell Design for Both the p-on-n Case and the n-on-p Case

	p-on-n Cell		n-on-p Cell	
	μm	cm^{-3}	μm	cm^{-3}
GaAs cap layer	0.6	$p = 5 \times 10^{19}$	0.6	$n = 2 \times 10^{18}$
$(Al_{0.9}Ga_{0.1})As$ window	0.04	$p \approx 10^{19}$	0.04	$n \approx 10^{18}$
GaAs emitter	0.5	$p = 2 \times 10^{18}$	0.2	$n = 1 \times 10^{18}$
GaAs base	3.5	$n = 2 \times 10^{17}$	3.8	$p = 5 \times 10^{17}$
$(Al_{0.2}Ga_{0.8})As$ window	0.2	$n = 1 \times 10^{18}$	0.2	$p = 2 \times 10^{18}$
GaAs buffer	0.6	$n = 1 \times 10^{18}$	0.6	$p = 2 \times 10^{18}$

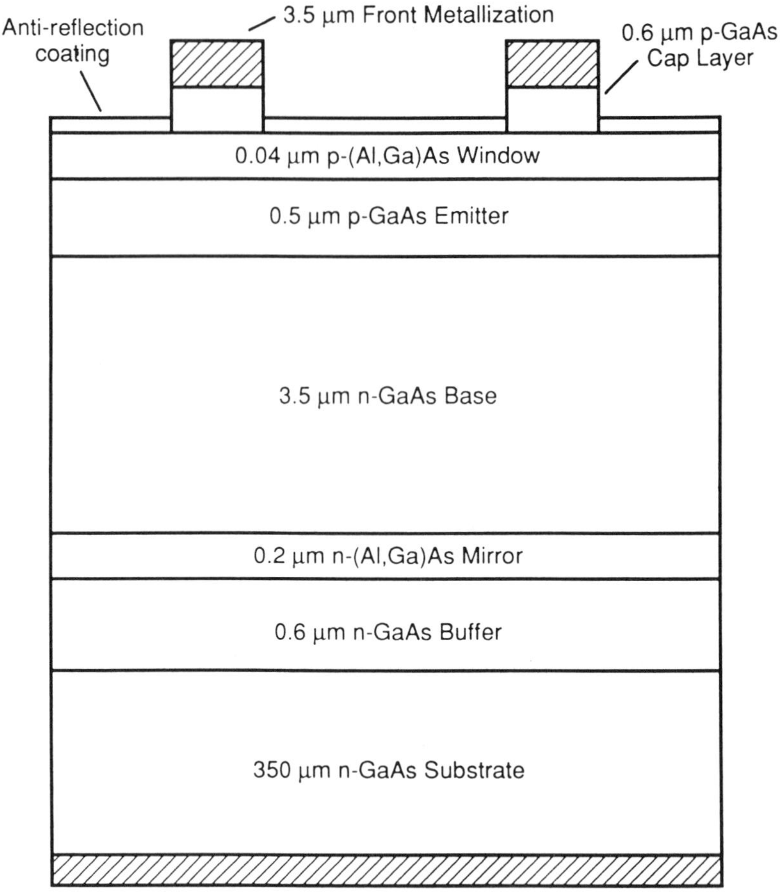

FIGURE 5.1 Schematic diagram of the cross-sectional structure of a typical GaAs concentrator cell design.

using either n-on-p or p-on-n configurations appear to have similar potential for high performance (DeMoulin and Lundstrom, 1987; MacMillan et al., 1988; Wojtczuk et al., 1990). The growth of the thin layers and abrupt heterojunctions employed in this design requires the use of advanced epitaxial growth techniques, such as MOVPE or MBE. The 400 Å $(Al_{0.9}Ga_{0.1})As$ front window is nearly transparent to photons below 3 eV in energy; it prevents minority carriers in the emitter from diffusing to the front surface of the cell, where they would recombine at a high rate. The thin emitter is heavily doped, $\geqslant 2 \times 10^{18}\,cm^{-3}$, to reduce sheet resistance. Because GaAs is a direct band gap material with high absorption coefficients, the 4 μm thickness of the emitter and base is sufficient to absorb nearly all the photons with energy greater than the 1.42 eV band gap (Casey et al., 1975). Both $(Al_xGa_{1-x})As$ and heavily doped

GaAs have been used successfully as a minority-carrier mirror at the rear of the cell. However, band gap narrowing reduces the effectiveness of p^+-GaAs as a minority-carrier mirror (Lundstrom et al., 1990). Also, both n-on-p and p-on-n designs have been used to produce high-efficiency cells (MacMillan et al., 1988).

GaAs has a larger band gap than Si (1.42 vs. 1.12 eV); in finished cells, this results in a relatively higher open circuit voltage for GaAs, but lower short circuit current. In Figure 5.2, the current–voltage characteristics of one sun, high-efficiency GaAs, Si, and $CuInSe_2$ cells are compared (Kurtz et al., 1991; Zhao et al., 1990; Basol et al., 1991).

Radiative recombination is a significant or dominant portion of the total bulk recombination in typical GaAs solar cells, resulting in bulk lifetimes on the order of 1–100 ns (Lush et al., 1992). Because of the relatively large band gap of GaAs (relative to Si), recombination in the space-charge region ($n = 2$ dark current) of the cell can be important as well. Also, the unpassivated surface of GaAs exhibits a relatively large recombination velocity of approximately 10^7 cm/s. A related phenomenon is the large perimeter-related, $n \approx 2$ current observed in GaAs diodes. At the one sun operating voltage, the $n \approx 2$ dark current makes up as much as one-half of the total recombination current (DeMoulin et al., 1987; Tobin et al., 1988a; Stellwag et al., 1990). Finally, it should be noted that heavily doped GaAs, such as is used in the cell design

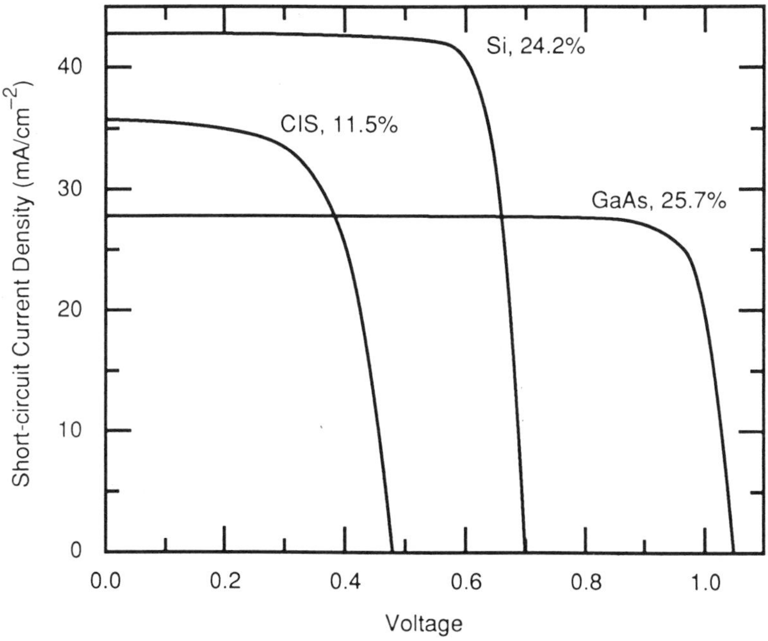

FIGURE 5.2 One sun, AM1.5G current–voltage characteristic for high-efficiency cells made from GaAs (Kurtz et al., 1991), Si (Zhao et al., 1990), and $CuInSe_2$ (Basol et al., 1991).

shown in Figure 5.1, does not go into high-level injection for optical concentration ratios less than about 10,000.

3.2. Current Flow and Resistance

It is important to minimize series resistance in a concentrator cell because of the high currents that flow laterally in the emitter, vertically under the grid lines in the GaAs cap layer, and along the grid lines themselves. The current produced in a $0.2\,\mathrm{cm}^2$ concentrator cell can exceed 5 A at 1,000 ×. Figure 5.3 depicts current flow near the surface of a cell and illustrates key resistance points.

The emitter in a GaAs cell is thin, and the sheet resistance of the emitter, R_S, limits the separation of the grid lines to 50–100 μm. After carriers travel laterally across the emitter, they must move vertically across the two heterojunction boundaries, at each edge of the emitter. This resistance, designated R_{HJ} in Figure 5.3, is not normally dominant. The carriers must next traverse the boundary between the metal grid line and the GaAs cap. (In the case of a p-type emitter, some might prefer to picture electrons flowing across the boundary to annihilate holes in the cap layer.) This contact resistance, R_C, limits the minimum grid shadowing that can be achieved. Values of R_C less

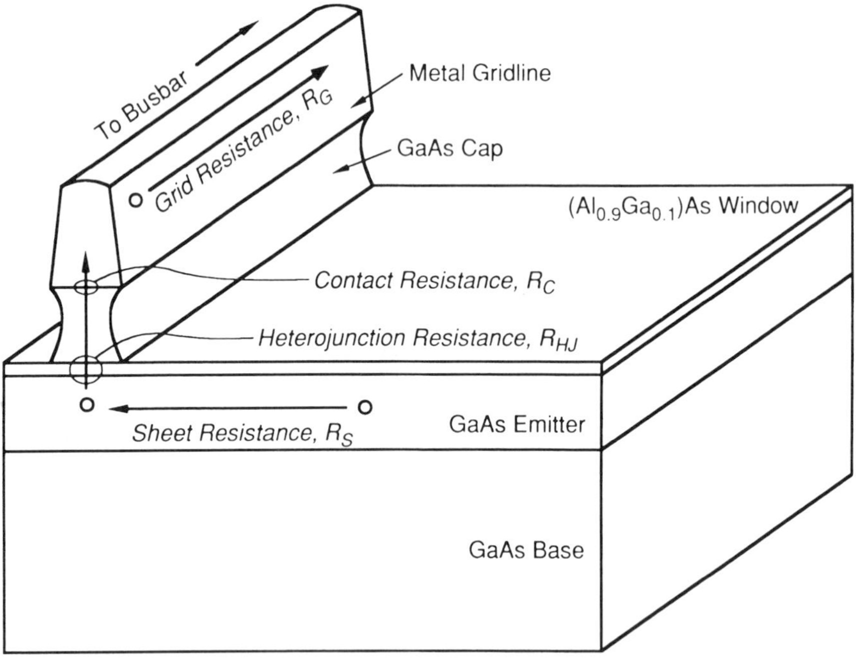

FIGURE 5.3 Schematic diagram of top of concentrator cell, with metal grid line, showing paths of current flow and key resistance points. The antireflection coating is not shown.

than $10^{-5}\,\Omega\text{-cm}^2$ are usually acceptable and can be realized using a thin alloyed layer of Au/Zn/Au (for a p-type emitter) or Au/Ge/Ni/Au (for an n-type emitter). Several other contact layer schemes have been explored for improved reliability and high temperature stability (Spitzer et al., 1988).

Finally, current must flow down the grid lines to the busbar. In state-of-the-art cells, Ag or Al grid lines are used, with a cross section of $3 \times 3\,\mu m$. To deposit grid lines with the profile shown in Figure 5.3, a lift-off procedure would have to be used (Tobin et al., 1990). Use of a plating technique would cause the line to be wider at the top than at the bottom, which would increase shadowing significantly. Figure 5.4 shows a schematic plan view of a typical concentrator cell grid pattern (the actual grid line density would be much higher); the area of the illuminated circular active region is typically 0.1–0.2 cm^2. For 1,000× operation the grid shadowing in the active region would be 4%–8%, depending on grid metal thickness and contact resistance. Note that the portion of the diode junction covered by the busbar contributes to the dark current, but not to the short circuit current.

Figure 5.5 shows simulated current versus voltage characteristics for an n-on-p GaAs concentrator cell, with varying frontside contact resistances, for 1,000×, AM1.5D conditions. These data were calculated using the quasi-three-dimensional computer program CELLOPT, which simulates the operation of the antireflection and semiconductor layers in one dimension and the emitter and front contact grid resistances in two dimensions (Hamaker, 1985). The simulated cell design employed $3 \times 3\,\mu m$ grid lines with 6% shadowing and a $0.3\,\mu m$, n-type emitter doped $2 \times 10^{18}\,cm^{-3}$. The data show that there is a severe degradation of cell performance for contact resistance greater than $1 \times 10^{-4}\,\Omega\text{-cm}^2$, due to poor fill factor. There is little improvement in the simulated efficiency of this cell design for $R_C < 1 \times 10^{-5}\,\Omega\text{-cm}^2$.

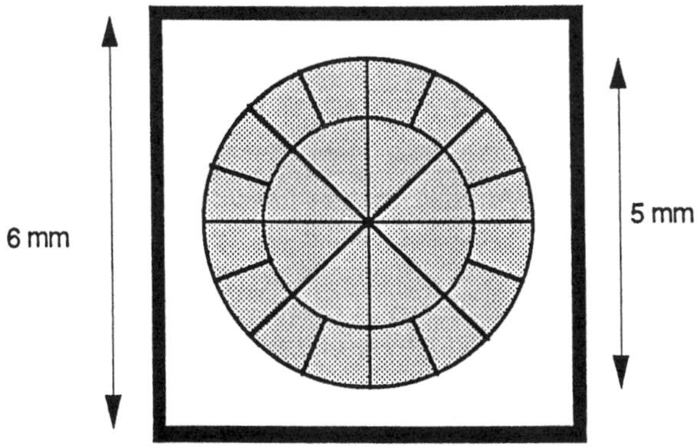

FIGURE 5.4 Schematic plan view of the grid line pattern for a GaAs concentrator cell. The area of the circular active region is typically 0.1–0.2 cm^2. The actual grid line density would be much higher than is shown here, with a shading factor of 4%–8% for 1,000× operation.

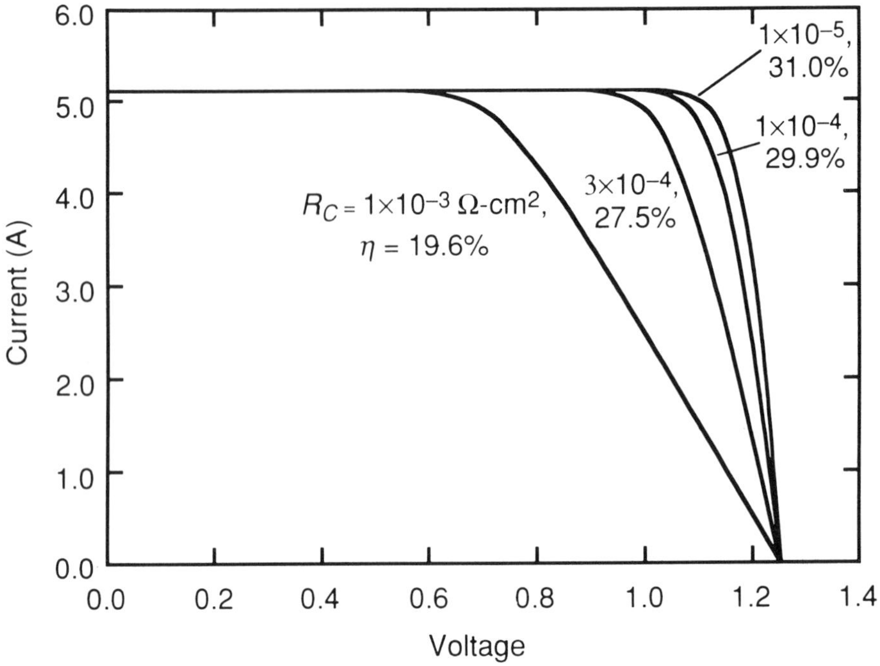

FIGURE 5.5 Simulated current versus voltage characteristics for a 0.2 cm^2, n-on-p GaAs concentrator cell, assuming different front contact resistances (R_C), under 1,000 ×, AM1.5D conditions.

3.3. Cell Efficiency at Elevated Temperature and Fluence

GaAs cells operating at high concentration absorb a large amount of heat, from infrared radiation, from thermalization of recombined carriers, and from flow of high currents in resistive emitters and grid lines. The junction temperature can easily exceed 50°C under realistic operating conditions (Kuryla et al., 1991). Figure 5.6 shows the simulated AM1.5D efficiency of an n-on-p cell as a function of cell temperature at 400 × (Klausmeier-Brown et al., 1991). The efficiency decreases approximately linearly with increasing temperature. The energy gap also decreases at higher temperatures. The reduction in the gap increases absorption slightly, but not enough to make up for the loss in V_{oc}. In general, this sensitivity of the band gap of semiconductors to temperature favors the use of higher band gap materials, such as GaAs, in solar concentrator applications.

The measured efficiency for two different cells is plotted as a function of concentration ratio in Figure 5.7 (MacMillan et al., 1988; uncorrected for 1991 Sandia recalibration, see Chapter 1). One is an n-on-p cell with a front contact grid designed for 100–150 × operation. The one sun short circuit current of this cell is 27.7 mA/cm^2; its fill factor peaks at about 150 ×, while its open

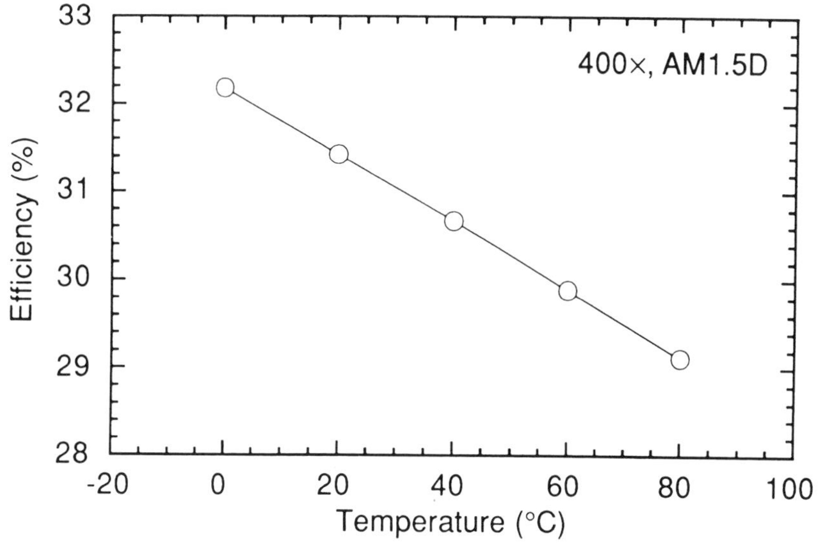

FIGURE 5.6 Simulated cell efficiency versus junction temperature for an n-on-p GaAs concentrator cell under 1000 ×, AM1.5D conditions (Klausmeier-Brown et al., 1991).

circuit voltage continues to increase to 1.19 V at 1,000 ×. These two competing effects cause the efficiency of the cell to peak at about 400 ×. The second cell is a more heavily gridded p-on-n cell designed for 1,000 × operation; it has a one sun J_{sc} of 25.9 mA/cm^2. The V_{oc} of this cell is nearly identical to that of the

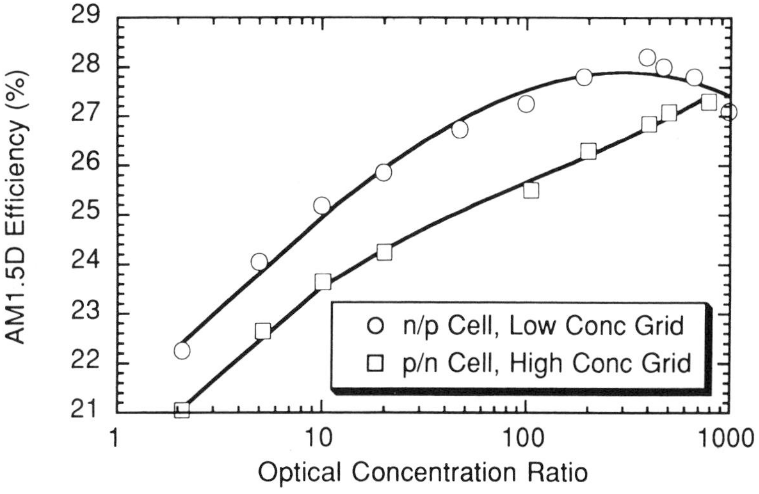

FIGURE 5.7 Measured cell efficiency versus concentration at 28°C for an n-on-p cell with a grid designed for 100 suns and a p-on-n cell with a grid designed for 1,000 suns (MacMillan et al., 1988).

more lightly gridded cell, but its fill factor increases with concentration to 0.89 at 1,000 ×. This shows how grid design is a key element of good concentrator cell design. The importance of minimizing grid series resistance tends to favor the use of smaller cell diameters for a given concentration ratio, with cost considerations setting a lower limit on cell size (Kaminar and Hamaker, 1985).

4. CONCENTRATOR SYSTEM CONSIDERATIONS

Photovoltaic modules for high concentration are quite different from one sun modules. The entire front surface of a concentrator module is covered with lenses that concentrate light on the cells beneath them. An array of concentrator modules is not fixed in place, but must use a two axis solar tracking system to maintain an accurate focus.

4.1. Concentrator System Components

The front lens panel is typically made using square Fresnel lenses, each with an area of $\sim 200\,\text{cm}^2$. These lenses would probably be stamped from plastic for production modules. Beneath the primary Fresnel lens is a secondary lens/collector. A good secondary element relaxes the requirements for pointing accuracy of the tracking system and may also reduce nonuniformity of illumination of the cell's surface. Inexpensive secondaries have been made from conically shaped metallic reflectors. Glass lens secondaries give better performance and can be made relatively cheaply if they are molded. Figure 5.8 shows a 12 cell, 1,000 × module and a cell assembly with glass secondary lens (Kuryla et al., 1991). The annular ring on the secondary lens would facilitate automated handling and placement.

Shadowing by the front grid lines on a concentrator cell reduce the maximum achievable short circuit by 4%–8%. This loss can be nearly eliminated through the use of prismatic cover slides, which direct light around the grid lines (Kaminar et al., 1988; see Chapter 10). However, it might be difficult to attach such coverslides in a cost-effective manner in a production environment.

4.2. Adjustments for Concentrator Optics

Compared with a one sun, flat-plate solar array, a concentrator array has additional optical losses. The overall optical efficiency (light entering the cell divided by light incident normally on the Fresnel lens) of the module depicted in Figure 5.8 is approximately 85%. The efficiency of the Fresnel lenses is ~93% in the active pass band. Reflection at the front surface of the secondary can be limited to less than 2% by using an antireflection coating; reflection at the back of the secondary is also about 2%. To minimize reflection at the front surface of the cell, the cell's antireflection coating should be designed to match the index of the adhesive used to attach the secondary lens to the cell. (Grid

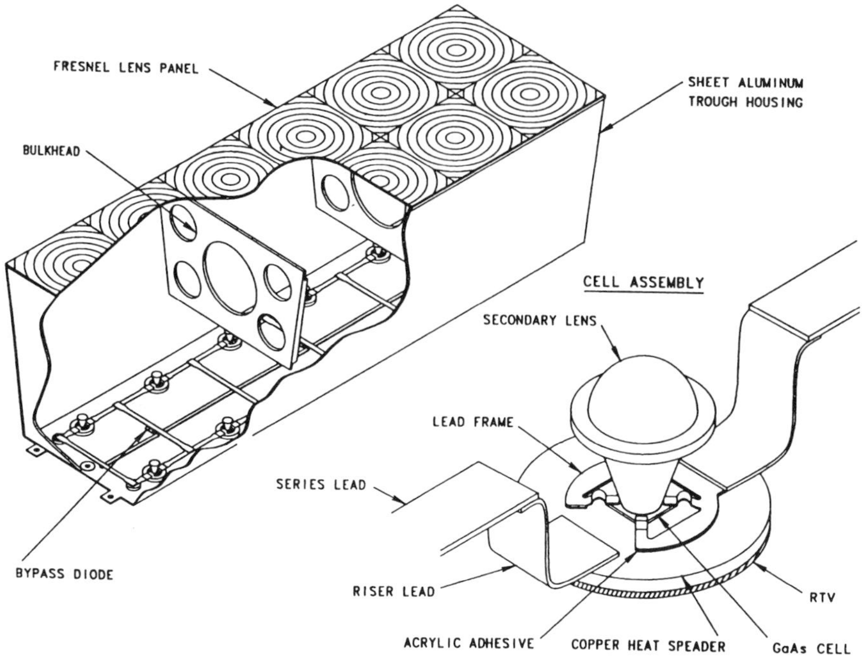

FIGURE 5.8 Twelve cell, 1,000 × module and cell assembly with glass secondary lens (Kuryla et al., 1991).

line shadowing would further reduce the optical efficiency by 4% or more.)

A final, additional "loss" must be considered when comparing flat-plate and concentrator systems. A typical high-concentration system responds almost exclusively to normally incident light. At AM1.5, only about 85% of the available light is normally incident (i.e., global vs. normal/direct insolation; measurements of the solar spectrum are discussed in Chapter 20). Some of the obliquely incident rays can be collected by a flat-plate system, albeit with decreasing efficiency as the angle of incidence increases. Note that the efficiency of concentrator cells and systems is usually referenced to the direct spectrum (AM1.5D), normalized to the incident power of the global spectrum (AM1.5G; a value of $100\,mW/cm^2$ is typically used for the incident power). Taking into account the optical losses discussed above, a concentrator cell efficiency of 28% (AM1.5D) would translate to a module efficiency of 24%–25% (at 28°C). In addition, all solar photovoltaic systems suffer from losses due to external wiring and inverter inefficiency, which would lower the final system efficiency to 21%–22%.

4.3. Cell Efficiency/Cost Targets

When evaluating the merits of various photovoltaic systems, one must consider not only efficiency but also cost. In 1990, Sandia National Laboratories issued a request for proposals for the development of collector systems that identified a near-term goal for energy cost (EC) of \$0.12/kW-hr and a long-term goal of \$0.06/kW-hr. A calculation of the energy cost takes into account the module efficiency, module cost and area, average insolation, cost of the tracker, cost of external wiring and power conditioning, indirect cost, operation and maintenance, and depreciation (Department of Energy, 1987).

The cost, in volume production, for a $44\,m^2$ array that would use a $1{,}000\times$, 36 cell version of the module shown in Figure 5.8 has been calculated to be \$530/m^2, with the module itself at \$350/m^2, external wiring at \$120/m^2, and the rest of the array, including the tracking system, at \$60/m^2 (figures are from the text of a proposal by Varian Associates, Inc., in response to the Sandia Collector RFP, 1990; the DOE formula [Department of Energy, 1987] assumes the use of a somewhat more expensive tracking system than is assumed here). The energy cost of such a system would be

$$\mathrm{EC} = \frac{0.0317}{\text{Cell efficiency at } 25^{\circ}\mathrm{C}} + 0.0053$$

assuming an indirect cost factor of 0.5. Thus, for a cell efficiency of 28%, the energy cost would be EC = \$0.118/kW-hr. If an indirect cost factor of 0.3 is used, which may be appropriate for long-term cost goals, and the concentration ratio is increased to $2{,}000\times$, the formula becomes:

$$\mathrm{EC} = \frac{0.0210}{\text{Cell efficiency at } 25^{\circ}\mathrm{C}} + 0.0053$$

Using this formula, one finds that a cell efficiency of 35% would result in an EC of \$0.065/kW-hr, close to the long-term goal of \$0.06/kW-hr.

4.4. System Cost, Manufacturability, and Safety

It is incorrect to assume that high-efficiency, high-concentration options for photovoltaic power are inherently more expensive than lower efficiency, flat-plate systems. In general, if one system is twice as efficient as another, then its allowed cost is twice as high. The cost of the site must be considered also; a system that is twice as efficient as another would use only half as much land for the same energy output.

Benner (1991) identifies substrates, epitaxy, and cell fabrication as three potential areas for cost reduction. Substrate cost can be reduced by achieving very high production volume, by separating the film from the substrate and reusing the substrate, or by using an alternative substrate, such as Ge or Si. Ge substrates are currently less expensive than GaAs, and Ge has almost the same lattice constant. With advances in heteroepitaxy, Si substrates may become more viable. The cost of GaAs epitaxy is fairly high due to its high cost of consumables and also because of the danger of transporting and storing compressed, highly toxic gases. Benner suggests that epitaxial process control, cost, and safety might be improved by generating gaseous consumables on site at a large photovoltaic plant or by using remote plasma generation of arsine (Pihlstrom et al., 1991).

The cost of GaAs or multijunction concentrator cell fabrication is minimized principally by maintaining a monolithic device structure and through incremental process refinement. Potential trouble spots in the fabrication process include maintaining good front contacts, adherence of the antireflection coating, the cap etch procedure, and, for some multijunction cells, maintaining a good tunnel junction interconnect. The integrity of the $(Al_xGa_{1-x})As$ window is critical; the cap layer must be completely etched away, and the window layer must be well protected by the AR coating. The window layer can oxidize via pin holes in the AR coating in a matter of months, rendering a cell useless. A cell with a stable, relatively low band gap window might eliminate the need for a cap layer altogether. A cell design demonstrated by Kurtz et al. (1991), which uses a 1.9 eV $(Ga_{0.5}In_{0.5})P$ window, would be a good candidate for elimination of the cap layer.

5. PROSPECTS FOR IMPROVED CELL EFFICIENCY

As discussed earlier in the chapter, the best GaAs concentrator cells have efficiencies of 28%. However, we have just seen that such cells can only meet near-term, developmental cost goals. To meet long-term cost goals, 35% efficient cells will be required. This section examines the prospects for improving GaAs concentrator cell efficiency beyond 30%.

5.1. Extrapolation from the Best Laboratory Cells

In their paper describing their 28.7% efficiency (200 ×, AM1.5D), Tobin et al. (1990) project that the one sun, AM1.5G efficiency of the cell could be increased from 24.8% to 26.5% in the near term by making small improvements in cell design and processing. This would improve the 200 ×, AM1.5D efficiency to ~30.7%. In an analysis of this same cell, Lush and Lundstrom (1991) project that its radiative limit one sun, AM1.5G efficiency is 27.2%, which translates to a 200 ×, AM1.5D efficiency of ~31.5%. This is still significantly less than the proposed 35% long-term goal; to reach it, other approaches must be considered.

5.2. Light-Trapping Techniques

High-efficiency silicon cells commonly make use of light-trapping techniques to increase the effective cell thickness for absorption of photons. Using such techniques, a given short circuit current can be achieved in a thinner cell, which reduces the recombination volume and thus increases the open circuit voltage. One method of implementing light trapping is to add texture to the front surface of the cell. This has recently been demonstrated for GaAs cells, using liquid epitaxial lateral overgrowth (Honsberg and Barnett, 1991). One might also make use of the various nonplanar etchants of GaAs to provide texturing (Bailey, 1991). In general, however, it may prove difficult both to texturize and to passivate the front surface of a GaAs cell. An alternative to texturing the cell itself is to use a light-trapping coverslide (Landis, 1991).

To provide reflection at the back of the cell, several researchers have proposed a Bragg reflector structure composed, for example, of thin alternating layers of AlAs and $(Al_xGa_{1-x})As$. Such a structure can reflect nearly 100% of long wavelength photons. Implementation of a Bragg reflector in an otherwise conventional heteroface cell would enhance the efficiency by approximately 1% (absolute).

5.3. Taking Advantage of Photon Recycling

Another possibility for enhancing GaAs cell performance is to design a cell that takes advantage of the photon recycling effect (Asbeck, 1977). Because the absorption length at the band edge is only 5–10 μm, radiative recombination events give rise to secondary photons that may be reabsorbed, leading to enhanced diffusion lengths. Lush and Lundstrom (1991) have calculated that an ideal cell that combines light trapping, a very thin active region, and photon recycling effects would be as much as 3% more efficient (absolute) than a conventional heteroface concentrator cell; the projected efficiency limit is slightly over 35% for 1,000 ×, AM1.5D conditions.

5.4. Other Approaches

Several cell designs that do not conform to the conventional heteroface model have been demonstrated or suggested. One of the most notable is a cell in which the $(Al_xGa_{1-x})As$ window is replaced by 1.9 eV $(Ga_{0.5}In_{0.5})P$. This cell currently claims the record single junction, one sun, AM1.5G efficiency of 25.7% (Kurtz et al., 1991). Another proposed structure is similar to the point contact design that has been instrumental in setting record efficiencies for Si cells. However, detailed modeling has shown that this design probably would not bring comparable benefit in GaAs (DeMoulin and Lundstrom, 1987).

It is also possible to improve the cost and efficiency performance of a concentrator system by reducing cell cost rather than increasing efficiency. A design that aims to accomplish this is the diffused junction cell, which is discussed in Chapter 6.

A discussion of alternative approaches would not be complete without mention of some of the research on multijunction cells. Two-terminal, current-matched, monolithic multijunction cell configurations based on GaAs include 1.9 eV $(Al_xGa_{1-x})As$ or (GaIn)P on GaAs; 1.75 eV $(Al_xGa_{1-x})As$ on

1.15 eV (In_yGa_{1-y})As; and GaAs on Ge. For comparison purposes, the 35% longterm goal discussed earlier in this chapter in the context of single junction cells should be used for these two terminal combinations. Both designs that employ 1.9 eV upper cells have demonstrated excellent one sun efficiencies of over 27%. Another monolithic system that has generated very high efficiency is InP on 0.7 eV (In_yGa_{1-y})As (Wanlass et al., 1991). Finally, the system that has demonstrated the highest efficiency of all is the voltage-matched combination of GaAs on GaSb. This nonmonolithic combination is discussed in Chapter 6. It appears that a multijunction concentrator cell design is more likely to meet the long-term energy cost target of $0.06/kW-hr than is single junction GaAs.

6. SUMMARY

The efficiency of the best GaAs concentrator cells has increased to approximately 29% at high concentration levels. The efficiency of the best module, a 1,000 ×, passively cooled design, exceeded 22% using 25%–26% efficient cells. These performance points are approximately adequate to meet the near- to medium-term DOE energy cost target of $0.12/kW-hr.

To meet the long-term energy cost goal of $0.06/kW-hr, higher performance is needed: cells with efficiencies over 35% at concentration and module efficiencies of about 30%. It has been projected that an ideally configured cell, which would make use of yet-to-be-developed light-trapping and photon-recycling enhancement techniques, could reach 35%. However, continued development of multijunction concentrator cells may provide a more certain path to reaching long-term targets.

REFERENCES

Arvizu, D.E. (1985), *18th IEEE Photovoltaic Specialists Conf. Record*, IEEE, New York, pp. 1529–1537.

Asbeck, P. (1977), *J. Appl. Phys.* **48**, 820–822.

Bailey, S.G. (1991), *22th IEEE Photovoltaic Specialists Conf. Record,* IEEE, New York, pp. 235–239.

Basol, B.M., Kapur, V.K., and Halani, A. (1991), *22nd IEEE Photovoltaic Specialists Conf. Record,* IEEE, New York, pp. 893–902.

Benner, J.P. (1991), *22nd IEEE Photovoltaic Specialists Conf. Record,* IEEE, New York, pp. 7–11.

Casey, H.C., Jr., Sell, D.D., and Wecht, K.W. (1975), *J. Appl. Phys.* **46**, 250–257.

DeMoulin, P.D., Kyono, C.S., Lundstrom, M.S., and Melloch, M.R. (1987), *18th IEEE Photovoltaic Specialists Conf. Record,* IEEE, New York, pp. 93–97.

DeMoulin, P.D., and Lundstrom, M.S. (1987), *18th IEEE Photovoltaic Specialists Conf. Record,* IEEE, New York, pp. 925–930.

Department of Energy, U.S. (1987), *Five Year Research Plan 1987–1991*, DOE/CH10093-7, pp. 27–31.

Fan, J.C.C., Bozler, C.O., and Chapman, R.L. (1978), *Appl. Phys. Lett.* **32**, 390–392.

Gale, R.P., McClelland, R.W., King, B.D., and Gormley, J.V. (1988), *20th IEEE Photovoltaic Specialists Conf. Record,* IEEE, New York, pp. 446–450.

Gray, J.L., Lundstrom, M.S., and Schwartz, R.J. (1987), *Sandia Report 86-7048*, Sandia National Laboratories, Albuquerque, NM, pp. 7–8.

Hamaker, H.C. (1985), *J. Appl. Phys.* **58**, 2344–2351.

Hamaker, H.C., Ford, C.W., Werthen, J.G., Virshup, G.F., Kaminar, N.R., King, D.L., and Gee, J.M. (1985), *18th IEEE Photovoltaic Specialists Conf. Record,* IEEE, New York, pp. 327–331.

Henry, C.H. (1980), *J. Appl. Phys.* **51**, 4494–4499.

Honsberg, C.B., and Barnett, A.M. (1991), *22nd IEEE Photovoltaic Specialists Conf. Record,* IEEE, New York, pp. 262–267.

Iles, P.A., Yeh, Y.-C.M., Ho, F.H., Chu, C.-L., and Cheng, C. (1990), *IEEE Electron Device Lett.* **11**, 140–142.

James, L.W., and Moon, R.L. (1975), *11th IEEE Photovoltaic Specialists Conf. Record,* IEEE, New York, pp. 402–408.

Kaminar, N.R., and Hamaker, H.C. (1985), *Sandia Report 85-7021*, Sandia National Laboratories, Albuquerque, NM, pp. 5–7.

Kaminar, N.R., Liu, D.D., MacMillan, H.F., Partain, L.D., Ladle Ristow, M., Virshup, G.F., and Gee, J.M. (1988), *20th IEEE Photovoltaic Specialists Conf. Record,* IEEE, New York, pp. 766–768.

Klausmeier-Brown, M.E., Partain, L.D., and Werthen, J.G. (1991), *22nd IEEE Photovoltaic Specialists Conf. Record,* IEEE, New York, pp. 58–62.

Kurtz, S.R., Oolson, J.M., and Kibbler, A. (1991), *21st IEEE Photovoltaic Specialists Conf. Record,* IEEE, New York, pp. 138–140.

Kuryla, M.S., Ladle Ristow, M., Partain, L.D., and Bigger, J.E. (1991), *22nd IEEE Photovoltaic Specialists Conf. Record,* IEEE, New York, pp. 506–511.

Landis, G.A. (1991), *22nd IEEE Photovoltaic Specialists Conf. Record,* IEEE, New York, pp. 1304–1307.

Lundstrom, M.S., Klausmeier-Brown, M.E., Melloch, M.R., Ahrenkiel, R.K., and Keyes, B.M. (1990), *Solid-St. Electron.* **33**, 693–704.

Lush, G.B., Levi, D.H., MacMillan, H.F., Ahrenkiel, R.K., Melloch, M.R., and Lundstrom, M.S. (1992), *Appl. Phys. Lett.* **61**, 2440–2442.

Lush, G.B., and Lundstrom, M.S. (1991), *Solar Cells* **30**, 337–344.

MacMillan, H.F., Hamaker, H.C., Kaminar, N.R., Kuryla, M.S., Ladle Ristow, M., Liu, D.D., Virshup, G.F., and Gee, J.M. (1988), *20th IEEE Photovoltaic Specialists Conf. Record,* IEEE, New York, pp. 462–468.

Mathers, C.D. (1977), *J. Appl. Phys.* **48**, 3181–3182.

Melloch, M.R., Tobin, S.P., Bajgar, C., Stellwag, T.B., Keshavarzi, A., Lundstrom, M.S., and Emery, K. (1990), *21st IEEE Photovoltaic Specialists Conf. Record,* IEEE, New York, pp. 163–167.

Owen, R.J., and Borden, P.G. (1981), *15th IEEE Photovoltaic Specialists Conf. Record,* IEEE, New York, pp. 329–335.

Pihlstrom, B.G., Thompson, L.R., and Collins, G.J. (1991), *Solar Cells*, **30**, 415–421.

Shockley, W., and Queisser, H.J. (1961), *J. Appl. Phys.* **32**, 510–514.

Sinton, R.A., and Swanson, R.M. (1987), *IEEE Trans. Electron Devices* **34**, 2116–2123.

Spitzer, M.B., Dingle, J.E., Gale, R.P., Zavracky, P., Boden, M., and Doyle, D.H. (1988), *20th IEEE Photovoltaic Specialists Conf. Record,* IEEE, New York, pp. 930–933.

Stellwag, T.B., Dodd, P.E., Carpenter, M.S., Lundstrom, M.S., Pierret, R.F., Melloch, M.R., Yablonovitch, E., and Gmitter, T.J. (1990), *21st IEEE Photovoltaic Specialists Conf. Record,* IEEE, New York, pp. 442–447.

Tobin, S.P., Vernon, S.M., Bajgar, C., Geoffroy, L.M., Keavney, C.J., Sanfacon, M.M., and Haven, V.E. (1988a), *Solar Cells* **24**, 103–115.

Tobin, S.P., Vernon, S.M., Bajgar, C., Haven, V.E., Geoffroy, L.M., and Lillington, D.R. (1988b), *IEEE Electron Device Lett.* **9**, 256–258.

Tobin, S.P., Vernon, S.M., Wojtczuk, S.J., Bajgar, C., Sanfacon, M.M., and Dixon, T.M. (1990), *21st IEEE Photovoltaic Specialists Conf. Record,* IEEE, New York, pp. 158–162.

Vander Plas, H.A., James, L.W., Moon, R.L., and Nelson, N.J. (1978), *13th IEEE Photovoltaic Specialists Conf. Record,* IEEE, New York, pp. 934–940.

Wanlass, M.W., Ward, J.S., Emery, K.A., Gessert, T.A., Osterwald, C.R., and Coutts, T.J. (1991), *Solar Cells* **30**, 363–371.

Wojtczuk, S.J., Tobin, S.P., Keavney, C.J., Bajgar, C., Sanfacon, M.M., Geoffroy, L.M., Dixon, T.M., Vernon, S.M., Scofield, J.D., and Ruby, D.S. (1990), *IEEE Trans. Electron Device* **37**, 455–462.

Woodall, J.M., and Hovel, H.J. (1977), *Appl. Phys. Lett.* **30**, 492–493.

Zhao, J., Wang, A., and Green, M.A. (1990), *21st IEEE Photovoltaic Specialists Conf. Record,* IEEE, New York, pp. 333–335.

CHAPTER SIX

High-Efficiency III–V Multijunction Solar Cells

LEWIS M. FRAAS, JX Crystals, Inc., Issaquah, WA 98027

1. INTRODUCTION

An energy conversion efficiency of 28% for a GaAs cell is within 10% of the theoretical limit efficiency (approximately 30%) for a single junction solar cell. There is very little room for improvement for single junction devices. However, there is considerable opportunity for improvement with multijunction devices because there are intrinsic losses in single junction devices that result from the fact that the sun's spectrum is not monochromatic. For example, it is known that when a GaAs photovoltaic cell is illuminated with near band gap monochromatic light with a wavelength of 8,500 A and a corresponding photon energy of 1.45 V, nearly all of the photons are converted to current and the device generates a voltage of 1.1 V with a fill factor of 0.85 so that the energy conversion efficiency is 1.1 × 0.85./1.45 or 64% (Spitzer et al., 1991). Can solar cells be made with conversion efficiencies over 50%?

Single junction solar cell energy conversion efficiencies are typically less than 30% because, for any given material, the sun's spectrum contains photons with both too much and too little energy for optimal conversion. This is shown in Figure 6.1. The sun contains photons in the visible, near-infrared, and infrared (IR) regions of the spectrum. The visible photons generate electron–hole pairs with energy well above the minimal excitation energy. The excess energy is lost as heat as minority carriers thermolize down to the band edge. The IR photons have too little energy to be absorbed. Only the near-IR photons are converted efficiently to electric power by the cell. If, however, one were to split the solar spectrum into two or more parts, one might pick materials tailored to convert

Solar Cells and Their Applications, Edited by Larry D. Partain.
ISBN 0-471-57420-1

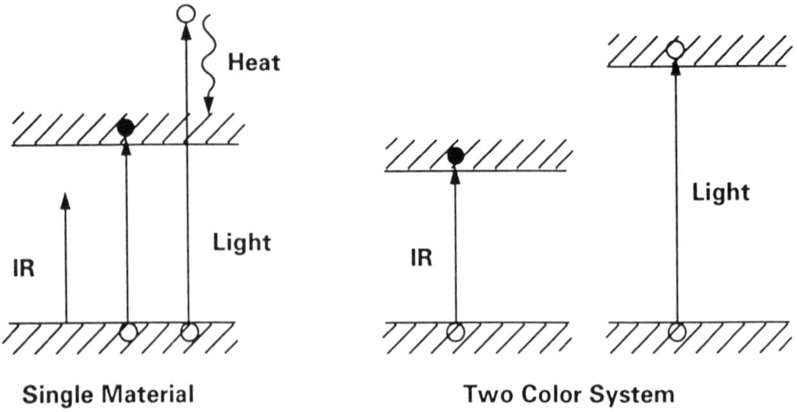

FIGURE 6.1 Schematic diagram showing how a two-color solar cell utilizes the various wavelengths of solar radiation. In a single material (left), the low-energy photons have too little energy to be absorbed and the high-energy photons are absorbed with considerable excess energy. The excess energy is wasted as heat. Using two materials (right), these energy losses can be decreased.

each part optimally into electric power, as is shown in Figure 6.1 for two materials. In the limit of a large enough number of materials, the sun's spectrum would be broken into monochromatic parts and the conversion efficiency might approach 60%.

Theoretical calculations have been done by Bennett and Olsen (1978) for multijunction cells containing two, three, four and more active junctions with ideally optimized band gaps. They considered both one sun and sunlight concentrated to 1,000 suns. At a concentration of 1,000 suns of terrestrial illumination, they found that the theoretical energy conversion efficiency for a two junction cell stack is 42.5%, while for a three junction cell stack the theoretical energy conversion efficiency is indeed 50%.

Can this 50% solar energy conversion efficiency actually be achieved? No, not yet, but an efficiency of 35% has now been achieved with a two junction GaAs/GaSb cell stack (Fraas and Avery, 1990). Relative to expectations several years ago (Fraas, 1985), this is quite an accomplishment. Indeed, perhaps 40%–50% efficient cells will be demonstrated before the end of the century.

However, this chapter focuses on a review of the recent successful work on two junction solar cells. It begins with a brief review of the theory of two junction cell performance, followed by a brief summary of the six cell pairs with experimentally demonstrated efficiencies exceeding the efficiency of a single junction GaAs cell. Since several two junction cells have now been demonstrated with efficiencies over 30%, a major part of this chapter deals with two junction cell manufacturability and economic viability. Does higher efficiency pay, or is it simply a scientific curiosity? We find that the monolithic $GaInP_2$/GaAs on Ge two junction cell can be viable for one sun flat-plate applications in space while the GaAs/GaSb mechanical cell stack is potentially viable with concentrated sunlight for both space and terrestrial applications.

2. CHOOSING IDEAL BAND GAPS AND REAL MATERIALS

The III–V semiconducting alloys offer the ability to tune band gaps over a very wide range from less than 0.5 V to over 2.0 V, and there are several binary III–V compounds (e.g., GaAs, InP, GaSb) in this range with well-characterized materials properties. Using generalized material properties for this family of alloys, a given solar spectrum, and appropriate computer programs, several authors have calculated iso-efficiency plots as a function of upper and lower cell band gap energies. Figure 6.2 shows two such iso-efficiency plots for realistic concentration ratios and cell operating temperatures for both space and terrestrial spectra (Wanlass et al., 1991).

The first salient feature noteworthy in these plots is that the point of maximum efficiency for either space or terrestrial application occurs at a lower band gap energy near 0.95 V with an upper band gap energy near 1.65 V. This fact guided researchers for ten years in attempts to make ideal component cells with ternary and quaternary III–V alloys such as AlGaAs, $GaInP_2$, GaAsP, GaInAs, and GaAsSb with band gap energies near 1.65 or 0.95 V. This work has been without notable success in terms of actual two junction cells with efficiencies over 30% using the prescribed band gaps of 0.95 and 1.65 V. Materials problems with lattice mismatch, ternary composition control, source material purity, epitaxy, and so forth abound.

After gaining a healthy respect for the materials problems involved with alloys containing three or more elements, Fraas, et al. (1987) proposed fabricating a GaAs/GaSb two junction cell stack because it would be composed of two simple binary compounds and, therefore, could be rapidly developed. Subsequently, Fraas and Avery (1990) demonstrated a GaAs/GaSb cell stack with an efficiency over 35%. The iso-efficiency curves in Figure 6.2 were then published and behold—the GaAs/GaSb cell pair energies of 1.42/0.72 fall right on top of a secondary maximum for the terrestrial case. The peak efficiency of 41% for this secondary maximum is only one percentage point below the efficiency of 42% at the primary maximum.

As of this writing, there are now six cell pairs with demonstrated conversion efficiencies over the best efficiency achieved with a single junction GaAs device. These cell pairs are AlGaAs/GaAs (Chang et al., 1991), $GaInP_2$/GaAs (Olson and Kurtz, 1990), GaAs/Si (Gee and Virshup, 1988), GaAs/GaSb (Fraas and Avery, 1990), GaAs/$CuInSe_2$ (Kim et al., 1991), and InP/$GaInAs_2$ (Wanlass et al., 1991). Of these, all six involve at least one binary compound, and five use GaAs as one of the component cells. This illustrates the importance of using simple, well-developed materials. Referring to Figure 6.2, the two that use GaAs as the bottom cell (AlGaAs/GaAs and $GaInP_2$/GaAs) are well off the theoretical maximum efficiency because they fail to absorb the IR portion of

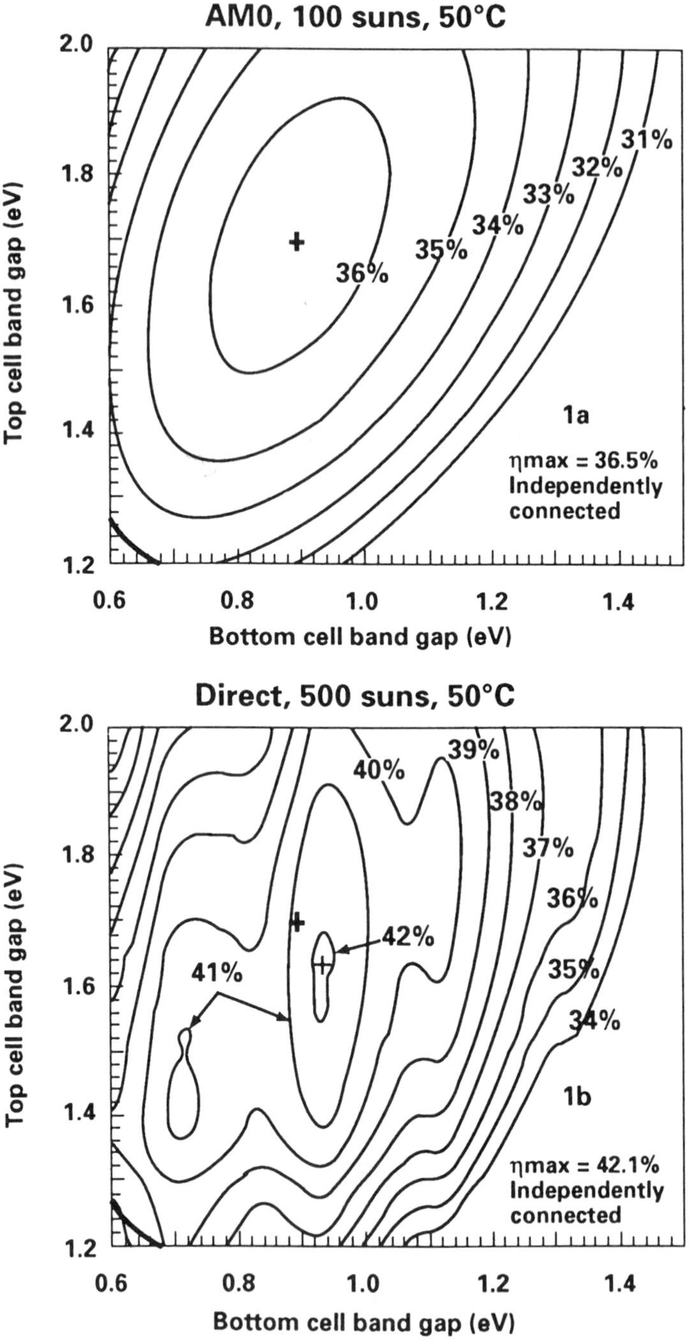

FIGURE 6.2 Computer-modeled iso-efficiency contours for two junction concentrator tandem cells. The two cases considered represent relevant operating conditions for terrestrial and space applications.

the sun's spectrum. Furthermore, the GaAs/Si and GaAs/$CuInSe_2$ cell pairs suffer because only a small increment of the infrared is absorbed in the bottom cell. Hence these pairs have theoretical efficiencies well off the maximum possible efficiency for a two junction pair. These facts indicate that materials considerations (and small budgets) have driven the respective sponsoring groups to compromise. Of the above six pairs, only the GaAs/GaSb and InP/$GaInAs_2$ pairs fall near an efficiency maximum, in both cases near the secondary maximum.

In addition to materials compositions, it is also important to describe physical cell-stack configurations. In this regard, two physical cell-stack configurations are being developed: the monolithic two junction cell and the mechanically stacked tandem cell. The $GaInP_2$/GaAs cell pair shown in Figure 6.3 is typical of a monolithic cell-stack design while the GaAs/GaSb cell pair shown in Figure 6.4 is typical of a mechanically stacked design.

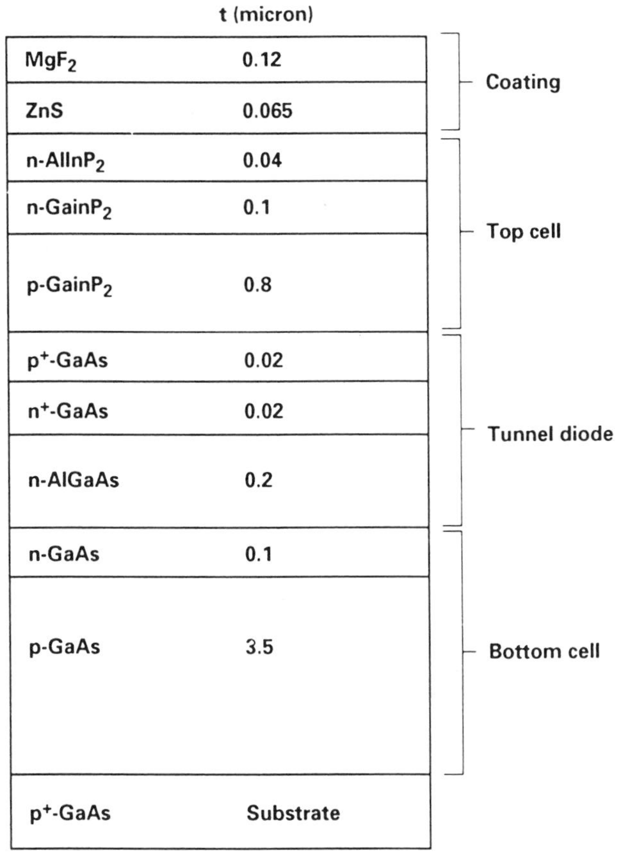

FIGURE 6.3 A schematic cross section of the $GalnP_2$/GaAs tandem cell. A GaAs contacting layer and metalization are not shown.

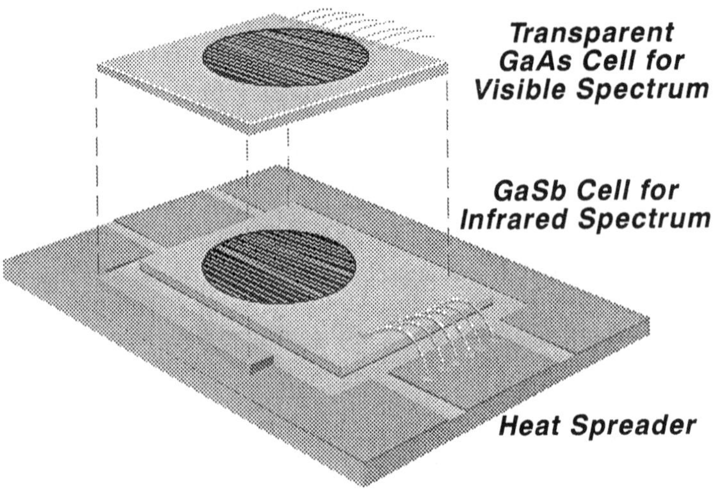

FIGURE 6.4 GaAs/GaSb cell stack in wire bonded form on ceramic substrate.

The structure in Figure 6.3 is grown epitaxially using metal organic chemical vapor deposition (MOCVD) equipment. Referring to Figure 6.3, note that an n-on-p $GaInP_2$ top cell is grown on top of an n-on-p GaAs bottom cell with an intermediate very thin p^+-on-n^+ GaAs tunnel junction connecting these two cell together in series. Counting the two AlGaAs and $AlInP_2$ window layers, eight layers are grown in sequence incorporating five elements in total, not counting dopants. Three different ternary alloys are involved (GaAlAs, $GaInP_2$, and $AlInP_2$). The result is a two terminal monolithic solar cell with two light-sensitive junctions grown on a single wafer. Since the two cells are series connected, it is important that the light-generated currents in both cells be equal. Otherwise, the cell with the lowest current will limit the current in the other cell and drag down the energy conversion efficiency for the cell stack. Careful control of the $GaInP_2$ top cell composition and thickness is essential for current matching for this monolithic cell stack. The most attractive feature of this monolithic cell-stack configuration is that it looks just like a single junction cell to the user.

From a materials point of view, the GaAs/GaSb mechanically stacked tandem cell shown in Figure 6.4 is much simpler. It uses two rectangular binary compound chips with off-center illuminated areas. The top GaAs chip has grid-metal patterns on both top and bottom to allow nonabsorbed IR to pass through to the bottom GaSb chip. The two rectangular chips are rotated 90° with respect to each other and glued together with a transparent adhesive. The result is a four terminal device where the two junctions can be independently accessed. An additional advantage from a materials point of view is that both junctions in GaAs and GaSb cells can be formed by diffusions, thereby avoiding the use of the toxic gases commonly used in MOCVD epitaxy. Although this structure is obviously simpler from a materials point of view, upon initial inspection this structure appears to have two disadvantages. First, because it is a four terminal device, wiring into modules appears to be more complex, and, second, because two wafers are used instead of one, the structure appears to be more costly than that in Figure 6.3. In reality, however, the four terminal configuration has several performance advantages associated with a voltage-matched wiring configuration, and the cost of the second wafer is offset by the high cost of epitaxy for the structure of Figure 6.3. These issues are discussed in more detail in subsequent sections of this chapter.

3. SYSTEMS APPLICATIONS—CONCENTRATOR VERSUS ONE SUN

Although the two junction cells out perform the simple single junction silicon cell, these high-efficiency cell stacks are obviously more complex and more expensive. It will be necessary to leverage the cost of these cell stacks by incorporating them in concentrated sunlight modules or by using them in applications in space where high cost is acceptable. Since tandem cell stacks are probably most appropriately used with concentrated sunlight, this discussion begins with the advantages of concentration from the cell point of view and then continues onto some of the limits imposed on tandem cells by the use of concentration. Finally, the potential of two junction cells for one sun space applications is discussed.

The most obvious advantage of concentration is the leverage that it applies to single crystal wafer cost. It substitutes inexpensive glass or plastic lens material for very expensive single crystal (gem stone) material. In the terrestrial case, for example at 1,000 suns, one 45 × 45 mm GaAs wafer will provide enough cells to populate a 1 m^2 module. The cells from that single wafer will produce 200 W of electric power. This leverage on cost associated with concentration can also be important for space applications. For example, with concentration, it is conceivable that a GaAs-based module can beat a silicon-based module in both performance and cost.

In addition to this cost leverage advantage associated with concentration, there are three additional more subtle advantages inherent in concentration. The first is illustrated in Figure 6.5. Higher yields occur when smaller concentrator cells are fabricated than when large one sun cells are fabricated. The second subtle advantage is shown in Figure 6.6. The open circuit voltage and hence efficiency of a solar cell increases with the higher current densities associated with higher concentration ratios. For two junction cells, this benefit is doubled. The third subtle advantage is that since the cells are at the focal plane behind the lens collector plane, there is plenty of room for wiring and one has a very high lens packing factor. Room for wiring is important for the mechanically stacked cells.

Concentration does, however, impose some constraints on cell fabrication. The main constraint is a very low series resistance at high current densities. In practice to date, this has been a problem for the monolithic cell stack structures

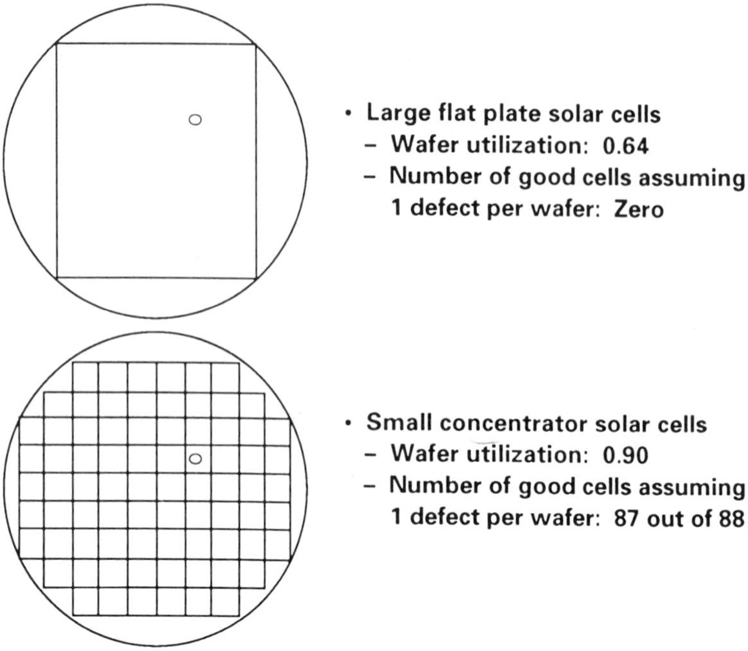

FIGURE 6.5 Small concentrator cells have higher manufacturing yields.

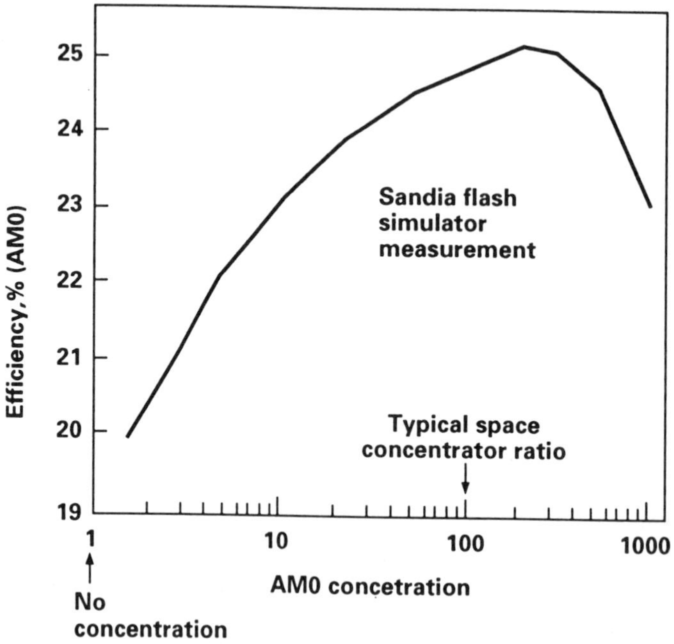

FIGURE 6.6 Efficiency versus concentration ratio for GaAs cells.

(Fig. 6.3). The tunnel junction in the high-performance $GaInP_2$/GaAs cells fabricated to date limits the cell's performance to low concentrations. Although good low resistance tunnel junctions have been separately fabricated, they have not yet been successfully incorporated in this monolithic cell structure to allow high performance at over 100 suns. The GaAs/GaSb mechanically stacked structure in Figure 6.4 does not suffer from this limitation and can be incorporated in both space and terrestrial concentrator modules (see Chapter 14).

Although concentrators are promising for space applications, they are not yet widely used. One sun arrays are still the main source of power in space. For this reason, the CLEFT GaAs/$CuInSe_2$ mechanically stacked tandem cell has been developed for space applications. Advantages of this cell in addition to its high performance include high radiation resistance and light weight.

More recently, the $GaInP_2$/GaAs on Ge cell has been proposed (Cavicchi et al., 1991) and may well be a natural upgrade replacement for the presently used GaAs on Ge flat-plate cell. The $GaInP_2$/GaAs on Ge cell evolves nicely from the GaAs on Ge cell in that both cells would use thin Ge wafers and MOCVD epitaxy. The best $GaInP_2$/GaAs cells are more efficient. The most immediate question: How reproducible is the epitaxial growth sequence? Assuming good reproducibility, this tandem cell could become a commercial quantity for one sun space applications.

4. THE STATUS OF TWO JUNCTION CELL STACKS TODAY

Table 6.1 summarizes the current verified energy conversion efficiencies at both one sun and under concentration for both space and terrestrial illumination conditions for six cell pairs. These six cell pairs are the only cell pairs that have efficiencies higher than a single junction GaAs cell. The efficiency of a GaAs cell at one sun and under concentrated sunlight is listed in Table 6.1 as well for purposes of comparison.

The underlined efficiency numbers in Table 6.1 highlight the cells that in the author's opinion could become commercially viable in the not too distant future. Thus at one sun, high III–V cell cost will limit applications to space. In this area, the $GaInP_2$/GaAs on Ge cell evolves nicely from the GaAs on Ge cell. Based on reported experimental results, the efficiency could thereby increase by 2 percentage points from 21.7% to 23.6%. Theoretically, this gain can be higher. In combination with sunlight concentration, the GaAs/GaSb cell stack can be viable for both space and terrestrial applications. Table 6.1 (Benner, 1991) lists the verified performances levels for this cell stack at 100 suns as 34.2%. However, for the terrestrial application, use at 1,000 suns would be more appropriate, at which point the cell-pair efficiency will go over 35%. The InP/$GaInAs_2$ cell pair could also be viable in space under concentration given a viable tunnel junction. A good tunnel junction is more easily fabricated in low band gap materials (Fraas and Knechtli, 1978), suggesting that

TABLE 6.1 Performance Status for Two Junction Solar Cells at One Sun and Under Concentration With GaAs Single Junction Cell Included for Comparison

Cell Type	Concentration (AM1.5)	AM1.5D Efficiency	AM0 Efficiency	Manufacturers
GaAs	200	28.7%[a]	24.5%	Spire
GaAs/GaSb	100	34.2%	30.5%	Boeing
GaAs/Si	500	31.0%		Sandia
$InP/GaInAs_2$ (three terminal)	50	31.8%		NREL
GaAs	1		21.7%	Spire
$GaInP_2/GaAs$	1	27.5%	23.6%	NREL
AlGaAs/GaAs	1	27.6%	23.6%	Varian
CLEFT/CIS	1		23.1%	Kopin

[a]The underlined efficiencies highlight the options for which an argument for economic viability is made in this chapter.

a tunnel junction for the $InP/GaInAs_2$ cell is feasible. However, in this event, a GaInAsP quaternary bottom cell would be required for good current matching.

5. MONOLITHIC VERSUS MECHANICALLY STACKED TWO JUNCTION CELLS

As noted in Table 6.1, the two junction cell with the highest performance is the GaAs/GaSb mechanical cell stack. From Figure 6.2, this is also the cell pair with the highest theoretical efficiency. However, the packaging and circuit wiring problems associated with the mechanically stacked cell appear to be more complex than for the monolithic cell and need to be addressed before proceeding with a cost comparison of the different cell options.

The advantage of the monolithic approach is that the final cell is a two terminal device and no creativity is required in incorporating this two terminal cell into modules. However, as has been discussed in previous sections, those monolithic cells that have been fabricated to date have materials compositions well off the maximum possible efficiency achievable with two junction cells. Furthermore, there are major problems with fabricating the successful two terminal cell stacks with high yields because of the material complexities. Finally, as concentrator cells, they have high internal resistance associated with the tunnel junction. These problems lead us to a thoughtful investigation of how mechanical cell stacks might be assembled and wired together into a module.

Simple circuit wiring and stack assembly procedures have now been developed for the GaAs/GaSb cell stack. The circuit wiring is accomplished with a prefabricated flex circuit designed for a voltage-matched triplet as a cell group, and the cell stack is built into a TAB (tape automated bonding) frame. After stack qualification testing in the TAB tape, concentrator module focal plane assembly is accomplished by excising the stacks and die and outer-lead bonding them onto the flex circuit. The module focal plane assembly is described in Chapter 14. In the following paragraphs, the voltage-matched triplet concept and the cell stack assembly into the TAB sites are briefly described. These concepts and procedures are described in more detail by Fraas et al. (1992).

The circuit wiring issue is now addressed. A mechanically stacked tandem cell is a four terminal device possessing a + and − terminal for each of the two cells. For the GaAs/GaSb tandem cell, the maximum power voltage of the GaSb cell is approximately one-third of the maximum power voltage of the GaAs cell. This voltage ratio allows us to construct the voltage-matched triplet circuit shown in Figure 6.7a. In this circuit, the three GaAs top cells from three cells pairs are wired in parallel, while the three GaSb bottom cells are wired in series. A simple flex circuit version of this two terminal voltage-matched triplet concept is shown in Figure 6.7b. The result is a two terminal voltage-matched triplet cell group that is modular and can be extended in series or parallel as if it were a single two terminal solar cell. Thus, the solution to the module wiring problem for mechanically stacked cells is quite simple.

What about mechanical stack assembly? Figure 6.8 shows schematically a TAB process for the production of testable cell stacks. In this TAB process, the spacers on the GaSb chips are applied at the wafer level by photolithography or screen printing. As the two types of wafers are diced up, die of each type are glued together to create cell stacks (see Figs. 6.9 and 6.10). These cell stacks are then inner lead bonded into a three-beam-set TAB tape (Figure 6.11). These lead sets contact the top and bottom of the top cell and the top of the

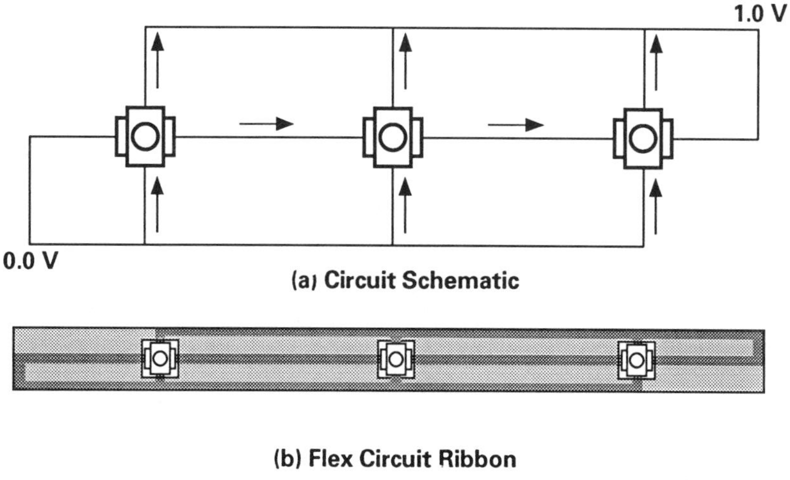

FIGURE 6.7 Triplet tandem cell circuit.

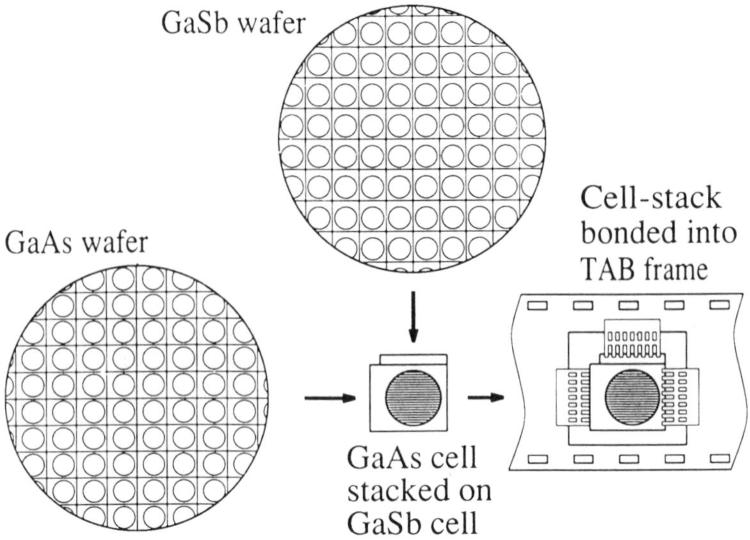

FIGURE 6.8 Tape automated bonding (TAB) of a cell stack: A GaAs cell is first adhesive bonded to a GaSb cell and then inserted into a tape site for inner lead bonding.

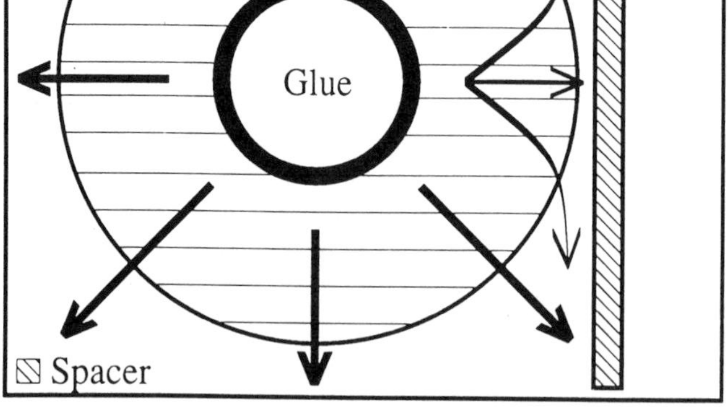

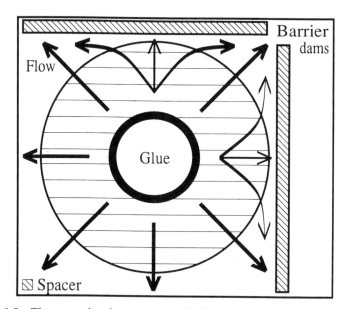

FIGURE 6.9 The spacer-barrier pattern on the GaSb chip provides electrical stand-off as well as a thin line for adequate thermal conductivity. The barrier dams also serve to keep glue away from the TAB contact surfaces.

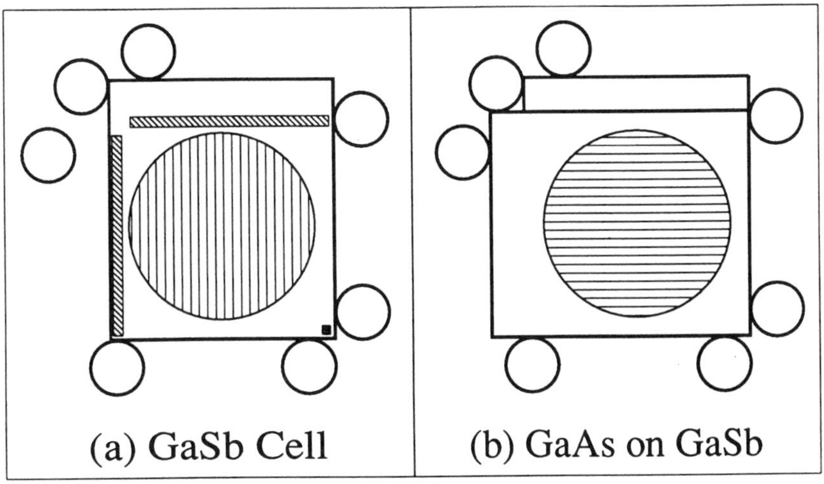

FIGURE 6.10 Cell-stack pin alignment jig is used during adhesive bonding of the GaAs cell to the GaSb cell. The alignment fixture accommodates multiple stacking sites.

bottom cell. The stack can then be tested in the tape by probing pads on the tape and the bottom of the bottom cell. A photograph of a high-efficiency GaAs/GaSb cell stack in a TAB frame is shown in Figure 6.12. This assembly process is quite straightforward and has the additional advantage over the monolithic stack that both cells in the stack can be independently measured.

6. FABRICATING III–V SOLAR CELLS

The relative cost advantages of different III–V cell fabrication schemes are often debated but seldom analyzed. In this section, I attempt to provide some quantitative cost comparisons that will prove useful in the next section of this chapter.

To quantitatively compare the different potential two junction cell stacks (e.g., Fig. 6.3 vs. Fig. 6.4), it is important to understand the fabrication processes used and their relative costs. In this section, the cell fabrication steps are broken down into three categories: crystal growth, junction formation, and photolithographic processing. Under the "crystal growth" label, I include the actual crystal growth, wafer slicing, and polishing. Under "junction formation", I consider both diffused junctions and epitaxially grown-in junctions. Under the "photolithographic processing" label, I include photolithography, dielectrics, metalizations, and all processes for cell fabrication up through dicing.

I am interested in addressing two types of cost questions. First, what are the relative costs? For example, what about crystal growth versus MOCVD epitaxy or MOCVD epitaxy versus diffusion? Second, what is the projected cell total fabrication cost? This section deals primarily with the relative cost trade

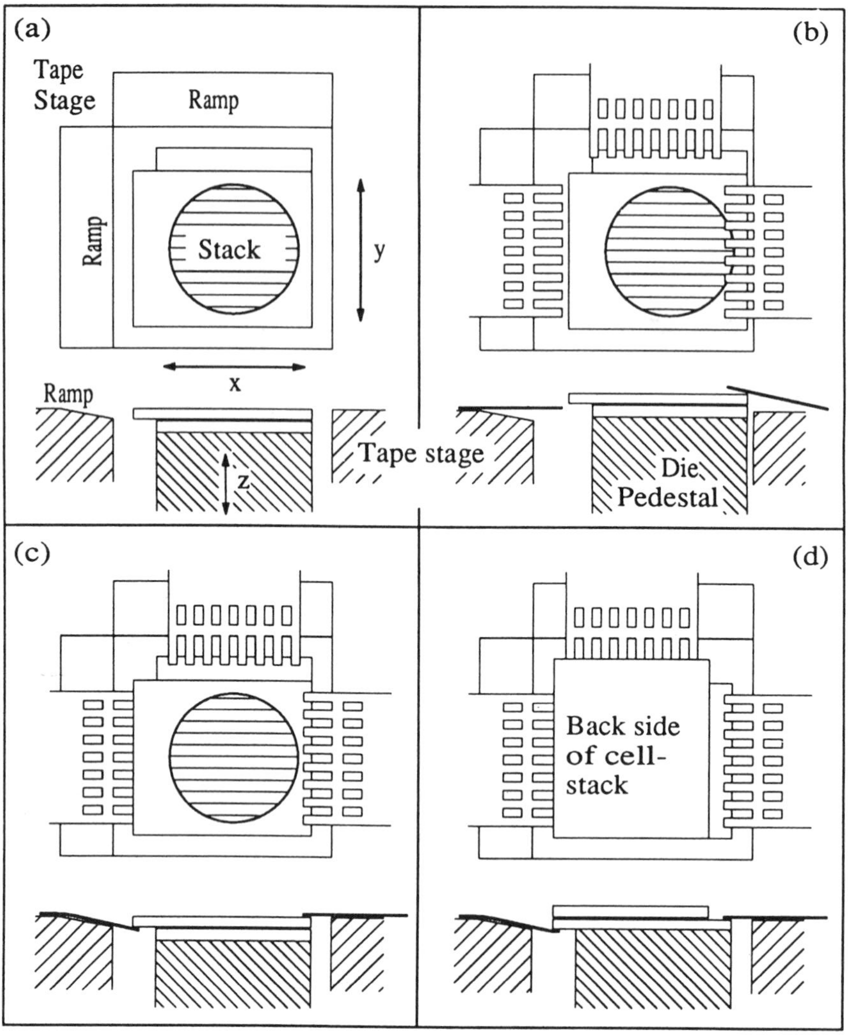

FIGURE 6.11 Three-dimensional TAB sequence is shown **(a–d)** for inner lead bonding of a cell stack into a tape site. The cell stack is mounted on a pedestal that is free to move within a square opening in the tape stage.

offs. The numbers presented will be in the context of the potential large-scale terrestrial utility application for concentrator tandem cells. In terms of relative cost, the cost trends will also apply to the space application. Section 7 of this chapter deals with the total cost versus allowed cost for the terrestrial utility application.

Table 6.2 presents our cost summary for the hypothetical assignment to process GaAs cells at a volume of 200,000 wafers (45 × 45 mm) per year. This corresponds to a production volume of 30 MW per year of GaAs cells

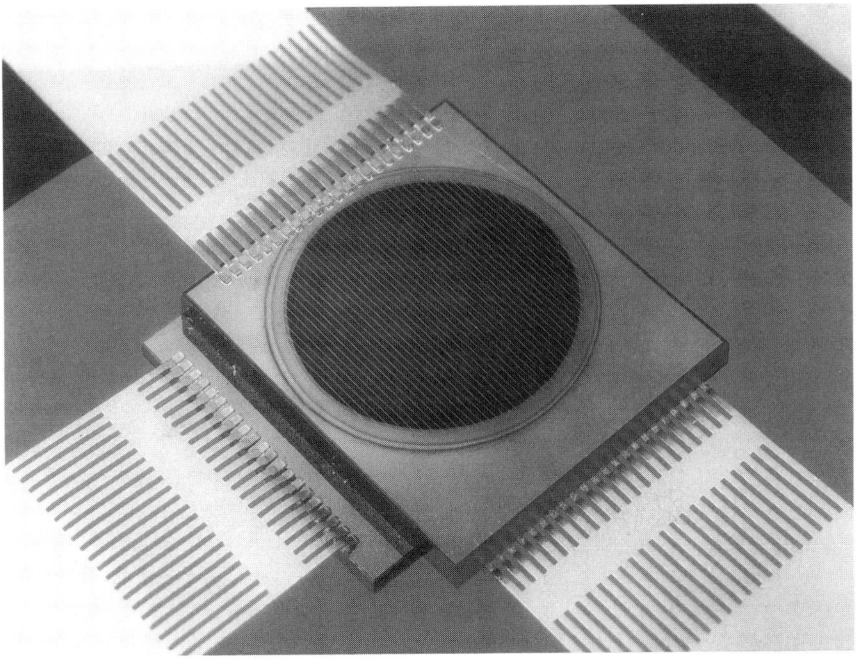

FIGURE 6.12 Cell stack packaging.

operating at 1,000 suns concentration. I begin by discussing Table 6.2 in the context of processing GaAs cells using the traditional MOCVD grown-in junction process discussed in Chapter 5.

Beginning with the crystal growth operation, we have assumed horizontal Bridgeman as the growth method and identified the cost of the equipment, labor, materials, and labor overhead required. We arrive at an in-house wafer cost of $17.60/wafer. We have also obtained an outside quote of $30/wafer (from Laser Diodes, 1992) for wafers produced at a rate of 60,000 per year. These numbers are roughly consistent given the lower vendor volume and a customary vendor markup.

Next, we evaluated the cost of growing a 5 μm thick epitaxial GaAs layer using MOCVD. We assumed a Cambridge Instruments barrel reactor as is used at Applied Solar Energy Corp. Again, cost of equipment, labor, and materials are listed. The final cost is $12.40/wafer. It is interesting that about half of this cost is in the MOCVD source materials. Also note that 700 pounds of arsine would be used per year. Given this safety negative, the real cost of epitaxy is then about the same as the cost of a wafer.

Table 6.2 also includes the cost of two typical process steps, i.e., 5 μm of evaporated silver metalization and a photolithography step. Unlike integrated circuit processing, the photolithography assumes mechanical alignment for the cell macrofeatures. The cost of either of these steps is low at $2 or less per

TABLE 6.2 Costs of Various III–V Wafer Fabrication Steps at Volume of 200,000 Wafers (45 × 45 mm)/ Year

	Crystal Growth	MOCVD (5 μm)	VCE (0.2 μm)	Diffusion	Metal Evaporation (5 μm)	Photo-lithography
Primary equipment						
Unit price	100 K	600 K	250 K	150 K	300 K	100 K
× Number ×	10	5	5	1	2	1
Depreciation	0.2	0.2	0.2	0.2	0.2	0.2
=Cost/year	200 K	600 K	250 K	30 K	120 K	20 K
Secondary equipment						
Total price ×	1,000 K	1,100 K	250 K	0	0	0
Depreciation	0.2	0.2	0.2			
=Cost/year	200 K	220 K	50 K			
Clean Laboratory						
Square feet	300	0	0	0	300	300
Cost/year	60 K				60 K	60 K
Materials						
1. Item	Ga+As	Reactants	Reactants	0	Silver	Resist
Cost/Wafer	5.50	5.00	0.20		0.35	0.75
2. Item	Quartz					
Cost/Wafer	1.50					
Direct Labor						
No. of operators	8	4	2	1	1	1
× No. of shifts	3	2	3	2	2	2
× Salary/year	25 K	25 K	25 K	25 K	25 K	25 K
Direct labor/year	600 K	200 K	150 K	50 K	50 K	50 K
Maintenance/year	150 K	150 K	50 K	25 K	25 K	15 K
Overhead labor						
Rate (%)	150	150	150	150	150	150
O.H./year	900 K	300 K	225 K	75 K	75 K	75 K
Total	$2.11 M	$1.47 M	$0.73 M	$180 K	$330 K	$220 K
$/Wafer	$17.60	$12.40	$3.85	$0.90	$2.00	$1.85

wafer. A curious result is that the photoresist costs more than the silver after accounting for the quantities required.

Next, since the junction in a GaSb IR cell is fabricated using diffusion (Fraas and Avery, 1990), we have evaluated the cost of diffusion in Table 6.1 as well. Since diffusion equipment is relatively inexpensive and has very high throughput, the cost per wafer is very low at $0.90/wafer. This low cost of diffusion has led us to explore a diffusion process for fabricating GaAs cells as well.

Our diffused junction GaAs cell fabrication process is shown in Figures 6.13 and 6.14. Our process begins with the assumption that an AlGaAs window layer is still required for low GaAs emitter surface recombination velocity. However, this 500 A thick window layer can be thought of as a surface

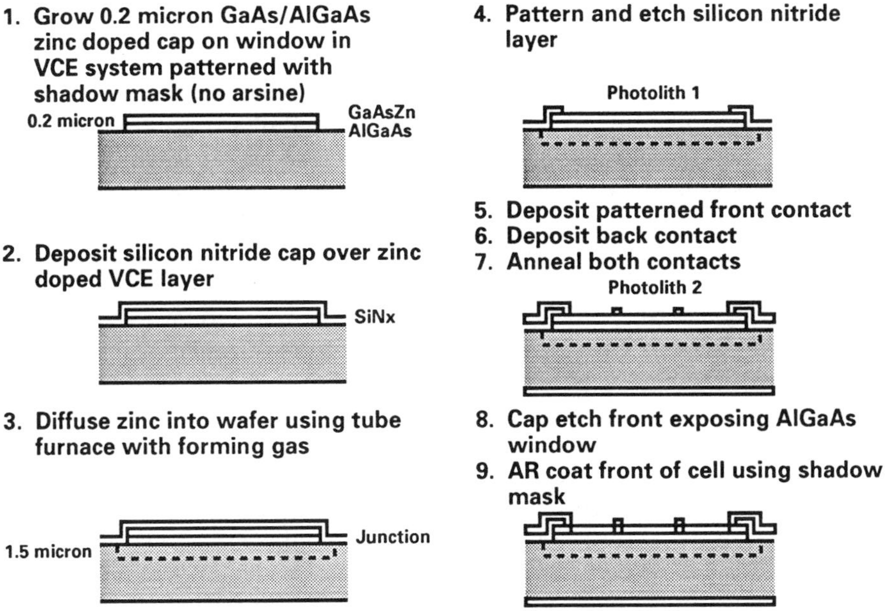
1. Grow 0.2 micron GaAs/AlGaAs zinc doped cap on window in VCE system patterned with shadow mask (no arsine)
0.2 micron
GaAsZn
AlGaAs
2. Deposit silicon nitride cap over zinc doped VCE layer
SiNx
3. Diffuse zinc into wafer using tube furnace with forming gas
1.5 micron
Junction
4. Pattern and etch silicon nitride layer
Photolith 1
5. Deposit patterned front contact
6. Deposit back contact
7. Anneal both contacts
Photolith 2
8. Cap etch front exposing AlGaAs window
9. AR coat front of cell using shadow mask

FIGURE 6.13 Process for zinc-diffused GaAs solar cell.

passivation layer, and, like an oxide on silicon, it can be rapidly formed in a simpler deposition system. A single zone vacuum chemical epitaxy (VCE) system (Fraas et al., 1990) using triethyl gallium, trimethyl aluminum, and evaporated arsenic, thereby avoiding arsine, would be ideal. It is further customary to grow a 1,500 A thick GaAs cap for making ohmic contact, and this can be accomplished in the same VCE process step. The crystalline quality of this layer is not critical. Although the growth of these two layers is required in addition to diffusion, the point is that the sum of these layers is only 0.2 μm, much less than the 5 μm thick sandwich required for the conventional MOCVD cell. Using a load-locked VCE machine, the growth time (including heat up and cool down times, for 0.2 μm is approximately one-fourth the time

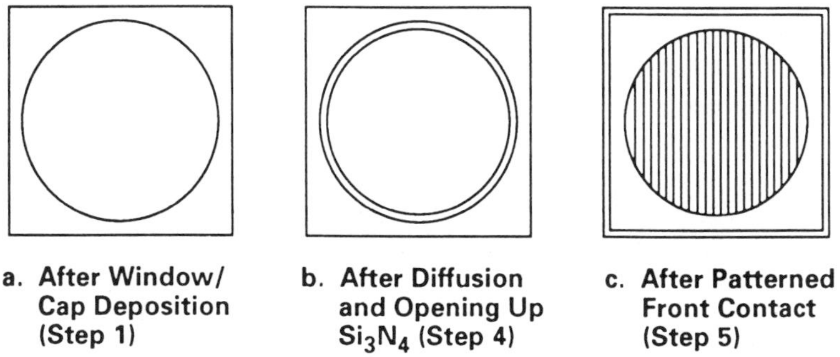
a. After Window/ Cap Deposition (Step 1)
b. After Diffusion and Opening Up Si_3N_4 (Step 4)
c. After Patterned Front Contact (Step 5)

FIGURE 6.14 Top view of GaAs cell through process.

for the conventional structure. The wafer throughput is therefore increased in less expensive, safer equipment.

Referring to Figures 6.13 and 6.14, this alternate process begins with the growth of a window-cap layer (step 1). It is important to note that the cap layer is heavily zinc doped and that the window-cap layer is deposited through a shadow mask only in the circular region to be illuminated in the center of each cell (Fig. 6.14a). The heavily zinc-doped patterned cap serves as the dopant source for the diffusion. In step 2, a silicon nitride layer is deposited over the front of the wafer. This layer serves to hold in the zinc and arsenic during the next diffusion step where the zinc is driven into the wafer to form the emitter and junction (step 3). Steps 4 and 5 are the two photolithography steps. In step 4, a hole is opened in the nitride (Fig. 6.14b), and in step 5 the patterned front metal is deposited (Fig. 6.14c). The nitride dielectric is left in place under the bus metal contact. Steps 4 through 9 are identical to the conventional MOCVD cell fabrication process.

Table 6.2 lists the cost associated with VCE and diffusion for this revised process. The sum cost is \$4.75/wafer. In addition to totally avoiding toxic gases, this process is considerably cheaper than the traditional MOCVD process.

The demonstrating of a 35% efficient solar cell is exciting, but is this cell intrinsically too expensive to be economically viable for terrestrial utility applications? To answer this question, it is first necessary to calculate allowable module and cell cost given the amount of electricity produced and the price of electricity. This calculation is done in Chapter 14 with the result that the tandem cell stacks can cost \$160/m^2 of optical collector area. Chapter 14 also notes that one 45 × 45 mm wafer of each type will populate a 1 m^2 module. Therefore, we have an allowance of \$80 per wafer.

From the information derived in Section 5 and summarized in Table 6.2, the cost of fabricating GaAs and GaSb solar cell wafers in quantities of 200,000 per year can now be calculated approximately. The results are presented in Table 6.3. The first column in this table refers to the fabrication of a GaAs cell wafer using MOCVD process. The \$23 cost for the bare wafer comes from the calculated wafer cost of \$17.60 (Table 6.2) divided by a 0.75 yield through the process. Similarly, the \$15 per wafer for MOCVD comes from the \$12.40 Table 6.2 value divided by a 0.8 yield. Since there is not enough information in Table 6.2 to calculate the photolithographic processing cost, we use the following estimation procedure. Yoo and Calaniss (1992) have estimated the cost of processing a silicon concentrator solar cell wafer through a five mask level process to be \$45 per wafer. Since the GaAs cell process involves three mask levels, I estimate the process cost to be 3/5 × 45 = \$27 per wafer. Summing these three cost contributions gives an estimated GaAs solar cell wafer cost of \$65, which is less than the \$80 allowance. Column 2 in Table 6.3 is again for a GaAs solar cell wafer. However, in this case the vendor-quoted wafer price of \$30 is used with the 0.75 yield assumption. The result is a GaAs processed solar cell wafer cost of \$82, still close enough to the target value. The independent calculated result presented in Chapter 5 is \$74 per wafer.

Column 3 in Table 6.3 presents the results for a GaAs solar cell fabrication process in which the safer VCE/diffusion process is substituted for the MOCVD process. The \$5 cost for junction formation comes from the \$3.85 Table 6.2 value divided by an 0.8 yield factor. The resultant GaAs solar cell wafer cost is then calculated to be \$55.

Column 4 in Table 6.3 is the calculated process cost for a GaSb solar cell wafer. Since no epitaxy is required, the junction formation process is cheaper so that the total GaSb solar cell wafer cost is projected to be \$51.

The cost of the two chips in a GaAs/GaSb cell stack is then projected to be \$55 + \$51 = \$106, which is well below the \$160 allowance and not much above the cost of an MOCVD GaAs cell (\$82) using purchased wafers. Note that MOCVD is a necessity for the $GaInP_2$/GaAs cell so that the cost of this cell wafer in high volume would be approximately \$82 as well. However, without a 1,000x tunnel junction, this cell is not presently a terrestrial concentrator cell option. There are, however, no technical obstacles for the GaAs/GaSb cell stack for terrestrial concentrator applications, and *our conclusion is that the two chip mechanical stack can be economical for terrestrial application.*

The difficulty today is that these costs can only be achieved with high volume production. In the above case, a 30 MW module business at \$1.50/W represents annual sales of \$45 million. Considerable investment in equipment and development will be required to build up this business. Although the commitment would be small compared with the F22 advanced tactical fighter program, it is still large for a new solar industry. Meanwhile, GaAs wafer costs are dropping as the number of wafers sold annually is growing through sales into the photonics and microwave industries.

8. CONCLUSIONS

Considerable progress has been made with two junction cell stacks over the last several years. GaAs/GaSb cell stacks have now been fabricated with

TABLE 6.3 Processed Solar Cell Wafer Cost (45 × 45 mm)

	Cases[a]			
	1	2	3	4
Crystal growth	23	40	23	23
Junction formation	15	15	5	1
Photolith processing	27	27	27	27
Total	\$65	\$82	\$55	\$51

[a] 1, GaAs with in-house crystal growth and MOCVD junction; 2, GaAs with purchased wafers and MOCVD junction; 3, GaAs with in-house growth and diffusion; 4, GaSb with in-house crystal growth and diffusion.

efficiencies over 35%. A 50% energy conversion efficiency is theoretically possible using three junction cells. The GaAs/GaSb cell stack is now being incorporated in space concentrator modules (see Chapter 14). Calculated cell costs suggest that these GaAs/GaSb cell stacks can be incorporated into terrestrial concentrator modules to produce economical utility scale electric power. Unfortunately, the financial commitment required to develop this market is on the order of $50 million. Meanwhile, the GaAs industry is evolving with photonic and microwave applications so that eventually a lower cost technology base might be able to address this electric power application.

REFERENCES

Benner, J.P. (1991), *22nd IEEE Photovoltaic Specialists Conf. Record,* IEEE, New York, pp. 7–11.

Bennett, A., and Olsen, L.C. (1978), *13th IEEE Photovoltaic Specialists Conf. Record,* IEEE New York, pp. 868–873.

Cavicchi, B.T., Krut, D.D., Lillington, D.R., Kurtz, S.R., and Olson, J.M. (1991), *22nd IEEE Photovoltaic Specialists Conf. Record,* IEEE, New York, pp. 63–67.

Chang, B.C., Virshop, G.F., Klausmeier-Brown, M., Ladle-Ristow, M., Wanlass, M.W. (1991), *22nd IEEE Photovoltaic Specialists Conf. Record,* IEEE, New York, pp. 54–57.

Fraas, L.M. (1985), *Current Topics in Photovoltaics,* Academic Press, New York, pp. 169–221.

Fraas, L.M., and Avery, J.E. (1990), *Optoelectronics Devices Technol.* **5**, 297–310.

Fraas, L.M., and Knechtli, R.C. (1978), *13th IEEE Photovoltaic Specialists Conf. Record,* IEEE, New York, pp. 886–891.

Fraas, L.M., Kuryla, M.S., Bigger, J.E. (1992), in *AIP 268, Photovoltaic Advanced Research and Development Project,* R. Noufi, Ed., pp. 39–46.

Fraas, L.M., Malocsay, E., Girard, G., Sundaram, V.S., Thompson, A. (1990), *J. Crystal Growth* **305**, 35–45.

Fraas, L.M., Partain, L.D., Cape, S.A., McLeod, P. (1987), *Solar Cells* **19**, 73–83.

Gee, J.M., and Virshup, G.F. (1988), *20th IEEE Photovoltaic Specialists Conf. Record,* IEEE, New York, p. 754.

Kim, N.P., Stewart, J.M., Stanbery, B.J., Mickelsen, R.A., Devaney, W.E., Chen, W.S., and Burgess, R.M. (1991), *22nd IEEE Photovoltaic Specialists Conf. Record,* IEEE, New York, pp. 68–72.

Olson, J.M., and Kurtz, S.R. (1990), *Appl. Phys. Lett.* **56**, 623–626.

Spitzer, M.B., McClelland, R.W., Dingle, B.D., Dingle, J.E., Hill, D.S., and Rose, B.H. (1991), *22nd IEEE Photovoltaic Specialists Conf. Record,* IEEE, New York, pp. 142–146.

Wanlass, M.W., Coutts, T.J., Ward, J.S., Emery, K.A., Gessert, T.A., and Osterwald, C.R. (1991), *22nd IEEE Photovoltaic Specialists Conf. Record,* IEEE, New York, pp. 38–45.

Yoo, H., and Calaniss, S. (1992), *Sandia Report SAND92-1454,* Sandia National Laboratories, Albuquerque, NM, p. 81.

CHAPTER SEVEN

a-Si:H-Based Solar Cells

A. CATALANO, National Renewable Energy Laboratory,
Golden, CO 80401

1. INTRODUCTION AND BACKGROUND

Although hydrogenated amorphous silicon (a-Si:H) is being used in a number of mature and emerging applications, ranging from active matrix flat-panel displays to neural networks, it owes its present popularity to the first practical use of the material in solar cells (Carlson and Wronski, 1976). The first solar cells were p/i/n devices prepared by the glow discharge decomposition of silane—a preparation method whose present widespread use, one might speculate, is also attributable to the success of the a-Si:H solar cell.

It is useful as a preamble to further discussion to enumerate the key elements that make a-Si:H an attractive photovoltaic device material, as well as some of the problems that require further development. Among the positive attributes of a-Si:H we may include

- A very high optical absorption coefficient ($> 10^5\,cm^{-1}$) over the visible spectrum, making submicron thick devices practical and economical
- A very high optical absorption coefficient ($> 10^5\,cm^{-1}$) over the visible spectrum, making submicron thick devices practical and economical
- A simple low temperature (approximately 250°C) deposition process that can be used to coat uniformly even highly textured surfaces with extremely thin (100 Å) films (the overall process is easily adapted to either continuous or batch processing)
- In keeping with the spirit of renewable energy, a process that is environmentally benign and creates virtually no hazardous wastes

Solar Cells and Their Applications, Edited by Larry D. Partain.
ISBN 0-471-57420-1

- Virtually limitless starting materials
- Easy monolithic integration and patterning of the devices
- Easy to make alloys with other group IV elements, allowing the optical gap to be raised or lowered.

Offsetting these substantial attractive features are several liabilities:

- a-Si:H is subject to a gradual degradation in performance as a result of light exposure or electrical (forward) bias. The degradation is easily reversed by heating at temperatures above 100°C
- The electrical transport properties are inferior to comparable crystalline materials but adequate for use in solar cells and many other applications
- Degenerate doping to produce highly conducting films has not been possible as a result of a mechanism by which compensating defects are generated. Resistivities in the range of 10^2–10^3 Ω-cm are, however, possible.

The first a-Si:H–based solar cells were p/i/n devices of low efficiency. Substantial gains in performance were made by temporarily discarding the original p/i/n device in favor of the simpler Schottky diode that enabled conversion efficiencies to reach 6% (Carlson, 1977). Improvements in the process of doping a-Si:H, however, permitted p/i/n homojunction devices to surpass the performance of the Schottky diode (Carlson, 1980). Efficiencies in the range of about 8% are attributed to the development of a heterojunction p/i/n device that utilized an amorphous silicon carbide (a-SiC:H) p-layer to form the light-incident front junction (Hamakawa et al., 1982). The important psychological barrier of 10% conversion efficiency was obtained by preparing heterojunction devices on textured substrates to yield a markedly higher current (Catalano et al., 1982). Grading the optical band gap of the p/i interface to reduce interface recombination has led to a substantial increase in open circuit voltage, short wavelength response, and fill factor and enabled efficiencies in the range of 12% to be demonstrated (Catalano et al., 1987a, b). Although efficiencies as high as 12.65% have been reported, the published quantum efficiencies integrated over the global AM1.5 spectra do not yield the reported current, and these results may be questionable (Arai et al., 1991). Substantial improvements in conversion efficiency and stabilized performance appear possible through the use of multijunction, multi-band-gap devices. Although this work is at an early stage, impressive results, including conversion efficiencies as high as 13.7%, have been reported for triple junction, multi-band-gap devices (Yang et al., 1988; Catalano et al., 1989; Ichikawa et al., 1991). Reports of these high efficiencies and improvements in the stabilized performance have largely been responsible for the efforts to develop wide and narrow band gap alloys based on a-Si:H. Figure 7.1 illustrates the progress in improving the initial efficiency of a-Si:–

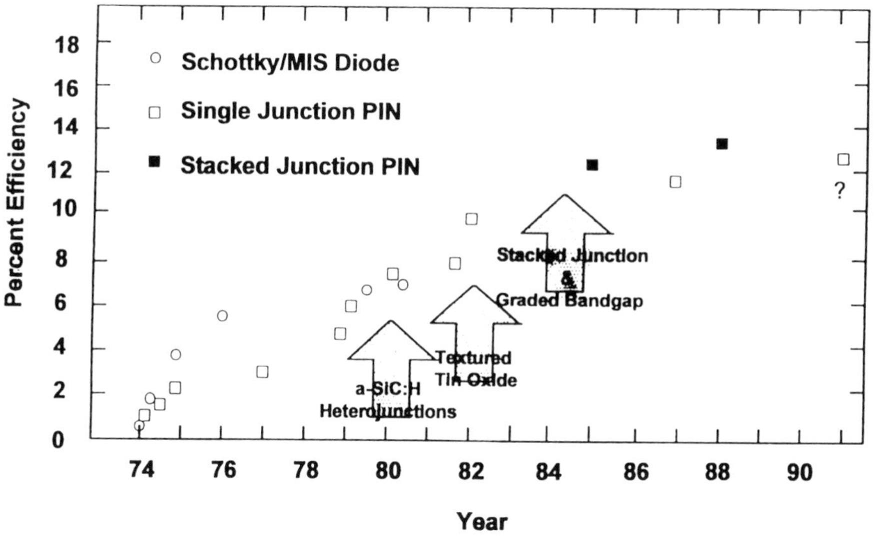

FIGURE 7.1 Progress in the initial efficiency of amorphous silicon solar cells.

based solar cells. More recently, emphasis has been placed on *stabilized* efficiency, which is a less well-defined term.

The first commercial solar cell products based on a-Si:H arrived in the marketplace in the early 1980s in the form of small (a few square centimeters) monolithically integrated modules that were incorporated in handheld calculators. The availability of these small modules, the advent of the low-power CMOS microprocessor, and the twisted nematic liquid crystal display combined to form a unique product that today is both low in cost and ubiquitous. By some estimates, sales of a-Si:H solar cells in these and related products may have, at its peak, pushed the sales volume of these cells beyond that of their crystalline counterpart. Since that auspicious beginning a-Si:H has begun to compete with crystalline silicon for small area consumer electronics applications dominated by battery charging. Although large area a-Si:H arrays have been deployed, the technology utilized has been relatively immature, and the low array efficiency (approximately 3%–4%) make them unattractive. The emerging multijunction, multi-band-gap technology, which is discussed in more detail in this chapter, has the potential to repeat for terrestrial power generation what has been done for calculators.

2. AMORPHOUS SILICON FUNDAMENTALS

Amorphous semiconductors, for the most part, have not been part of the solid state device revolution because of their poor electrical transport properties and the difficulty in changing the conductivity type by doping. The former problem may be ascribed to the heavy concentration of defects present in amorphous materials. These high concentrations of defects originate due to the relaxation of the constraint of a periodic lattice with rigidly assigned atomic positions. "Errors" in arrangement translate into strained, weak, or nonexistent bonds with the consequent wide array of corresponding defects that frustrate transport and encourage carrier trapping and recombination. Difficulty in doping arises from the so-called 8-N rule of Mott (1967), which anticipates that the lattice will adjust to satisfy the needed bonding of added impurities. a-Si:H is successful as a semiconductor material because it avoids these limitations by virtue of the addition of hydrogen to complex with the defects implicit in the consequences of disorder. The limitations imposed by disorder, however, are not entirely avoided because the ability of hydrogen to complex defects also affects those desirable unsatisfied bonds arising from substitutional doping. The light-induced degradation likely emerges from the same root cause. Hydrogen incorporated over a range of bond strengths arising from disorder gives rise to bonds with a distribution of strengths. Recombination of electron–hole pairs at higher energy may release their energy locally, breaking these relatively weak bonds.

The incorporation of hydrogen into the growing film is essential to obtaining a device quality film. Films deposited by means such as electron beam deposition yield extremely high defect levels as a consequence of the lack of hydrogen (Lecomber and Spear, 1979). A variety of methods have been used to prepare hydrogenated a-Si, and in the following section several of these are reviewed.

2.1. Preparation

The deposition of device quality, a-Si:H requires, as a minimum, a method that permits low impurity levels by such contaminants as Na, O_2, C, and N_2, allows control of the critical film thickness appropriate for devices (10–100 Å), and allows excellent uniformity for the large area solar cells ($<1\,m^2$) required. We take as an absolute requirement that hydrogen be incorporated in the proper fashion and to the required degree. Few methods, besides the plasma-enhanced chemical vapor deposition (PECVD) or "glow discharge" (GD) deposition have proven capable of meeting all these requirements. Table 7.1 gives a summary of the preparation methods that have been utilized, as well as the resulting device performance obtained for solar cells.

Although photo-CVD and sputtering have proven to be attractive alternatives to the GD process, the methods impose added limitations. Photo-CVD requires significant engineering to overcome the difficulties associated with admitting the ultraviolet light through a window into the vacuum system and in forming uniform films. Sputtering, which permits high rate deposition and adjustment of the hydrogen content via control of its partial pressure during film formation, generally requires higher hydrogen content and results in wider band gap material than the GD process. This inevitably leads to poorer device

TABLE 7.1 Deposition Methods Employed for a-Si:H Solar Cells

Method	References	Temperature	Rate	Comments
RF, DC glow discharge	Madan (1989) and Catalano et al. (1987a)	~250°C	18 Å/s	Excellent uniformity: 9.74% at 18 Å/s, 12% at 4.4 Å/s
Microwave glow discharge	Watanabe et al. (1987a, b)	200°C	15 Å/s	Films only
ECR glow discharge	Hanaki et al. (1987), Hattori et al. (1987), Ichikawa et al. (1987), and Kitagawa et al. (1987a, b)	300°C	High	Doped layers, μx films
Controlled plasma magnetron CPM	Nishikuni et al. (1988)	200°C	15 Å/s	8.2% (100 cm^2)
Photolytic decomposition	Saitoh (1982), Yoshikawa and Yamaga (1987), Konagai (1987), and Konagai et al. (1988)	250°C	1–3 Å/s	11.5% μx films prepared
Sputtering	Muller et al. (1984) and Rudder et al. (1983, 1984)	250°–300°C	~3 Å/s	5.5%
Cluster beam evaporation	Anderson et al. (1987)			H ambient, electrical bias, films only
Catalytic chemical vapor deposition	Matsumura (1987)	300°C	30 Å/s	Films only
Chemical vapor deposition	Hegedus et al. (1986) and Scott et al. (1986)	150°C–350°C	~1 Å/s	4%

performance. Recent improvements in the quality of sputtered amorphous films have been reported. (Pinarbasi et al., 1990), suggesting that film properties and stability of such material may be superior to those prepared by glow discharge. However, these conclusions are based on measurements of mid-gap state density, not device measurement, and thus must be considered tentative.

Another novel method of preparing a-Si:H, termed the *hot-wire* (HW) *assisted chemical vapor deposition method*, has shown promising results as indicated by measurements of film properties (Mahan et al., 1991). In this method, a high temperature tungsten filament is used to dissociate silane gas at low pressure. The method is unique in that minority-carrier diffusion lengths of 1,000–2,000 Å are obtained in films containing 1% hydrogen compared with the 10% hydrogen usually found in high-quality films. Unfortunately, these impressive properties have not, as of this writing, been confirmed by device measurement.

Variations of the GD method, such as remote plasma deposition, electron-cyclotron resonance (ECR) deposition, and other magnetically confined methods, have also provided additional control of the deposition process. In the remote plasma deposition process, an active species, usually atomic hydrogen, is produced remote from the substrate onto which film growth occurs. The reactive species is then mixed with silane gas or other silicon-containing gaseous media to form the a:Si film. Studies indicate that the hydrogen content of the samples can be controlled to a greater degree than the conventional GD method and that the resulting films have superior properties as evidenced by measurement of infrared absorption and photoconductivity (Parsons et al., 1987; Lucovsky and Tsu, 1987). Unfortunately, little information on devices is available to support these conclusions, which like so many, must remain in question until more information is obtained.

ECR deposition enables a high rate of deposition due to the high plasma power density afforded by confinement of electrons to a circular path in the plasma. Although high rates of deposition of amorphous films have been obtained with magnetic confinement (Nakano et al., 1988; Shing and Pool, 1990), the use of the method has also permitted the preparation of microcrystalline silicon carbide p-type films. The demonstration of high-voltage p/i/n devices (0.965 V) has been attributed to the use of this method (Hattori et al., 1987). Other magnetically confined methods have shown promise in preparing high-efficiency devices.

2.2. Defects and Doping

As discussed earlier, the relaxation of the requirement of a periodic lattice permits the local bonding to change to accommodate impurities such as dopants. Thus, the majority of amorphous semiconductors cannot be doped. Although a-Si:H is exceptional in this matter, it cannot be degenerately doped. Street (1982) and Street et al. (1985) have provided a quantitative explanation for the phenomenon that is based on the equilibrium between compensating acceptor-like states and the ionized dopant in, for example n-type phosphorous doped material. The neutral threefold coordinated phosphorous atom (P_3^0) becomes ionized in a fourfold site (P_4^+) expressed as

$$P_3^0 = P_4^+ + e^- \tag{1}$$

The resulting electron, however, can be accommodated at a lower energy via a reaction with fourfold coordinated silicon yielding a compensating defect according to

$$e^- + Si_4^0 \rightleftharpoons D^- \tag{2}$$

Assuming that charge neutrality must be maintained by the system so that

$$(P_4^+) = (e^-) + (D^-) \tag{3}$$

one can readily show that the defect density D^- varies with $[P_3^0]^{1/2}$ according to the relation

$$(P_3^0) = \alpha'\beta[1 + \beta/(Si_4^0)](D^-)^2/(Si_4^0) \tag{4}$$

where α' and β are the rate constants for Eqs. (1) and (2), respectively. The electron density N_e is then given by

$$N_e = \beta(D^-)/(Si_4^0) = (P_3^0)^{1/2}\{\beta/\alpha'[(Si_4^0) + \beta]\}^{1/2} \tag{5}$$

At high concentrations $P_4^0 \ll P_3^0$, and the defect concentration should increase as the square root of the concentration. This has been verified by experiment. This self-compensating mechanism fixes the Fermi level between the D^- and the donor level and frustrates degenerative doping. A similar argument can be made to explain the difficulty in doping the system p-type. In general, the conductivity activation energy and resistivity are both higher for the p-type material. This is believed to arise from the lower hole mobility and the larger tail density of states associated with the valance band (Dressnandt and Rothwarf, 1984).

2.3. Theory

The theoretical treatment of a-Si:H–based devices is far more complicated than the corresponding crystalline device. The complication arises primarily due to the difficulty in simulating mathematically the somewhat uncertain nature of energy states in the material. Unlike crystalline material in which localized energy states occupy a well-defined energy and band edges are sharp, amorphous semiconductor energy states are more often better described by distributions. To make matters more complicated, the occupancy states and the dimensions of these contributions vary with Fermi level position and doping level, not to mention the preparation conditions (Moustakas, 1984; Street and Winer, 1989). A schematic illustration of the density of states profile is given in Figure 7.2.

Swartz (1982) was among the first to propose a comprehensive model of the most important amorphous silicon solar cell, the p/i/n diode. The model assumed a constant lifetime throughout the i region and a single level for recombination. Crandall (1983) proposed a model based on drift length assuming a uniform electric field. The current collection was determined by an effective collection length, L_c,

$$L_c = L_e + L_h \tag{6}$$

where L_e and L_h are the electron and hole drift lengths, respectively. This deceptively simple approach was shown to be useful in predicting the fill factor of devices (Faughnan and Crandall, 1984) and has proved useful on an ad hoc basis for establishing a link between defect density and device performance (Smith and Wagner, 1985).

Hack and Shur (1985) advanced a more comprehensive model that included a continuous exponential distribution of deep localized donor- and acceptor-like states in the gap, using a minimum density of states of $10^{16}\,cm^{-3}\,eV^{-1}$.

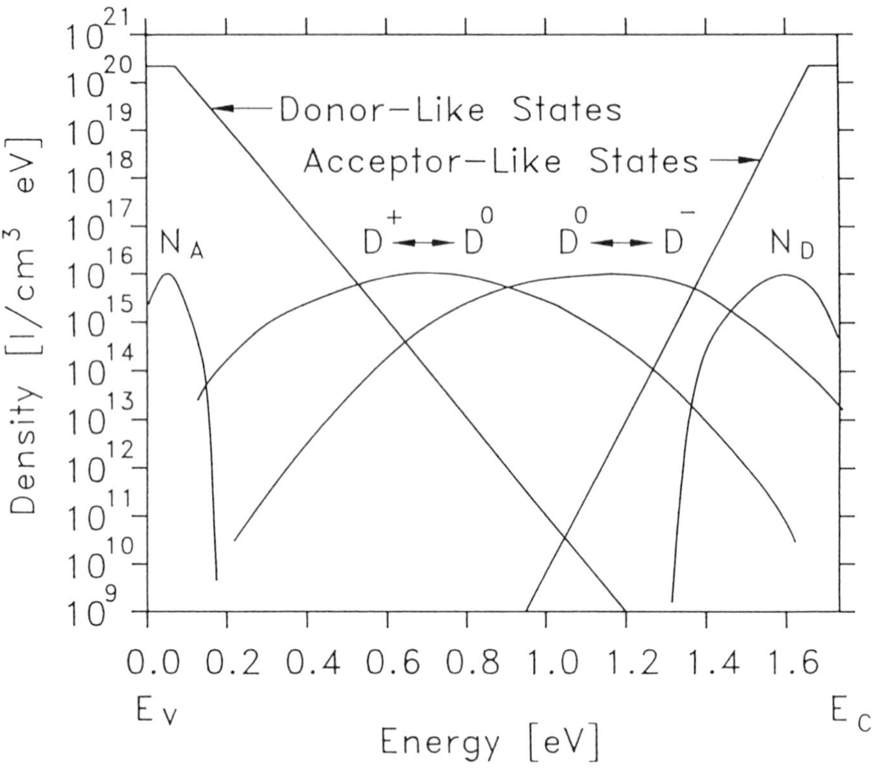

FIGURE 7.2 Plot of the energy state distribution in amorphous silicon.

The doping levels of the p and n layers and the effect of nonuniform optical excitation were also included. The band mobilities were assumed to be spatially invariant in this model. Based on these calculations Hack and Shur (1985) concluded that the principal loss mechanism in the device is bulk recombination determined by the free carrier densities. Among the especially significant conclusion of this work was the demonstration that transport is determined by the "limiting" carrier, in this case holes rather than the sum of drift lengths. This conclusion provided the explanation for the commonly observed phenomenon of lower collection efficiency measured via illumination through the n layer compared with the p layer of a p/i/n structure. Unfortunately, interface recombination was not explicitly treated in the device. More recently, Schwartz et al. (1989) proposed a numerical model that included both recombination via "dangling" bond states in the midgap region and via interface states. The distribution of midgap states was assumed to follow a gaussian distribution. Among the significant conclusions of their work is that bulk recombination through midgap states is the dominant mechanism through which minority carriers are lost, followed by interface recombination.

This model has been further refined (McElheny et al., 1988; Chatterjie et al., 1990) to include a more general set of boundary conditions. The model is referred to as *analysis of microelectronics of photonic structures* (AMPS). It is especially noteworthy in that it permits almost unlimited flexibility to the user in terms of grading materials properties and other parameters. Unlike previous models AMPS includes contact effects on device performance and explicitly includes tunneling and direct band-to-band recombination, although Auger recombination is not accounted for. It has also been extended to permit treatment of multijunction devices (Hou and Fonash, 1991). Although the use of this model has only just begun, it has already predicted unanticipated observations confirmed by later experiments. For example, Hou and Fonash predicted a photogating effect that would lead to greater than unity effective quantum efficiencies. This was explained as due to trapped charge generated with AC light modulating the flow of current generated by a DC bias. Experiments have verified this phenomenon.

3. AMORPHOUS SILICON-BASED ALLOYS

The lack of well-defined crystal structure in amorphous materials relaxes, to a large extent, the requirement of stoichiometry. The concept of well-defined phases in the solid state also disappears. Hence, although crystalline silicon nitride forms a single phase only at the composition Si_3N_4 and silicon carbide at SiC, the amorphous silicon-based alloys encompass the range a-Si_{1-x}:C_x:H and a-Si_{1-x}:N_x, where $0 \leqslant x \leqslant 1$. The hydrogen containing alloys have interesting and important potential device applications. For solar cells, the ability to alloy silicon with nitrogen or carbon to achieve a wide band gap and with germanium to obtain a lower band gap alloy permits the fabrication of multijunction, multi-band-gap devices.

Substantially higher conversion efficiency can be obtained by using multi-band-gap, multijunction devices. Kuwano et al. (1982) have calculated that a two terminal tandem device with maximum conversion efficiency of 21% should be possible using a low gap (1.1 eV) alloy. With a triple junction structure employing 1.7 eV a-Si:H in the middle junction the same authors calculated that an efficiency of 24% could be obtained by combining ordinary a-Si:H with both high and low band gap cells. These results are summarized graphically in Figures 7.3 and 7.4.

Although the optical band gap of a-Si:H can be adjusted over a narrow range near 1.7 eV by varying substrate temperature to adjust hydrogen content, alloys with germanium and carbon offer far greater flexibility to obtain lower and higher band gaps, respectively. a-Si_{1-x}:C_x:H alloys incorporated as the i layer in p/i/n devices exhibit open circuit voltages (V_{oc}) that increase monotonically with optical band gap. Unfortunately, the transport properties become worse as the carbon content increases, leading to a loss in fill factor and eventually even V_{oc} as shown in Figure 7.5. Nonetheless, V_{oc} as high as

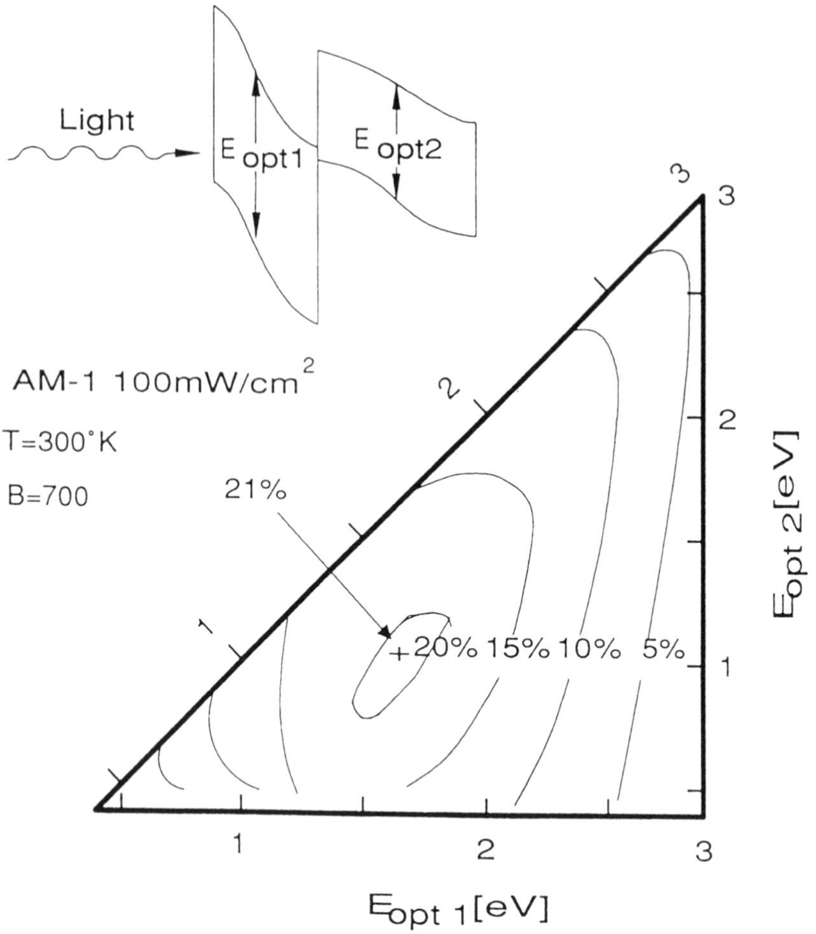

FIGURE 7.3 Theoretical conversion efficiency of a two terminal, dual junction, dual band gap solar cell. The maximum conversion efficiency of 21% is obtained for a 1.75 eV/1.15 eV band gap combination.

1.05V have been obtained (Catalano et al., 1987b). The loss in fill factor is accompanied by other physical findings that suggest their origin. For example, the Urbach energy, which is a measure of the width of the conduction and valence band tail state density, increases with increasing carbon content. Li et al. (1991) have shown that the structure of the a-SiC:H plays an important role in determining the electrical and optical processes and have related changes in atomic structure to process changes. Figure 7.6 shows the infrared absorption spectra of a-SiC:H films prepared from several sources. Films grown from an undiluted mixture of silane and methane, for example, show absorption in the infrared characteristic of CH_2 and CH_3 bonds, as well as SiH_2 bonds. Such films also exhibit high rates of light-induced degradation. The use of hydrogen dilution substantially reduces the occurrence of such bonds leading to an

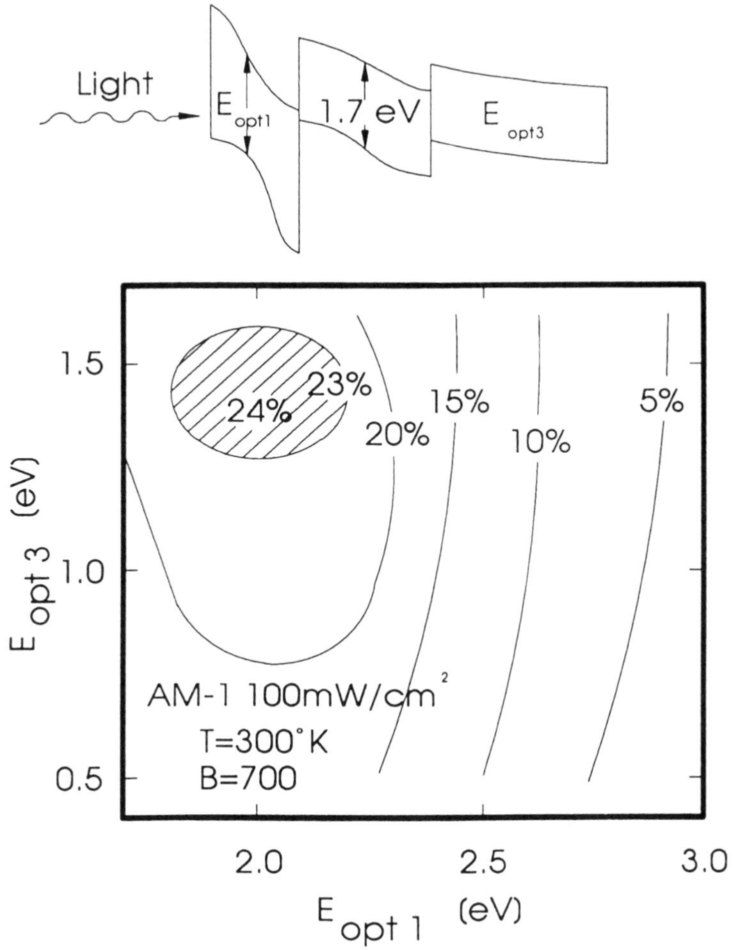

FIGURE 7.4 Theoretical conversion effciency of a two terminal, triple junction cell with 1.7 band gap middle junction cell. The optimum conversionefficiency of 24% is obtained for $E_{g1} = 2.0\,eV$ and $E_{g3} = 1.45\,eV$.

improvement in optical and electrical properties as shown in Figures 7.7 and 7.8. Given the obvious importance of structure in determining both electrical and optical properties, novel feedstocks such as disilyl and trisilyl methane (DSM, TSM) have been used to "build in" the desired Si–C bonds in the gas-phase precursor to solid formation. The evidence that supports the hypothesis that such precursors lead to improved electrical transport and optical properties is also shown in Figures 7.7 and 7.8 for the Urbach energy and minority-carrier diffusion length.

The low band gap alloy a-Si_{1-x}:Ge_x:H exhibits a similar structure–property relationship, and dilution of the feedstocks GeH_4 and SiH_4 with hydrogen also plays a strong role in improving quality. Yang et al. (1991) have

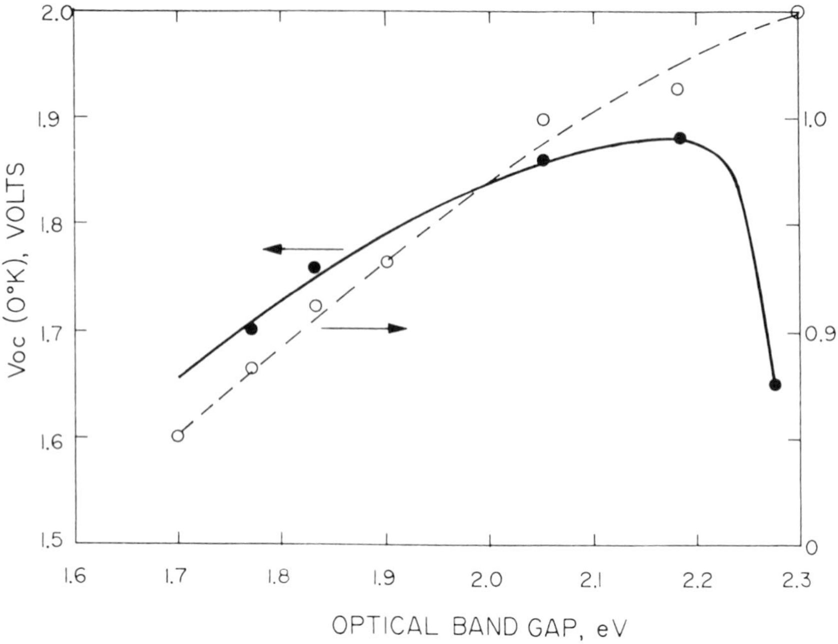

FIGURE 7.5 Open circuit voltage at T = 0°K (extrapolated, closed circles) and T = 298°K (open circles) versus optical band gap of a-SiC:H devices.

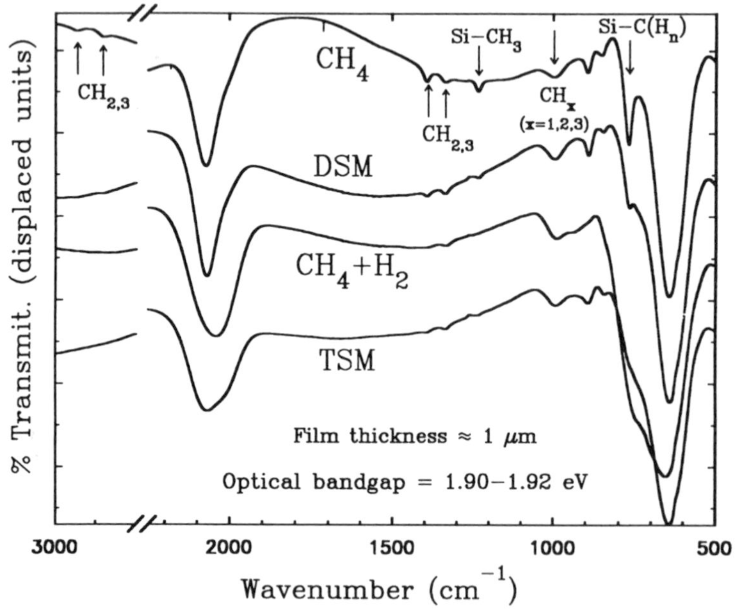

FIGURE 7.6 Infrared transmission spectra of a-SiC:H alloys prepared from various feedstock gases. All films have an optical band gap near 1.90–1.92 eV. DSM and TSM refer to disilyl methane and trisilyl methane, respectively.

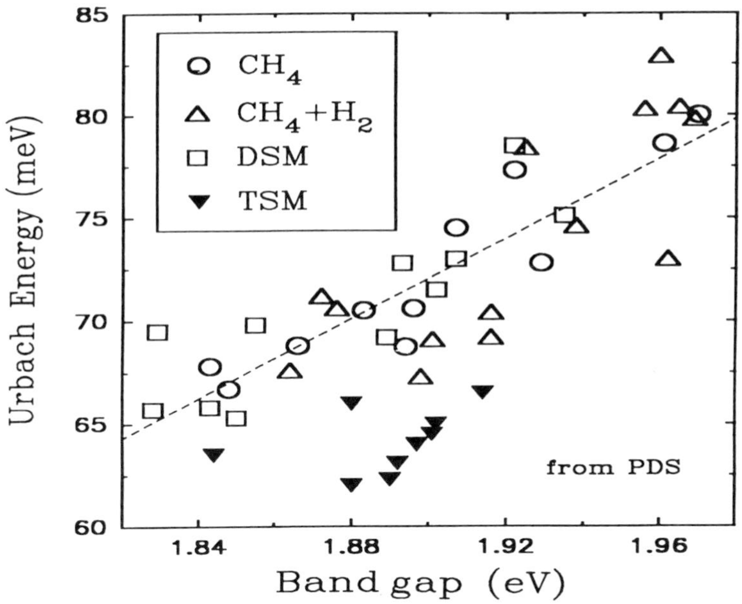

FIGURE 7.7 Urbach energy versus band gap for a-SiC:H films based on different feedstocks.

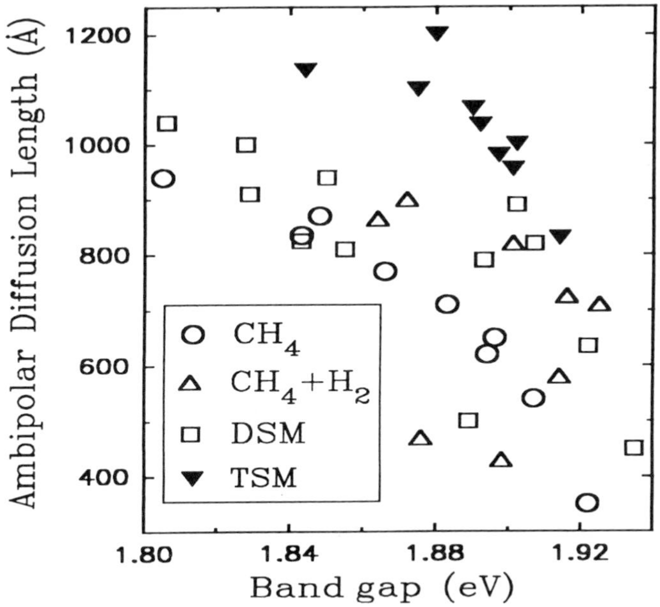

FIGURE 7.8 Hole diffusion length versus band gap for a-SiC:H alloys from different feedstocks.

shown, for example, that the addition of GeH_4 to a silane glow discharge produces films whose infrared spectra are indicative of substantial silicon dihydride bonding. The use of hydrogen reduces the concentration of dihydride bonds and substantially improves transport properties as shown in Figures 7.9 and 7.10, which plot the $\mu\tau$ product derived from photoconductivity measurements and the ambipolar diffusion length. Raman spectroscopy of a-Si_{1-x}:Ge_x:H films reveals that without hydrogen dilution there is a higher fraction of Si–Si and Ge–Ge bonds than would be expected for a homogeneous alloy. This preferential bonding may arise from the much greater ease in dissociating GeH_4 than SiH_4.

Devices of a-SiGe:H appear to be somewhat more complicated than ordinary a-Si:H p/i/n devices. Guha et al. (1989), for example, have suggested that the conventional wisdom of grading the p/i/n diode from high to low band gap is inappropriate for a-Si_{1-x}:Ge_x:H devices. Instead, it is suggested that grading the band gap from high to low from both p/i and n/i interfaces is necessary to obtain the highest efficiency. Computer simulation of the device indicated that such profiles effectively maintained a uniform electric field throughout the i region, thereby enhancing carrier collection. Arya et al. (1989) also studied the effect of grading and obtained similar results.

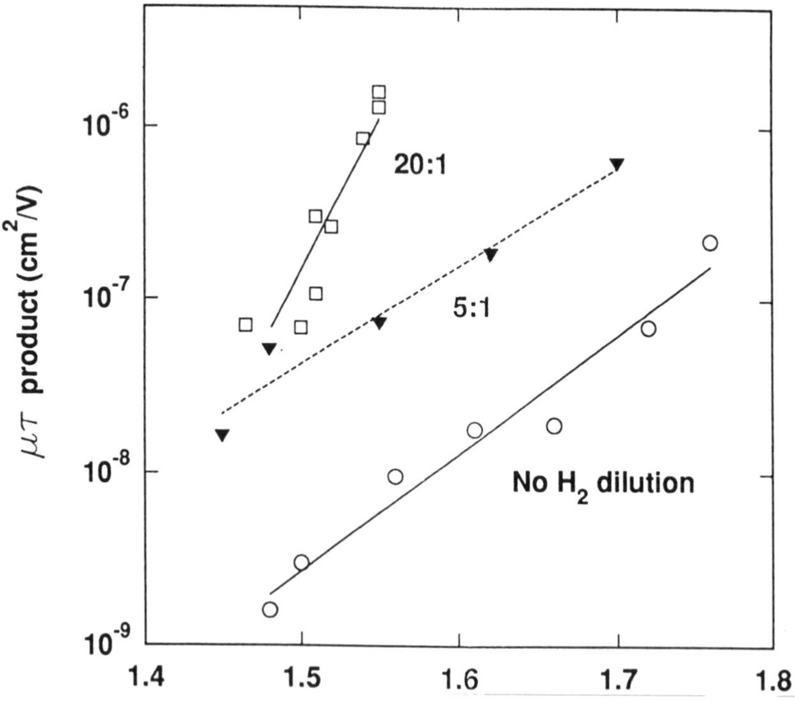

FIGURE 7.9 $\mu\tau$ product of a-SiGe alloys prepared with various hydrogen dilution ratios.

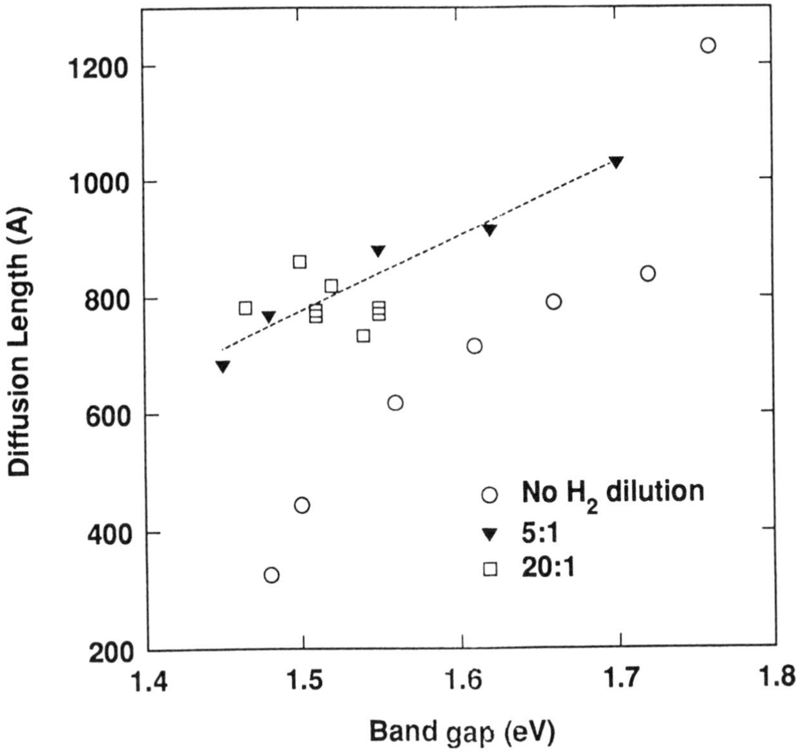

FIGURE 7.10 Minority-carrier diffusion length of a-SiGe:H alloys prepared with various hydrogen dilution ratios.

Incorporating the alloys in multijunction devices has indeed led to high conversion efficiency. For example, Yang et al. (1988), reporting on small area diodes, mentioned a record conversion efficiency of 13.7% for a triple junction dual band gap device comprised of a-Si:H for the top two junctions and a-Si_{1-x}:Ge_x:H for the bottom cell. In this device, the amorphous semiconductor layers were deposited onto a textured, highly reflecting stainless steel substrate overcoated with silver and ZnO, using a transparent conducting top contact. Large area triple junction modules with laser scribed interconnects have also been fabricated that exhibit a conversion efficiency of 10.85% (Catalano et al., 1992).

3.1. Stability of a-Si:H

Prolonged light exposure induces metastable changes in a-Si:H films and devices usually referred to as the Staebler-Wronski effect (SWE) (Staebler and Wronski, 1977). Although device performance and other transport properties return to their original values by heating above 100°C, this still presents the most serious limitation to the widespread use of a-Si:H solar cells. Light exposure has been shown to increase the density of dangling bonds via electron spin resonance (ESR) and reduce minority-carrier diffusion length (Dresner et al., 1981). A quantitative treatment of the kinetics of light-induced formation of defects has been presented by Stutzmann et al. (1984, 1985), who also reported on the effects of impurities, light exposure and so forth. The process of light-induced defect formation is viewed as one in which light provides the necessary energy to reconfigure the system at a higher energy level that is essentially "frozen in" at room temperature but that relaxes to its ground state via annealing. The process can also take place via carrier recombination when electrical stress is applied, e.g., via forward bias of a p/i/n diode. It is likely that carrier recombination and the damaging effect of the energy released are the root cause of defect formation. Stutzmann et al. suggested that the concentration of defects N_s, determined via ESR, increases quantitatively with time, t, and generation rate, G, according to the equation

$$N_s \propto t^{1/3} G^{2/3} \tag{7}$$

The process appeared to be self-limiting, and it was suggested that recombination of the excess carriers via the photoinduced carriers may successfully compete with the degradation process or be limited by the number of "defective" sites available. The kinetics of the annealing of the defects has also been studied (Stutzmann et al., 1986) and found to be a monomolecular process in which the rate R is activated according to the relationship

$$R = \nu_0 \exp(-E_a/kT_a) \tag{8}$$

where ν_0 is the attempt to anneal prefactor, T_a is the annealing temperature, k is Boltzmann's constant, and E_a is the activation energy. The lack of an exact fit suggested a distribution of energies rather than a single value.

Redfield and Bube (1989, 1990), in a series of papers, argue that a more accurate quantitative treatment can be obtained by a stretched exponential expression that implicitly recognizes the self-limiting nature of the process according to the relation

$$N(t) = N_{ss} - (N_{ss} - N_0)\exp[-(t/J_0)^{\beta}] \tag{9}$$

where N(t) is the defect density at time t, N_{ss} is the saturated defect density, N_0 is the initial defect density, J_0 is an effective time constant, and β is the stretch parameter. Unlike the $t^{1/3}$ model, in whch saturation was assumed to occur because the direct recombination process responsible for degradation was frustrated by other recombination routes, the stretched exponential model explicitly includes saturation.

Smith and Wagner (1985) combined the defect creation model of Stutzmann et al. with the device model of Faughnan and Crandall (1984) to predict the dependence of device efficiency on time. Also included in that treatment was a prediction of the effect of the i-layer thickness of the p/i/n device on the rate of degradation. Unfortunately, this did not yield results comparable to experiment, and an ad hoc model was produced to cover the wide range of device design parameters needed for practical applications (Catalano et al., 1985). The latter model predicted that for exposure times up to about 1,000 hours the device efficiency η at time t decreased from its initial value according to

$$\eta/\eta_0 = 1.1 - K \log t/t_0 \tag{10}$$

where the rate constant K depended on the thickness of the i layer of the p/i/n device,

$$K = 0.165\, d^{0.54} \tag{11}$$

Most recently, Yang and Chen (1991), using high intensity illumination, showed that the stretched exponential model can be used to describe device degradation and verified that saturation also occurs in devices. Figure 7.11a plots the normalized efficiency versus time for devices light soaked at 140 suns at various temperatures, showing explicitly that saturation occurs. Figure 7.11b plots the normalized efficiency at 50°C and several light intensities. The efficiency was related to the defect density N via the expression

$$\eta = A \log N \tag{12}$$

where A is a constant. Figure 7.12 plots the relationship between N_s/N_0 and exposure intensity for devices exposed at several temperatures. At high temperatures, saturation occurs at low defect levels, relative to the initial value N_0, while at 50°C, close to the value normally encountered under ambient conditions, the saturation value is highly dependent on intensity. This sensitivity to intensity and temperature is probably responsible for the wide variation in experimental results reported and the difficulty in observing saturation of devices.

Yang and Chen, in a work yet to be published, have evidence suggesting that degradation at high intensity populates metastable states that are more easily annealed than those accessed by degradation at low intensity, which anneal more slowly. The occurrence of a range of annealing energies is not consistent with the stretched exponential model that effectively combines the various energy states in a single distribution. These experiments suggest that the energy levels accessed during degradation and the rate of annealing are different for each intensity. Hence, the value β that contains the information on this distribution may oversimplify the actual relationship process occurring during degradation and hide potentially important phenomenology.

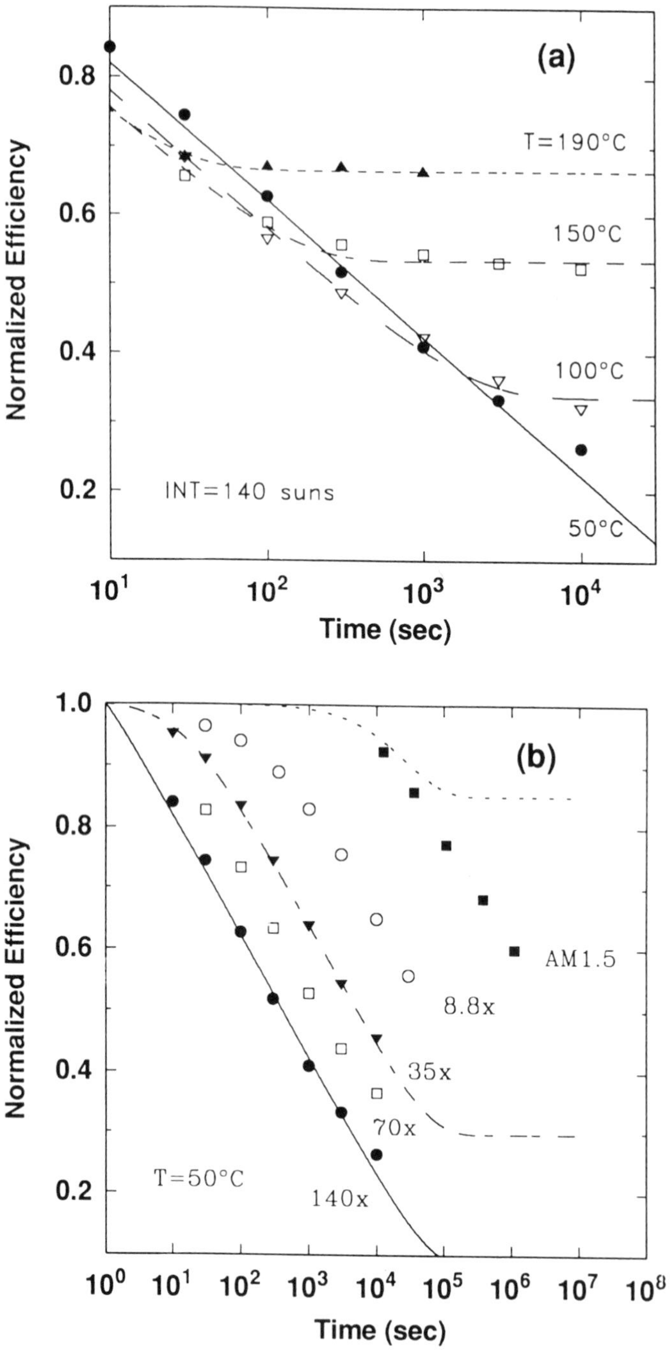

FIGURE 7.11 Time dependence of cell degradation at various temperatures and 140 suns **(a)** and at various light intensities and 50°C **(b)**. Curves are fits using the SJT models.

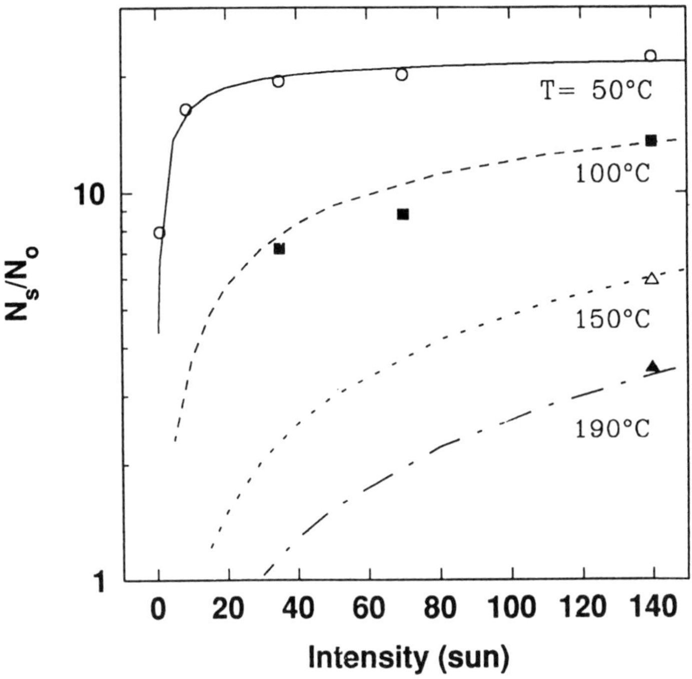

FIGURE 7.12 Saturated defect density as a function of light intensity for various temperatures.

ACKNOWLEDGMENTS

Many thanks are due to the scientists at Solarex for their contributions to this work. I also expecially thank Sharon Martino and Wendy Larsen for preparing the manuscript. Partial support from NREL under Subcontract No. ZM-0-19033-1 is also gratefully acknowledged.

REFERENCES

Anderson, J.C., Biswas, S., and Guo, H. (1987), *J. Appl. Phys.* **61**, 604–613.

Arai, Y., Ishii, M., Shinohara, H., and Yamazaki, S. (1991), *IEEE Electronic Device* **12**, 460–461.

Arya, R., Bennett, M.S., Rajan, K., and Catalano, A. (1989), *Proceedings of the 9th E.C. PVSEC*, Freiburg, Germany, pp. 251–254.

Carlson, D.E. (1977), *IEEE Trans. Electronic Device ED-24*, 449–453.

Carlson, D.E. (1980), *Solar Energy Mater.* **3**, 503–518.

Carlson, D.E., and Wronski, C.R. (1976), *Appl. Phys. Lett.* **28**, 671–673.

Catalano, A., Arya, R.R., Fortmann, C., Morris, J., Newton, J., and O'Dowd, J.G. (1987a), *19th IEEE Photovoltaic Specialists Conf. Record*, IEEE, New York, pp. 1506–1507.

Catalano, A., Arya, R.R., Yang, L., Bennett, M., Newtown, J., and Wiedeman, S. (1989), *Tech. Dig. PVSEC-4*, Sydney, Australia, pp. 421–427.

Catalano, A., Bennett, M., Arya, R.R., Rajan, K., and Newton, J. (1985), *18th IEEE Photovoltaic Specialists Conf. Record*, IEEE, New York, pp. 1378–1382.

Catalano, A., Bennett, M., Newton, J., Yang, L., Li, Y.-M., Fieselmann, B., Wiedeman, S., and D'Aiello, R.V. (1992), *NREL PVAR&D 11th Review Meeting*, Denver, Colorado.

Catalano, A., D'Aiello, R.V., Dresner, J., Faughnan, B., Firester, A., Kane, J., Schade, H., Smith, Z.E., Schwartz, G., and Triano, A. (1982), *16th IEEE Photovoltaic Specialists Conf. Record*, IEEE, New York, pp. 1421–1422.

Catalano, A., Fortmann, C., Newtown, J., Arya, R.R., and Wood, G. (1987b), *Tech. Dig. PVSEC-3,* Tokyo, Japan, pp. 705–708.

Chatterjie, P., McElheny, P.J., and Fonash, S.J. (1990), *J. Appl. Phys.* **67**, 3803–3809.

Crandall, R.S. (1983), *J. Appl. Phys.* **54**, 7176–7178.

Dresner, J., Goldstein, B., and Szostak, D. (1981), *Appl. Phys. Lett.* **38**, 998–999.

Dressnandt, N.C., and Rothwarf, A. (1984), *17th IEEE Photovoltaic Specialists Conf. Record*, IEEE, New York, pp. 364–368.

Faughnan, B.W., and Crandall, R.S. (1984), *Appl. Phys. Lett.* **44**, 537–539.

Guha, S., Yang, J., Pawlikiewicz, A., Glatfelter, T., Ross, R., and Ovshinsky, S.R. (1989), *Appl. Phys. Lett.* **54**, 2330–2332.

Hack, M., and Shur, M., (1985), *J. Appl. Phys.* **58**, 997–1020.

Hamakawa, Y., Okamoto, H., and Tawada, Y. (1982), *Int. J. Solar Energy* **1**, 125.

Hanaki, K., Hattori, T., Hamakawa, Y. (1987), *Tech. Dig. PVSEC-3*, Tokyo, Japan, pp. 49–52.

Hattori, Y., Kruangram, D., Toyaman, T., Okamoto, H., and Hamakawa, Y. (1987), *Tech. Dig. PVSEC-3*, Tokyo, Japan, pp. 171–174.

Hegedus, S., Rocheleau, R., Cebulka, J.M., and Baron, B.N. (1986), *J. Appl. Phys.* **60**, 1046–1054.

Hou, J.Y., Arch, J.K., and Fonash, S.J. (1991), *22nd IEEE Photovoltaic Specialists Conf. Record*, IEEE, New York, pp. 1260–1264.

Ichikawa, Y., Aizawa, H., Shimabukuro, H., Nagao, Y., and Sakai, H. (1987), *Tech. Dig. PVSEC-3*, Tokyo, Japan, pp. 29–32.

Ichikawa, Y., Fujikake, S., Ohta, H., Sasaki, T., and Sakai, H. (1991), *22nd IEEE Photovoltaic Specialists Conf. Record*, IEEE, New York, pp. 1296–1301.

Kitagawa, M., Ishihara, S., Setsune, K., Manabe, Y., and Hirao, Y. (1987a), *Jpn. J. Appl. Phys.* **26**, L231–L233.

Kitagawa, M., Setsune, K., Manabe, Y., and Hirao, T. (1987b), *J. Appl. Phys.* **61**, 2084–2087.

Konagai, M. (1987), *Tech. Dig. PVSEC-3*, Tokyo, Japan, pp. 15–20.

Konagai, M., Kim, W.Y., Shibata, A., Kazama, Y., Seki, K., Tsukuda, S., Yamanaka, S., and Takahashi, K. (1988), *Tech. Dig. PVSEC-4*, Sydney, Australia, pp. 197–204.

Kuwano, Y., Ohnishi, M., Nishiwaki, H., Tsuda, S., Fukatsu, T., Enomoto, K., Nakashima, Y., and Tarui, H. (1982), *16th IEEE Photovoltaic Specialists Conf. Record*, IEEE, New York, pp. 1338–1343.

Lecomber, P.G., and Speak, W.E. (1979), *Top. Appl. Phys.* **36**, 251–285.

Li, Y.-M., Fieselman, B.F., and Catalano, A. (1991), *22nd IEEE Photovoltaic Specialists Conf. Record*, IEEE, New York, 1231–1235.

Lucovsky, G., and Tsu, D.V. (1987), *J. Non-Cryst. Sol.* **97/98**, 265–268.

Madan, A. (1989), *Paper Presented at the SERI PV AR&D Meeting*, Lakewood, California.

Mahan, A.H., Nelson, B.P., Salamon, S., and R.S. Crandall (1991), *Mater. Res. Soc. Symp. Proc.* **219**, 673–678.

Matsumura, H. (1987), *Tech. Dig. PVSEC-3*, Tokyo, Japan, pp. 33–36.

McElheny, P.J., Arch, J.K., Lin, H.-S. and Fonash, S.J. (1988), *J. Appl. Phys.* **64**, 1254–1265.

Mott, N.F. (1967), *Adv. Phys.* **16**, 49.

Moustakas, T.D. (1984), *Semiconductors Semimetals* **21 (Part A)**, 65.

Muller, W., Pirrung, J., Schroder, B., and Geiger, J. (1984), *Solar Energy Mater.* **10**, 171–186.

Nakano, S., Nakamura, N., Ninomiya, K., Tarui, H., Matsuyama, T., Nishikuni, M., Kiyama, S., Hishikawa, Y., Dohjo, H., Tsuda, S., Ohnishi, M., Kishi, Y., and Kuwano, Y. (1988), *20th IEEE Photovoltaic Specialists Conf. Record*, IEEE, New York, pp. 123–128.

Nishikuni, M., Ninomiya, K., Tanaka, M., Matsuoka, T, Nakam, S., Okuda, N., Shibuya, H., Ohnishi, M., Kishi, Y., Kumano, Y., and Ohara, S. (1988), *Tech. Dig. PVSEC-4*, Sydney, Australia, pp. 91–96.

Parsons, G.N., Tsu, D.V., and Lucovsky, G. (1987), *J. Non-Cryst. Sol.* **97 98**, 1375–1378.

Pinarbasi, M., Kushner, M.J., and Abelson, J. (1990), *J. Appl. Phys.* **68**, 2255–2264.

Redfield, D., and Bube, R.H. (1989), *Appl. Phys. Lett.* **54**, 1037.

Redfield, D., and Bube, R.H. (1990), *Phys. Rev. Lett.* **65**, 464.

Rudder, R.A., Cook, J.W., and Lucovsky, G. (1983), *Appl. Phys. Lett.* **43**, 871–873.

Rudder, R.A., Cook, J.W., Schetzina, F., and Lucovsky, G. (1984), *J. Vac. Sci. Technol.* **A2**, 326–329.

Saitoh, T. (1982), *Jpn. J. Appl. Phys.* **22 (Suppl. 22)**, 617.

Schwartz, R.J., Park, J.N., Gray, J.L., and Turner, G.B., (1989), *Tech. Dig. PVSEC-4*, Sydney, Australia, pp. 607–613.

Scott, B., Olbricht, W.L., Meyerson, B.S., Scott, B.A., and Piecenik, R.M. (1986), *5th E.C. PVSEC*, Athens, Greece, 788–793.

Shing, Y.H., and Pool, F.S., (1990), *21st IEEE Photovoltaic Specialists Conf. Record*, IEEE, New York, pp. 1574–1578.

Smith, Z.E., and Wagner, S. (1985), *Mater. Res. Soc. Symp. Proc.* **49**, 331–338.

Staebler, D.L., and Wronski, C.R., (1977), *Appl. Phys. Lett.* **31**, 292–294.

Street, R.A. (1982), *Phys. Rev. Lett.* **49**, 1187–1190.

Street, R.A., Biegelsen, D.K., Jackson, W.B., Jackson, N.M., and Stutzmann, M. (1985), *Phil. Mag. B.* **52**, 235–245.

Street, R.A., and Winer, K. (1989), *13th ICALS,* **Pt. II**, 645–647.

Stutzmann, M., Jackson, W.B., and Tsai, C.C. (1984), *Appl. Phys. Lett.* **45**, 1075–1077.

Stutzmann, M., Jackson, W.B., and Tsai, C.C. (1985), *Phys. Rev. B.* **32**, 23–47.

Stutzmann, M., Jackson, W.B., and Tsai, C.C. (1986), *Phys. Rev. B.* **34**, 63–72.

Swartz, G.A. (1982), *J. Appl. Phys.* **53**, 712–719.

Watanabe, T., Azuma, K., Tanaka, M., Nakatani, M., Sonobe, T., and Shimada, T. (1987a), *Tech. Dig. PVSEC-3*, Tokyo, Japan, pp. 57–60.

Watanabe, T., Tanaka, M., Azuma, K., Nakatani, M., Sonobe, T., and Shimada, T. (1987b), *Jpn. J. Appl. Phys.* **26**, L288–L290.

Yang, J., Ross, P., Glatfelter, T., Mohr, R., Hammond, G., Bernotaitis, C., Chen, E., Burdick, H., Hopson, J., and Guha, S. (1988), *20th IEEE Photovoltaic Specialists Conf. Record*, IEEE, New York, pp. 241–246.

Yang, L., and Chen, L. (1991), *J. Non-Cryst. Sol.* **137 138**, 1189–1192.

Yang, L., Chen, L., and Catalano, A. (1991), *Mater. Res. Soc. Symp. Proc.* **219**, 259–264.

Yang, L., Newton, J., and Fieselmann, B.F. (1989), *Mater. Res. Soc. Symp. Proc.* **149**, 497–502.

Yoshikawa, A., and Yamaga, S. (1984), *Jpn. J. Appl. Phys.* **23**, L91–L93.

CHAPTER EIGHT

$CuInSe_2$- and CdTe-Based Solar Cells

KIM MITCHELL, Siemens Solar, Camarillo, CA 93011

1. INTRODUCTION

Polycrystalline photovoltaic (PV) materials show substantial promise for achieving the Department of Energy (DOE) PV costs and performance goals. Cadmium telluride (CdTe) and $CuInSe_2$ (CIS) both recently demonstrated device efficiencies in the 15% range with promise of achieving 20%–25% efficiencies. This chapter reviews the materials properties, device options, cell performance characteristics, and future prospects for CdTe- and CIS-based PV materials.

2. BACKGROUND

To be a viable contributor to large-scale electrical supply, a PV materials technology must be manufacturable (1 million m^2 per year or greater), durable (20 year plus outdoor life), low cost (less than \$2/W_{peak}), and with the requisite module efficiency (15% or greater) to achieve the PV system cost goals. Polycrystalline materials offer several advantages for achieving the above PV materials requirements. In the late 1970s, the U.S. DOE PV Research and Development program investigated several relatively unexplored polycrystalline thin-film materials including CIS, CdTe, indium phosphide (InP), zinc phosphide (Zn_3P_2), copper selenide (Cu_2Se), tungsten diselenide (WSe_2), gallium arsenide (GaAs) and zinc silicon arsenide (ZnSiAs) (Mitchell, 1982).

Solar Cells and Their Applications, Edited by Larry D. Partain.
ISBN 0-471-57420-1

Two of these materials groups, namely, $CuInSe_2$-based alloys and CdTe, are approaching commercial demonstration. Workers at the University of South Florida recently achieved 15.8% efficiency on a 1 cm^2 CdTe device (Chu et al., 1992), and the EUROCIS project has shown 14.8% efficiency for CIS devices (Stolt et al., 1992) and 15.2% efficiency for 1.12 eV $CuIn(Se_{0.75}S_{0.25})_2$ (Walter et al., 1992). Siemens Solar has extended their 14%–15% efficient CIS device technology to 0.4 M^2, 10% aperture efficient modules. BP Solar, Solar Cells Incorporated, and Golden Photon (previously Photon Energy) are developing 0.4–0.7 M^2 size CdTe modules.

2.1. Materials Properties

An excellent review of the basic materials properties of CdTe has been published by Zanio (1978). Due to the many developments in CdTe, *Solar Cells* and the *International Journal of Solar Energy* have devoted special issues to CdTe thin-film solar cells (Coutts et al., 1988; Bonnet, 1992). Properties of CIS and other chalcopyrites are described by Shay and Wernick (1975). A special issue of *Solar Cells* has also been devoted to CIS (Coutts et al., 1986). Several conference reports on ternary and multinary compounds (e.g., Pamplin et al., 1984; Deb and Zunger, 1987), as well as a review paper on CIS by Rockett and Birkmire (1991), are useful. For PV applications, the key properties are the energy band gap (1.44 eV for CdTe and 0.95 eV for CIS), optical absorption coefficients, crystal structure (zinc blende for CdTe and chalcopyrite for CIS), and phase diagram and thermodynamic properties that define the film formation and defect chemistry.

Both CdTe and CIS are direct energy band gap materials, strongly absorbing the solar spectrum within a few micrometers thickness. Figure 8.1 compares the photon currents generated in CIS, CdTe, thin-film silicon:hydrogen alloys (a-Si:H), and crystalline silicon (c-Si) calculated from the optical absorption coefficients and assuming no reflection losses for the 100 mW/cm^2 ASTM air-mass 1.5 global reference spectrum (Tuttle et al., 1990; Mitchell, 1979; Eskenas et al., 1985; Runyan, 1965). The photon current increases as a function of optical path length (i.e., absorber layer thickness), and J_{max} is the maximum available photon current density for the material. CIS with a 0.95 eV optical band gap (51 mA/cm^2 J_{max}) and CdTe with a 1.44 eV band gap (31 mA/cm^2 J_{max}) both absorb 90% of the available photons in 1 μm thicknesses so that film thicknesses of only 1–3 μm are sufficient for thin-film PV applications.

Thin films translate to reduced PV materials requirements. To quantify this, the number of kilograms required per megawatt PV capacity equals the density of the material (g/cm^3) times the film thickness (micrometers) divided by the conversion efficiency. Given the respective CIS and CdTe densities of 5.77 and 6.2 g/cm^3, using 3 μm thick films and assuming a conservative 15% PV conversion efficiency implies that 115 kg (253 lb) of CIS or 124 kg (273 lb) of CdTe are required per megawatt of PV modules produced, minimal amounts compared with other energy conversion technologies.

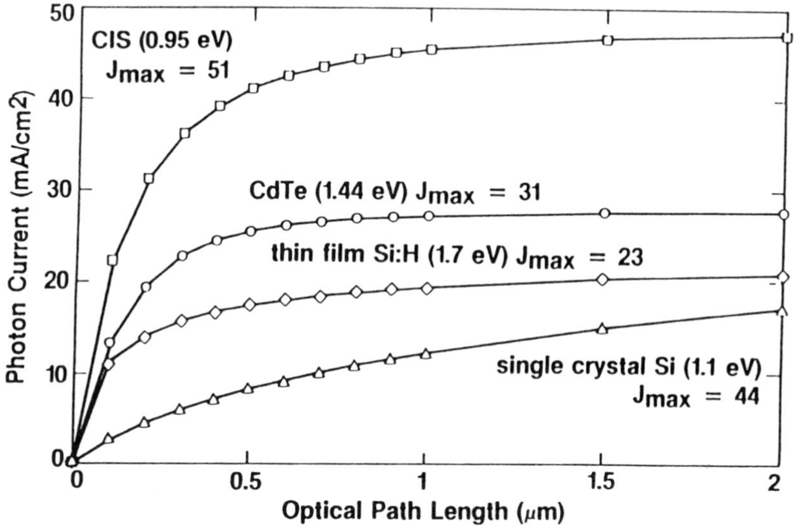

FIGURE 8.1 Photo current versus optical path length.

2.2. Device Types

The components of a polycrystalline solar cell, illustrated in Figure 8.2, are the substrate, the back electrical contact, the semiconductor absorber layer, the junction or window layer, the front grid or transparent electrode layer, and, finally, an antireflection (AR) layer. Thin-film solar cells are configured either as superstrate (the light enters the device through the substrate) or substrate (the light enters the cell through the layer opposite the substrate) devices. In either approach, glass is the predominant substrate due to its low cost and environmental durability, although alumina, metal foils, and plastics have been considered. A number of device structures have been explored for polycrystal-

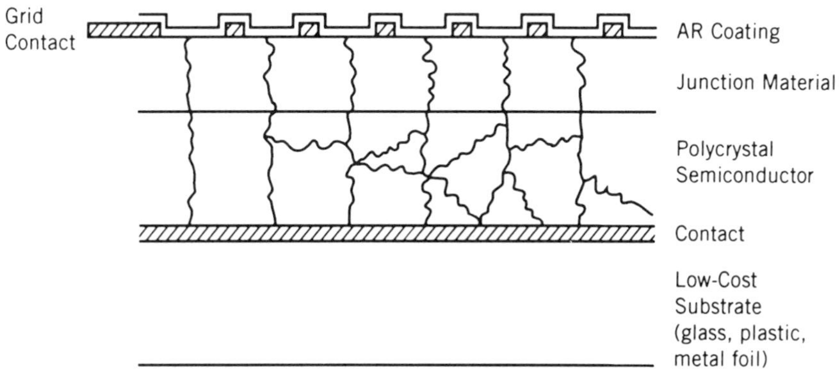

FIGURE 8.2 Cross section of a thin-film polycrystalline cell.

line materials: Schottky barriers, homojunctions, heterojunctions, and p/i/n junctions. The most successful have been heterojunctions and p/i/n junctions, where a wide band gap window layer allows the light to be absorbed in or close to the space charge region. In this case, the minority-carrier diffusion length can be 1 μm or less and the minority-carrier lifetime in the picosecond to nanosecond range to achieve 15% or greater conversion efficiencies. In contrast, c-Si requires tens of microseconds to millisecond carrier lifetimes to achieve similar efficiencies. Several books are available covering topics relevant to polycrystalline semiconductor device physics, including those of Sze (1969), Milnes and Feucht (1972), Hovel (1975), Mitchell (1979), Kazmerski (1980), Fonash (1981), Chopra and Das (1983), and Fahrenbruch and Bube (1983).

2.3. Performance Expectations

The efficiency of a single junction solar cell is limited to about 18%–24% by lack of absorption of photons below the band gap energy and thermal dissipation of the excess photon energy above the band gap energy (Wolf, 1960; Wysocki and Rappaport, 1960). The practical efficiency limit for a single junction is determined by the PV material and junction quality and the device configuration.

Sites (1988) presented a photocarrier and voltage loss analysis for thin-film solar cells as depicted in Figure 8.3, which projected respective CIS and CdTe maximum theoretical efficiencies of 23.5% and 27.5% and practical efficiencies of 15% and 18%. Improvements in material and device technologies continually increase the projected practical efficiencies. For example, efficiencies for CIS and CIS-related alloys are now expected to reach 20%–25% (Mitchell, 1989; Menner and Schock, 1992).

2.4. Multijunction Options

Multiple junction PV converters can achieve higher PV efficiencies than single junction ones. The solar spectrum is divided into wavelength ranges, and appropriate band gap semiconductors are selected to convert the photons efficiently in each wavelength range. The photons at short wavelengths are absorbed in a high band gap semiconductor with a resultant higher voltage output, and the longer wavelength photons are absorbed by a low band gap semiconductor. Present day multijunctions involve only two or three band gaps, although in theory an infinite number of band gaps are possible (Mitchell, 1978; Loferski, 1982).

For two band gap tandems, the junctions can be operated in four terminal mode, in which the maximum power points of each junction are tracked separately, or in two terminal mode, in which the two junctions are constrained to pass the same electrical current. Two terminal and four terminal tandem junction performances have been modeled for polycrystalline materials (Mitchell, 1984). Values assumed for the calculations were selected to best fit

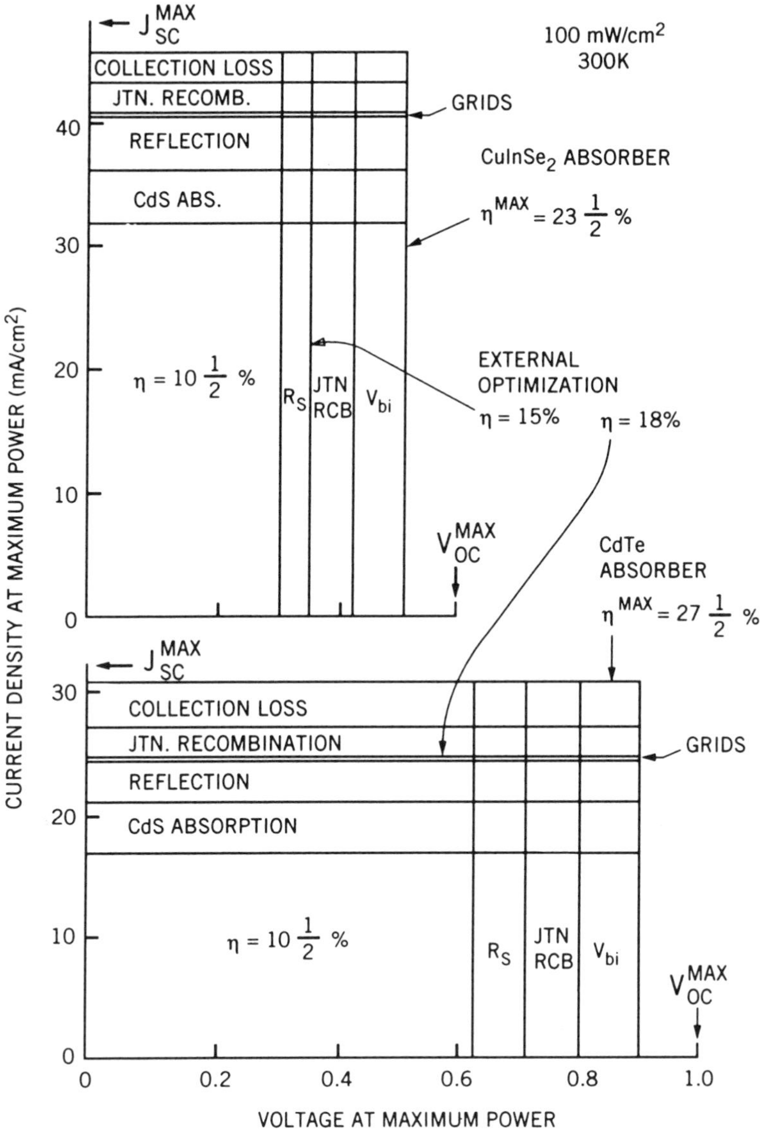

FIGURE 8.3 Maximum power current density and voltage for CdTe and CIS cells. Estimated magnitudes of the primary loss mechanisms as shown (After Sites, 1988).

the experimental data, with 18% single junction efficiencies projected. As shown in Figures 8.4 and 8.5, efficiencies above 24% were predicted for top cell band gaps of 1.8–1.9 eV and bottom cell band gaps in the 1.15–1.25 eV range. About two-thirds of the total efficiency is contributed by the top cell. For four terminal tandems, efficiencies above 20% are achievable for a broad range of top and bottom cell band gaps (see Fig. 8.4). Similar efficiencies are achievable

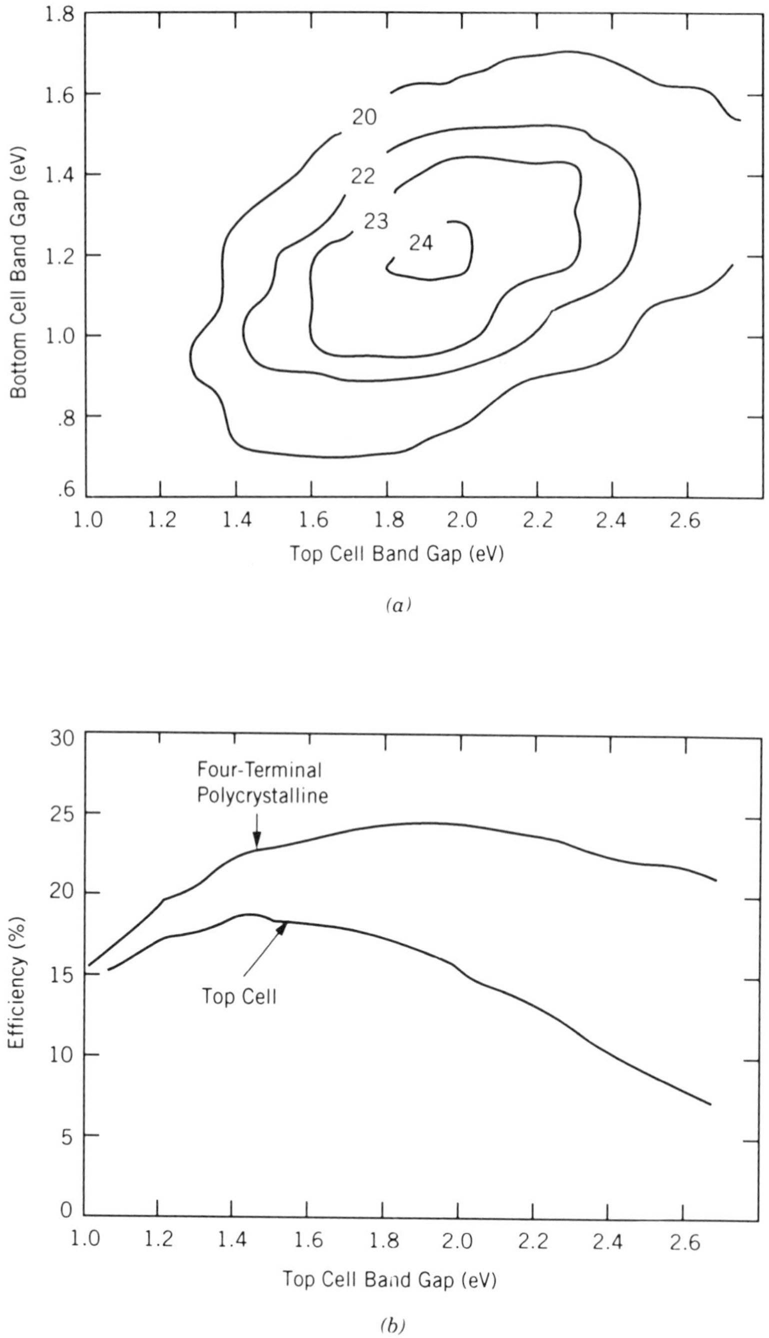

FIGURE 8.4 Four terminal AM1.5 global PV efficiency for polycrystalline thin-film cells: **(a)** efficiency vs. top and bottom cell band gaps: **(b)** top cell and total efficiency vs. top cell band gap. SOURCE: Mitchell et al. (1984).

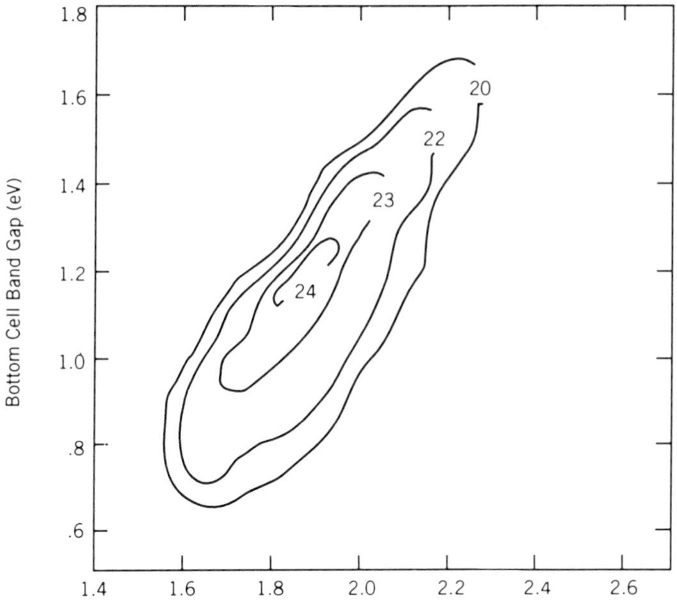

(a)

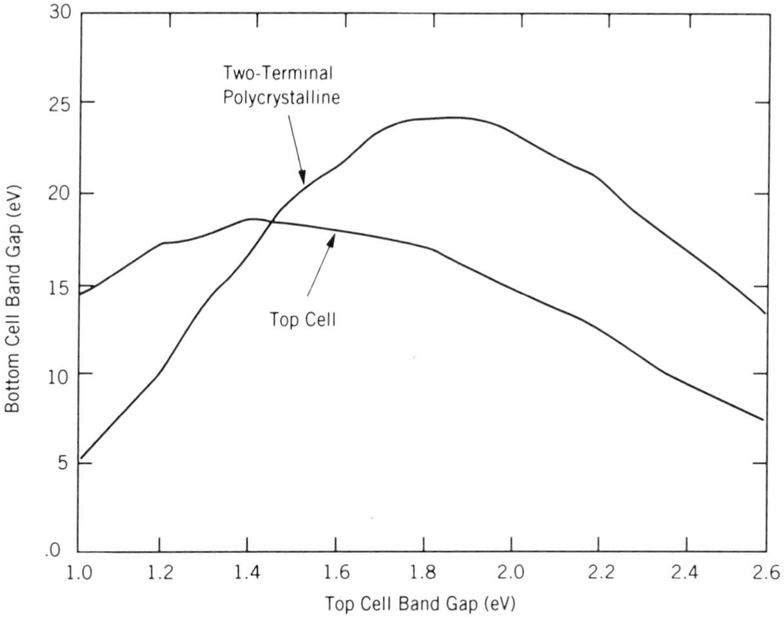

(b)

FIGURE 8.5 Two terminal AM1.5 global PV efficiency for polycrystalline thin-film cells: **(a)** efficiency vs. top and bottom cell band gaps; **(b)** top cell and total efficiency vs. top cell band gap. SOURCE: Mitchell et al. (1984).

for two-terminal tandems but over a narrower range of top and bottom cell band gaps (see Fig. 8.5).

As with single junction solar cells, improvements in material and device technologies continue to increase the projected practical efficiencies for tandems. Tandem efficiencies in the 30%–35% range are achievable based on 20%–25% single junction efficiencies.

3. DEPOSITION PROCESSES

One of the most remarkable results of the thin film polycrystalline PV research program has been the demonstration of the large spectrum of deposition processes possible for fabricating thin-film PV devices. These include vacuum evaporation, sputtering, chemical vapor deposition, plasma deposition, electrodeposition, chemical spray, and sintered film (Mitchell, 1982). Thin-film processing has been the subject of several books, for example, those by Maissel and Glang (1970), Lowenheim (1978), Vossen and Kern (1978), Bunshah et al. (1982), Chopra and Das (1983), Sherman (1987), and Schuegraf (1988). CIS and CdTe thin-film formation techniques are summarized in the next sections.

3.1. $CuInSe_2$ Alloys

CIS and related thin-film alloys have received substantial attention due to the high efficiencies demonstrated on cells and modules. Work at Boeing prompted early interest in the material, reporting 9.4%CdS/CIS cell efficiencies (Mickelsen and Chen, 1980, 1981). Thin-film CIS alloy formation technique explored include co-evaporation of the elements (Mickelsen and Chen, 1980, 1981; Dhere et al., 1984), flash evaporation (Romeo et al., 1986), laser synthesis (Joliet et al., 1985; Laude et al., 1986), MOCVD (Sagnes et al., 1992), sputtering (Thornton et al., 1984; Ermer et al., 1985; Thornton and Lommasson, 1986), spray pyrolysis (Mooney and Lamoreaux, 1986; Bougnot et al., 1986), electrodeposition (Bhattacharya and Rajeshwar, 1986; Hodes and Cahen, 1986; Thouin et al., 1992), screen printing (Arita et al., 1988), and sulfurization and selenization of metal layers (Grindle et al., 1979; Binsma and Van Der Linden, 1982; Chu et al., 1984; Kapur et al., 1984; Ermer and Love, 1986; Basol and Kapur, 1990; Verma et al., 1991; Bodegard and Stolt, 1992). Additional deposition approaches are referenced by Mitchell (1982) and Rockett and Birkmire (1991).

The highest efficiency devices have been fabricated using co-evaporation from the elements and sulfurization and selenization of metal layers. Figure 8.6 illustrates three source evaporation. The substrate is heated in the range of 350°–550°C. To prevent loss of Se from the films, the Se evaporation rate is maintained at about three times the Se incorporation rate into the film. The Cu and In rates are also controlled to achieve the desired Cu/In ratio.

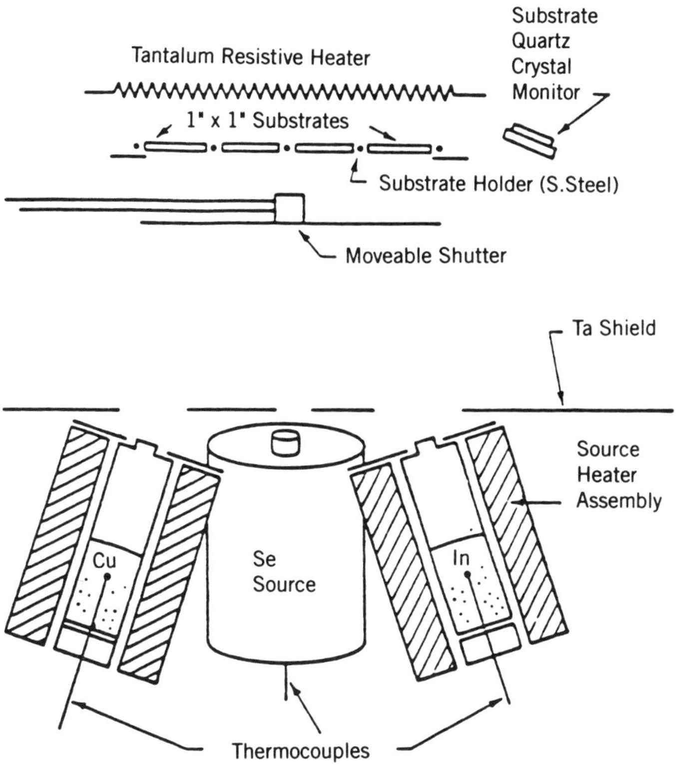

FIGURE 8.6 Diagram of a three source CIS evaporation system.

Selenization or sulfurization are carried out using either gas sources such as H_2Se and H_2S or elemental sources or layers of Se or S. The metal layers are deposited by a number of techniques, such as electrodeposition, evaporation, and sputtering. The initial reaction involves the metallurgical formation of the compound. Additional high-temperature annealing improves the electronic properties of the films.

3.2. CdTe Alloys

A number of deposition techniques have been used to form CdTe alloys: atomic layer epitaxy (Skarp et al., 1991), chemical spray (Serreze et al., 1981; Albright et al., 1990), chemical vapor deposition (Chu et al., 1981; Rohatgi et al., 1991), close-space sublimation (Tyan and Perez-Albuerne, 1982; Mitchell et al., 1975; Chu, 1988; Chu et al., 1992), electrodeposition (Basol et al., 1980; Basol and Tseng, 1986; Woodcock et al., 1991), sintered films (Nakayama et al., 1976, 1980; Matsumoto et al., 1984), stacked elemental layer (Bhatti et al., 1991), and vacuum evaporation (Uda et al., 1978; Huber et al., 1981).

Remarkably, cell efficiencies in excess of 10% have been achieved by atomic layer epitaxy, chemical spray, close-space sublimation, electrodeposition, MOCVD, and screen printing. CdTe has the ability to form large grain, oriented films independent of deposition technique. Two examples, close-space sublimation and screen-printing, are described.

A side view of a close-space sublimation system (also known as close-space vapor transport [CSVT]) is illustrated in Figure 8.7. The source-to-substrate spacing is usually only a few millimeters. Deposition ambient pressure ranges from 1 to 760 torr, with deposition rates dependent on the ambient pressure and the source temperature. Micrometer per minute growth rates are easily achievable (Mitchell et al., 1985).

The process sequence for screen-printed CdS/CdTe cells is shown in Figure 8.8. Each material to be printed is first ground into small particles and mixed with a binder and fluxing agent. It is then screen-printed onto the substrate, dried at 120°C to evolve the binder, and fired at a higher temperature (620°C) to sinter the film. The final steps of the process involve screen-printing the back contact, for example, with carbon paste with a small amount of Cu acceptor impurity, and annealing the entire device to improve the junction quality.

3.3. Window Layers

The window layer has two primary roles: (1) optically couple the light into the PV absorber layer with minimal optical reflection and absorption losses and (2) transport the photogenerated majority-carrier current to the outside circuit with minimal electrical resistance losses. For superstrate devices, MgF_2 has been used as an AR coating to reduce the 4% reflection loss at the air–glass interface. AR etched glass is a second alternative, reducing the air–glass reflection loss to less than 2%.

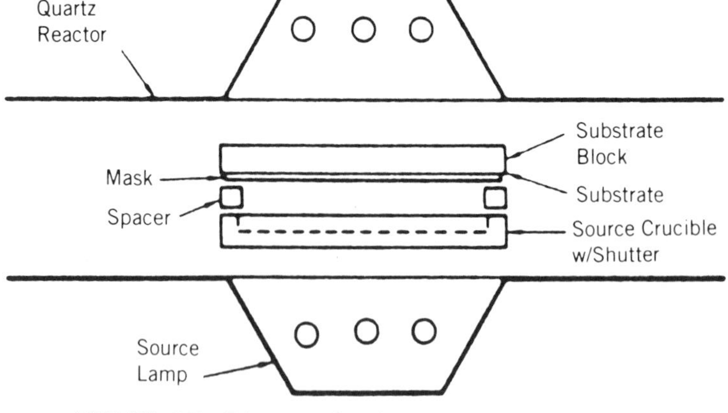

FIGURE 8.7 Diagram of a close-space sublimation system.

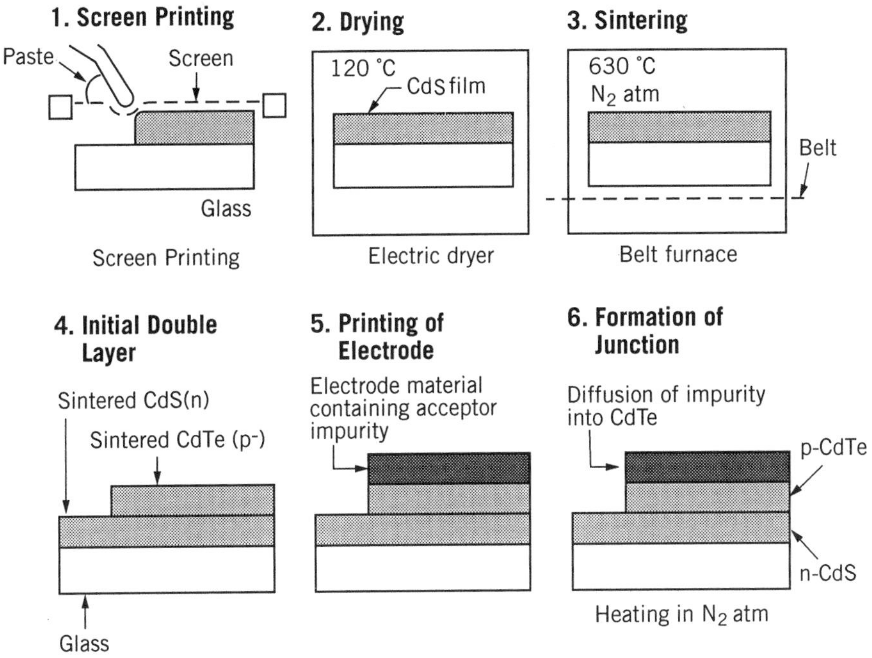

FIGURE 8.8 Sequence for formation of screen-printed CdTe cells.

Metal oxides play a major role as both a partial AR layer and a transparent conductive layer. Reviews of transparent conductor technologies include those by Jarzebski (1973), Haacke (1977), and Chopra et al. (1983). Tin oxide (SnO_2 or TO) and indium tin oxide (ITO) coated glass are commercially available and commonly used. ZnO is another candidate transparent conductor.

The highest efficiency CIS and CdTe cells have been fabricated with ZnO layers in conjunction with thin CdS (Mitchell et al., 1985; Potter et al., 1985; Mitchell et al., 1988a; Mauch et al., 1991). Techniques to deposit ZnO include spray pyrolysis (Aranovich et al., 1979; Major et al., 1983), sputtering (Barnes et al., 1980; Minami et al., 1984; Schropp et al., 1988), and chemical vapor deposition (Ghandi et al., 1980; Roth and Williams, 1981; Smith, 1983; Wenas et al., 1991; Hu and Gordon, 1992).

CdS has a strong affinity to grow oriented films independent of the substrate type. Deposition techniques include chemical spray (Roderick, 1980; Albin and Risbud, 1987), chemical deposition (Nagao and Watanabe, 1968; Lincot and Ortega-Borges, 1992; Lincot et al., 1992; Kessler et al., 1992), and evaporation (Hall et al., 1980). The chemical deposition technique has demonstrated the advantage that it not only deposits the CdS layer, but also chemically etches and passivates the substrate film. Chemical deposition has also been applied to deposit ZnS films (Ortega-Borges et al., 1992) and indium hydroxide and sulfide films (Velthaus et al., 1992).

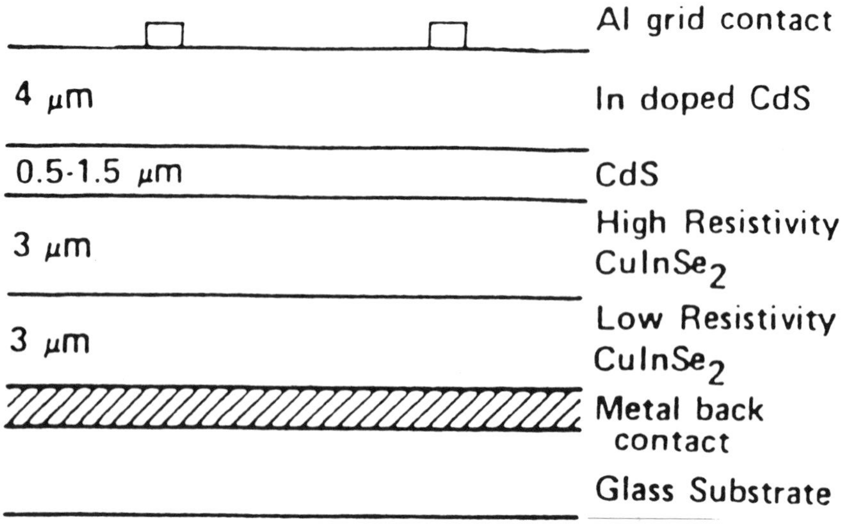

FIGURE 8.9 Cross section of a thick CdS/bi layer CIS cell.

4. CELL PERFORMANCE

CdTe and CIS both have recently demonstrated device efficiencies in the 15% range. As shown in the following tables, a number of CdTe and CIS device structures and deposition processes have achieved efficiencies above 10%. These results are described below.

4.1. $CuInSe_2$-Based Cells

Figures 8.9 and 8.10 illustrate typical CIS device structures both of which are substrate devices. The first one utilizes a thick doped CdS layer as the window layer and a bilayer CIS sequence where a Cu-rich layer is deposited first,

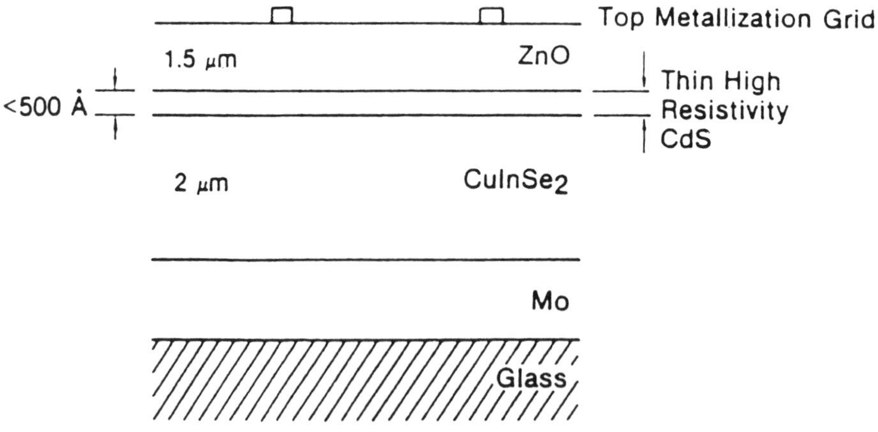

FIGURE 8.10 Cross section of a ZnO/thin CdS/CIS cell.

followed by an In-rich layer (Mickelsen and Chen, 1981). The second device achieves higher J_{sc} by using ZnO as the transparent conductor and a thin CdS layer at the junction (Potter et al., 1985). Figure 8.11 shows the spectral response for a 14.1% device of this type extending from the ZnO to the CIS band edge, resulting in a 41 mA/cm^2 J_{sc}.

The cell parameters for CdS/CIS thin-film cells over 10% efficiency are summarized in Table 8.1. The highest efficiencies have been achieved with devices using a ZnO transparent conductor with a thin CdS junction layer. The best device, 14.8% efficient, consists of soda–lime glass with a 1,500 nm Mo layer, a 2,500 nm co-evaporated CIS layer, a 10 nm chemically deposited CdS layer, a 500 nm RF magnetron sputtered Al-doped ZnO layer, and a final 120 nm MgF_2AR coating (Stolt et al., 1992; Schock et al., 1992). The substrate temperature is ramped from 350° to 550°C during the CIS deposition. After fabrication, the cell was annealed in air at 200°C. An additional anneal under illumination at open circuit for 20 min further improved efficiency. The dependence of efficiency on V_{oc} in Table 8.1 is evident. The best V_{oc} of 513 mV is the highest reported for CIS to date.

Table 8.2 shows that CIS cells fabricated with non-Cd–containing window layers can achieve efficiencies over 10%, with 13.3% the highest efficiency. Table 8.2 also reflects the dependence of efficiency on cell V_{oc}.

Table 8.3 presents the cell parameters for CdS/I–III–VI_2 multinary thin-film solar cells. The 15.2% efficiency reported for a 1.12 eV $CuIn(Se_{0.75}S_{0.25})_2$ cell is the first CIS-based alloy device to exceed 15% efficiency (Walter et al., 1992). With 613 mV V_{oc}, the device parameters are similar to those of a

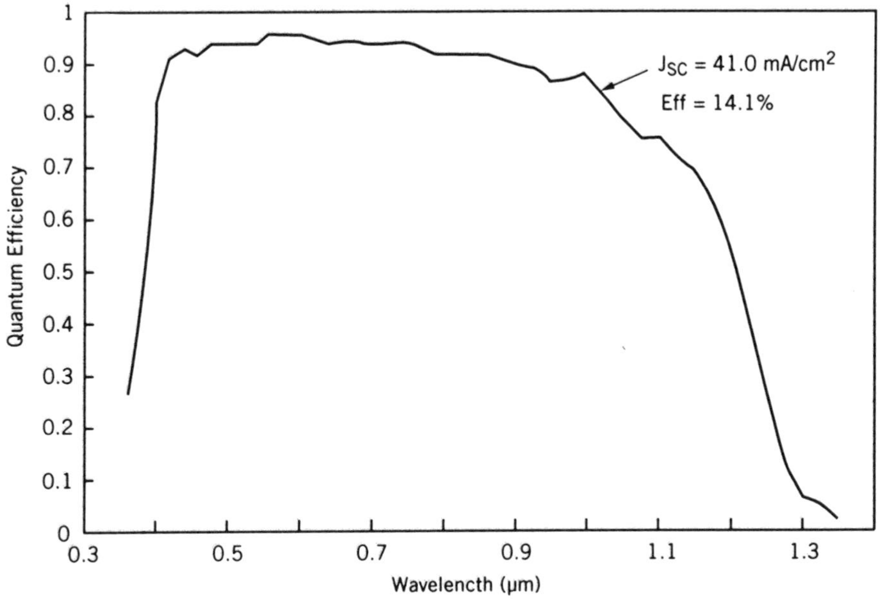

FIGURE 8.11 Spectral response of a 14% efficient ZnO/thin CdS/CIS cell.

Table 8.1 $CdS/CuInSe_2$ Thin-Film Solar Cells Over 10% Efficiency

Cell Structure	V_{oc} (mV)	J_{sc}[a] (mA/cm^2)	FF	Eff[a] (%)	Area[a] (cm^2)	Reference
MgF_2/ZnO/CdS/CIS	513	40.4	0.720	14.8	0.26	Stolt et al. (1992)
ZnO/CdS/CIS	508	41.0	0.677	14.1	3.5	Mitchell et al. (1988a)
ZnO/CdS/CIS	486	39.8	0.720	13.9	0.25	Schock and Pfisterer (1992)
ZnO/CdS/CIS	483	38.3	0.667	12.4	0.93	Basol et al. (1991)
AR/CdZnS/CIS	440	41.0	0.688	12.4	0.96	Devaney et al. (1985)
ITO/CdZnS/CIS	426	36.1[b]	0.637	11.2[b]	0.08	Birkmire and Phillips (1987)
ZnO/CdZnS/CIS	419	37.8	0.664	10.5	0.04	Urabe et al. (1991)
AR/CdS/CIS	444	35.3	0.653	10.3	—	Noufi et al. (1986)
ZnO/CdS/CIS	427	37.4	0.641	10.2	0.18	Arya et al. (1992)

[a]Active area, 100 mW/cm^2 ASTM air mass 1.5 global spectrum, 25°C.
[b]87.5 mW/cm^2, 32°C.

crystalline Si solar cell. This device also uses the ZnO transparent conductive and thin chemically deposited CdS layer. Walter et al. (1992) also describe the structural and optical property dependence on S content. One important finding is that, for In-rich films, the optical data follow the pseudobinary tie line whereas Cu-rich films do not.

The mechanisms causing the lower efficiencies at higher band gaps is unclear. For $CuInS_2$ (1.40 eV), the efficiency is 7.3% (Mitchell et al., 1988c). For $CuGaSe_2$ (1.68 eV), the best efficiency is 5.8% (Dimmler et al., 1987). Further research is necessary to understand the defect chemistry of these materials that determines the electronic properties. Table 8.3 also includes a comparison of a 1.0 eV I–II–VI_2 multinary cell with and without a CdS buffer layer. The 14.3% efficiency is the highest reported for a non-CdS window layer CIS alloy cell.

TABLE 8.2 Noncadmium Window $CuInSe_2$ Thin-Film Solar Cells

Cell Structure	V_{oc} (mV)	J_{sc}[a] (mA/cm^2)	FF	Eff[a] (%)	Area[a] (cm^2)	Reference
ZnO/CIS	477	39.5	0.704	13.3	3.5	Ermer et al. (1992)
ZnO/BF/CIS[b]	443	37.7	0.62	10.3	0.25	Schock and Pfisterer (1992)
ZnO/ZnSe/CIS	391	40.1	0.641	10.0	3.5	Mitchell et al. (1988a)
ZnO/ZnSe/CIS	420	33.7	0.604	8.5	1.2	Nouhi et al. (1987)

[a]Active area, 100 mW/cm^2 ASTM air mass 1.5 global spectrum, 25°C.
[b]BF, Cd-free buffer layer.

TABLE 8.3 CdS–III–VI_2 Multinary Thin-Film Solar Cells

Cell Structure	E_g (eV)	V_{oc} (mV)	J_{sc}[a] (mA/cm^2)	FF	Eff[a] (%)	Area[a] (cm^2)	Reference
MgF_2/ZnO/CdS/ $CuIn(Se_{0.75}S_{0.25})_2$	1.12	613	33.5	0.74	15.2	0.07	Walter et al. (1992)
ZnO/CdS/I–III–VI_2 alloy	0.98	534	39.2	0.718	15.1	3.4	J.E. Ermer, personal communication (1992)
ZnO/CdS/I–III–VI_2 alloy	1.00	539	38.1	0.706	14.5	3.4	Ermer et al. (1992)
ZnO/I–III–VI_2 alloy	1.00	544	37.0	0.712	14.3	3.4	J.E. Ermer, private communication (1992)
MgF_2/ZnO/CdS/ $CuIn(Se_{0.45}S_{0.55})_2$	1.27	663	28.9	0.74	14.2	0.07	Walter et al. (1992)
ZnO/ZnCdS/ $CuIn_{0.74}Ga_{0.26}Se_2$	1.11	546	36.7	0.684	13.7	0.99	B.J. Stanbery, private communication (1992)
Al_2O_3/ZnCdS/ $CuIn_{0.63}Ga_{0.37}Se_2$	1.27	658	28.0	0.680	12.4	0.38	Walter et al. (1991)
ZnO/CdS/I–III–VI_2 alloy	1.43	728	23.3	0.656	11.1	3.5	Ermer et al. (1992)
ZnCdS $CuIn_{0.37}Ga_{0.63}Se_2$	1.44	580	20[b]	0.55	7.5[b]	0.3	Dimmler et al. (1988), Bloss and Schock (1988)
ZnO/CdS/$CuInS_2$	1.40	592	22.7	0.546	7.3	3.3	Mitchell et al. (1988c)
$Zn_{0.35}Cd_{0.65}S/CuGaSe_2$	1.68	845	11.6	0.50	5.8	0.2	Dimmler et al. (1987), Bloss and Schock (1988)

[a]Active area, 100 mW/cm^2 ASTM air mass 1.5 global spectrum, 25°C.
[b]85 mW/cm^2 illumination.

4.2. CdTe-Based Cells

The CdTe thin-film solar cell has achieved a world record 15.8% efficiency (Chu et al., 1992). As shown in Table 8.4, at least 12 groups have achieved CdTe efficiencies in excess of 10% using different processes and device structures. The majority of CdTe devices are deposited on a glass/transparent conductor substrate in a superstrate configuration.

The 15.2% cell was fabricated on fluorine-doped tin oxide–coated 7059 glass. A 50–80 nm thick CdS is formed by chemical deposition from an aqueous solution containing Cd acetate followed by a 3–5 μm thick CdTe layer deposited by close-space sublimation. The back contact is formed by first etching the CdTe surface and then depositing HgTe. The completed device is annealed in an inert atmosphere and finally a MgF_2 AR coating on the front glass (Chu et al., 1992).

TABLE 8.4 CdTe Thin-Film Solar Cells over 10% Efficiency

Cell Structure	V_{oc} (mV)	J_{sc}[a] (mA/cm^2)	FF	Eff[a] (%)	Area[a] (cm^2)	Reference
MgF_2/glass/SnO_2 CdS/CdTe	843	25.1	0.745	15.8	1.05	Chu et al. (1992)
Glass/SnO_2/CdS/ CdTe	819 807	23.5[b] 23.8	0.74 0.66	14.2[b] 12.7	0.02 1.0	Woodcock et al. (1991)
Glass/SnO_2/CdS/ CdTe	804	23.8	0.73	14.0	0.12	Skarp et al. (1991)
Glass/ITO/CdS/ CdTe	720	27.9	0.65	13.1	0.03	Morris et al. (1990)
Glass/CdS/CdTe	754	27.9	0.606	12.8	0.78	Matsumoto et al (1984)
Glass/SnO_2/CdS/ CdTe	788 783	26.2 25.0	0.614 0.627	12.7 12.3	0.30 0.31	Kazmerski and Emery (1992) Albright et al. (1990)
Glass/SnO_2/CdS/ CdTe/ZnTe/Ni	767 740	20.9 22.0	0.696 0.64	11.2 10.4	1.07 4.0	Kazmerski and Emery (1992) Meyers (1986)
Glass/ITO/CdS/ CdTe	790 764	20.1 18.0	0.694 0.734	11.0 10.1	0.19 —	Kazmerski and Emery (1992), Shafarman et al. (1991)
Glass/SnO_2/CdS/ MOCVD CdTe	745 704	22.1 22.6	0.660 0.627	10.9 9.9	0.08 —	Kazmerski and Emery (1992), Rohatgi et al. (1991)
Glass/ITO/CdS/ CdHgTe	620	27.0	0.63	10.6	1.48	Basol and Tseng (1986)
Glass/In_2O_3/CdS/ CdTe	750	17.0[c]	0.62	10.5[c]	0.1	Tyan and Perez-Albuerne (1982)
Glass/SnO_2/CdTe	663	28.1	0.563	10.5	4.0	Mitchell et al. (1985)

[a]Active area, 100 mW/cm^2 ASTM air mass 1.5 global spectrum, 25°C.
[b]84 mW/cm^2 ELH illumination.
[c]75 mW/cm^2 illumination.

Electrodeposited cells and modules are under development by BP Solar resulting in small area efficiencies up to 14.2% and a 10.1% efficient, 706 cm^2 aperture area module (Woodcock et al., 1991). Electrodeposition of CdTe has been demonstrated on areas up to 60 × 100 cm.

Atomic layer epitaxy (ALE) has also achieved 14% cell efficiencies. ALE deposits the CdTe from elemental Cd, S, and Te sources at a 420°C substrate temperature via sequential chemical reactions on the deposition surface. In addition to the typical glass/TC/CdS/CdTe structure, the device includes a 250 nm thick graded CdS_xTe_{1-x} between the 50 nm thick CdS and the 3 μm thick CdTe layers.

The CdTe cell performances achieved above are remarkably independent of processing approach, although important distinctions exist. Most devices required a postformation heat treatment to achieve high efficiencies. For example, deposition temperature is significant. A high melting point glass, Corning 7059 borosilicate glass, was used for the close-space sublimation depositions.

4.3. Multijunction Performance

Multijunctions, as presented in Section 2.4, provide the opportunity to achieve higher efficiencies than single junctions. This was especially important during the 1980s, when there was uncertainty that single junction thin-film devices could reach the U.S. Department of Energy goal of 15% efficiency. The most advanced thin film options during that time period were thin-film a-Si:H alloys (TFS) with band gaps near 1.7 eV and CIS. Not only were the TFS and CIS band gaps optimum for multijunctions, but the manufacturing technologies were complementary. TFS and CIS were superstrate and substrate device technologies, respectively, combining to form an environmentally durable glass/glass package. 15.6% efficiency was demonstrated in a 4 cm^2 four-terminal test device (Mitchell et al., 1988b). Subsequently, large area (0.4 m^2) tandem TFS/CIS modules achieved 41.5 W or 10.5% aperature efficiency (Mitchell et al., 1990).

The demonstration of high-efficiency all-polycrystalline tandems requires improvement in the performance of the high band gap junctions. CdTe presently is the highest efficiency top cell candidate (15%), but its 1.44 eV band gap limits the current into the bottom cell unless the CdTe is made very thin. A 9.9% efficient CdTe/CIS mechanically stacked solar cell was reported by Meyers et al. (1988). Adding Ga and S to CIS increases the band gap of CIS as shown in Table 8.3, but for band gaps above 1.3 eV, cell efficiencies are below 12%, requiring further development. In the future, single junction efficiencies above 15% should be achievable for 1.6–1.8 eV band gap polycrystalline materials, providing a basis for 20% tandem PV technologies.

For space applications, high-efficiency thin single crystal GaAs-based cells have been combined with CIS to achieve 23.1%AM0 efficiency. Efficiencies over 25%AM0 with specific module powers of 700 W/kg are projected, incorporating advanced GaAs heterostructures (Stanbery et al., 1987, 1990; Kim et al., 1991).

5. PROSPECTS

On a relative scale compared with other PV technologies, thin-film polycrystalline solar cells with total device thicknesses of 5 μm or less can tolerate significant levels of chemical, structural, and electronic impurities, inhomogeneities, and defects. While polycrystalline materials are compatible with several large-scale deposition techniques, including vacuum evaporation, sputtering, chemical vapor deposition, plasma deposition, electrodeposition, chemical spray, and sintered film, the spatial nonuniformities in thickness and composition and physical and chemical interactions at the interfaces do impact the resultant device performance. Different from crystalline Si and III–V compounds, polycrystalline thin-film deposition with glass as the typical substrate involves heterogeneous nucleation and growth, thermal expansion coefficient differences, lattice mismatch, chemical interactions at interfaces, and grain boundaries. Elimination of defects in large-scale, low-cost PV manufacturing is difficult. Determining how to minimize their impact on device performance through gettering or passivation is essential. The key issues are to define the specific trade offs between cell and module PV performance and the underlying layer and interface material and electronic properties and how to manipulate the deposition techniques to achieve these requisite material and electronics properties.

5.1. Key Research Topics

A performance loss analysis described in Section 2.3 and illustrated in Figure 8.3 implies maximum theoretical efficiencies of 23.5% and 27.5% for CdTe and CIS respectively (Sites, 1988). Subsequent improvements in the materials and devices suggest that practical single junction CdTe and CIS efficiencies can reach 20%–25% (Mitchell, 1989; Menner and Schock, 1992).

Pursuit of CIS and CdTe efficiencies beyond 15% requires understanding the primary mechanisms controlling device performance, namely, photocarrier collection, junction rectification, and electrical resistance losses. As noted in Section 2.1, for 100 mW/cm^2 ASTM AM1.5 G spectrum, the theoretical maximum photocurrents for CdTe and CIS are 31 and 51 mA/cm^2, respectively. Limitations to achieving theoretical J_{sc} arise from optical reflection, unwanted optical absorption in front layers, incomplete optical absorption in the absorber layer, and incomplete minority-carrier collection. Improved ZnO front transparent electrode texture and conductivity will reduce optical reflection and plasma absorption losses. Optical confinement by reducing the absorber thickness and using optical back reflectors will promote absorption of light in the active device region. In addition, improved minority-carrier lifetime and/or the use of minority-carrier mirrors will increase long wavelength response of the cell.

Analysis of the current–voltage behavior of CIS cells implies that the dominant junction loss mechanism is recombination (Mitchell and Liu, 1988; Mitchell, 1989). Microshunts (microscopic local areas of increased junction shunting) have been identified by optical beam–induced current (OBIC) and electron beam–induced current (EBIC) imaging (Chesarek et al., 1988). Substantial spatial variations in electronic response up to 0.1 mm in dimension are also evident in these films. Transmission electron microscopy (TEM) and energy dispersive X-ray analysis (EDX) have identified compositional and structral nonuniformities in the CIS layers. In addition, intragrain and grain surface properties and both the CdS/CIS and Mo/CIS interfaces contribute to junction recombination. The CIS defect chemistry within the grains, at the grain surfaces, and at the junction and contact interfaces ultimately determines the limit to recombination. All of these are related to the processing of the device either through the quality of the feedstock materials or through the processing conditions.

The additions of alloy elements, such as Zn and S to CdTe or Ga and S to CIS, can also increase V_{oc} by increasing the absorber layer band gap as long as the minority-carrier lifetime is not degraded. The nature of the heterojunction window layer also significantly impacts the V_{oc} and cell performance. High band gap II–VI and chalcopyrite window layers are options. Table 8.2 shows the promise of non-Cd window layers for CIS. For CIS, continued optimization of processing conditions is expected to achieve 550–570 mV V_{oc}, and modifications to the material and interfaces will achieve V_{oc} of 600 mV and higher. For CdTe, improved understanding of the defect chemistry and the role of oxygen and $CdCl_2$ annealing and better, durable electrical contacts are key research issues.

5.2. Prospects for Manufacturing

Strategies Unlimited, in a 1989 report assessing CIS PV technology, projected a CIS manufacturing cost of \$1.60/W_p for a 5 MW/year production level, dropping to \$0.72/W_p at a 25 MW/year production level. The primary figure of merit for PV technologies, \$/W, is derived from the manufacturing process parameters such as throughput, yield, and costs and the device performance parameters, namely, conversion efficiency, short circuit current J_{sc}, open circuit voltage V_{oc}, and fill factor FF.

Both CIS and CdTe are in the piloting phases of thin-film module development. Several papers have been presented on CIS module development (Mitchell et al., 1987, 1990; Pier et al., 1992). Table 8.5, a summary of module performance data, shows that module plate aperture efficiencies of 11.2% and 9.7% have been achieved on areas of 938 and 3,890 cm^2. An encapsulated and framed CIS module measures 37.7 W or 9.7% aperture efficiency.

TABLE 8.5 Siemens Solar Industries CIS Performance Achievements

Module Structure	Power (W)	Area (cm^2)	Eff (%)	I_{sc} (mA)	J_{sc} (mA/cm^2)	V_{oc} (V)	FF
1 × 1 ft module plate	10.5	938	11.2	641	39.3	25.5	0.64
1 × 4 ft module plate	40.8	3890	10.5	2550	37.0	24.5	0.65
1 × 4 ft module	37.7	3883	9.7	2440	35.5	24.0	0.64

Commercial screen-printed CdTe devices for consumer electronics are now manufactured by Matsushita Electric. BP Solar is in the pilot-production phase of manufacture using a CdTe electrodeposition process. Golden Photon (formerly Photon Energy) is fabricating CdTe modules for the PVUSA demonstration project by a chemical vapor process. Solar Cells Incorporated has been extending a technique similar to close-space sublimation technique to fabricate large area modules.

CdTe and CIS alloys have recently demonstrated device efficiencies in the 15% range, with promise of achieving 20%–25% efficiencies. Substantial challenges and opportunities in terrestrial applications are ahead. In addition, high-power densities and radiation hardness make CdTe and CIS thin-film PV technologies particularly attractive for space applications (Gay et al., 1984; Mickelsen et al., 1985; Burgess et al., 1988; Landis et al., 1989). Several thin-film CdTe and CIS large-scale module manufacturing technologies are being developed, capable of achieving less than $1 per watt.

REFERENCES

Albin, D.S., and Risbud, S.H. (1987), *Thin Solid Films* **147**, 203–212.

Albright, S.P., Ackerman, B., and Jordan, J.F. (1990), *IEEE Trans. Electron Devices* **ED-37**, 434–437.

Aranovich, J., Ortiz, A., and Bube, R. H. (1979), *J. Vac. Sci. Technol.* **16**, 994–1003.

Arita, T., Suyama, N., Kita, Y., Kitamura, S., Hibino, T., Takada, H., Omura, K., Ueno, N., and Murozono, M. (1988), *20th IEEE Photovoltaic Specialists Conf. Record*, IEEE, New York, pp. 1650–1655.

Arya, R.R., Lommasson, T., Russell, L., Wiedman, S., Skibo, S., and Catalano, A. (1992), *11th Photovoltaic Adv. Res. Dev. Mtg.*, pp. 112–115.

Barnes, J.O., Leary, D.J., and Jordan, A.G. (1980), *J. Electrochem. Soc.* 127, 1636–1640.

Basol, B.M., and Kapur, V.K. (1990), *21st IEEE Photovoltaic Specialists Conf. Record*, IEEE, New York, pp. 546–549.

Basol, B.M., Kapur, V.K., and Halani, A. (1991), *22nd IEEE Photovoltaic Specialists Conf. Record*, IEEE New York, pp. 893–897.

Basol, B.M., Stafsudd, O.M., Rod, R.C., and Tseng, E.S.F. (1980), *Proc. 3rd Eur. Solar Energy Conf.*, Reidel, Boston, pp. 878–881.

Basol, B.M., and Tseng, E.S. (1986), *Appl. Phyys. Lett.* **48**, 946–948.

Bhattacharya, R.N., and Rajeshwar, K. (1986), *Solar Cells* **16**, 237–244.

Bhatti, M.T., Groake, E.P., Miles, R.W., Carter, M.J., and Hill, R. (1991), *Proc. 10th Eur. Solar Energy Conf.,* Kluwer Academic, Boston, pp. 574–576.

Binsma, J.J., and Van Der Linden, H.A. (1992), *Thin Solid Films* **97**, pp. 237–243.

Birkmire, R.W., and Phillips, J.E. (1987), *Stable High-Efficiency $CuInSe_2$-Based Polycrystalline Thin Film Tandem Solar Cells: Final Contract Report*, Solar Energy Research Institute Report No. SERI/STR-211-3249.

Bloss, W. H., and Schock, H.W. (1988), *Proc. 8th Eur. Solar Energy Conf.,* Kluwer Academic, Boston, pp. 1571–1577.

Bodegard, M., and Stolt, L. (1992), *Proc. 11th Eur. Solar Energy Conf.*, pp. 878–881.

Bonnet, D., Ed. (1992), *Int. J. Solar Energy* **12**.

Bougnot, J., Duchemin, S., and Savelli, M. (1986), *Solar Cells* **16**, 221–236.

Bunshah, R.F., et al. (1982), *Deposition Technologies for Films and Coatings*, Noyes Publications, Park Ridge, NJ.

Burgess, R.M., Chen, W.S., Devaney, W.E., Doyle, D.H., Kim, N.P., and Stanbery, B.J. (1988), *20th IEEE Photovoltaic Specialists Conf. Record,* IEEE, New York, 908–912.

Chesarek, W., Mitchell, K., Mason, A., and Fabick, L. (1988), *Solar Cells* **24**, 263–270.

Chopra, K.L., and Das, S.R. (1983), *Thin Film Solar Cells*, Plenum Press, New York.

Chopra, K.L., Major, S., and Pandya, D.K. (1983), *Thin Solid Films* **109**, 1–45.

Chu, T.L. (1988), *Solar Cells* **23**, 31–48.

Chu, T.L., Chu, S.S., Britt, J., Chen, G., Ferekides, C., Schultz, N., Wang, C., and Wu, C.Q. (1992), *11th Eur. Solar Energy Conf.*, pp. 988–990.

Chu, T.L., Chu, S.S., Lin, S.C., and Yue, J. (1984), *J. Electrochem. Soc.* **127**, 2128–2185.

Chu, T.L., Chu, S.S., Pauleau, Y., Stokes, E.D., Jiang, C.L., Murthy, K., and Abderrassoul, R. (1981), *15th IEEE Photovoltaic Specialists Conf. Record,* IEEE, New York, pp. 1271–1276.

Coutts, T., Kazmerski, L., and Wagner, S., Eds. (1986), *Solar Cells* **16**.

Coutts, T., Kazmerski, L., and Wagner, S., Eds. (1988), *Solar Cells* **23**, 1–126.

Deb, S.K., and Zunger, A., Eds. (1987), *Proc. 7th Int. Conf. Ternary Multinary Compounds*, Materials Research Society, Pittsburgh.

Devaney, W.E., Mickelsen, R.A., and Chen, W.S. (1985), *18th IEEE Photovoltaic Specialists Conf. Record*, IEEE, New York, 1733–1734.

Dhere, N.G., Lourenco, M.C., Dhere, R.G., and Kazmerski, L.L. (1984), *Solar Cells* **13**, 59–65.

Dimmler, B., Dittrich, H., Klenk, R., Mauch, R.H., Menner, R., and Schock, H.W. (1988), *Proc. 8th Eur. Solar Energy Conf.*, Kluwer Academic, Boston, pp. 1583–1587.

Dimmler, B., Dittrich, H., Menner, R., and Schock, H.W. (1987), *19th IEEE Photovoltaic Specialists Conf. Record*, IEEE, New York, 1454–1460.

Ermer, J., Gay, R., Pier, D., and Tarrant, D. (1993), *J. Vacuum Science Technology* **A11**, 1888–1895. American Vacuum Society, New York.

Ermer, J.H., and Love, B.E. (1986), *Method for Forming $CuInSe_2$ Films*, U.S. Patent No. 4,798,660, filed Dec. 22, 1986, issued Jan. 17, 1989.

Ermer, J.E., Love, R.B., Khanna, A.K., Lewis, S.C., and Cohen, F. (1985), *18th IEEE Photovoltaic Specialists Conf. Record*, IEEE, New York, pp. 1655–1658.

Eskenas, K.L., and Mitchell, K.W. (1985), *18th IEEE Photovoltaic Specialists Conf. Record*, IEEE, New York, pp. 720–725.

Fahrenbruch, A.L., and Bube, R.H. (1983), *Fundamentals of Solar Cells*, Academic Press, New York.

Fonash, S.J. (1981), *Solar Cell Device Physics*, Academic Press, New York.

Gabor, A.M., Tuttle, J.R., Contreras, M., Albin, D.S., Franz, A., Niles, D.W., and Noufi, R. (1994), *Proc. 12th Eur. Solar Energy Conf.*, H.S. Stephens & Assoc., Bedford, U.K., pp. 939–943.

Gay, C.F., Potter, R.R., Tanner, D.P., and Anspaugh, B.E. (1984), *17th IEEE Photovoltaic Specialists Conf. Record*, IEEE, New York, pp. 151–154.

Ghandhi, S.K., Field, R.J., and Shealy, J.R. (1980), *Appl. Phys. Lett.* **37**, 449–451.

Grindle, S.P., Smith, C.W., and Mittleman, S.D. (1979), *Appl. Phys. Lett.* **35**, 24–26.

Haacke, G. (1977), *Annu, Rev. Mater. Sci.* **7**, 73–93.

Hall, R.B., Birkmire, R.W., Eser, E., Hench, T.L., and Meakin, J.D. (1980), *14th IEEE Photovoltaic Specialists Conf. Record*, IEEE, New York, pp. 706–711.

Hedstrom, J., Ohlsen, H., Bodegard, M., Kylner, A., Stolt, L., Hariskos, D., Ruckh, M., and Schock, H.W. (1993), *23rd IEEE Photovoltaic Specialists Conf. Record*, IEEE, New York, pp. 364–371.

Hodes, G., and Cahen, D. (1986), *Solar Cells* **16**, 245–254.

Hovel, H.J. (1975), *Solar Cells* **11**, Academic Press, New York.

Hu, J., and Gordon, R.G. (1992), *J. Appl. Phys.* **71**, 880–890.

Huber, W., Lopez-Otero, A., Formann, C., Fahrenbruch, A.L., and Bube, R.H. (1981), *15th IEEE Photovoltaic Specialists Conf. Record*, IEEE, New York, pp. 1062–1067.

Jarzebski, Z.M. (1973), *Oxide Semiconductors*, Pergamon, New York.

Joliet, M.C., Antoniadis, C., Andrew, R., and Laude, L.D. (1985), *Appl. Phys. Lett.* **46**, 266–267.

Kapur, V.K., Choudary, U.V., and Chu, A.K. (1984), *Process of Forming A Compound Semiconductive Material*, U.S. Patent No. 4,581,108, filed Jan. 6, 1984, issued Apr. 8, 1986.

Kazmerski, L.L. Ed. (1980), *Polycrystalline and Amorphous Thin Films and Devices*, Academic Press, New York.

Kazmerski, L., and Emery, K.A. (1992), *Proc. 2nd World Renewable Energy Congress*, Reading, United Kingdom, in press.

Kessler, J., Velthaus, K.O., Ruckh, M., Laichinger, R., Schock, H.W., Lincot, D., Ortega, R., and Vedel, J. (1992), *Proc. 6th Int. PV Science and Engineering Conf.*, IBH Publishing, New Delhi, pp. 1005–1010.

Kim, N.P., Stewart, J.M., Stanbery, B.J., Mickelsen, R.A., Devaney, W.E., Chen, W.S., Burgess, R.M., McClelland, R.W., Shastry, S., Dingle, J., Hill, D.S., Dingle, B.D., and Gale, R.P. (1991), *22nd IEEE Photovoltaic Specialists Conf. Record*, IEEE, New York, pp. 68–72.

Landis, G.A., Bailey, S.G., and Flood, D.J. (1989), *Advances in Thin-Film Solar Cells for Lightweight Space Photovoltaic Power*, NASA Technical Memorandum 102017, National Technical Information Service, Springfield, Virginia.

Laude, L.D., Joliet, M.C., and Antoniadis, C. (1986), *Solar Cells* **16**, 199–209.

Lincot, D., and Ortega-Borges, R. (1992), *J. Electrochem. Soc.* **139**, 1880–1889.

Lincot, D., Ortega-Borges, R., Vedel, J., Ruckh, M., Kessler, J., Velthaus, K.O., Hariskkos, D., and Schock, H.W. (1992), *Proc. 11th Eur. Solar Energy Conf.*, pp. 870–873.

Loferski, J.J. (1982), *16th IEEE Photovoltaic Specialists Conf. Record*, IEEE, New York, pp. 648–654.

Lowenheim, F.A. (1978), *Electroplating*, McGraw-Hill, New York.

Maissel, L.I., and Glang, R. (1970), *Handbook of Thin Film Technology*, McGraw-Hill, New York.

Major, S., Banerjee, A., and Chopra, K.L. (1983), *Thin Solid Films* **108**, pp. 333–340.

Matsumoto, H., Kuribayashi, K., Uda, H., Komatsu, Y., Nakano, A., and Ikegami, S. (1984), *Solar Cells* **11**, 367–373.

Mauch, R.H., Hedstrom, J., Lincot, D., Ruckh, M., Kessler, J., Klinger, R., Stolt, L., Vedel, J., and Schock, H.W. (1991), *22nd IEEE Photovoltaic Specialists Conf. Record*, IEEE, New York, pp. 898–902.

Menner, R., and Schock, H.W. (1992), *Proc. 11th Eur. Solar Energy Conf.*, pp. 834–837.

Meyers, P.V. (1986), *Proc. 7th Eur. Solar Energy Conf.*, Kluwer Academic, Boston, 1211–1213.

Meyers, P.V., Liu, C.H., Russell, L., Ramanathan, V., Birkmire, R.W., McCandless, B.E., and Phillips, J.E. (1988), *20th IEEE Photovoltaic Specialists Conf. Record*, IEEE, New York, pp. 1448–1451.

Mickelsen, R.A., and Chen, W.S. (1980), *Appl. Phys. Lett.* **36**, 371–373.

Mickelsen, R.A., and Chen, W.S. (1981), *15th IEEE Photovoltaic Specialists Conf. Record*, IEEE, New York, pp. 800–804.

Mickelsen, R.A., Chen, W.S., Stanbery, B.J., Dursch, H., Stewart, J.M., Hsiao, Y.R., and Devaney, W. (1985), *18th IEEE Photovoltaic Specialists Conf. Record*, IEEE, New York, pp. 1069–1073.

Milnes, A.G., and Feucht, D.L. (1972), *Heterojunctions and Metal-Semiconductor Junctions*, Academic Press, New York.

Minami, T., Nanto, H., Shooji, S., and Tanata, S. (1984), *Japan. J. Appl. Phys.* **23**, L280–ML282.

Mitchell, K., Eberspacher, C., Ermer, J., and Pier, D. (1988a), *20th IEEE Photovoltaic Specialists Conf. Record*, IEEE, New York, pp. 1384–1389.

Mitchell, K., Eberspacher, C., Ermer, J., Pier, D., and Milla, P. (1988b), *Proc. 8th Eur. Solar Energy Conf.*, Kluwer Academic, Boston, pp. 1578–1582.

Mitchell, K., Potter, R., Ermer, J., Wieting, R., Eberspacher, C., Tanner, D.P., Knopp, K., and Gay, R. (1987), *19th IEEE Photovoltaic Specialists Conf. Record*, IEEE, New York, pp. 13–18.

Mitchell, K.W. (1978), *Proc. Int. Electron Device Meeting*, 254–M257.

Mitchell, K.W. (1979), *Evaluation of the CdS/CdTe Heterojunction Solar Cell*, Garland Publishing, New York.

Mitchell, K.W. (1982). *Annu. Rev. Mater. Sci.* **12**, 401–415.

Mitchell, K.W. (1984), *Tech. Digest 1st International Photovoltaic Science and Engineering Conf.*, Japan Convention Services, Tokyo, pp. 691–694.

Mitchell, K.W. (1989), *Proc. 9th Eur. Solar Energy Conf.*, Kluwer Academic, Boston, pp. 292–293.

Mitchell, K.W., Eberspacher, C., Cohen, F., Avery, J., Duran, G., and Bottenberg, W. (1985), *18th IEEE Photovoltaic Specialists Conf. Record*, IEEE, New York, pp. 1359–1364.

Mitchell, K.W., Fahrenbruch, A.L., and Bube, R.H. (1975), *J. Vac. Sci. Technol.* **12**, 909–911.

Mitchell, K.W., and Liu, H.I. (1988), *20th IEEE Photovoltaic Specialists Conf. Record*, IEEE, New York, pp. 1461–1468.

Mitchell, K.W., Pollock, G.A., and Mason, A.V. (1988c), *20th IEEE Photovoltaic Specialists Conf. Record*, IEEE, New York, pp. 1542–1544.

Mitchell, K.W., Willett, D., Eberspacher, C., Ermer, J., Pier, D., and Pauls, K. (1990), *21st IEEE Photovoltaic Specialists Conf. Record*, IEEE, New York, pp. 1481–1486.

Mooney, J.B., and Lamoreaux, R.H. (1986), *Solar Cells* **16**, 211–220.

Morris, G.C., Tanner, P.G., and Tottszer, A. (1990), *21st IEEE Photovoltaic Specialists Conf. Record*, IEEE, New York, pp. 575–580.

Nagao, M., and Watanabe, S. (1968). *Jpn. J. Appl. Phys.* **7**, 684–685.

Nakayama, N., Matsumoto, H., Nakano, A., Ikegami, S., Uda, H., Yamashita, T. (1980), *Jpn. J. Appl. Phys.* **19**, 703–712,

Nakayama, N., Matsumoto, H., Yamaguchi, K., and Ikegami, S. (1976), *Jpn. J. Appl. Phys.* **15**, pp. 2281–2282.

Noufi, R., Powell, R.C., Ramanathan, V., and Matson, R.J. (1986), *Proc. 7th Int. Conf. Ternary Multinary Compounds*, Materials Research Society, Pittsburgh, pp. 453–458.

Nouhi, A., Stirn, R.J., and Hermann, A. (1987), *19th IEEE Photovoltaic Specialists Conf. Record*, IEEE, New York, pp. 1461–1465.

Ortega-Borges, R., Lincot, D., and Vedal, J. (1992), *Proc. 11th Eur. Solar Energy Conf.*, pp. 862–865.

Pamplin, B.R., Joshi, N.V., and Schwab, C., Eds. (1984), *Proc. 6th Int. Conf. Ternary Multinary Compounds*, Pergamon, New York.

Pier, D., Chesarek, W., Dietrich, M., Kuriyagawa, S., and Gay, R. (1992), *Proc. 11th Photovoltaic Adv. Res. Dev. Meeting*, pp. 107–111.

Potter, R.R., Eberspacher, C.E., and Fabick, L.B. (1985), *19th IEEE Photovoltaic Specialists Conf. Record*, IEEE, New York, pp. 1659–1664.

Rockett, A., and Birkmire, R.W. (1991), *J. Appl. Phys.* **70**, R81–R97.

Roderick, G.A. (1980), *Proc. 3rd Eur. Solar Energy Conf.*, Reidel Publishing, Boston, pp. 327–334.

Rohatgi, A., Sudharsanan, R., Ringel, S.A., and MacDougal, M.H. (1991), *Solar Cells* **30**, pp. 109–122.

Romeo, N., Canevari, V., Sberveglieri, G., Bosio, A., and Zanotti, L. (1986), *Solar Cells* **16**, 155–164.

Roth, A.P., and Williams, D.F. (1981), *J. Appl. Phys.* **52**, pp. 6685–6692.

Runyan, W.R. (1965), *Silicon Semiconductor Technology*, McGraw-Hill, New York.

Sagnes, B., Salesse, A., Artaud, M.C., Duchemin, S., and Bougnot, J. (1992), *Proc. 11th Eur. Solar Energy Conf.*, pp. 854–857.

Schock, H.W. (1994), *Proc. 12th Eur. Solar Energy Conf.,* H.S. Stephens & Assoc., Bedford, U.K., pp. 944–947.

Schock, H.W., Burgelman, M., Carter, M., Stolt, L., and Vedel, J. (1992), *Proc. 11th Eur. Solar Energy Conf.*, pp. 116–119.

Schock, H.W., and Pfisterer, F. (1992), *Proc. 11th Photovoltaic Adv. Res. Dev. Meeting*, American Inst. Physics, New York, pp. 511–519.

Schropp, R.E.I., Matovich, C.E., Bhat, P.K., and Madan, A. (1988), *20th IEEE Photovoltaic Specialists Conf. Record*, IEEE, New York, pp. 273–276.

Schuegraf, K.K. (1988), *Handbook of Thin-Film Deposition Processes and Techniques*, Noyes Publications, Park Ridge, NJ.

Serreze, H.B., Lis, S., Squillante, M.R., Turcotte, R., Talbot, M., and Entine, G. (1981), *15th IEEE Photovoltaic Specialists Conf. Record*, IEEE, New York, pp. 1068–1072.

Shafarman, W.N., Birkmire, R.W., Fardig, D.A., McCandless, B.E., Mondal, A., Phillips, J.E., and Varrin, R.D. (1991), *Solar Cells* **30**, 61–67.

Shay, J.L., and Wernick, J.H. (1975), *Ternary Chalcopyrite Semiconductors: Growth, Electronic Properties, and Applications*, Pergamon, New York.

Sherman, A. (1987), *Chemical Vapor Deposition for Microelectronics*, Noyes Publications, Park Ridge, NJ.

Sites, J.R., (1988), *20th IEEE Photovoltaic Specialists Conf. Record*, IEEE, New York, pp. 1604–1607.

Skarp, J., Koskinen, Y., Lindfors, S., Rautiainen, A., and Suntola, T. (1991), *Proc. 10th Eur. Solar Energy Conf.*, Kluwer Academic, Boston, pp. 567–569.

Smith, R.T. (1983), *Appl. Phys. Lett.* **43**, pp. 1108–1110.

Stanbery, B.J., Avery, J.E., Burgess, R.M., Chen, W.S., Devaney, W.E., Doyle, D.H., Mickelsen, R.A., McClelland, R.W., King, B.D., Gale, R.P., and Fan, J.C. (1987), *19th IEEE Photovoltaic Specialists Conf. Record*, IEEE, New York, pp. 280–284.

Stanbery, B.J., King, B.D., Burgess, R.M., McClelland, R.W., Kim, N.P., Gale, R.P., and Mickelson, R.A. (1990), *IEEE Trans. Electronic Devices* **Ed-37**, 438–442.

Stolt, L., Bodegard, M., Hedstrom, J., Kessler, J., Ruckh, M., Velthaus, K.O., and Schock, H.W. (1992), *Proc. 11th Eur. Solar Energy Conf.*, pp. 120–123.

Strategies Unlimited (1989), *Copper Indium Diselenide Photovoltaic Technology, Report T-8*, Strategies Unlimited, Mountain View, CA.

Sze, S.M. (1969), *Physics of Semiconductor Devices*, Wiley Interscience, New York.

Thornton, J.A., Cornog, D.G., Hall, R.B., Shea, S.P., and Meakin, J.D. (1984), *Proc. 17th IEEE Photovoltaic Specialists Conf. Record*, IEEE, New York, pp. 781–785.

Thornton, J.A., and Lommasson, T.C. (1986), *Solar Cells* **16**, 165–180.

Thouin, L., Guillemoles, J.F., Massaccesi, S., Ortega, R., Cowache, P., Rouquette-Sanchez, S., Lincot, D., and Vedel, J. (1992), *Proc. 11th Eur. Solar Energy Conf.*, pp. 866–869.

Tuttle, J.R., Noufi, R., and Dhere, R.G. (1990), *IEEE Trans. Electronic Devices* **ED-37**, 1494–1495.

Tyan, Y., and Perez-Albuerne, E.A. (1992), *16th IEEE Photovoltaic Specialists Conf. Record*, IEEE, New York, pp. 794–800.

Uda, H., Taniguchi, H., Yoshida, M., and Yamashita, T. (1978), *Jpn. J. Appl. Phys.* **17**, pp. 585–586.

Urabe, K., Hama, T., Roy, M., Sato, H., Fujisawa, H., Ohsawa, M., Ichikawa, Y., and Sakai, H. (1991), *22nd IEEE Photovoltaic Specialists Conf. Record*, IEEE, New York, pp. 1082–1087.

Velthaus, K.O., Kessler, J., Ruckh, M., Hariskos, D., Schmid, D., and Schock, H.W. (1992), *Proc. 11th Eur. Solar Energy Conf.*, pp. 842–845.

Verma, S., Varrin, R.D., Birkmire, R.W., and Russell, T.W.F. (1991), *22nd IEEE Photovoltaic Specialists Conf. Record*, IEEE, New York, 914–929.

Vossen, J.L., and Kern, W. (1978), *Thin Film Processes*, Academic Press, New York.

Walter, T., Menner, R., Koble, Ch., and Schock, H.W. (1994), *Proc. 12th Eur. Solar Energy Conf.*, H.S. Stephens & Assoc., Bedford, U.K., pp. 1755–1758.

Walter, T., Menner, R., Ruckh, M., Kaser, L., and Schock, H.W. (1991), *22nd IEEE Photovoltaic Specialists Conf. Record*, IEEE, New York, pp. 924–927.

Walter, T., Ruckh, M., Velthaus, K.O., and Schock, H.W. (1992), *Proc. 11th Eur. Solar Energy Conf.*, pp. 124–127.

Wenas, W.W., Yamada, A., Konagi, M., and Takahashi, K. (1991), *Jpn. J. Appl. Phys.* **30**, L441–L443.

Wolf, M. (1960), *Proc. IRE* **48**, 1246–1263.

Woodcock, J.M., Turner, A.K., Ozsan, M.E., and Summers, J.G. (1991), *22nd IEEE Photovoltaic Specialists Conf. Record*, IEEE, New York, pp. 842–847.

Wysocki, J.J., and Rappaport, P. (1960), *J. Appl. Phys.* **31**, 571–578.

Zanio, K. (1978), *Semiconductors and Semimetals, Vol. 13, Cadmium Telluride*, Academic Press, New York.

NOTE ADDED IN PROOF

Since the original manuscript was written, the performance of CIS-based PV technologies continues to advance with cell efficiencies approaching 17%. With a MgF_2 antireflection coating, a ZnO/CdS/Cu(In,Ga)Se_2 cell produced 16.9% active area efficiency with a 35.8 mA/cm^2 J_{sc}, 641 mV V_{oc}, and 0.735 fill factor (Hedstrom et al., 1993). A corresponding CIS cell had 15.4% efficiency with a 41.2 mA/cm^2 J_{sc}, 515 mV V_{oc}, and 0.726 fill factor. Subsequently, the CIS efficiency was improved to 16.0% with a 41.2 mA/cm^2 J_{sc}, 519 mV V_{oc}, and 0.75 fill factor (Schock, 1994). 12.0% efficient ZnO/CdS/$CuInS_2$ cells were also reported (Walter et al., 1994). Independently, a group at the National Renewable Energy Laboratory reported a 0.43 cm^2, 16.4% total area efficient ZnO/CdS/Cu(In,Ga)Se_2 cell (Gabor et al., 1994). These results demonstrate the rapid progress being made and support the analysis presented in the chapter, projecting thin-film polycrystalline PV efficiencies in the 20%–25% range.

PART TWO

MODULE TECHNOLOGIES

CHAPTER NINE

Silicon One Sun, Terrestrial Modules

STEPHEN J. HOGAN, Spire Corporation, Bedford, MA 01730-2396

1. INTRODUCTION

The interconnection and assembly of photovoltaic cells into a safe and stable package is an important aspect in the manufacture of terrestrial photovoltaic modules. Solar cells allow versatile system design opportunities, as they may be combined in a wide range of series and parallel configurations to obtain the desired voltages and currents. It is important that this combination of cells is packaged in a reliable structure to ensure achievement of the desired lifetime when exposed to environmental stresses. While presently the "typical" manufacturing process is expected to provide package lifetimes in excess of 20 years, there are a variety of techniques that may also be used for shorter life requirements. It is the intent of this chapter to address only the combining and packaging techniques commonly used on "power-producing" silicon-based modules. These modules typically require a stable package, as they are expected to deliver near full power over a maximum time frame. Later chapters will deal with other technologies and applications.

2. MODULE FABRICATION METHODS

The development of module fabrication has progressed over the last 15 years to yield a highly reliable and durable package for the interconnected cells. Much of the technology used today was originated in the efforts of the U.S. Department of Energy (DOE) program active in the early 1980s. This program,

Solar Cells and Their Applications, Edited by Larry D. Partain.
ISBN 0-471-57420-1

administered by the Jet Propulsion Laboratory (JPL), was aimed at addressing the manufacturing issues of photovoltaics. Efforts have also been carried out in other countries, but the current method of manufacture by all silicon cell manufacturers is predominantly similar.

The original techniques used in the manufacture of photovoltaic modules included highly labor intensive methods of interconnecting cells and encapsulating them. The cells had more variety in manufacture than today's techniques, as described in Chapter 2, and were metalized in several ways. The connecting means was determined largely by the cell metalization, which originally was evaporated contacts based on space cell technology. This often required a large-scale bonder similar to those used in the semiconductor industry. As the cell metalization progressed to either a plated or screen-printed system, the interconnect method became more standard. While some manufacturers used ultrasonic bonding of ribbon materials onto plated contacts, the majority of manufacturers have now resorted to reflow soldering thin, solder-plated copper ribbons onto the cells. These ribbons, or tabs, were (and often still are) manually soldered on to the cells by operators using soldering irons. Once the cell's tabs are soldered to their front contacts, they are interconnected to each other by soldering the tabs of one cell onto the back of another cell. This interconnection procedure is continued to form a string of cells. The strings are connected together using a heavier gauge bus ribbon to achieve the desired voltage and current. A typical module consists of 36 cells in series to generate about 15 volts at peak power operation. This voltage is appropriate for battery charging, while modules can be interconnected to generate higher voltages. The module current, directly related to the number of strings in parallel, is typically two to three amperes per string (depending on the cell size and efficiency).

The evolution of the module encapsulation method has a similarly interesting history. Original encapsulation techniques were also labor intensive, using silicone potting materials to embed the cell strings between two sheets of glass. This method, which proved to give a long product life, was expensive, messy, and slow. Curing of the silicone pottant compound often took several days. The JPL effort looked at alternate materials for both the encapsulation and glass processes. An early alternative to liquid encapsulation materials was the development and use of polyvinyl butyral (PVB) in sheet form placed between the cells and the glass. This material is used in the manufacture of laminated safety glass for automotive windshields. However, PVB requires special storage and handling and has a high durometer and a high moisture transmission rate. As a result, PVB was subsequently replaced by a formulation based on ethylene vinyl acetate (EVA). The EVA was less expensive, easier to store, softer (lower stress on fragile cells), less transmissive of moisture, and more stable. Recent investigations, however, suggest limitations on the applications of modules fabricated with this material (Wohlgemuth and Petersen, 1992; Pern and Czanderna, 1992), as discussed later. Other advancements in module design involve the materials used for the front and back covering.

While glass remained the material of choice for the front surface, the back glass was replaced by a durable plastic film. Research on the front glass showed that tempered, low-iron content glass is preferred. Tempering provides reliability against impact damage from such sources as hail and rocks. Preference is given to low-iron glass based on the higher transmission exhibited by this glass over the cell's spectral response band. This superior transmission improves the current from the module by up to 5%. The back surface of the modules was replaced by a film of one or more pre-laminated layers, including such materials as polyvinyl fluoride (PVF) film, aluminum foil, or polyester film (Nowlan, 1990). This rugged film resists mechanical abrasion, is much lighter in weight, and is easier to laminate than the glass it replaced. The research over the past 15 years has led to a durable, reliable package for module encapsulation.

A typical module design is shown in Figure 9.1. This diagram shows the cross-sectional structure, including the front glass (superstrate) material. The module production process begins with the electrical sorting of the solar cells under bias lighting. This sorting procedure groups cells with similar currents at a fixed voltage so that similar cells may be used in a module string. If nonsimilar cells are used, the current from the string is only as great as the current from the lowest current-producing cell in the string, thereby creating the possibility of significant power loss from higher current output cells. The cells have tabs applied to their front surface using a reflow soldering technique. The cells are then placed face down with the tabs of the adjacent cell over the back surface. A soldering operation connects the tabs from one cell to the

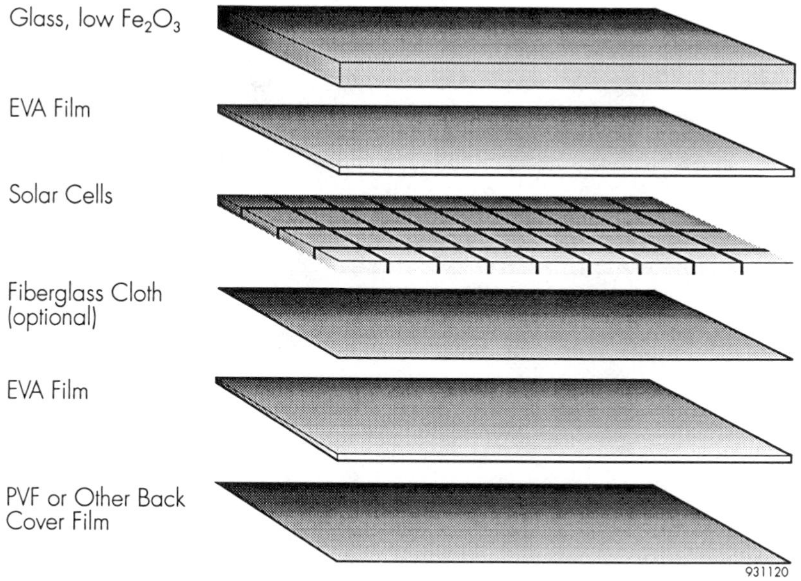

FIGURE 9.1 Superstrate module design.

other. Often an automated machine such as that shown in Figure 9.2 is used. The cell strings are then interconnected with bus ribbon into the desired series and parallel configuration.

The glass is cleaned, and a thin (450 μm) sheet of EVA is placed onto the surface of the glass. The solar cell strings are placed onto this assembly, followed by another layer of EVA, sometimes a porous fiberglass sheet, and finally the PVF or other film back covering material. This laminate assembly is often laid up with the back sheet being cut to allow the output bus ribbons to exit the laminate assembly prior to lamination.

The laminate assembly is placed in a laminator, where the assembly is subjected to a vacuum under a carefully heated environment. The time and temperature of this process varies with the EVA formulation and is the most crucial step in the formation of a reliable final laminate. A typical processing profile for conventional EVA includes a 5 min evacuation time followed by a 15 min cure time. The temperature during laminate and cure is around 150°C, at a vacuum of around 500 mTorr. The laminator has a diaphragm that provides the capabilities both to evacuate behind the back material and to press on it, forcing EVA to flow and fill any gaps in the module. Figure 9.3 is a cross-sectional depiction of the type of equipment used for the lamination process.

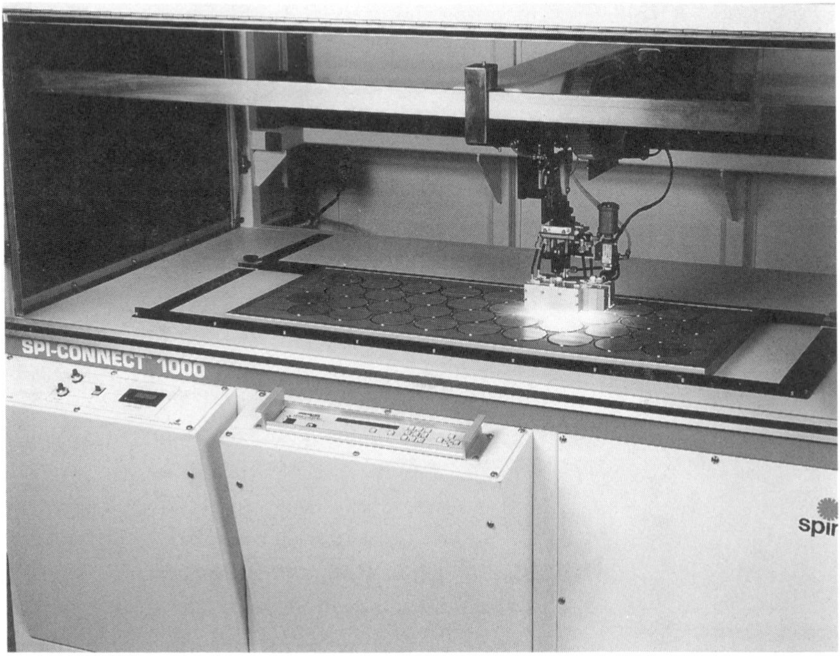

FIGURE 9.2 Automated soldering equipment.

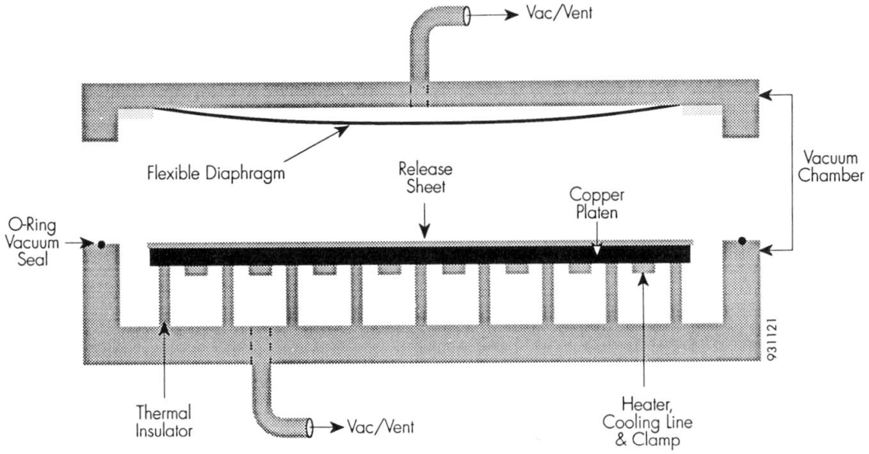

FIGURE 9.3 Lamination equipment process chamber.

The heat of the lamination process melts the EVA and activates a curing agent forming cross-linked bonds with exceptional adhesion. Peel strength and gel tests are done on the final product to verify that the EVA has cured sufficiently and created strong bonds between the glass, cells, and back cover material. Junction boxes are attached over the output bus ribbons with a weatherproof adhesive. Frames are placed around the edges of the glass using a rubber edge seal, completing the module fabrication process. The completed module is tested with a solar simulator, and labels are attached prior to packing the modules for shipment.

3. ECONOMIC ANALYSIS OF MANUFACTURING COSTS

Analyses and examinations of manufacturing costs of photovoltaic production have been reported for many years (Callaghan et al., 1985; Whisnant et al., 1985; Strategies Unlimited, 1987). Major government programs around the world have been focused on reducing the production costs associated with crystalline silicon module production and were instrumental in financing programs in thin-film technologies. Today the predominant technology of power module production throughout the world remains based on crystalline silicon techniques. It is therefore important to understand and appreciate the cost drivers of the technology.

The silicon technology of today is dominated by either cast (multicrystalline) or CZ (single crystal) materials. Each has inherent advantages, but both typically yield a similar cost per watt for the finished product. While the cast techniques can take advantage of lower growth costs, the efficiency of the final cells is typically less than single crystal material. This reduced efficiency raises the cost per watt. Presently the production cost for modules manufactured from either cast or CZ cells is around $5.00 (USD) per watt at the 1 MW per year production level. This cost is primarily driven by the production level, and early in 1985 an assessment was made of production at higher volumes (Hogan and Little, 1985). More recent analysis have focused on production quantities at 10 MW per year (Hogan et al., 1991a, b; Darkazalli et al., 1991) and production factors of conversion efficiency, labor rates, materials cost, and yield. These analyses predict the cost of manufacture for the 10 MW per year case to be in the range of $3.00/Watt.

The $3.00/Watt production cost at the 10 MW level makes several assumptions. The analysis is usually done for high labor rate areas (such as $7 per hour), a modest cell efficiency (14%), thick silicon wafers (400 μm), and moderate yields (90%). Figure 9.4 gives a bar chart of the costs associated with both cast and CZ module production.

These costs are based on such techniques as inner diameter blade wafering of the cast or CZ silicon, screen printing of contacts, and encapsulation using the glass/EVA/PVF systems. The capital equipment cost required to produce the wafers, cells, and modules is around $18 million. The analysis uses a 7 year depreciation schedule and a 100% overhead rate on an average $7.00 per hour labor rate on 160 employees. Cost analysis has used several tools, including the JPL developed IPEG model, the EPRI developed IMCAP model, and a Spire Corporation internal model called IACM. All models delivered production costs within 5% of each other. Figure 9.5 shows the breakdown of the cost for both CZ and cast silicon in categories of materials, labor, overhead, depreci-

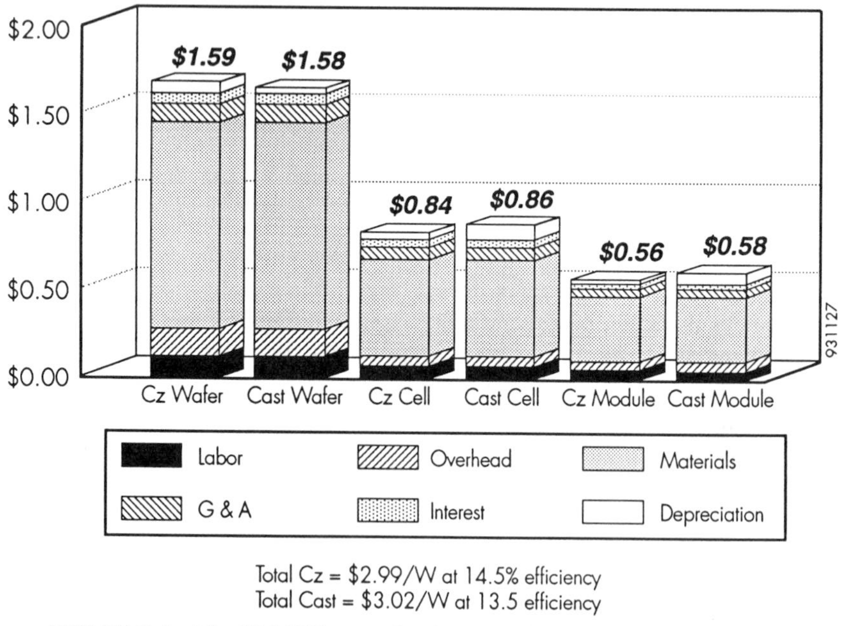

FIGURE 9.4 The 10 MW/yr production costs for different technologies.

"L"

Luminance = candela/sq.cm.

ϵ/cm^2

(E) Illuminance

photon : lumen/cm^2

lm/cm^2

steridian

Sun's power density { 1366 W/m

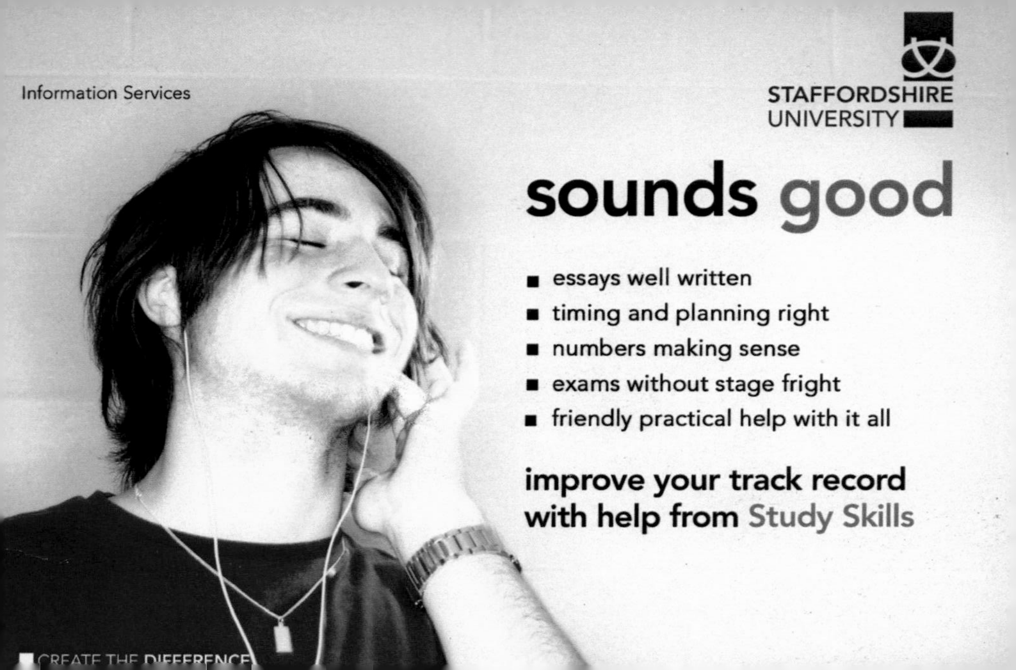
Information Services
STAFFORDSHIRE
UNIVERSITY
sounds good
essays well written
timing and planning right
numbers making sense
exams without stage fright
friendly practical help with it all
improve your track record
with help from Study Skills
CREATE THE DIFFERENCE

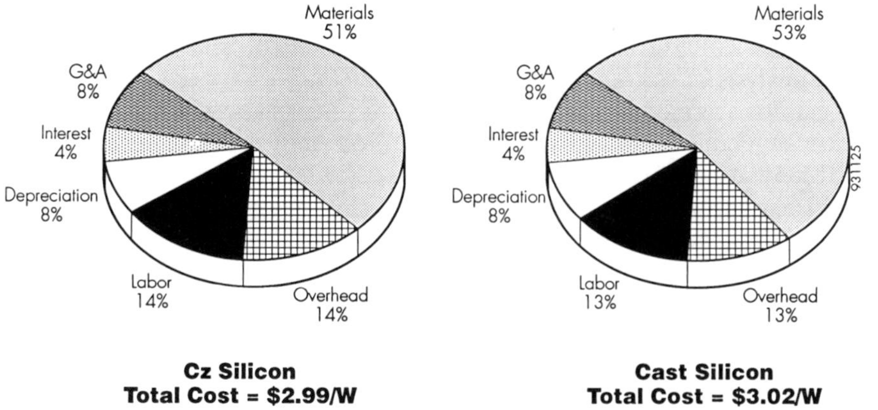

FIGURE 9.5 PV manufacturing cost categories.

ation, general and administrative (G&A), and interest. The analysis shows that the driving cost is associated with the materials required to fabricate the silicon cell and module.

When examining the future direction of the technology to determine the most promising cost reduction areas, it is helpful to look at the sensitivity of production costs to varying parameters. Figure 9.6 presents a cost sensitivity curve to several factors in the production assumptions and shows that cell efficiency and production level play major roles in the cost of the end product. For example, a 25% increase in cell efficiency over the baseline of 14.5% will reduce the manufacturing cost of the module by \$0.50/Watt. This type of analysis is useful in determining the emphasis of future activities to produce the greatest return on activities.

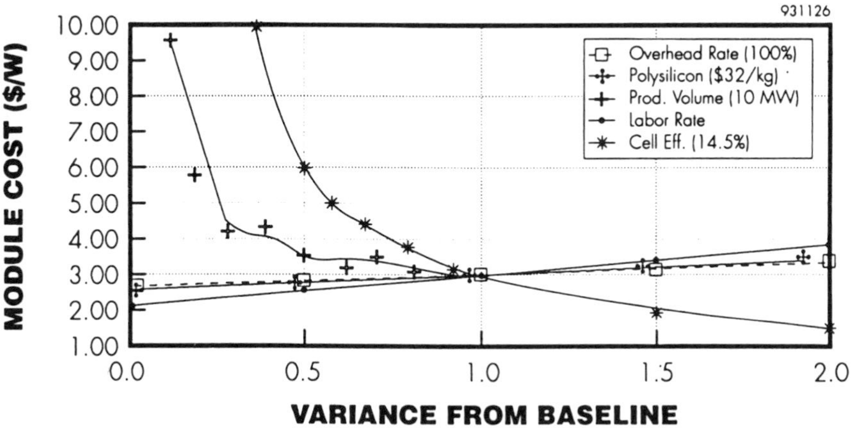

FIGURE 9.6 Cost sensitivity analysis.

4. FUTURE TRENDS

The analysis of cost sensitivity in the previous section suggests the directions of emphasis in module production. These directions may be summarized in the areas of module efficiency, materials, and production techniques.

Many techniques have been suggested for improving cell and module efficiency and were discussed in prior chapters. While many of these techniques may greatly increase the conversion efficiency of the cells and modules, the increased power from the modules must be weighed against the higher processing costs associated with efficiency enhancements. These trade offs have been discussed for cast silicon (Wohlgemuth et al., 1990) and other silicon technologies (Hogan et al., 1991a, b). These studies demonstrated that improved efficiencies often required very expensive capital equipment or greatly reduced throughput by adding slow steps. The future efforts to improve efficiencies must balance the power output gain against increased module production costs. It is likely that a 20% reduction in module production costs may be realized if reasonable cell efficiency improvements are made.

Module materials are also being examined to reduce costs and increase reliability. In particular, recent studies (Pern and Czanderna, 1992) have indicated possible long-term stability concerns with EVA as an encapsulant. While other reports (Wohlgemuth and Petersen, 1992) indicate that the stability question with EVA is only related to exposure to higher temperatures than normal, there is still a strong interest in resolving any questions regarding stability. Other materials being studied are alternate back cover materials and flexible module systems that do away with the glass front. To reduce module costs further, the elimination of the metal frame around the module is being considered, especially in applications where many modules may be mounted side by side such as in utility applications.

The economic analysis also indicates the strong relationship between cost and production levels. Recent programs have been implemented by both private industries and government agencies to stimulate the growth of producers. While such programs as the DOE PVMaT encourage the growth of the industry, the increased capacity also will reduce module cost. Increased production will allow the development of automation techniques and production processes that are affordable at these higher production levels. There is strong reason to believe that production prices for silicon-based technologies will fall below the $2.00/Watt level with the realization of advanced processing techniques and increased production levels.

REFERENCES

Callaghan, W.T., Henry, P.K., and McGuire, P.A. (1985), *18th IEEE Photovoltaic Specialists Conf. Record,* IEEE, New York, pp. 214–220.

Darkazalli, G.D., Hogan, S.J., and Nowlan, M.J. (1991), *22nd IEEE Photovoltaic Specialists Conf. Record,* IEEE, New York, pp. 818–821.

Hogan, S.J., Darkazalli, G.D., and Little, R. (1991a), *ISES Solar World Congr.*, Boulder, CO, 41–45.

Hogan, S.J., Darkazalli, G.D., and Wolfson, R. (1991b), *10th Eur. PVSEC.* WIP, Munich, pp. 276–280.

Hogan, S.J., and Little, R. (1985), *6th Eur. PVSEC,* WIP, Munich, pp. 1060–1068.

Nowlan, M.J. (1990), *Photovoltaic Module Reliability Conference*, NREL SERI/CP-4079, Golden, CO, pp. 259–278.

Pern, F.J., and Czanderna, A.W. (1992), *Solar Energy Mater. Solar Cells,* **25**, 3–23.

Strategies Unlimited (1987), *Crystalline Silicon Photovoltaic Manufacturing Cost Analysis,* PVIIS Report T-7, Mountain View, CA.

Whisnant, R.A., Champagne P.T., Wright, S.R., Brookshire, K.C., and Zuckerman, G.J. (1985), *18th IEEE Photovoltaic Specialists Conf. Record,* IEEE, New York, pp. 1537–1544.

Wohlgemuth, J.H., Narayanan, S., and Brenneman, R. (1990), *21st IEEE Photovoltaic Specialists Conf. Record,* IEEE, New York, pp. 221–226.

Wohlgemuth, J.H., and Petersen, R. (1992), *Photovoltaic Module Reliability Workshop,* NREL, Golden, CO, pp. 313–327.

CHAPTER TEN

Silicon Low-Concentration, Line-Focus, Terrestrial Modules

MARK J. O'NEILL, Entech, Inc., DFW Airport, TX 75261-2246

I. INTRODUCTION

Low-concentration, line-focus modules offer a straightforward path to near-term economic viability for terrestrial photovoltaic systems. At the present time, excellent-quality, proven-lifetime, one sun silicon cell modules are being mass produced by numerous companies around the world, as described in Chapter 9. However, the price of such modules is currently about $\$4/W_{Peak}$, which is several times too costly to compete with conventional energy sources. This one-sun module price equates to about $\$500/m^2$ on a price per unit area basis. In contrast, excellent-quality, proven-lifetime, acrylic plastic Fresnel lenses are being mass produced at prices of about $\$50/m^2$. When such lenses are used to intercept and focus incident sunlight onto one sun silicon cells, the total module cost is much lower than for one sun modules. For a 20 sun concentrator module, the cell area and cost are reduced to about $\$0.20/W_{Peak}$, while the lens cost equates to about $\$0.30/W_{Peak}$. In addition to these obvious cost advantages, low-concentration modules also provide excellent manufacturability, unequaled field performance, and demonstrated durability in the outdoor environment. The following sections further describe this simple, low-cost, robust photovoltaic technology.

Solar Cells and Their Applications, Edited by Larry D. Partain.
ISBN 0-471-57420-1

2. BRIEF DESCRIPTION OF A LOW-CONCENTRATION, LINE-FOCUS MODULE

Figure 10.1 shows the three key elements in a line-focus terrestrial concentrator module, as discussed below.

1. An optical *concentrator* focuses incident direct sunlight into a linear image. In Figure 10.1, the arch-shaped linear Fresnel lens is the optical concentrator.
2. A photovoltaic energy *receiver* is placed along the focal line of the optical concentrator. The receiver includes a number of series-connected silicon cells (the cell string) bonded to a waste heat dissipator. In Figure 10.1, the extruded aluminum finned heat sink provides natural convection air-cooling of the receiver. The receiver generally includes protective bypass diodes and is fully encapsulated for environmental durability.
3. A structural *housing* supports the concentrator and receiver. In Figure 10.1, the V-shaped housing is made of thin-gauge, marine-grade aluminum. The housing also encloses the air volume between concentrator and receiver, thereby minimizing moisture and dirt infiltration.

Thus the line-focus module is simple in construction and contains very few parts, making it amenable to low-cost, high-volume production.

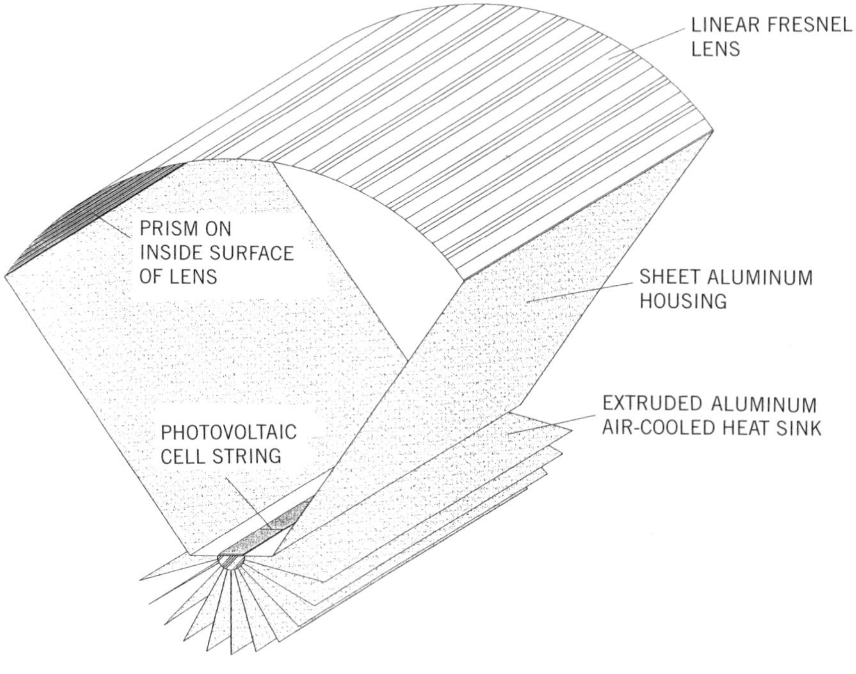

FIGURE 10.1 Low-concentration, line-focus module with air-cooled receiver.

3. RATIONALE FOR THE LOW-CONCENTRATION, LINE-FOCUS MODULE APPROACH

As decribed in Chapters 2 and 9, high-quality, one sun silicon cells and modules are currently being mass produced by a number of respected companies around the world. The market price for one sun silicon modules (including cells, electrical interconnections, front glazing, rear encapsulation, and perimeter framing) is presently about $\$4/W_{Peak}$ (Maycock, 1992–93). By adapting this one sun cell technology to low-concentration, line-focus modules, the cost content of the photovoltaic receiver (excluding the heat sink) becomes about $\$0.40/W_{peak}$ at a 10 sun concentration or $\$0.20/W_{peak}$ at a 20 sun concentration. Simply stated, a 10 sun concentrator saves about 90% of the required cell area (and cost), while a 20 sun concentrator saves about 95% of the required cell area (and cost), both compared with a one sun module.

To estimate the total cost of a line-focus module, the cost of the photovoltaic receiver (excluding heat sink) discussed above must be added to the costs of the optical concentrator (lens or mirror), the heat sink, and the structural housing elements. Fortunately, high-quality, long-life, acrylic plastic Fresnel lenses are already being mass produced by 3M at volume prices corresponding to about $\$0.30/W_{Peak}$ (Jaster, 1993). Similarly, natural convection, air-cooled heat sinks are currently being mass produced by aluminum extruders for a price equivalent to about $\$0.30/W_{Peak}$ (O'Neill and McDanal, 1993). Finally, aluminum housings are currently being mass produced by sheet-metal fabricators for prices equivalent to about $\$0.30/W_{Peak}$ (O'Neill and McDanal, 1993). Clearly, the sum of these mass-produced parts (about $\$1.10/W_{Peak}$ for a 20 sun module, or $\$1.30/W_{Peak}$ for a 10 sun module) is about 70% less than the current selling price for one sun photovoltaic modules.

Compared with higher concentration, point-focus modules, the line-focus approach is far more robust in terms of allowable tolerances during fabrication, installation, and operation (as discussed in later sections of this chapter). Furthermore, production runs of line-focus modules in the 20 sun range have already demonstrated excellent manufacturability, coupled with over 16% lot-average module efficiency (O'Neill et al., 1991a). Laboratory 20 sun modules have already exceeded 21% peak efficiency (O'Neill and McDanal, 1989). Independent, side-by-side array field tests of 20 sun concentrator modules versus one sun modules (mono-, poly-, and amorphous silicon types) have shown much better performance for the low-concentration technology (Candelario et al., 1992). Thus no technical breakthroughs are required to reach excellent performance levels at very attractive prices for low-concentration, line-focus modules. The technology is already in hand.

4. OPTICAL CONSIDERATIONS

Theoretically, either a reflective mirror or a refractive lens could be used to make the optical concentrator in a low-concentration, line-focus module. In the late 1970s and early 1980s, a number of substantial research, development, and demonstration (RD&D) programs were funded by the U.S. Department of Energy (DOE), related to reflective line-focus concentrator technology. Several large installations resulted from these multimillion dollar RD&D programs (Burgess and Walker, 1979). However, none of these reflective concentrator systems provided performance levels approaching expectations. Indeed, the highest long-term solar-to-electric conversion efficiency achieved by the best of these reflective concentrator systems was about 1%, despite the use of excellent concentrator cells (Thomas, 1984). The fundamental problem with reflective concentrators for photovoltaic applications is their sensitivity to mirror shape errors. Very small shape errors cause significant local image defocusing effects. For series-connected solar cells in the focal line of the mirror, the total current is limited to that produced by the least-illuminated cell in the string. Even with thermally sagged silvered glass mirrors, which have worked well in solar heat collector applications, these effects resulted in very low photovoltaic system performance. Other unique problems identified with these reflective systems included highly nonuniform irradiance over the cell, dirt accumulation on both mirrors and exposed photovoltaic receivers, and single-axis tracking losses (receiver end shading) for horizontally mounted concentrators.

Under the same DOE-sponsored RD&D program, a first-generation refractive line-focus concentrator system was also deployed in 1982 (O'Neill and Muzzy, 1985). In contrast to the negative experiences with reflective systems, the refractive system, using an optimized arched acrylic Fresnel lens, achieved an annual solar-to-electric conversion efficiency of about 8%, as anticipated. Furthermore, the refractive system attained a 5-year, long-term efficiency of about 7%, despite considerable power conditioning unit downtime (O'Neill, 1988). The refractive system used solar cells of the same type from the same manufacturer as those used in the reflective systems. Clearly, the difference in performance was in the optical concentrators. Based on these conclusive results, DOE and its cognizant National Laboratory, Sandia, scrapped any further reflective photovoltaic concentrator development work in the early 1980s. Fresnel lenses were selected as the concentrators of choice for future photovoltaic systems.

Fresnel lenses can be configured with prisms on either the inner or outer surface of the lens. To avoid substantial performance losses, the prisms should be on the inner lens surface (facing the receiver), not the outer lens surface (facing the sun). Figure 10.2 compares the solar ray paths for a portion of a lens with prisms on the inner surface versus a lens with prisms on the outer surface. Note that numerous rays are lost for the outer-prism lens configuration due to interceptions with the nonworking prism surfaces. Such losses are unavoidable for Fresnel lenses with prisms on the outer surface. In addition to avoiding such optical losses, the inner-prism lens configuration also minimizes dirt collection on the prismatic surface.

Fresnel lenses can also be configured in flat form or in arched form. Basic ray trace analyses easily demonstrate that the lens should have a convex outer surface for superior concentration performance. Figure 10.3 shows several

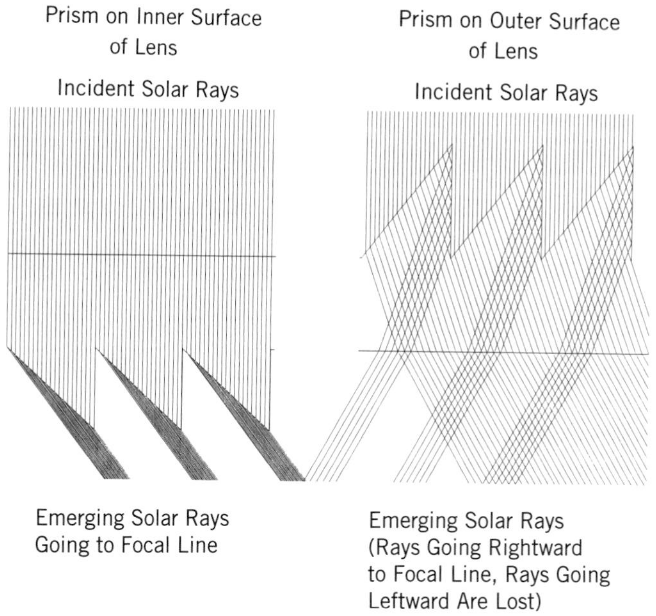

FIGURE 10.2 Ray trace showing why prisms should be on the inner lens surface.

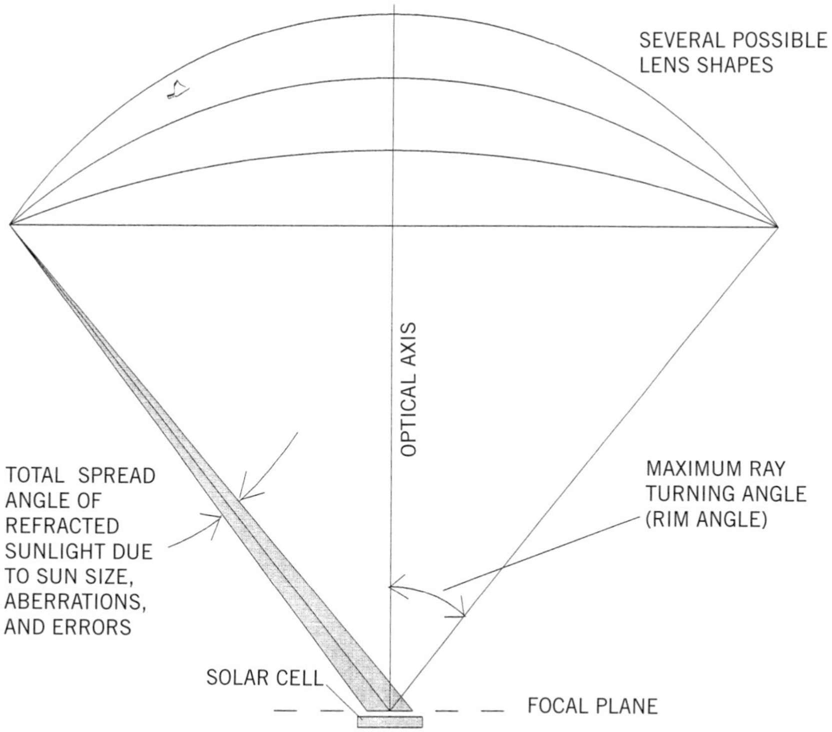

FIGURE 10.3 Possible lens shapes for a line-focus module.

possible lens shapes (of an infinite number of possibilities), each with prisms on the inner surface. Two important angles are shown in Figure 10.3: (1) the lens rim angle (ϕ_{rim}) which corresponds to the largest ray turning angle and which sets the aperture size of the lens; and, (2) the total spread angle of the refracted sunlight (θ_{spread}), which determines the width of the focal line. From basic trigonometry, for small values of the spread angle, one can easily show that the maximum value of the geometric concentration ratio (GCR = lens aperture width/cell active width) of the lens/receiver combination is

$$GCR = \sin(2\phi_{rim})/\tan(\theta_{spread}) \tag{1}$$

To maximize the GCR, a large rim angle and a small spread angle are clearly desirable. Table 10.1 compares the spread angle as a function of rim angle for two limiting-case lens designs. The first lens is a conventional *flat* Fresnel lens, with prisms on its inner surface. The second lens is a *transmittance-optimized* lens, which corresponds to a specific arched shape and prismatic pattern, which together ensure refraction symmetry for all parts of the lens. This refraction symmetry dictates that the solar ray angle of incidence at the smooth outer lens surface is equal to the solar ray angle of emergence at the prismatic inner lens surface for every location in the lens. From basic physical optics considerations, one can show that this symmetrical refraction lens design minimizes the

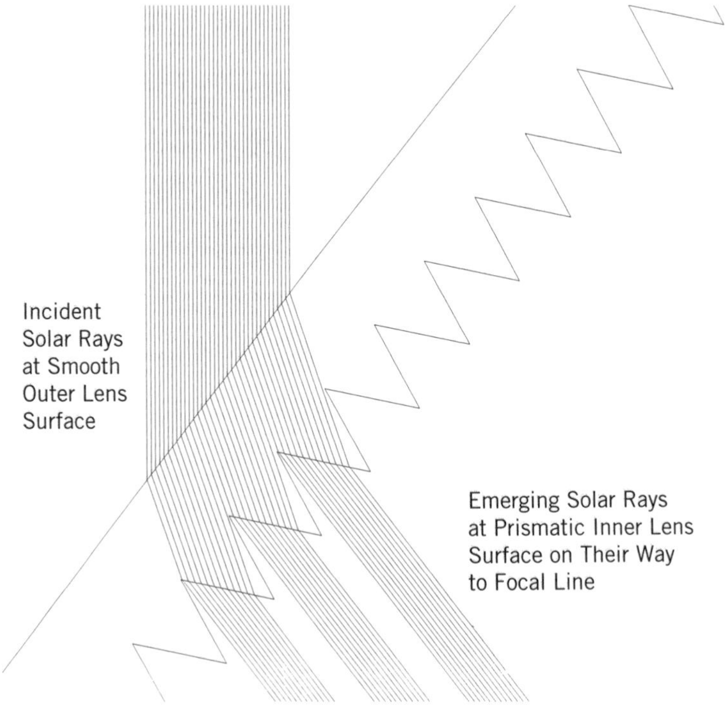

FIGURE 10.4 Ray trace for a portion of the symmetrical refraction lens.

reflection losses for every prism in the lens, thereby maximizing lens transmittance (O'Neill, 1978). Figure 10.4 shows the solar ray paths for a portion of a symmetrical refraction lens. Note that the angles of incidence and emergence are equal for each ray passing through the optimal lens. Note also that the tips of the prisms and the nonworking prism surfaces are well outside of the ray paths and therefore cause no optical losses.

In Table 10.1, six components of the solar ray spread angle are calculated (from Snell's refraction law and basic diffraction theory) for each lens type for a variety of ray turning angles (0°–39°). The first spread angle component is due to the finite angular diameter of the solar disk (about 0.53°) This angle is magnified by refraction through the flat lens, but is not changed in size by the optimal lens. The second spread angle component is due to shape errors, i.e., a local lens slope angle slightly different from the theoretical value. Slope errors of at least ±1° will be caused by temperature–humidity expansion–contraction effects alone, as further discussed in Section 8. The spread angle due to slope errors is large for the flat lens at large turning angles, but is negligible for the optimal lens, due to refraction symmetry. While not shown in Table 10.1, the spread angle for a reflective concentrator subject to the same slope error is 4° regardless of ray turning angle. Thus the shape error tolerance of the optimal lens is *hundreds of times better* than for flat lenses or reflective concentrators.

The third spread angle component in Table 10.1 is due to manufacturing inaccuracies in forming the prism angles. For acrylic lenses made by several sophisticated approaches (casting, compression molding, or 3M's Lensfilm process), small prism angle errors of ±0.125° are typical. For extruded lenses, the errors are much larger and the present results do not apply.

The fourth spread angle is due to dispersion (chromatic aberration). Acrylic plastic's refractive index varies from about 1.50 in the near-ultraviolet portion of the solar spectrum to about 1.48 in the near-infrared portion, causing this ray spread effect. The fifth spread angle component is due to diffraction. Individual beams of sunlight passing through the individual prisms are diffracted in the same manner as beams passing through narrow slits. The diffractive spread depends on the beam width and on the wavelength of the light. For commercially available lenses from 3M, the prisms have a constant depth of about 0.23 mm. The beam widths for both types of lenses were calculated based on this constant prism depth. The total spread angle containing 95% of the diffracted photons was then calculated for Table 10.1. The larger spread angle for the flat lens is due to the narrower beam width for the flat lens compared with the optimal lens.

The final spread angle component in Table 10.1 is due to a sun-pointing error of ±0.5°, a typical value for present sun-tracking arrays. Table 10.1 is truncated at 39° ray turning angle, because a flat lens begins to reflect rays totally internally at higher turning angles, resulting in zero transmittance. The optimal lens can operate at much larger turning angles, but all of the image spread functions continue to increase, resulting in unacceptably large image sizes for much larger turning angles.

TABLE 10.1 Optical Considerations: Spread Angle Functions (in Degrees) for Two Types of Fresnel Lenses

Lens Type	Ray Turning Angle	Solar Disk[a] Spread	Slope Error[b] Spread	Prism Error[c] Spread	Dispersive[d] Spread	Diffractive[e] Spread	Sun-Pointing[f] Spread
Flat	0	0.53	0.00	0.12	0.00	0.22	1.00
Optimal	0	0.53	0.00	0.12	0.00	0.22	1.00
Flat	3	0.53	0.01	0.13	0.12	0.22	1.01
Optimal	3	0.53	0.00	0.12	0.12	0.22	1.00
Flat	6	0.55	0.06	0.13	0.25	0.22	1.03
Optimal	6	0.53	0.00	0.13	0.25	0.22	1.00
Flat	9	0.56	0.13	0.15	0.38	0.28	1.07
Optimal	9	0.53	0.00	0.13	0.37	0.27	1.00
Flat	12	0.59	0.24	0.17	0.52	0.38	1.12
Optimal	12	0.53	0.00	0.13	0.50	0.36	1.00
Flat	15	0.63	0.39	0.19	0.68	0.50	1.19
Optimal	15	0.53	0.00	0.14	0.63	0.46	1.00
Flat	18	0.68	0.59	0.23	0.85	0.62	1.29
Optimal	18	0.53	0.01	0.15	0.76	0.56	1.00

Flat	24	0.85	1.20	0.34	1.29	0.95	1.60
Optimal	24	0.53	0.01	0.17	1.04	0.77	1.00
Flat	27	0.97	1.66	0.43	1.59	1.16	1.83
Optimal	27	0.53	0.01	0.18	1.19	0.89	1.00
Flat	30	1.14	2.32	0.55	1.98	1.45	2.15
Optimal	30	0.53	0.01	0.19	1.35	1.01	1.00
Flat	33	1.38	3.28	0.72	2.51	1.83	2.62
Optimal	33	0.53	0.01	0.21	1.52	1.15	1.00
Flat	36	1.76	4.85	0.99	3.30	2.40	3.34
Optimal	36	0.53	0.01	0.23	1.70	1.30	1.00
Flat	39	2.41	8.96	1.44	4.71	3.35	4.65
Optimal	39	0.53	0.02	0.25	1.89	1.46	1.00

[a]Solar disk spread angle is calculated based on the annual average solar disk angular diameter of 0.53°.

[b]Slope error spread angle assumes a $\pm 1°$ lens contour slope error, which is consistent with acrylic lens expansion/contraction due to temperature and humidity variations.

[c]Prism error spread angle corresponds to a prism apex angle error of $\pm 0.12°$, which is consistent with 3M's Lensfilm production process, which is the current state of the art in Fresnel lens manufacture.

[d]Dispersive (chromatic aberration) spread angle is based on acrylic plastic's refractive index variation from 1.50 in the ultraviolet to 1.48 in the infrared portions of the solar spectrum.

[e]Diffractive spread angle is calculated for a median wavelength of 0.7 μm and for a constant prism height of 230 μm, corresponding to the prisms produced by the continuous 3M Lensfilm process. The stated diffractive spread angle contains 95% of the energy in the diffracted beam passing through the prism.

[f]Sun-pointing spread angle assumes a $\pm 0.5°$ sun-tracking error about the critical axis, typical of present trackers.

The achievable GCR of the two lenses can be estimated by combining the various spread angles in Table 10.1 into a total spread angle. Although all six components are not truly random, a root-sum-of-squares (RSS) total is physically more meaningful than a simple algebraic summation. For the 39° turning angle case, the RSS total spread angle for the optimal lens is about 2.65°. Using Eq. (1), the achievable GCR is 21 ×. This 21 × GCR value is in fact being used in one commercially available line-focus concentrator, which employs a 40° rim angle optimal lens (O'Neill and McDanal, 1992). In contrast, a flat lens with a 39° rim angle could only achieve a GCR of 5 ×, based on the RSS total spread angle in Table 10.1.

For 27° rim angles, the optimal lens could achieve a GCR of about 25 ×, while the flat lens could achieve about 14 ×. However, for equal aperture width, the module depth is about 60% greater for the 27° rim angle than the 39° rim angle. Greater module depth is detrimental in terms of higher housing costs (more material) and increased wind loads (more drag area). Thus, the higher rim angle is generally preferred for the optimal lens, despite a slight reduction in achievable concentration ratio. For flat lenses, small rim angles and their inherent disadvantages must be tolerated to achieve reasonable concentration levels.

The most important optical performance parameter is net optical efficiency, which is defined for photovoltaic applications as the ratio of the number of photons focussed onto the solar cell per unit time divided by the number of photons incident on the outer lens surface per unit time. This definition is equivalent to the focussed photon flux integrated over the cell area divided by the one-sun photon flux integrated over the lens aperture area. Commercially available optimal lenses made by 3M, using a continuous Lensfilm process provide typical optical efficiency values of 90% ± 2% at 21 × GCR (O'Neill and McDanal, 1992). Optimal lenses made by E-Systems and ENTECH in the early 1980s, using a much less precise extrusion–embossing process, had much lower optical efficiency levels (e.g., 70% at 25 ×) due to malformed prisms (O'Neill, 1984a). Developmental arched lenses made by SEACorp, using the profile extrusion process have provided optical efficiency levels above 80% at 10 × (Kaminar et al., 1991, 1992). Due to their lower optical performance, flat lenses are no longer in use for line-focus concentrators. However, point-focus photovoltaic concentrators using flat lenses are still under development, as discussed in Chapter 11.

Another key optical consideration is the irradiance profile over the photovoltaic cell in the focal line of the lens. With a Fresnel lens, this profile can be tailored by design. Each individual prism's angle can be selected to place its individual image contribution at any desired position in the focal plane. The final irradiance profile over the cell can thus be tailored by selecting the angles for the thousands of individual prisms to provide overlapping images that add up to any desired pattern. Figure 10.5 shows the results of such image tailoring for two commercially available optimal lenses (O'Neill and McDanal, 1993). These two lenses have the same rim angle (40°), aperture width (85 cm), focal length (73 cm), and optical efficiency (90% at 21 ×). However, the prism angle

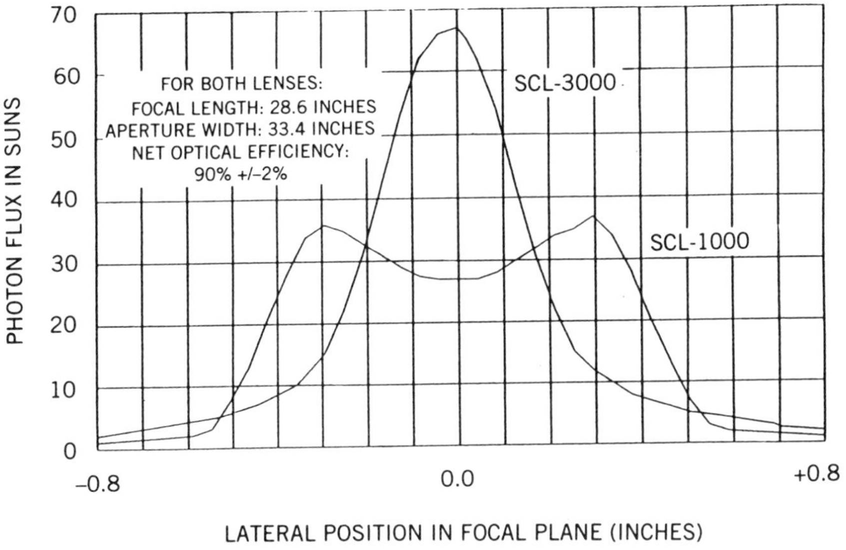

FIGURE 10.5 Measured focal plane photon flux profiles for two optimal arched Fresnel lens optical concentrators.

distributions are slightly different for the two lenses, one providing a milder focus (SCL-1000) and the other providing a sharper focus (SCL-3000), as shown in Figure 10.5. Figure 10.6 shows the sun-pointing error tolerance of the

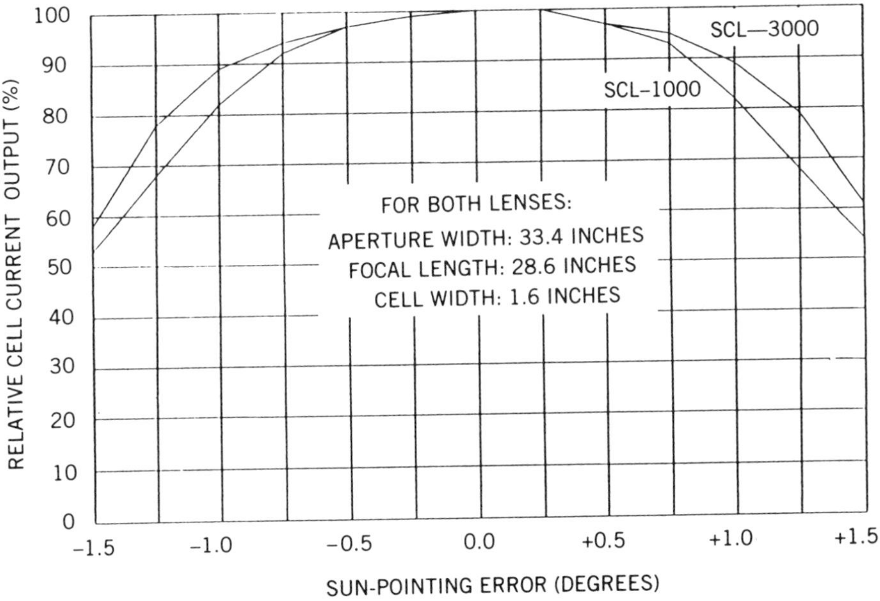

FIGURE 10.6 Measured sun-pointing error tolerance for two optimal arched Fresnel lens optical concentrators at 21 × geometric concentration ratio.

two lenses when they are coupled with a solar cell sized for 21× geometric concentration. The milder focus lens provides good (i.e., less than 10% loss) off-track performance up to about 0.8°, while the sharper focus lens provides good off-track performance up to about 1.0°. Figure 10.6 thus demonstrates the trade off between irradiance uniformity over the cell and sun-tracking error tolerance.

5. ELECTRICAL CONSIDERATIONS

Figure 10.7 shows a typical silicon cell used in a commercial line-focus concentrator module. The cell area represents one-half of a standard 10 cm square one sun type cell. Dual busbars allow the current to be efficiently extracted from both sides of the cell. The basic electrical characteristics of silicon cells have been fully discussed in Sections I and II of this book. The key consideration for cells to be used in low-concentration, line-focus modules relates to series-resistance minimization. Figure 10.8 shows the impact of total cell series resistance on cell efficiency for the cell shown in Figure 10.7, when that cell is operated at 20 suns irradiance. At this irradiance level, the current output of the cell is 25–30 A. Note that each milli-ohm of added resistance results in about one percentage point drop in cell efficiency.

Total cell series resistance is due to several components, which can be individually analyzed. Starting at the back of the cell, the current must flow

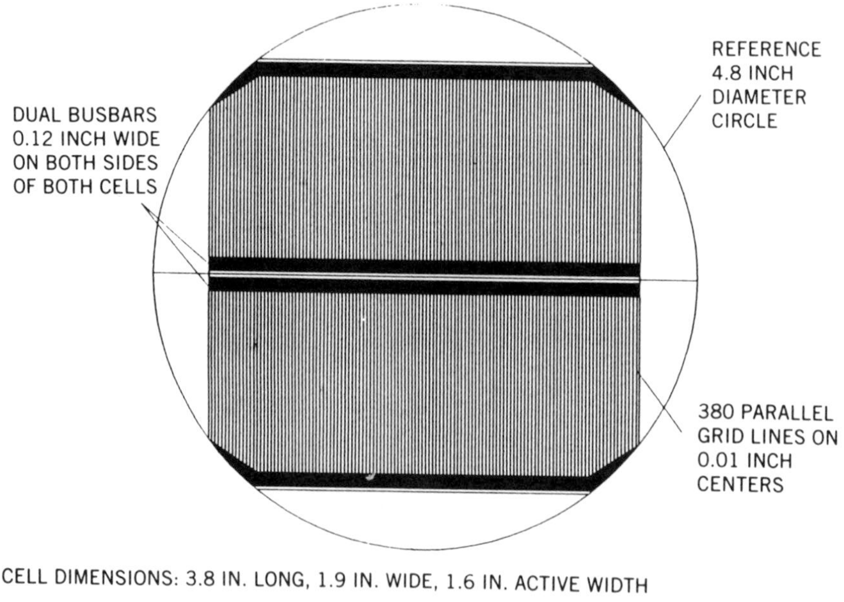

FIGURE 10.7 Two 20 sun concentrator cells on a single silicon wafer.

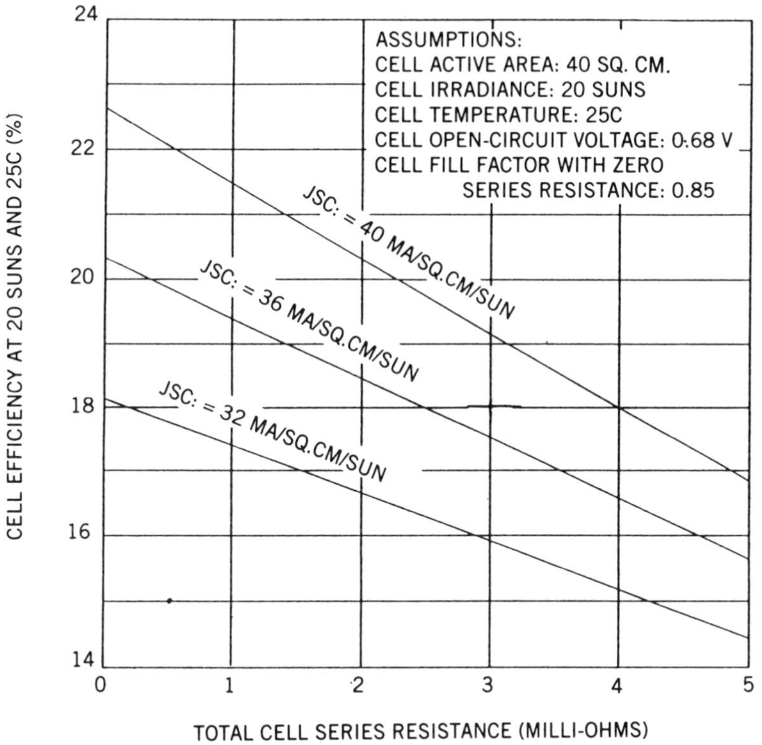

FIGURE 10.8 Calculated 20 sun cell efficiency versus total cell series resistance.

laterally across the back metalization, encountering normal metalization resistance. The current must then flow from the back metal into the bulk silicon material, encountering contact resistance between the back metal and the underlying semiconductor. The current must then flow from the backside of the cell to the emitter at the topside of the cell through the bulk silicon material, encountering bulk resistance. The current must then flow laterally in the emitter layer of the cell toward the topside metal gridlines, encountering sheet resistance. The current must then flow from the emitter layer into the grid lines, encountering contact resistance between the emitter and metal grid lines. The current must then flow along the grid lines to the busbars, encountering normal metalization resistance. Approximate equations for the key components of series resistance for line-focus concentrator cells can be derived by integrating the voltage drop (or power loss) along the various current flow paths (e.g., O'Neill, 1990).

To minimize the cell series resistance, very heavy metalization coverage is desirable on both the top and bottom surfaces of the solar cell. However, heavy metalization coverage on the topside of the cell normally carries a penalty relating to incident light blockage (grid shadowing loss). This potential loss can fortunately be overcome by using a prismatic cell cover, as shown in Figure 10.9 (O'Neill, 1987). The prismatic cover is made of transparent silicone rubber and includes small refractive elements that direct incident sunlight away from grid lines and onto active silicon material instead. By employing such prismatic cell covers, production cells of the type shown in Figure 10.7 have been manufactured in 20,000 cell quantities, with average efficiency levels of 18% at 20 suns irradiance and 25°C cell temperature (O'Neill et al., 1991a). These cells were made on low-cost, 0.5–1.0 Ω-cm Czochralski silicon wafers and employed 20%–40% topside metalization coverage to reduce total cell series resistance to about 1 mΩ.

To maximize system performance, cells that are series connected into circuits should be current matched. This is generally done by flash-testing individual cells at the design irradiance level, grouping them according to peak-power current, and ensuring that all of the cells in each receiver are from the same current group. Thereafter, modules that are series connected in the field should all employ cells that are likewise current matched.

Bypass diodes are used to prevent reverse-bias power dissipation overheating of cells that become shaded or cracked. Typically, no more than five cells are interconnected in series without a bypass diode protecting them.

All conductors and connectors in a receiver should obviously be capable of handling the highest anticipated module current, which corresponds to short circuit current under hot conditions with peak direct normal irradiance.

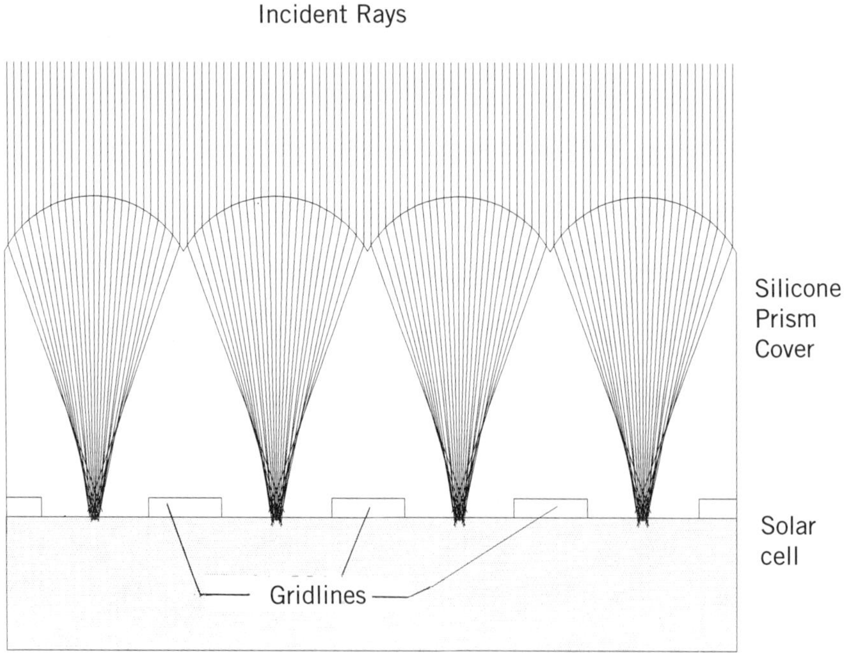

FIGURE 10.9 Ray trace of the prismatic cell cover.

Normally, conductors and connectors are selected based on a performance/cost trade off. For example, a conductor can be made larger to reduce resistive power loss, but this increases the conductor cost. The best conductor size is the one that minimizes system cost per unit of system energy delivered.

All electrically live parts of the receiver (including cells, diodes, conductors, and connectors) must be dielectrically isolated from the electrically grounded parts of the module (including heat sink and housing). Typically, the resistance between the live parts and the grounded parts of a receiver should be at least several hundred million ohms (several hundred megohms).

The complete photovoltaic receiver should be well encapsulated to prevent ground faults (current leakage from the electrically live parts of the receiver to the grounded parts of the module) under both dry and moist conditions (Barlow and Richards, 1988). Condensation routinely occurs inside concentrator modules, and small amounts of rain infiltration are not uncommon. Drain holes should be provided at appropriate locations in the module.

6. THERMAL CONSIDERATIONS

Since the operational conversion efficiency of silicon cells is typically less than 20%, more than 80% of the radiant energy focused onto the photovoltaic receiver must be dissipated as waste heat. The most common methods of rejecting waste heat are air cooling and water cooling. Figure 10.1 shows an air-cooled receiver, while Figure 10.10 shows a water-cooled receiver. Air

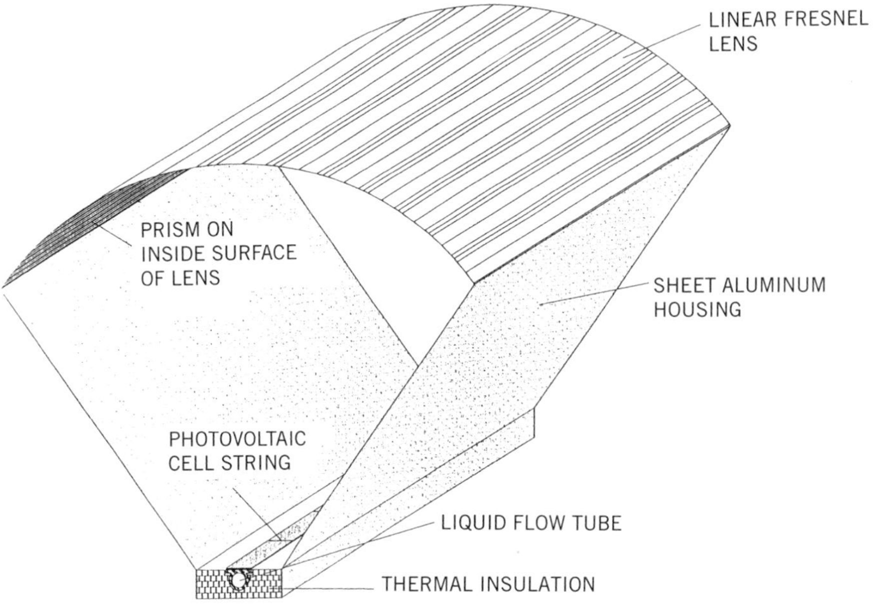

FIGURE 10.10 Low-concentration, line-focus module with water-cooled receiver.

cooling typically relies on passive, atmospheric natural convection cooling (with wind enhancement) of a finned heat sink, to which the cells are mounted. Water cooling generally relies on active, pumped-water (with antifreeze if needed) forced-convection cooling of a plate/tube heat sink to which the cells are mounted. Air cooling is simple and reliable. Water cooling is more complicated, but allows the thermal energy to be efficiently collected and used (e.g., for domestic water heating).

For either approach, the cell must be bonded to the heat sink with a thermally conductive, electrically insulating adhesive system. Silicone rubber, loaded with metal oxides to enhance thermal conductivity, is a popular and effective adhesive (e.g., alumina-loaded GE RTV-615). Dielectric isolation of 2,000–3,000 V can be obtained with an adhesive layer a few hundred microns thick, resulting in a typical cell-to-heat sink temperature difference of 10°–20°C for a 20 sun concentrator under peak sunlight conditions.

For air-cooled heat sinks, the heat sink-to-ambient temperature difference varies with wind speed, wind direction, and module orientation, as well as sunlight conditions. A critical parameter for any air-cooled heat sink is the ratio of total fin heat transfer area to total lens aperture area. The aperture area determines the quantity of waste heat that must be dissipated, while the fin area determines the effectiveness of convective heat removal. A direct analogy exists with one sun photovoltaic modules, which are also air cooled. Roof-mounted one sun modules generally only have their upper surface exposed for cooling, while frame-mounted one sun modules generally have both their upper and lower surfaces exposed for cooling. The ratio of heat transfer area to aperture area is about 1.0 for the former and about 2.0 for the latter. The temperature difference between module and air is much lower for the latter because of this better ratio. Commercially available line-focus modules have just begun to use very large heat sinks with about four times as much fin area as lens aperture area. With these new, large, extruded aluminum heat sinks, the heat sink-to-ambient temperature difference is typically only 15°–20°C for a 20 sun concentrator under peak sunlight conditions. Combining this temperature difference with the cell-to-heat sink difference results in a total cell-to-ambient temperature difference of about 25°–40°C, which is about the same as for a one sun module.

For a well-designed water-cooled heat sink, the heat sink-to-water temperature difference is generally less than 5°C. Thus, the overall cell-to-water difference is typically 15°–25°C for this approach. Obviously, the heat absorbed by the receiver cooling water must be removed at a remote thermal sink (e.g., by a heat exchanger in a hot water tank). This remote thermal sink generally dictates the temperature range of the receiver cooling water.

For water-cooled receivers, protection must be provided against overheating in the event of a flow loss (e.g., pump failure or hose leak). One common method of protection is to sense receiver temperature and to rotate the module away from the sun if the set-point temperature is exceeded.

7. SUN-TRACKING CONSIDERATIONS

As discussed in Section 4, line-focus concentrator modules in the 20 sun class must typically track the sun within $\pm 0.5^\circ - 1.0^\circ$ in the critical (lateral) plane to prevent significant light spillage beyond the cell string. In the orthogonal, less critical (longitudinal) plane, the same modules must track the sun within $\pm 10^\circ - 15^\circ$ to prevent significant lens defocusing effects. For lower concentration modules in the 10 sun class, these tracking error bands can be increased. In fact, single-axis tracking about a polar axis (parallel to the earth's spin axis) becomes practical for 10 sun and lower concentration ratios (Kaminar et al., 1992).

A variety of different approaches have been used for two axis sun tracking of line-focus modules. Several groups have experimented with large "heliostat" sun trackers, which employ rectangular structural platforms mounted on top of tall posts. The platform generally rotates in a bearing on top of the post, about a fixed vertical axis, for azimuth tracking. The platform is then tilted with a jack, about a moving orthogonal axis, for elevation tracking. Line-focus modules are mounted to the large platform, which can measure 10 m square or larger. This approach is also common for developmental point-focus concentrator systems, as discussed in Chapter 11.

Other groups have successfully employed low-profile "tilt–roll" sun trackers (O'Neill, 1984b). These units employ picture-frame-like structures, supported on a series of posts aligned in the east–west direction. The frames tilt from north to south in bearings mounted on top of the posts. The line-focus modules are mounted within the frames and individually rotate from east to west in Venetian-blind fashion. Tilt–roll trackers measuring 100 m long in the east–west direction, populated with 60 modules and providing over 167 m^2 of total lens aperture, have been successfully deployed and operated in Texas and California (O'Neill et al., 1991a). Incidentally, if the tilt angle is held constant at the local latitude angle, the tilt–roll tracker approach becomes the single-axis polar-tracker approach.

The tilt–roll approach provides better land use efficiency than the heliostat approach. The lens aperture–to–land area ratio is typically 20%–30% for a tilt–roll tracker compared with only 10%–15% for a heliostat tracker. The tilt–roll tracker is also closer to the ground than the heliostat tracker, simplifying module installation and alignment.

Various tracking control approaches have also been employed for line-focus concentrator systems. Early systems in the 1970s and 1980s primarily used sun-sensor–based, closed-loop methods, while more recent systems deployed in the 1990s generally use microprocessor-based, open-loop technology. The latter approach is now preferred, based on both reliability and economics.

8. PRACTICAL FIELD CONSIDERATIONS

Line-focus systems of the type described in this chapter have now been in the field for more than a decade. A number of practical lessons have been learned from these fielded systems, as further described below.

Acrylic plastic is the preferred material for the line-focus Fresnel lens. Acrylic samples have been exposed to the outdoor environment for more than 35 years, with very little optical or mechanical degradation (Mahoney et al., 1993). However, acrylic undergoes significant expansion and contraction with variations in both temperature and humidity. The coefficient of thermal expansion (CTE) is about 0.0001/°C. In the harshest environments, the lens could experience temperature swings from −40°C in winter to 60°C in summer, corresponding to the temperature cycling range adopted by Sandia National Labs for concentrator module testing (Barlow and Richards, 1988). Every dimension of the lens will change by 1% due to this temperature variation. In addition, acrylic absorbs water and swells, increasing every dimension by another 0.6% from dry conditions to saturated conditions. Combining these two effects, *overall dimensional changes of 1.6%* can occur between dry, cold winter conditions and humid, hot summer conditions. For a 3.7 m long lens, which is currently used in a commercially available concentrator, this equates to a 6 cm change in lens length. Similarly, for a typical 1 m wide lens, the width increases by nearly 2 cm. Designing a module to accommodate this enormous lens growth is no trivial problem. The most successful approaches employed to date have allowed the straight edges of the lens to slide in slots formed in the housing sidewalls to accommodate lengthwise growth and have allowed the arched lens to bow up and down to accommodate lateral growth.

Hail resistance is another important lens consideration. In many regions of the world, 2.5 cm diameter hail stones can be anticipated at least once in a 20–30 year field lifetime for a photovoltaic system. Sandia has performed numerous hail tests on acrylic lens materials over the years (Barlow and Richards, 1988). For a normal outdoor-grade acrylic sheet, a 3 mm thickness is required to withstand 2.5 cm hail stones at terminal velocity with no damage. For newer impact-resistant grades of acrylic (e.g., Rohm and Haas Implex), 1.5 mm thickness is adequate to survive the same hail impact.

The structural support aspects of module and tracker design are also crucial to long field lifetime. Generally, concentrator modules are designed to withstand survival wind loads in the 80–100 mph range without permanent damage. Similarly, tracking structures are generally designed to withstand survival wind loads in the 80–100 mph range without permanent damage. With a properly designed control system, including a reliable wind sensor and a backup drive motor power supply, the tracking structure can be placed in a safe "stow" position during periods of high wind, thereby minimizing the total area of modules and structures exposed to the wind. In addition to survival wind load considerations, the effect of operational wind loads must also be analyzed to ensure that deflections will not impact normal system performance.

Shading losses between adjacent modules and/or adjacent sun trackers must be carefully considered. Computerized hourly simulations of system performance throughout the year provide the easiest method of optimizing field layout. Generally, as the ground cover ratio (aperture area divided by land area) increases, shading losses increase, while land-related costs decrease. The optimal layout usually corresponds to the lowest system cost per unit energy delivered. However, for some applications, energy delivered at certain times (e.g., utility on-peak-load periods) is far more valuable than energy delivered at other times (utility off-peak-load periods), which must be factored into the optimization.

Performance degradation due to dust and dirt effects must also be considered. Generally, line-focus concentrator modules are constructed with weather seals between the lens and housing, and between the housing and receiver, to minimize dust and dirt infiltration into the module. However, external lens surfaces are subject to dirt accumulation. Data from actual systems in the field indicate that typical losses due to dirt buildup over several months of exposure are in the range of 5% (O'Neill and Muzzy, 1985). The frequency of lens washing should be established for a particular application, based on a trade off of the cost of washing versus the value of the energy lost between cleanings.

System lifetime experiences for line-focus modules have been positive. For each of the first 5 years of system operation (1982–1987), Sandia personnel carefully measured the electrical performance characteristics (IV curves) of every source circuit in an experimental 25 kW system at the Dallas–Ft. Worth Airport. No measurable performance degradation was detected over this period (O'Neill, 1988). Later, this experimental system was dismantled after 10 years of operation, and two modules were taken apart for component autopsies. After cleaning, the lenses were tested and found to be within 3% of their initial optical efficiency. Similarly, after cleaning, the receivers were tested and found to be within 5% of their initial electrical power output. Furthermore, the electrically live portions of the receivers (cells, diodes, and connectors) were also found still to be dielectrically isolated from the heat sinks ($>100\,M\Omega$).

9. SYSTEM PERFORMANCE CONSIDERATIONS

For a line-focus concentrator module, the module efficiency can be simply and accurately estimated as the product of lens optical efficiency times cell electrical conversion efficiency. As described in previous sections of this chapter, commercially available line-focus modules are being produced with 90% efficient lenses and 18% efficient cells (at 25°C cell temperature). These modules are therefore about 16% efficient at 25°C cell temperature. Figure 10.11 shows how the module efficiency varies with cell temperature. At a more typical operating cell temperature of around 60°C, the module efficiency is about 14%. When such modules are installed into source circuits in the field, wiring and mismatch losses account for about a 5% relative reduction in performance. Thus, an operational field efficiency of about 13% is typical for present line-focus systems. Future systems will undoubtedly perform at much higher levels, as high-efficiency laboratory cells make their way to commercial production.

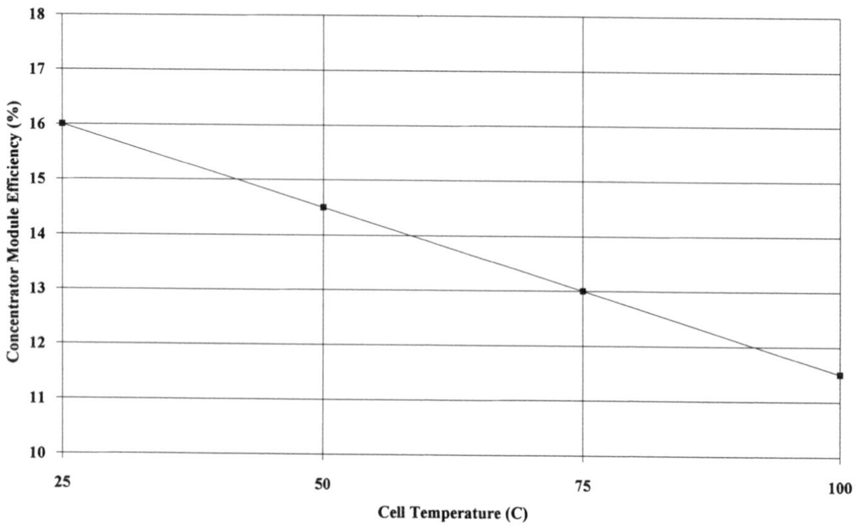

FIGURE 10.11 Typical production line-focus concentrator module efficiency.

The relative performance of line-focus concentrators versus other photovoltaic technologies has been independently measured by the PhotoVoltaics for Utility Scale Applications (PVUSA) Project, which has been jointly funded by the U.S. DOE, Pacific Gas and Electric Company, and other participants (Candelario et al., 1992). Located at Davis, California, the PVUSA Project includes side-by-side utility-scale (>20 kW each) photovoltaic systems of many different technologies. One sun modules using monocrystalline silicon, polycrystalline silicon, and amorphous silicon have been tested side by side with a 20 sun line-focus concentrator array. Figure 10.12 shows the relative system efficiencies through 1991. Note that the various systems became operational at different times (e.g., the concentrator system has operated since April 1991). Clearly, the line-focus concentrator system has provided the highest efficiency of all the technologies.

Energy production is more important than array efficiency. Figure 10.13 shows the measured monthly capacity factor for all of the PVUSA systems for 1991 (Candelario et al., 1992). Monthly capacity factor is defined by PVUSA as the electrical energy produced during the month divided by the kW rating of the system and also divided by the total number of hours in the month. Alternatively, capacity factor is equivalent to the percentage of total time (including nights) for the month that the photovoltaic system produced its rated power. (Utilities commonly use capacity factor as a measure of performance of conventional plants as well.) Note that the concentrator system has greatly outperformed the fixed one sun arrays in terms of capacity factor, especially so for the summer peak-load months, when the value of the electricity is the greatest (Hoff and Iannucci, 1990). Thus, in terms of both efficiency and energy production, the line-focus concentrator technology has been the performance leader at PVUSA.

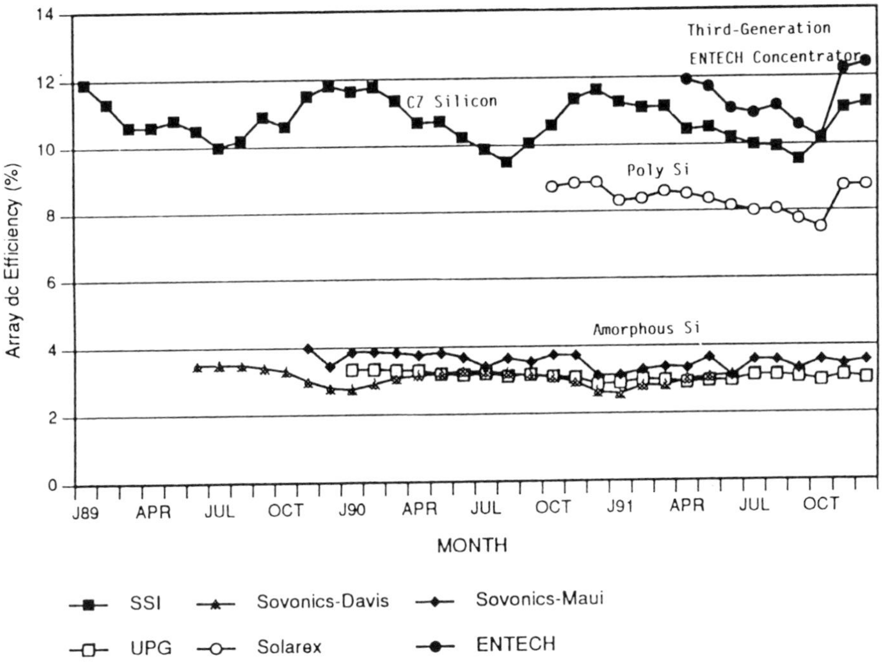

FIGURE 10.12 Photovoltaic array efficiencies at PVUSA.

The energy production of a line-focus photovoltaic system can be accurately predicted with an hour-by-hour computer simulation that treats variations in direct irradiance, ambient temperature, wind speed, module-to-module shading, sun tracker–to–sun tracker shading, and non-normal incidence angle effects. For a well-designed system, tracking and shading losses can be limited to about 10% on an annual basis, resulting in an annual sun-tracking efficiency of 90%.

A first-order approximation to the annual energy production of a line-focus concentrator system can be calculated as the product of the following five parameters:

1. Total collector field lens aperture area (m^2)
2. Annual total direct normal irradiance (DNI) at the site (kWh/m^2)
3. Annual sun-tracking efficiency (typically 90%)
4. Array operational efficiency (presently about 13%)
5. Power conversion unit (DC/AC inverter) efficiency (typically 95%)

The product of the first two items defines the annual solar energy available to the system. The annual direct normal irradiance varies widely with location

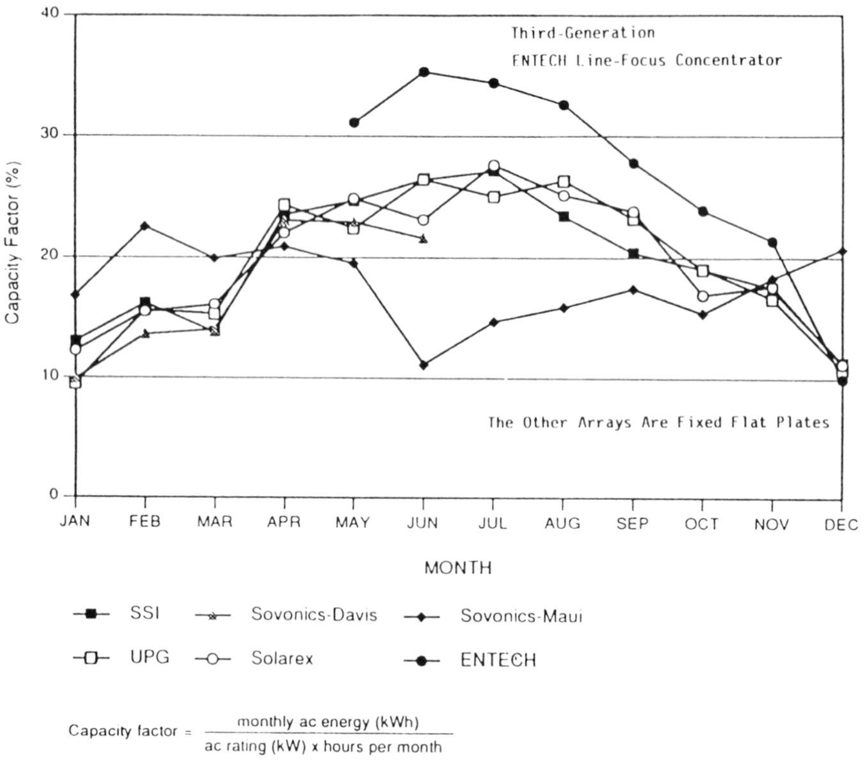

FIGURE 10.13 Photovoltaic array capacity factors at PVUSA.

(e.g., 2,600 kWh/m^2 in Albuquerque, 2,200 kWh/m^2 in Davis, and 1,600 kWh/m^2 in Dallas). The product of the last three items defines the system's annual conversion efficiency, which is presently about 11%.

10. SYSTEM ECONOMICS

While the basic economic rationale for the line-focus concentrator module approach was briefly discussed in Section 3, results of a more detailed system economics analysis are presented in Figure 10.14 (O'Neill et al., 1991a). The system price in $/W$_{Peak}$ is shown as the right-hand ordinate. This system price includes installed modules, sun trackers, and power conversion units. The corresponding levelized energy price in ¢/kWh is shown as the left-hand ordinate. This energy price is based on a clear desert site (2,600 kWh/m^2 annual direct irradiance) and a 10% capital recovery factor. The system and energy prices are plotted as functions of annual production volume of line-focus concentrator systems. Two curves are shown. The upper curve corresponds to the best present line-focus production module performance levels (17% peak module efficiency). The lower curve corresponds to anticipated future line-focus module performance levels (21% peak module efficiency). The 21% efficiency has already been demonstrated for a laboratory minimodule (O'Neill and McDanal, 1989). Note in Figure 10.14 that a production rate of 10 MW/year will result in a system price of \$2–3/W_{Peak} and an energy price of 10–15 ¢/kWh. Clearly, these values are approaching parity with some conventional energy sources (e.g., nuclear power plants).

In summary, low-concentration, line-focus concentrator technology offers a direct path to economic viability for photovoltaic systems. The key elements of the technology, lenses and cells, are already in mass production, at acceptable performance levels and prices. More than a decade of experience has shown that line-focus concentrators are manufacturable, rugged, efficient, and long lived. Side-by-side tests with other photovoltaic options have shown the line-focus approach to be unsurpassed in performance. No breakthroughs in efficiency, materials, stability, or manufacturing processes are needed for line-focus concentrators to reach economic viability. The simple line-focus concentrator technology is in hand, but it needs to be mass produced (at 10 MW/year or more) to achieve its full potential.

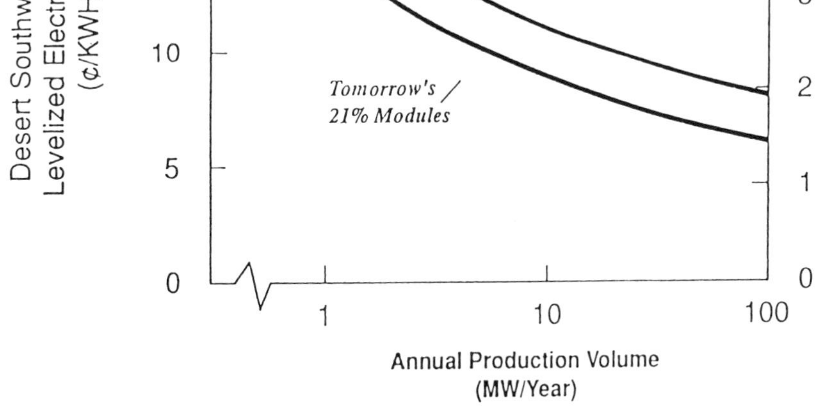

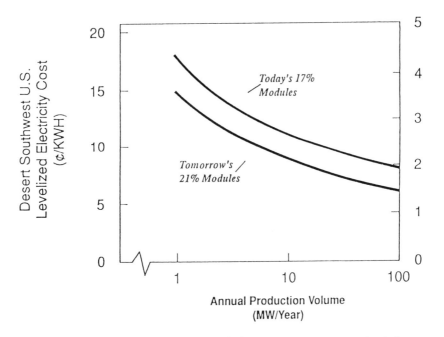

FIGURE 10.14 Linear Fresnel lens photovoltaic concentrator system estimated energy economics.

REFERENCES

Barlow, R.S., and Richards, E.H. (1988), *Qualification Tests for Photovoltaic Concentrator Cell Assemblies and Modules*, SAND86-2743, Sandia, Albuquerque, NM.

Burgess, E.L., and Walker, E.A., Eds. (1979), *Photovoltaic Application Experiment Designs*, U.S. DOE ALO-71, Sandia, Albuquerque, NM.

Candelario, T., et al. (1992), *1991 PVUSA Progress Report*, San Ramon, CA.

Hoff, T., and Iannucci, J.J. (1990), *21st IEEE Photovoltaic Specialists Conf. Record*, IEEE, New York.

Jaster, P. (1993), *3M Price List for 21× Solar Concentrator Lenses*, 3M, St. Paul, MN.

Kaminar, N., et al. (1991), *22nd IEEE Photovoltaic Specialists Conf. Record*, IEEE, New York.

Kaminar, N., et al. (1992), 1992 DOE/Sandia Crystalline PV Technology Project Review Meeting, Sandia, Albuquerque, NM.

Mahoney, A.R., et al. (1992), 23rd IEEE Photovoltaic Specialists Conf. Record, IEEE, New York.

Maycock, Paul (1992–93), *PV News*, monthly editions for 1992 and 1993.

O'Neill, M.J. (1978), *Solar Concentrator and Energy Collection System*, U.S. Patent No. 4,069,812.

O'Neill, M.J. (1984a), *Development of a Low-Cost Extrusion/Embossing Process for a Line-Focus Fresnel Lens Solar Concentrator*, ENTECH, Dallas–Ft. Worth Airport.

O'Neill, M.J. (1984b), *Linear Fresnel Lens Photovoltaic Concentrators*, 7th World Energy Engineering Conference, Atlanta, Georgia.

O'Neill, M.J. (1987), *Photovoltaic Cell Cover for Use With a Primary Optical Concentrator in a Solar Energy Collector*, U.S. Patent No. 4,711,972.

O'Neill, M.J. (1988), *ASHRAE Trans.* **96**.

O'Neill, M.J. (1990), *Development of a 22× Photovoltaic Concentrator Module and Deployment of a 10 kW Array of the New Modules*, ENTECH, Dallas–Ft. Worth Airport.

O'Neill, M.J., et al. (1991), *22nd IEEE Photovoltaic Specialists Conf. Record*, IEEE, New York.

O'Neill, M.J., et al. (1991b), *Photovoltaic Manufacturing Technology Improvements for ENTECH's Concentrator Module*, NREL/TP-214-4486, ENTECH, Dallas–Ft. Worth Airport.

O'Neill, M.J., and McDanal, A.J. (1989), *4th PVSEC*, Sydney, Australia.

O'Neill, M.J., and McDanal, A.J. (1992), 1992 DOE/Sandia Crystalline PV Technology Project Review Meeting, Sandia, Albuquerque, NM.

O'Neill, M.J., and McDanal, A.J. (1993), *23rd IEEE Photovoltaic Specialists Conf. Record*, IEEE, New York.

O'Neill, M.J., and Muzzy, D.B. (1985), *Fabrication, Installation, and Two-Year Evaluation of a 245 m² Linear Fresnel Lens Photovoltaic and Thermal Concentrator System at Dallas–Ft. Worth Airport, Texas*, DOE/ET/20626-1, Dallas–Ft. Worth Airport.

Thomas, M. (1984), *Monthly Performance Reports for Photovoltaic Application Experiments*, monthly editions for 1984, Sandia, Albuquerque, NM.

CHAPTER ELEVEN

Silicon and Gallium Arsenide Point-Focus Terrestrial Concentrator Modules

F.R. GOODMAN, JR, J.C. SCHAEFER, F.J. DOSTALEK, and E.A. DEMEO,
Electric Power Research Institute, Palo Alto, CA 94303

1. INTRODUCTION

Concentrating photovoltaic (PV) technology offers the potential for electricity production at lower cost than from flat-plate technologies principally because concentrators require less area devoted to high-cost material like pure silicon or gallium arsenide. This advantage is offset somewhat by the additional complexity required for tracking, which renders concentrators relatively less attractive than flat-plate PV for small installations. Concentrating PV has been successfully utilized in space, where weight and reliability are more important than cost, and it appeared that for terrestrial applications only cost reduction was necessary. This promise has attracted interest for more than a decade, and in fact costs have fallen as the technology has improved. But more than simply cost reduction was needed.

This chapter addresses the important issues in terrestrial, point-focus module design; the history in the United States of the more successful modules; issues in testing concentrator modules for high reliability; and, finally, the economics of electricity generation from concentrator PV. Point-focus modules concentrate light in both axes of the cell's surface, whereas line-focus modules focus it in only one axis.

Solar Cells and Their Applications, Edited by Larry D. Partain.
ISBN 0-471-57420-1

For terrestrial concentrators, module design involves a complex set of related issues and trade offs. Designers strive for high efficiency and reliability, but the fundamental objective is always low-cost electricity, which in turn requires high-efficiency cells with relatively low cost. After much work during the 1980s, which is summarized in Chapters 3 and 5, hundreds of cells with efficiencies in the 22%–26% range have been produced (private communication, Amonix, Sun Power and Sandia National Laboratories). The challenge during the 1990s is to produce them economically and to build economical, reliable arrays and systems in which to mount them.

2. MODULE DESIGN ISSUES

Solar concentration can be achieved either with reflective optics, such as with heliostats or parabolic troughs or dishes, or with refractive optics in the form of molded acrylic Fresnel lenses or of ground glass or molded lenses. Only the direct normal portion of the total irradiance, that which comes in parallel rays directly from the sun's disk, can be focused or concentrated, and as a result concentrating PV is primarily intended for sites with frequently clear and sunny conditions.

Fresnel lenses are believed by some to offer several advantages over reflective optics, such as parabolic troughs or dishes: the acrylic materials are cheaper per square meter than mirrored glass, optical efficiency can be greater because light must pass through only one weathered or dirty surface, and the geometry is such that the tracking error can be twice what it must be for reflective surfaces to provide the same imaging results. As a result, most concentrator modules in the past decade have been built with Fresnel lenses.

One issue in all concentrator systems is tracking error. To attain the aperture area necessary for cost-effective commercial power generation, concentrator modules will be assembled into large arrays, as shown in Figure 11.1. Under ideal circumstances all modules in an array would be oriented directly at the sun, so that the axis of orientation from the center of the cell through the center of the lens points directly at the center of the sun's disk. However, mechanical loads during windy periods will distort the array structure so that at least some modules in the array will experience tracking error. Modules must therefore be designed to accommodate it. For point-focus systems with concentration ratios as low as 35, the optics to concentrate sunlight consist of simply the Fresnel lenses in the front of the modules. In higher concentration designs a secondary optical element has generally been employed to provide more uniform flux on the cell with a range of off-axis tracking conditions. Figure 11.1 shows the placement of a reflective secondary optical element above a solar cell, as one component in a Fresnel lens PV concentrating system.

One indicator of a module's successful energy production in actual application is the angular tracking error, measured in degrees, that the module can

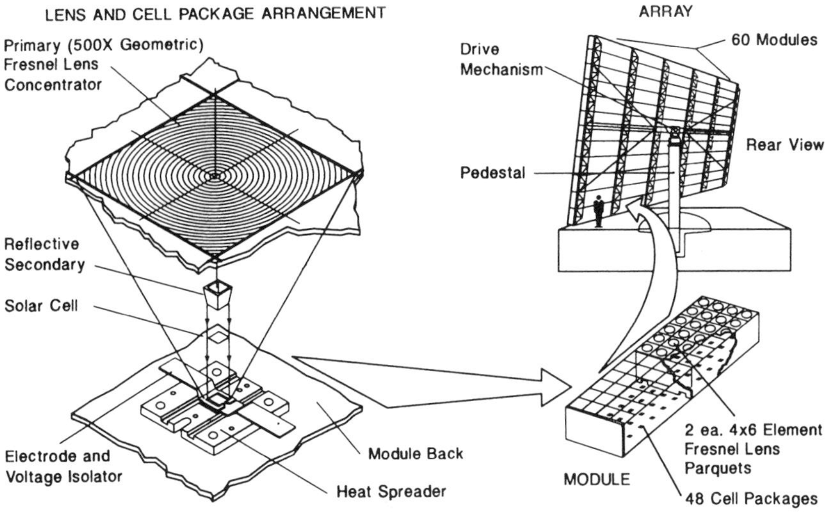

FIGURE 11.1 Components of a Fresnel lens concentration PV system. SOURCE: McNaughton and Richman (1992a, p. 2–3).

experience without significant deterioration in power output or efficiency. Figure 11.2 illustrates output as a function of tracking error; for this particular high-concentration module a tracking error of 1.2° reduces the module efficiency from 22.7% to about 21%.

Although other structures have been tested, concentrator arrays have typically been mounted on a two axis tracking gear box atop a concrete or steel post. Backlash in the gear box therefore forms a portion of the tracking error that must be accommodated. During the 1980s, little attention was paid to the use of gear boxes other than those developed for solar thermal heliostats, but new designs have been reported (Uroshevich et al., 1991).

Up to a point, higher concentration results in higher efficiency (see, for example, Figs. 1.8, 3.4 and 6.6), as well as lower cost, so concentration ratio (sometimes measured as the number of suns followed by a multiplication sign, ×) forms the basis for many other design choices. There are two measures in common use. Geometric concentration is the ratio of the Fresnel lens' aperture area to the illuminated cell area; it serves as an indicator of actual flux, but, because the optical components exhibit different efficiencies in different designs, the true optical concentration or actual irradiance on the cell needs to be measured rather than simply calculated. Efficiency also depends on the uniformity of flux distribution across the cell, which in turn is a function of the quality of the optical and tracking systems. Uniform flux is best but is never perfectly achieved even with the best of optical components (McNaughton and Richman, 1992b).

Cells operate more efficiently if they are cooler, so designers attempt to keep cell temperatures as low as possible. In the early 1980s some concentrator

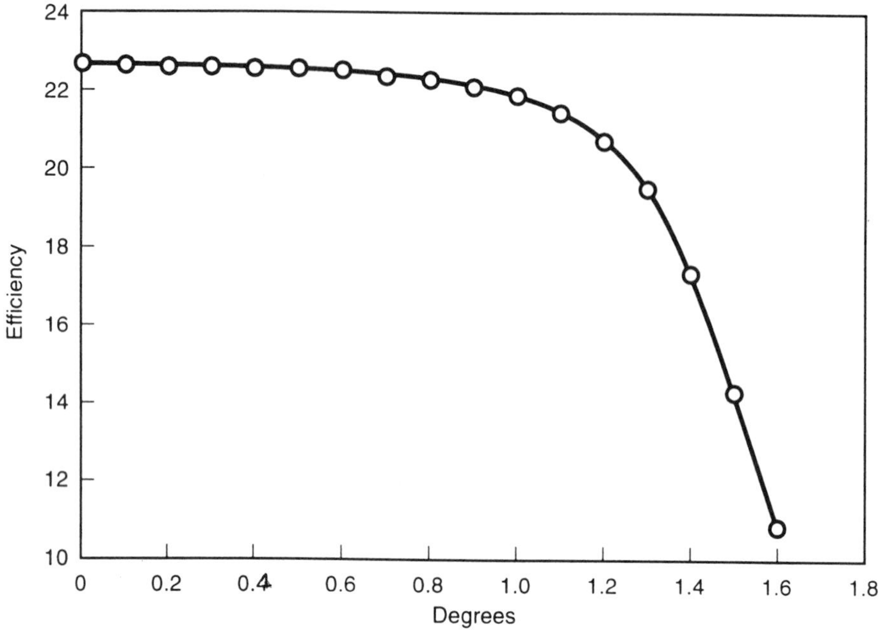

FIGURE 11.2 Module efficiency as a function of tracking error. The system under test was a 2×1 Fresnel lens parquet on an aluminum housing with 942 geometric concentration, GaAs cells each having about $0.2\,cm^2$ illuminated area, and solid crown glass secondary lenses. SOURCE: Kuryla et al. (1991).

modules and arrays were built with active circulation of a coolant-like water or antifreeze, but the additional complication drove costs higher. Most such early installations were unsuccessful for other reasons. Active cooling is now employed only for small test installations and for one prospective commercial design described below. Most commercial point-focus designs utilize passive cooling in the form of heat sinks to dissipate the excess heat developed in the cells, and the requirement for low cell temperature places a premium on high thermal conductivity between the cell and the heat sink. In some designs now under test, the metal module housing also serves as the heat sink. One measure commonly employed for the quality of cell cooling is the temperature differential between the cell and ambient. Figure 11.3 illustrates how temperatures in the cell and other components in one module change during a thermal test; the greatest differential occurs between ambient and the module back plane.

Soldering has been the most common means of attaching the backside cell metalization to the electrode assemblies and to the heat spreader, which carries away the heat. Because the cell material has a different thermal coefficient of expansion from that of the metal to which it is soldered, there will be some strain on this solder bond. The higher the cell temperature and the larger the cell, the greater this strain will be. Experience demonstrates that it is difficult

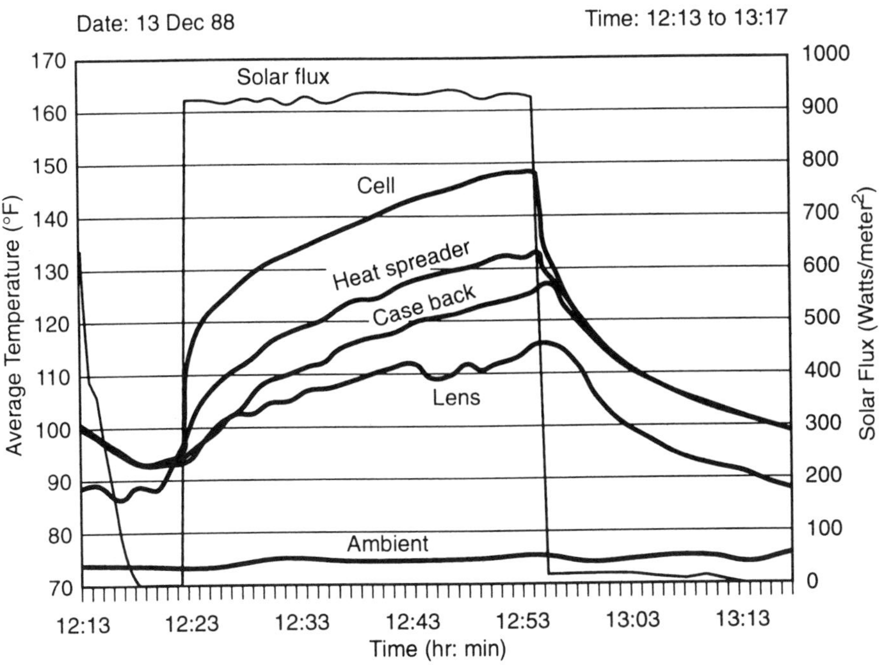

FIGURE 11.3 Module component temperatures. The measurements were taken on a 0.06 inch thick aluminum module with a 4×6 Fresnel lens parquet having a geometric concentration ratio of 500 and with $0.64\,cm^2$ Si cells soldered to ceramic voltage isolators on aluminum heat spreaders. SOURCE: McNaughton and Richman (1992b, p. 5–14).

to make this bond without voids in the solder and that under repeated heating and cooling voids and cracks tend to develop and grow. Voids of large enough size under a cell degrade its output, so they should be avoided or minimized. Solder bonds cannot be seen because they are under the cell. However, a variety of nonintrusive testing procedures have been developed as quality control measures. These include infrared scanning, ultrasonics, x-ray radiography, and measurement of open circuit voltages (McNaughton and Richman, 1992b; Richards and Chiang, 1990; Carroll et al., 1990). Testing procedures have been calibrated against actual field experience to ensure that they adequately reflect the damage that these solder bonds will experience in 30 years of operation (Chiang and Blankenau, 1988).

Safety requires that exposed metal parts such as heat sinks be grounded, so that an electrical insulator is also required between the cells, which operate at high voltage at the upper ends of the strings, and the heat sinks at ground potential. A variety of materials with high thermal conductivity and high electrical resistivity have been tested: ceramics, kapton, anodized aluminum, and electrophoretically coated acrylics (Richards et al., 1990).

Although modules are enclosed to protect electrical components from the elements, they must also "breathe" to prevent pressure differentials from mechanically distorting the module shape. The entry of outside air has enabled moisture entry, either by condensation or leakage or, in at least one case, both. The presence of moisture complicates the choice of materials to isolate electrically the cells and other components from ground. Electrical insulating components inside the modules, including cell surfaces, must therefore function even with some moisture on them.

A problem that has recurred at one time or another with most concentrator designs is the burning of module components because of severe off-axis tracking. Under ideal circumstances with arrays always oriented directly at the sun this would not be a problem, but under real operating conditions with wind stows (orientation of a tracker to reduce mechanical loads during periods of high wind) and occasional tracker failures, off-axis contingencies must be protected against with a reflective barrier to cover the electrical components. Usually this is an aluminum sheet.

Finally, in order for concentrator modules to be commercially successful for power generation, they must be manufacturable at a reasonable cost.

3. THE CONCENTRATOR DEVELOPER'S DILEMMA

The issues noted above pose a complex set of design and manufacturing process choices that module developers must make, often in the absence of complete information. They must weigh the need for testing to confirm engineering calculations against the temptation to proceed with incomplete information and hope to demonstrate high-performance modules. Successful developers will choose an appropriate mix of these two paths. The history of PV development in general, however, suggests the wisdom of caution and the need to test designs and materials before committing to them on a large scale. This dilemma is exacerbated by the absence of an early commercial market for concentrators, requiring a longer period of unrewarded research and development (DeMeo et al., 1990).

4. CONCENTRATOR PERFORMANCE DATA

The definition of conditions under which ratings are measured in concentrator modules is even more important than in flat-plate PV, as noted in Chapter 16, because of the wider temperature ranges over which cells operate. Laboratory cell measurements are made at room temperature, 25°C, and typically with flash testers. In actual operation, concentrator cells operate at much higher temperatures. To assess how a concentrator module will operate in a hot, desert setting, it is more logical to report module efficiencies as a function of ambient temperature. Therefore, wherever possible module efficiencies are reported below as measured in outdoor settings, without adjusting to 25°C cell temperature. Although 20°C is commonly used as a reference for ambient temperature, some utility companies have suggested that 40°C more accurately reflects the conditions they experience when they most need power (Lepley, 1990). Furthermore, while efficiency is an important indicator, the real measures of a successful module design are the amount and cost of energy it generates.

The efficiency data presented below should not be used to compare one module with another because they were recorded under different conditions. Some efficiencies are instantaneous under ideal conditions, some are averages, some are ranges, and one was taken with active cooling to hold cells at 25°C. Comparable, complete performance data have not been reported for all these modules.

4.1. History of Concentrator Module Testing

Recognizing the potential advantages inherent in concentrators and the need for test experience, early sponsors and designers built and installed at least six large PV systems in the early 1980s. Four of them utilized reflective parabolic trough designs with line-focus concentration. Experience showed either that they did not function at all or that they failed after short periods of time; none demonstrated efficiencies above 4% (Inglis, 1983). In all four of these cases the designers apparently built entire systems without first testing individual modules. None of the modules or systems, which ranged up to 240 kW in expected output, attained their anticipated ratings, and none of these organizations continues in the PV business. Since that time it has become more common for developers to test individual modules first to ensure proper system operation.

Almost all concentrator systems built since the early 1980s have been based on Fresnel lenses; one major exception that utilizes reflective optics for its primary concentration is described below.

4.2. Module Test Experience

Reflecting the advances made in concentrator cell efficiency during the late 1980s, a number of organizations are developing concentrator modules in the 1990s. Table 11.1 describes the major characteristics of modules tested earlier as well as those now under development.

Two of the early Fresnel lens systems, that built by Martin Marietta at the Phoenix–Sky Harbor Airport and that built by ENTECH at the Dallas–Fort Worth Airport, performed with efficiencies approaching 10% for a number of years. The Phoenix–Sky Harbor system operated with a 35× point-focus concentration ratio, whereas the Dallas–Fort Worth line-focus modules utilized a 25× concentration ratio. Two other line-focus systems built by ENTECH are now in operation at the 3M facility in Austin, Texas, and at PVUSA near Davis, California; they are discussed in Chapter 10.

TABLE 11.1 Concentrator Module Characteristics[a]

Developer and Model	Geo. Conc.	Cell	SOE	Module Features	Date of Test	No. of Modules Built	Reference
Martin Marietta							
Sky Harbor	33	Si	None	1 × 4	82–87		Eckert (1987)
Intersol	70	Si	Refl.	2 × 7	84–87		Boes (1990)
Alpha Solarco	492	Si	Glass silo	2 × 12	89–92	100	Carroll et al. (1990)
AESI	350	Si	Glass		88	1	Todoroff (1988)
EPRI Mods 1&2	500	Si BSC	Refl.	4 × 12	87–92	130	Whitaker and Dostalek (1991)
EPRI Mod 3	250	Si BSC	Refl.	4 × 6	93	3	
Midway	178	Si	Unknown	2 × 8	92	4+	Boes (1990)
SNL SBM3		Si UNSW	Refl., pr. cv.	2 × 12	90	2	Richards et al. (1990)
SNLConcept 90	146	Si BSC	Quartz silo	2 × 3 test	90	1	Chiang and Quintana (1990)
SNL Exptl	100	Si UNSW	Pr. cv.	2 × 6 test	90	1	Chiang and Richards (1990)
SunPower	250	Si BSC	See text		91	1	Verlinden et al. (1991)
Varian 1	1,000	GaAs	Glass TIR	6 × 6 test	85	1	Hester (1986)
Varian 2	942	GaAs	Glass TIR	2 × 6	90	1	Kuryla et al. (1991)

[a]SOE, secondary optical element; BSC, backside contact cell (described in Chapter 3); UNSW, University of New South Wales, described in Chapter 2; Refl, reflective secondary optical element; Pr. cv., prism cover (described in Chapter 10); TIR, glass secondary optical element with total internal reflection. Modules described as test modules are not protected against the elements sufficiently for continuous outdoor use.

Martin Marietta employed the same design for a 350 kW system in Saudi Arabia as it did for the 225 kW system at Phoenix–Sky Harbor. Modules were fabricated of molded acrylic, with the acrylic lenses cemented on the front. The large (more than 2 inches in diameter), round, single crystal silicon cells were soldered to copper conductors, which in turn were soldered to ceramic insulators. Ceramic insulators were attached to the external, grounded heat sinks to provide low resistance to heat flow.

Arizona Public Service Company's experience at Phoenix–Sky Harbor illuminated a variety of unforeseen difficulties (Eckert, 1987). Problems common to all PV systems—not just concentrating systems—are described in Chapter 16. Those specific to concentrators include water entry, electronic or mechanical tracking problems, and a variety of electrical and insulation failures. Water entry, either from leakage around the screw holes on the back of the modules or from condensation, has turned out to be an issue for most other concentrator modules also. Arizona Public Service drilled holes to allow the water to escape, faced the modules down at night—rather than up at the dark night sky—so that the cell mounts did not cool more rapidly than other module components or the air inside, and oriented the modules at an angle at night so rainwater could not collect on the back of the modules. Tracking problems included both mechanical failures in the gear boxes, which were difficult to repair because all the modules and the torque tube to which they were mounted had to be removed, and electronic failures in the controller that drives the motors and positions the array throughout the day. To provide repair parts between 1982 and 1987, modules were gradually cannibalized so that the capacity of the system declined to about 175 kW by 1987. That these problems arose with a first-generation design should not be a surprise, as there was little experience in 1982 on which designers could draw; the same is not true in the 1990s. Phoenix–Sky Harbor was dismantled later in 1987 because the land was committed to other uses; some of the modules were relocated to another site, where they are still in use.

From a performance perspective, an important aspect of the Phoenix–Sky Harbor experience is its capacity factor during the first half of 1987, which exceeded 29%. This suggests that in the future values of at least 30% can be attained with concentrators.

The Intersol Corporation of Colorado developed the Martin Marietta design further and supplied modules or systems to Pacific Gas and Electric Company, Arizona Public Service Company, Platte River Power Authority, Southern California Edison Company, and a pumping application in southern Colorado. These modules operated with smaller cells and initiated a trend toward higher concentration that has characterized development since the mid-1980s. However, Intersol has not supplied systems since the late 1980s.

Alpha Solarco installed a 125 m^2 proof-of-concept array with 100 modules near Las Vegas in 1989. One feature of its design was a silo-shaped glass secondary optical element, which provided more uniform flux than reflective secondaries, better tolerance to lens distortion and tracking error, and front surface encapsulation for the cell (Carroll et al., 1991; Maish et al., 1991). After the first 3 days of operation following installation, the array's measured output was found to be 10 kW, far less than the 15 kW expected. Analysis showed that the secondary optical elements were breaking because of increased absorption in the glass as it gradually discolored in the high ultraviolet flux. A new glass for the secondary optical elements that does not discolor has since been tested and found satisfactory (Carroll et al., 1991).

A module and array development program was begun at the Electric Power Research Institute (EPRI) in 1986 with the realization that a new, very high-efficiency silicon cell was going to be available. This program has progressed through the construction of some 130 modules (McNaughton and Richman, 1992b). Not all of these were populated with cells. Most were installed in two separate arrays with 60 modules each to test weatherability, structural deformations, response to wind loads, temperature variations, thermal issues, tracker performance, and general durability. Eleven other modules were built to test strength, thermal design, manufacturing error budgets, optical performance, solder joints, and electrical performance. Four modules are under long-term testing at utility sponsors' sites: near Atlanta with Georgia Power Company; in San Ramon, California, with Pacific Gas and Electric Company; in Phoenix, Arizona, with Arizona Public Service Company; and near Los Angeles with the Los Angeles Department of Water and Power.

Based on its experience in testing the first model or Mod 1, Cummings Enginering Company designed and tested a subsequent Mod 2 design. Electrical performance data from a Mod 1 module fully populated with cells show that actual efficiencies at operating temperatures range from 15% to nearly 18%, as shown in Table 11.2. An advanced system design that incorporates an integrated array backplane (like a printed circuit board) is currently under test (Moore, 1992).

Sandia National Laboratories (SNL) is cooperating with various industrial organizations to assist them in developing technology for cost-effective PV, including activities in cell development, soldering, module sealing, testing, refractive optics, and prismatic cell covers (Chamberlin and King, 1990). To demonstrate these technologies, SNL has also tested three concentrator modules using the leading two silicon concentrator cell designs: the Electric Power Research Institute (EPRI) and Stanford University BSC (back side contact, formerly known as the *point contact* and described in Chapter 3) (Sinton, 1992) and Martin Green's UNSW (University of New South Wales, described in Chapter 2) cells. Their performance is summarized in Table 11.2. The SNL Baseline Module 3 (SBM3) is a prototype for a manufacturable unit and incorporates the UNSW cell with prismatic covers (see Chapter 10) and electrophoretically coated, high-temperature acrylic on anodized aluminum as the electrical insulator. Module efficiency was measured as 16.7% at 1,000 W/m^2 direct normal (Richards et al., 1990). An experimental module with 12 cells and mounts like the SBM3, no reflective secondary optical element, a lower geometric concentration of 100, smaller lenses, and active cooling demonstrated a record silicon concentrator module efficiency of 20.4% (Chiang and

TABLE 11.2 Module Operating Results Since 1990

Developer and Model	Efficiency Observed	Test Conditions	Operational Period	Comments and Reference
EPRI Mod 1 Unit 3	16.3	851 W/m^2, T_c = 50C	12/5/90	Instantaneous
	14.9–17.6	tracking	10/26/90–12/5/90	Operational in Georgia (McNaughton and Richman, 1992a, b)
Sandia SBM3	16.7	1,000 W/m^2, T_c = 65C, T_a = 22	5/15/90	24 cell (Richards et al., 1990)
SNL Concept 90	19.0	800 W/m^2, T_c = 50C, T_a = 22	5/16/90	Test module, six cells (Chiang and Quintana, 1990)
SNL Exptl	20.4	800 W/m^2, T_c = 25C	4/19/90	Active cooling, test module, 12 cells (Chiang and Richards, 1990)
Varian 2	22.3	790 W/m^2, T_a = 11C	12/14/90	Instantaneous test
	20.0–22.0	Tracking	12/6/90–12/14/90	Cool, clean, in Palo Alto
	18.0–20.0	Tracking	4/19/91–4/21/91	Hot, dirty, in Phoenix (Kuryla et al., 1991)

Richards, 1990). A third module, the SNL Concept 90, with six BSC cells covered by a quartz secondary optical element and encapsulated in a clear epoxy resin developed a 19% efficiency (Chiang and Quintana, 1990).

The SunPower Corporation is developing a PV central receiver concept, utilizing multiple dense arrays of monolithically interconnected BSC cells. Each dense array is fabricated from a single wafer (Swanson, 1992). These dense arrays are each 6 cm on a side, the largest square area that can be sliced from a 4 inch round wafer. A dense array with its electrode and cold-plate assemblies forms a module. A variable number of modules is built into a receiver—the exact number depending on system size. This concept avoids the large parts count per watt in conventional Fresnel modules, enables economic active cooling because the coolant need not be pumped over a vast receiver area, and simplifies electrical connections. SunPower has also developed a flux homogenizer to mount in front of the receiver, so that the flux intensity does not vary across its surface (Verlinden et al., 1991). A one kW prototype system using parabolic dish concentration, similar to that illustrated in Figure 11.4, was tested for 2 months in 1992.

This central receiver concept at a larger scale requires some means of concentrating sunlight in a relatively uniform manner on a larger number of these modules, which could be done with a large field of heliostats in a multi-megawatt plant (Stolte, 1992). Worthy of note is Stolte's finding that a 50 MW central receiver plant produces electricity at about the same estimated

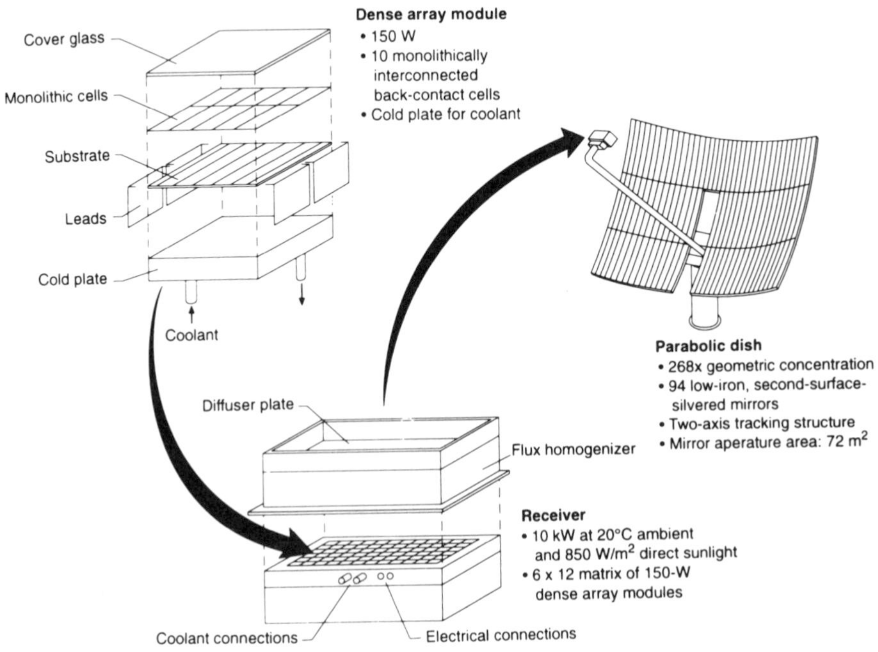

FIGURE 11.4 Parabolic dish concentrator PV system.

cost, approximately \$0.13/kW-h, as a Fresnel lens plant at \$0.12/kW-h. However, the actual cost for either of these approaches will not be known until commercial plants are built. Value engineering could result in lower costs than the estimates.

Tests of the most recent 942×, gallium arsenide Varian Associates design (Varian 2 in Tables 11.1 and 11.2, to distinguish it from the 6×6 module Varian built in 1985) in the form of a two cell minimodule version demonstrated in 1990 at a 22.7% conversion efficiency with an ambient temperature of 24°C and passive cooling. Over an 18 month period of outdoor service, a similar module with 12 cells in Phoenix developed actual efficiencies (not corrected to 25°C cell temperature) between 18% and 20%, as noted in Table 11.2 (Kuryla et al., 1991). This is the highest known efficiency for an operating, outdoor module and serves as a milestone for terrestrial photovoltaics.

A drop in the Varian 2's efficiency after a year of operation has been attributed to water entry in the form of condensation and improperly applied antireflective (AR) coatings. Preliminary results suggest that the AR coatings on one or two of the cells were affected. The two affected cell mounts were removed, and the other 10 cells continued to operate with no degradation.

The Wattsun Corporation's design utilizes a very short focal length (about 0.9 inches) and a correspondingly thin module, a 0.5 inch square aperture, convex rather than Fresnel refractive lenses, and silicon cells approximately 0.09 inches square. A module 12 inches square contains 576 cells that are mounted in a receiver that is assembled like a printed circuit board. Early test results demonstrated about 11% efficiency (Boes, 1990).

4.3. Qualification Testing of Concentrator Modules

The module failures observed during the 1980s prompted SNL to develop a multistep qualification test to simulate extreme conditions equivalent to 30 years of outdoor operation. Figure 11.5 shows the basic steps in the qualification test (Barlow and Richards, 1988). In this way failures can be identified early and in small quantities rather than after many modules have been built. Currently any concentrator developer would be foolhardy not to carry out this series of tests, as it illuminates failures in cell mounts and soldering, electrical leakage, and moisture-related problems.

5. COST OF ENERGY FROM CONCENTRATORS

The fundamental measure of PV's competitiveness with other forms of commercial power generation is levelized cost of energy (COE) per kW-h (DeMeo et al., 1992). Equation (7) in Chapter 16 summarizes this calculation, whose results, with concentrator values instead of the flat-plate estimates used in Chapter 16, are summarized in Figure 11.6. Sources are those listed in Chapter 16.

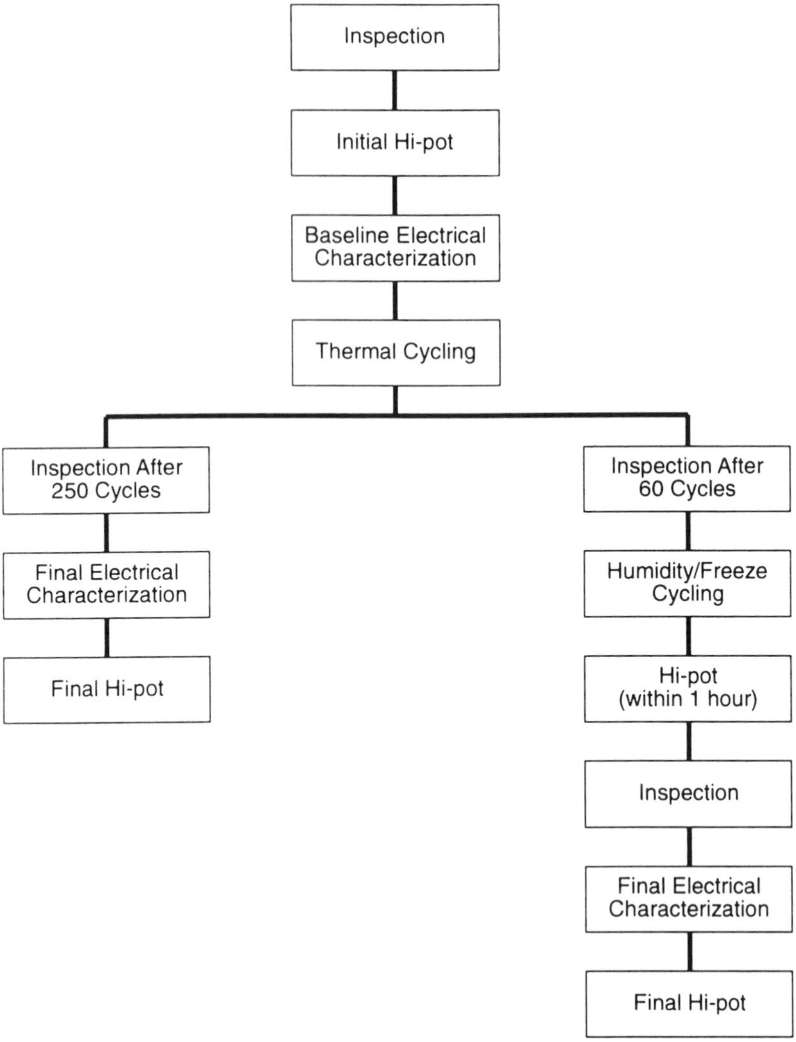

FIGURE 11.5 Sandia qualification test. SOURCE: Barlow and Richards (1988).

The lower portion of Figure 11.6 lists the baseline or most probable values for each of the key variables, along with probable upper and lower limits for each. Based on the histories of cell and module efficiencies, for example, it is reasonable to suppose that 22% concentrator modules can be manufactured in the foreseeable future (Stolte, 1992); in the best of all worlds it is possible to suppose that module efficiencies of 24% are attainable, whereas in the worst case it may be that no better than 16 percent will be achieved in mass production.

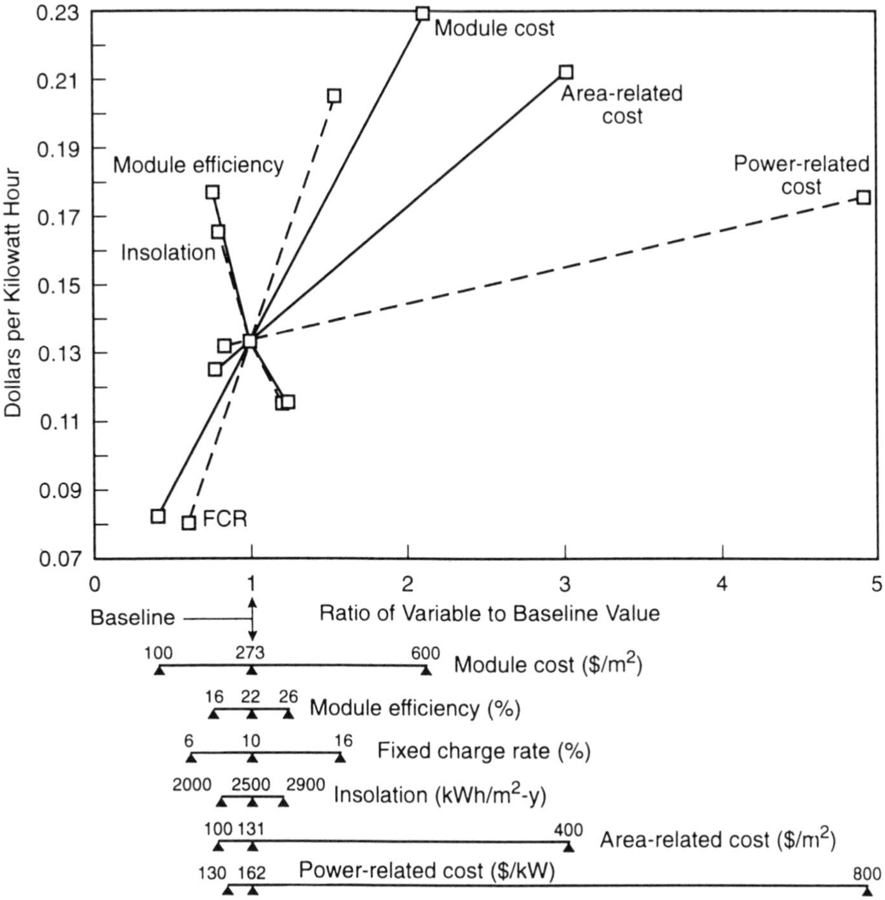

FIGURE 11.6 Concentrator COE as a function of key variables.

Figure 11.6 illustrates the effect that a variation in each variable alone has on the final COE. For example, the baseline COE is 13 cents, but if module efficiencies rise to 26% COE drops below 12 cents. As another example, with all other variables at their baseline values, a reduction in module cost from \$273 per square meter to \$100 per square meter reduces COE to about 8 cents per kW-h.

Figure 11.6 is only useful for calculating the effect on COE of one variable at a time. Consideration of more than one at a time requires the use of Eq. (7) in Chapter 16. For example, if the cost of modules falls to \$100 per square meter and direct normal insolation of 2,725 is attainable, the COE will decline to about 6 cents per kW-h.

6. SUMMARY

The major technical barrier to efficient concentrator technology, cells with adequate efficiency, has been overcome. In addition, module performance almost adequate for cost-effective commercial power generation has been demonstrated on a small scale. Although reliability problems and efficiency shortfalls have been illuminated by building and operating commercial-sized concentrator modules, these challenges can be met without further breakthroughs in technology. However, continued engineering effort is obviously necessary to eliminate the difficulties noted and to design modules that can be manufactured at a low enough cost. Adequate manufacturing volume should drive module cost down sufficiently to make concentrator PV cost effective.

The capital investment in this engineering and in the production facilities to manufacture modules, as well as cells, will only be forthcoming if the perspective market is of sufficient size. Issues in building this market are discussed in Chapters 16, 19, and 22.

REFERENCES

Barlow, R.S., and Richards, E.H. (1988), Qualification Tests for Photovoltaic Concentrator Cell Assemblies and Modules, Report SAND86-2743, Sandia National Laboratories, Albuquerque, NM.

Boes, E.C. (1990), *21st IEEE Photovoltaic Specialists Conf. Record*, IEEE, New York, pp. 944–951.

Carroll, D. (1991), *22nd IEEE Photovoltaic Specialists Conf. Record*, IEEE, New York, pp. 625–628.

Carroll, D., Bailor, B., and Schmidt, E. (1991), *22nd IEEE Photovoltaic Specialists Conf. Record*, IEEE, New York, pp. 518–522.

Carroll, D., Schmidt, E., and Bailor, B. (1990), *21st IEEE Photovoltaic Specialists Conf. Record*, IEEE, New York, pp. 1136–1141.

Chamberlin, J.L., and King, D.L. (1990), *21st IEEE Photovoltaic Specialists Conf. Record*, IEEE, New York, pp. 870–875.

Chiang, C.J., and Blankenau, S.J. (1988), *20th IEEE Photovoltaic Specialists Conf. Record*, IEEE, New York, pp. 1242–1245.

Chiang, C.J., and Quintana, M.A. (1990), *21st IEEE Photovoltaic Specialists Conf. Record*, IEEE, New York, pp. 887–891.

Chiang, C.J., and Richards, E.H. (1990), *21st IEEE Photovoltaic Specialists Conf. Record*, IEEE, New York, pp. 861–863.

DeMeo, E.A., Goodman, F.R. Jr., Peterson, T.M., and Schaefer, J.C. (1990), *21st IEEE Photovoltaic Specialists Conf. Record*, IEEE, New York, pp. 16–23.

DeMeo, E.A., Weinberg, C.J., and Tassiou, R. (1992), *Proc. IEA/ENEL Exec. Conf. Photovoltaic Systems Electric Utility Applications*, OECD/IEA, Paris.

Eckert, P. (1987), *Lessons Learned and Issues Raised at Sky Harbor—A Utility-Interactive Concentrator Photovoltaic Project, 1987*, Am. Soc. Mech. Eng. Solar Energy Conf., Honolulu.

Hester, S.L. (1986), *PG&E Photovoltaic Module Performance Assessment*, Report AP-4464, EPRI, Palo Alto.

Inglis, D.J. (1983), *Photovoltaic Field Test Performance Assessment: Technology Status Report Number 2*, Report AP-3244, EPRI, Palo Alto.

Kuryla, M.S., Ristow, M.L., Partain, L.D., and Bigger J.E. (1991), *22nd IEEE Photovoltaic Specialists Conf. Record*, IEEE, New York, pp. 506–511.

Lepley, T. (1990), *21st IEEE Photovoltaic Specialists Conf. Record*, IEEE, New York, upp. 903–908.

Maish, A.B., and Chamberlin, J.L. (1991), *10th Eur. Community Photovoltaic Solar Energy Conf.*, Kluwer Academic Publ., Dordrecht, The Netherlands, pp. 902–995.

Maish, A.B., Hund, T.D., Quintana, M.A., and Chiang, C.J. (1991), *10th Eur. Community Photovoltaic Solar Energy Conf.* Kluwer Academic Publ., Dordrecht, The Netherlands, pp. 988–991.

McNaughton, W.P., and Richman, R.H. (1992a), *A Summary of Recent Advances in the EPRI High-Concentration Photovoltaic Program*, Report TR-100392, EPRI, Palo Alto.

McNaughton, W.P., and Richman, R.H. (1992b), *Recent Advances in the EPRI High-Concentration Photovoltaic Program*, Report TR-100393, Vols. 1 and 2, EPRI, Palo Alto.

Moore, T. (1992), *EPRI J.*, **17**, 16–25.

Richards, E.H., and Chiang, C.J. (1990), *21st IEEE Photovoltaic Specialists Conf. Record*, IEEE, New York, pp. 1074–1079.

Richards, E.H., Chiang, C.J., and Quintana, M.A. (1990·), *21st IEEE Photovoltaic Specialists Conf. Record*, IEEE, New York, pp. 881–886.

Sinton, R.A. (1992), *Development Efforts on Silicon Solar Cells*, Report TR-10043, EPRI, Palo Alto.

Swanson, R.M. (1992), *Solar Eng.* **2**, 1067–1070.

Stolte, W.J., (1992), *Engineering and Economic Evaluation of Central-Station Photovoltaic Power Plants*, Report TR-101255, EPRI, Palo Alto.

Todoroff, B. (1988), *20th IEEE Photovoltaic Specialists Conf. Record*, IEEE, New York, pp. 1347–1352.

Uroshevich, M., Gron, M., and Tyjewski, P. (1991), *22nd IEEE Photovoltaic Specialists Conf. Record*, IEEE, New York, pp. 766–769.

Verlinden, P., Sinton, R.A., Swanson, R.M., and Crane, R.A. (1991), *22nd IEEE Photovoltaic Specialists Conf. Record*, IEEE, New York, pp. 739–743.

Whitaker, C.M., and Dostalek, F.J. (1991), *22nd IEEE Photovoltaic Specialists Conf. Record*, IEEE, New York, pp. 512–517.

CHAPTER TWELVE

Thin-Film Terrestrial Modules

YUKINORI KUWANO, Sanyo Electric Company, Ltd., Hashiridani, Hirakata, Osaka 573, Japan

1. INTRODUCTION

In this chapter, module technologies of thin-film solar cells, which hold promise as low-cost photovoltaics, are described. Concerning the techniques for producing these modules and their applications, we introduce some unique, recently developed solar cells and examples of new applications. In the last part, the prospects of a worldwide network of large-scale solar power systems and future topics are described.

2. TECHNOLOGIES FOR THIN-FILM SOLAR CELL MODULES

2.1. Materials

The materials used in thin-film solar cells include hydrogenated amorphous silicon (a-Si:H), polycrystalline silicon (poly-Si), and compound semiconductors.

2.1.1. a-Si:H. Solar cells that use a-Si:H (Spear and LeComber, 1975), a material developed relatively recently, were developed by Carlson and Wronski (1976) at RCA. They were subsequently put to first practical use as a power source for small calculators by Kuwano and Ohnishi (1981) at Sanyo. The a-Si:H solar cells are representative of the low-cost thin-film devices that are receiving much attention. The a-Si:H solar cells have the following features.

Solar Cells and Their Applications, Edited by Larry D. Partain.
ISBN 0-471-57420-1

- Fabrication is simple.
- Fabrication requires a low temperature (300°C or less).
- Little material is used (film thickness is 1 μm or less compared with about 300 μm of crystalline Si cells).
- The gas reaction is favorable for a large area.
- The low-temperature process allows the use of inexpensive substrates such as glasses.
- A high voltage can be easily obtained using integrated-type submodule structure.

Various applications that take advantage of these features have been developed. The conversion efficiency of these modules has exceeded 10% (Hamakawa, 1992), one target for practical-sized solar cells of at least 100 cm^2 for use as an electric power source.

2.1.2. Poly-Si. Recently, thin-film poly-Si gained attention for use in solar cells because of its physical properties. Specifically, it can be used as a narrow band gap material similar to the a-SiGe:H used in stacked a-Si:H solar cells (Marfaing, 1979) or for producing a low-cost poly-Si cell by using a thin-film poly-Si.

Currently, conversion efficiencies of 6% or more are being obtained with solar cells and thin-film poly-Si (Matsuyama et al., 1992). This material holds promise for use in thin-film solar cells or for use in the bottom cell of thin-film stacked solar cells.

2.1.3. Compound Semiconductors. The CdS/CdTe thin-film solar cells are being used because of their reliability and performance. They also offer hope for a low-cost, thin-film solar cell.

A typical solar cell using $CuInSe_2$ (CIS) is a CdS/$CuInSe_2$ structure developed by Wagner et al. (1974). Currently, module conversion efficiencies exceeding 10% are being obtained.

2.2. Module Structure

Solar cell modules for practical use consist of individual solar cells or submodules connected in parallel or in series in such a way as to provide the desired voltage and current. This structure is called a *solar cell module.*

2.2.1. Integrated-Type Submodule Structure. One feature of thin-film solar cells is that they are, as described above, easily processed. This feature allows them to be fabricated into unique module structures called *integrated-type structures*. Kuwano et al. (1980) developed the first integrated-type structure for a-Si:H solar cells.

An example of a module structure for consumer use that includes conventional crystalline solar cells is shown in Figure 12.1a. In this case, the cells are arranged and connected in series. In this way, the voltage required as an electric power source is obtained. However, this method involves complex processes, and the large amount of wiring reduces reliability. In contrast with this, many thin-film solar cells (such as the a-Si:H devices) can be cascade-connected on a single substrate so as to produce high voltages, as shown in Figure 12.1b, c. This entire structure is generally called a *submodule.*

The structure in Figure 12.1b is called a *type I integrated-type structure* and has all of the cells connected in series at the edge of the insulating substrate. The type II submodule structure shown in Figure 12.1c has adjacent cells connected in series at the boundary between them and has the merit of reducing the resistance loss of the transparent electrode. The features of integrated-type structure modules are

- High voltages suited to the intended use can be obtained from a single substrate.
- The transparent electrode is segmented into a number of electrodes, so resistance loss caused by enlarging the module size can be reduced.
- The module construction process can be simplified.
- It is possible to prevent the effects of pin holes from affecting the entire solar cell.

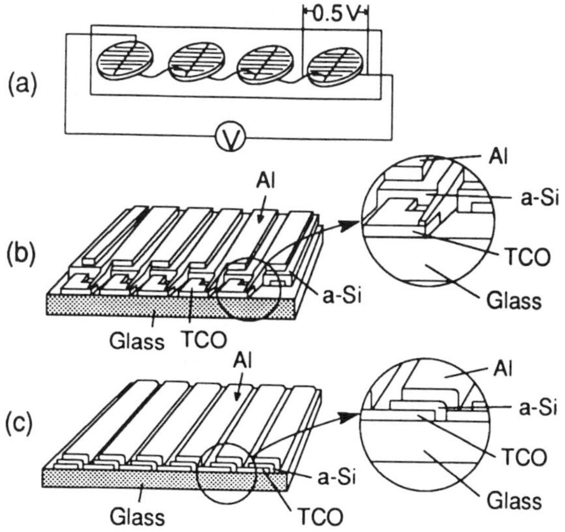

FIGURE 12.1 Conventional type solar cell module. **(a)** an integrated type a-Si : H solar cell module, **(b)** type I, and **(c)** type II.

The through-hole contact structure (Ohnishi et al., 1989a) has been recently developed. The new type of structure allows an even greater reduction in the electrical loss than the conventional integrated-type structure.

2.2.2. Module Structure. Figure 12.2 shows the thin-film solar cell module structures. A superstrate type module as shown in Figure 12.2a consists of submodules, transparent protective glass board, frame, filling, and metal sheet. A tempered glass substrate is usually used as the protective glass board. Polyvinyl butyral (PVB) or ethylene vinyl acetate (EVA) are usually used as the fillings. PVB has the good feature of low degradation of transparency by the ultraviolet light, and EVA provides good protection from moisture. The back side is coated using the plastic films and metal films for protection against moisture and for insulation. Figure 12.2b shows another type of module structure based on a large-sized submodule. The glass substrate, which is a component of the submodule, is directly used as the protective glass board, so that it is not necessary to connect the submodules by lead lines and is possible to simplify the assembling process of modules and to reduce the cost of modules. In the future, this type of module structure will become a leading one.

2.3. New Module Processes

2.3.1. a-Si:H Solar Cell Modules. The processes of making a-Si:H solar cells are described here. The process of manufacturing thin-film solar cells generally consists of thin-film fabrication processes and module fabrication processes,

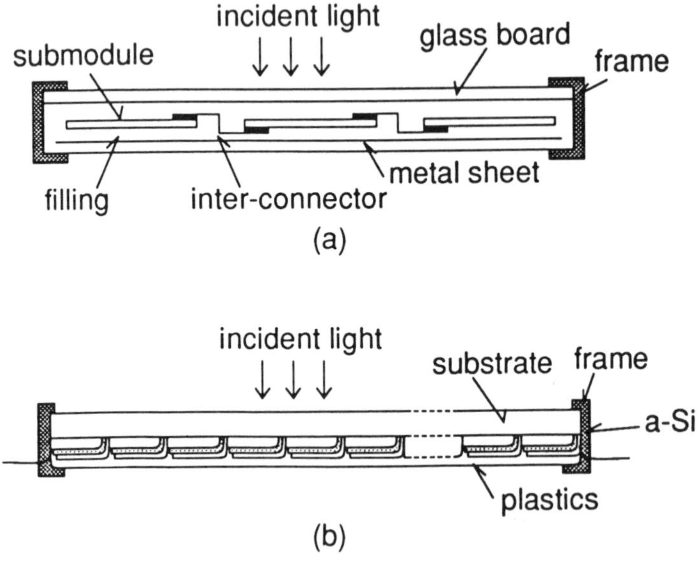

FIGURE 12.2 Module structures: **(a)** superstrate type and **(b)** large-area submodule type.

including patterning. The process of fabricating thin films for a-Si:H solar cells involves fabrication of a transparent, conductive-oxide thin film on a substrate of glass or other material, followed by fabrication of an a-Si:H p-type layer, i-type layer, and n-type layer, and, finally, the fabrication of a metal thin film to serve as the back electrode. The transparent conductive-oxide thin film and metal thin films are fabricated by evaporation and so on. RF-plasma CVD is widely used as an industrial process for fabricating the a-Si:H thin films. Conventionally, a single reaction chamber method has been used to fabricate the p, i, and n layers. However, the intermixing of the dopants in each layer cannot be avoided with this method, creating a problem in improving the solar cell characteristics. To solve this problem, Kuwano et al. (1982) developed the consecutive, separated reaction chamber method shown in Figure 12.3a. With this method, each a-Si:H layer is consecutively fabricated in a separate chamber, so it is possible to reduce intermixing of dopants remarkably better than previously available. This method is now becoming the standard method for fabricating a-Si:H solar cells.

The roll-to-roll method, shown in Figure 12.3b, is a mass production technique developed by the ECD company (Izu and Ovshinsky, 1983). It has the special feature of fabricating each a-Si:H layer on a continuous flexible substrate in separate reaction chambers divided by fine divisions. Other methods include the plasma box method of the Solems company (Fig. 12.3c) (Schmitt, 1989) and Fuji Electric's IVE method (Fig. 12.3d) (Sakai et al., 1985), in which the substrates are positioned vertically.

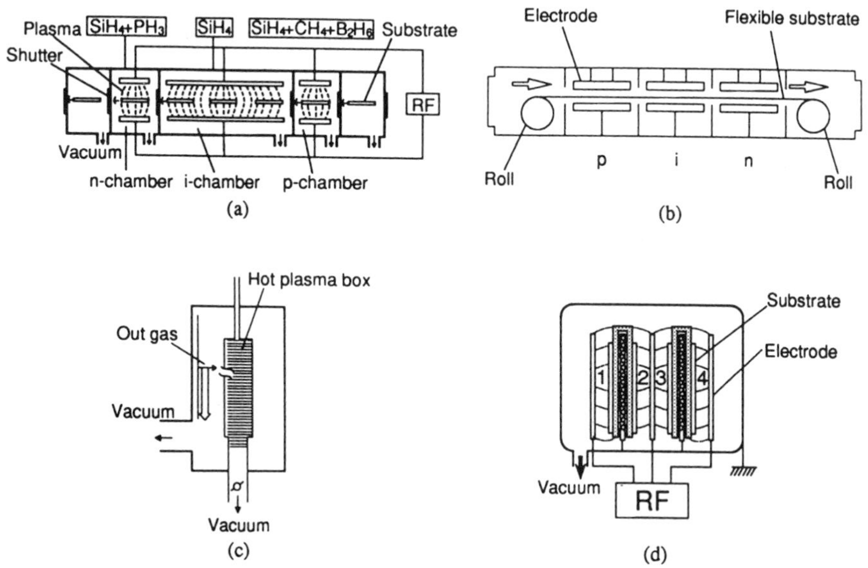

FIGURE 12.3 Fabrication methods for a-Si:H solar cells: **(a)** consecutive, separated reaction chamber method; **(b)** roll to roll method; **(c)** plasma box method; and **(d)** vertical arrangement of substrates method.

The integrated-type structure is formed by using the patterning process shown in Figure 12.4. Previously, the process involved wet or dry etching using a metal mask (Fig. 12.4a), but recently a new laser patterning method (Fig. 12.4b) was developed (Nakano et al., 1986). Laser patterning has a number of excellent advantages as an a-Si:H solar cell manufacturing process:

- Basically, there is no constraint on substrate size; large-area substrates can be processed, so it is efficient for high-power modules.
- Precision, high-density patterning makes it possible to increase the effective area.
- Processing is simplified, including completely dry processing and so on.
- All processes for the a-Si:H solar cell fabrication can be fully automated to one line.

This kind of laser patterning is an effective technique for the industrial production of a-Si:H solar cells.

2.3.2. Poly-Si Solar Cell Modules. At this time, most methods for fabricating poly-Si thin films ($<50\,\mu$m) are in the research stage. The ribbon or sheet poly-Si fabrication process is a leading method, as mentioned in Chapter 2, and their module technology is similar to that of crystalline-Si solar cells.

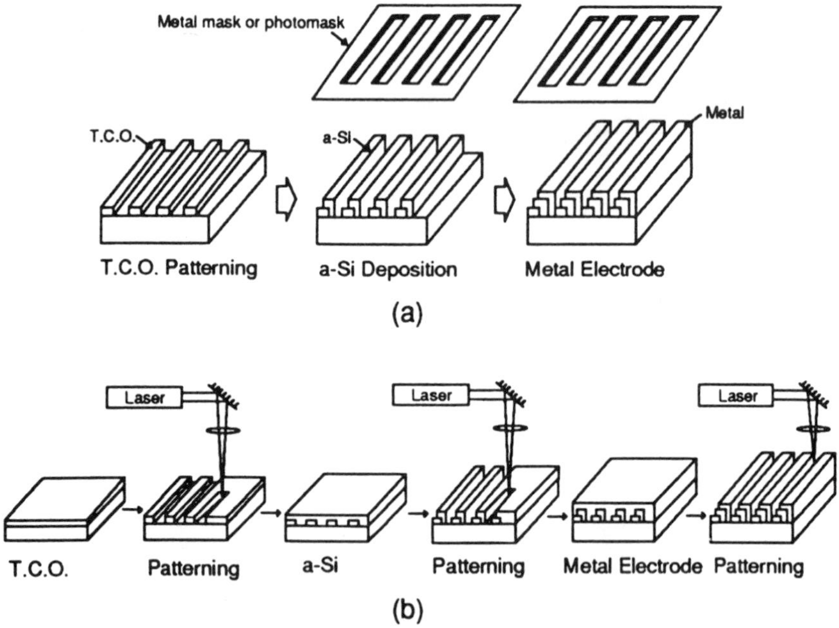

FIGURE 12.4 **(a)** Conventional patterning process and **(b)** laser patterning process.

2.3.3. Compound Semiconductors Solar Cell Modules. The methods of fabricating thin-film solar cells using compound semiconductors include vacuum evaporation techniques and screen printing techniques that use pastes. A CdS/CdTe solar cell manufacturing process that uses an all-printing method has been developed by Matsumoto et al. (1984) at Matsushita.

Of the CIS solar cells, CdS/$CuInSe_2$-type cells have the merit of higher efficiency. The methods for fabricating CIS cells include sputtering and vacuum evaporation methods, but the simpler methods of spraying (Duchemin et al., 1989) and screen-printing (Vervaet et al., 1989) are also being investigated.

2.4. Unique Solar Cell Modules

Here, focusing on a-Si:H solar cells, which are progressively being used as thin-film solar cells, we introduce some practical solar cell modules that have unique structures that take advantage of the merits of thin-film devices.

2.4.1. See-Through Solar Cell Module. The see-through structure transmits part of the incident light, so it can simultaneously provide lighting and produce electricity. This see-through structure is one type of module that makes the most of the special qualities of thin-film materials.

There are two types of see-through structures. The one shown in Figure 12.5 makes use of the ease of processing of thin film-materials by perforating part of the solar cell with many small holes so as to allow some of the light to pass through (Ohnishi et al., 1989b). The other type of structure makes use of the transparency of thin-film materials. Normally the light entering the solar cell is reflected by the metal back electrode and does not pass through. However, if the same type of transparent electrode is used for the back electrode is is used for the electrode where the light enters the cell, the solar cell will absorb part of the light striking it and allow the rest to pass through.

2.4.2. Ultralight Flexible Solar Cell Module. An example of another unique module is the flexible a-Si:H solar cell (Kishi et al., 1990). a:Si:H can be fabricated by low-temperature processes, as described earlier, so it is possible to use transparent plastic film as the substrate. Thus, by forming integrated-type a-Si:H solar cells directly on the plastic film, a super-lightweight,

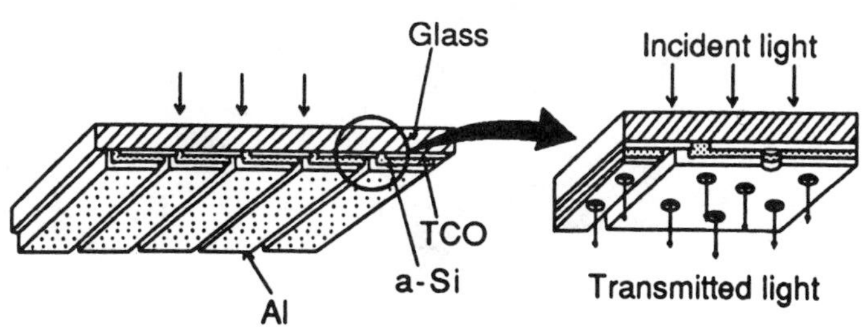

FIGURE 12.5 See-through solar cell module structure.

flexible a-Si:H solar cell module can be obtained. These solar cells can be bent and bonded to curved surfaces. What is more, they produce a 0.27 W/g output, which is larger than the 0.25 W/g output of the GaAs solar cell modules used in space applications. Thus, these solar cells are expected to find a wide range of applications.

2.4.3. Roofing Tiles With Solar Cell Modules. When solar cells are used to provide electric power, system design, including installation, is important. In particular, the installation method has a large effect on the cost of the solar cells. Thus, solar cells that are unified with building materials have been developed. The main merits of such integration are

- Installation does not require additional land.
- No structure is required to support the solar cells.
- The solar cells also serve as building materials.

Figure 12.6 shows a photograph of the external appearance of this type of module installed on a roof. These modules are made by fabricating the large-area integrated-type solar cell described above directly on a transparent Japanese-style glass roofing tile (Matsuoka et al., 1990). The roofing tile modules were jointly developed by Sanyo Electric and Asahi Glass. The dimensions of the glass tile that serves as the substrate is 305 × 305 mm. It weighs the same as a conventional tile of the same shape, but is three times as

FIGURE 12.6 External appearance of roofing tile modules installed on a roof.

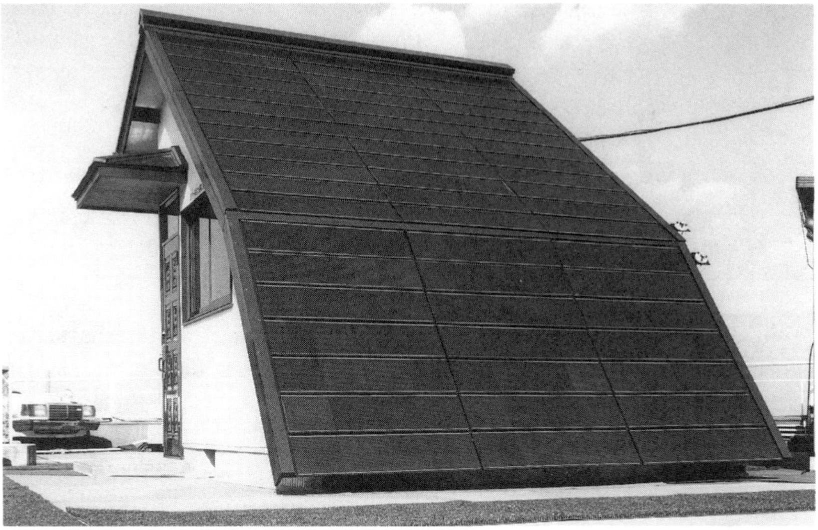

FIGURE 12.7 A model house using shingles with a-Si:H solar cell modules.

strong. The feasibility of direct fabrication on this kind of curved surface is a unique feature of thin-film solar cells. Solar shingles, which are solar cells unified with ordinary slate shingles, and their appearance as installed on a model house are shown in Figure 12.7 (Ohnishi et al., 1980b). Installation on this model house shows that the essential points in installation are about the same as for slate shingles, and the appearance is quite familiar.

2.5. Characteristics and Reliability

2.5.1. Characteristics and Reliability of a-Si:H Solar Cell Modules. The characteristics of a-Si:H solar cell modules are listed in Table 12.1. At this time, the maximum total conversion efficiency being obtained from cells of 100 cm^2 or larger is 12.0%. For large-area modules of 1,200 cm^2, efficiencies of 10% or more are being obtained.

Concerning the reliability of a-Si:H solar cells, from the aspects of both module fabrication techniques and data accumulation, a performance guarantee of from 20 to 30 years is needed. Module fabrication has attained a guaranteed durability of about 20 years. To ensure reliability for a longer period, it is necessary further to study filling materials, backing materials, and side shields with respect to moisture proofing and resistance to light-induced degradation.

The latest data on light-induced degradation for an a-Si:H solar cell is shown in Figure 12.8. A feature of this characteristic is that there is about 10%–20% degradation in the initial period after exposure to light, but after

TABLE 12.1 Characteristics of a-Si:H Solar Cell Modules

Junction Structure	Eff. (%)	V_{oc} (V)	J_{sc} (mA/cm^2)	FF	Area (cm^2)	Organization
Single junction cell						
p a-SiC/in a-Si:H	12.0	12.53	(130.1 mA)	0.735	100	Sanyo
p a-SiC/in a-Si:H	10.3	10.0	16.4/unit	0.691	100	Kaneka
Two terminal tandem cell						
a-Si:H/a-Si:H	10.1	15.8	8.3/unit	0.693	100	Kaneka
a-Si:H/a-Si:H	10.05	39.3	(441 mA)	0.696	1200	Fuji

that the characteristic becomes stable. Measures to improve this characteristic include decreasing the impurities in the a-Si:H layer (Tsuda et al., 1985), improving layer quality (Kuwano, 1987), and decreasing the thickness of the i-layer using stacked cell structures (Nakamura et al., 1985). By using stacked cell structures, the degradation is suppressed to less than 15% of the initial conversion efficiency (Ichikawa et al., 1990).

2.5.2. Characteristics and Reliability of poly-Si and Compound Semiconductor Thin-Film Solar Cell Modules. The characteristics of poly-Si thin-film solar cell and compound semiconductor solar cell (CdTe, CIS, etc.) modules are listed in Table 12.2. First, the Astro Power company has reported characteristics of large-area silicon thin-film solar cells. Concerning the compound semiconductor thin-film solar cells, BP Solar, Matsushita, and others have developed CdTe cells that have high conversion efficiency for a large area. Also, Siemens Solar has achieved conversion efficiency of near 10% with a large-area CIS solar cell.

Both the CdTe and CIS solar cells have high reliability. Particularly for the CIS cells, even exposure to AM1 light at 80°C for approximately 3,600 h does

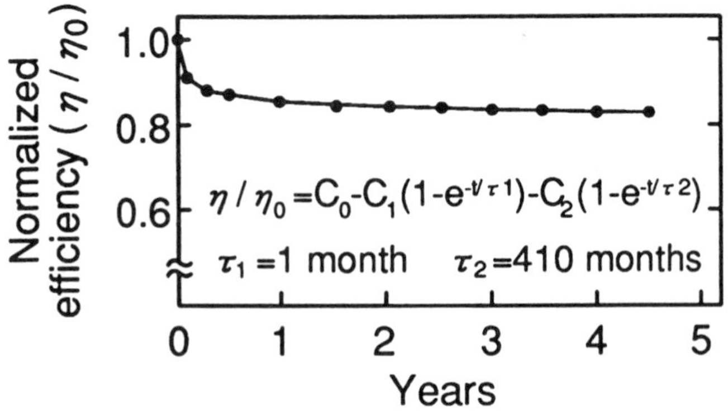

FIGURE 12.8 Light-soaked degradation data of an a-Si:H solar cell submodule.

TABLE 12.2 Characteristics of poly-Si and Compound Semiconductor Thin-Film Solar Cell Modules

Group	Material	Area (cm^2)	Eff (%)	Power (W)
Siemens Solar	$CuInSe_2$	3,883	9.7	37.8
Astro Power	Si-Film	3,984	9.5	34.2
Photon Energy	CdTe	3,323	6.4	21.3
Siemens Solar	$CuInSe_2$	938	11.1	10.4
Matsusha Battery	CdTe	1,200	8.1	9.73
BP Solar	CdTe	706	10.1	7.1
Photon Energy	CdTe	832	8.1	6.8

not change the characteristics. However, the thin-film properties of compound semiconductor are degraded by moisture, so this fact must be considered in the design of the solar cell modules.

3. APPLICATIONS OF THIN-FILM SOLAR CELL MODULES

Since the a-Si:H solar cell was developed as a cheap and mass produced solar cell, the application range of thin-film solar cells is quickly expanding from consumer products to portable applications and independent power sources.

3.1. Consumer Product Applications

The first application of thin-film solar cells came in 1980, when the new integrated-type a-Si:H solar cells appeared. With these integrated-type solar cells, solar power spread to calculators, radios, watches, and solar charging devices. Figure 12.9 shows examples of these consumer goods. Independent power source applications also continue to spread. Integrated-type solar cells are now used in solar "guide posts" (large electronic guide maps installed), water pumping systems, and portable devices (Fig. 12.10a–c).

Another recently developed consumer application that is getting a lot of attention is a solar-powered air conditioner (Fig. 12.11). Power consumption by air conditioners peaks in midsummer, and it is difficult to provide a stable supply of electricity. The generating power of solar cells and the air conditioner's power consumption both vary with time in much the same way, and so this system is expected to be able to reduce peak power demand in the summer. When the solar cells electric output is at its highest, it covers most of the air conditioner's power consumption. In addition, a bidirectional solar air conditioner that can send the electric output from the solar cells back into the commercial utility grid is being studied. With this, a consumer could take advantage of the electricity generated by the solar cells even when not using the air conditioner.

FIGURE 12.9 Consumer applications of a-Si:H solar cell modules.

(a)

FIGURE 12.10 Stand-alone applications of a-Si:H solar cell modules. **(a)** guide post.

(b)

(c)

FIGURE 12.10 Stand-alone applications of a-Si:H solar cell modules: **(b)** pump, and **(c)** portable power supply.

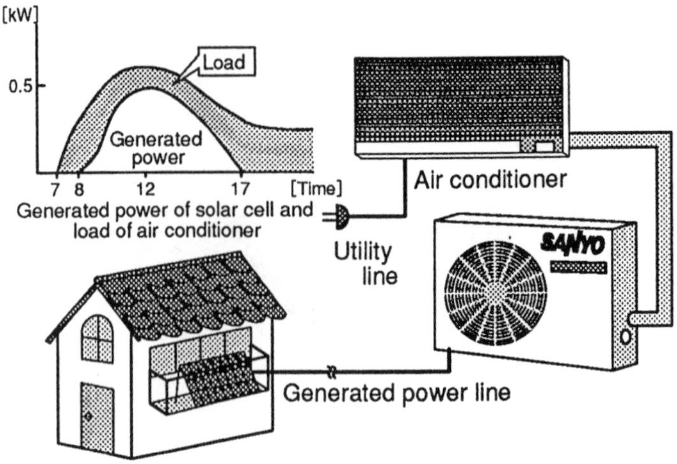

FIGURE 12.11 Solar air conditioner.

3.2. New Applications

Some unique applications of integrated-type a-Si:H solar cells and other unique a-Si:H solar cell modules are described here.

3.2.1. Solar Airplane. Figure 12.12 is a photograph of the Tampopo (Sunseeker), a solar airplane bearing the flexible a-Si:H solar cells described above. This is the first plane ever to use thin film solar cells to generate motive power, which in September 1990 made headlines by successfully crossing the North American continent.

3.2.2. Solar Airship. Another use for flexible solar cells is in airships. They are installed on the Solar Egg (Fig. 12.13), the first airship in the world to move by solar power.

3.2.3. Hybrid Solar Car. The car in Figure 12.14 is the "hybrid solar car" Mirai I (Future I), a next-generation electric car equipped with solar cells, small fuel cells, and NiCd secondary batteries. The motive power comes from the clean energy of the solar cells and other batteries, so the Mirai-I is getting much attention as a clean vehicle that does not emit exhaust gases or other noxious waste products. Furthermore, because it uses a hybrid power source, it can be driven any time, night or day.

3.3. Designs of Consumer Electronic Products With Solar Cells

Here, the designs of various devices that use solar cell modules are described. Figure 12.15a shows a circuit diagram for a calculator equipped with a solar

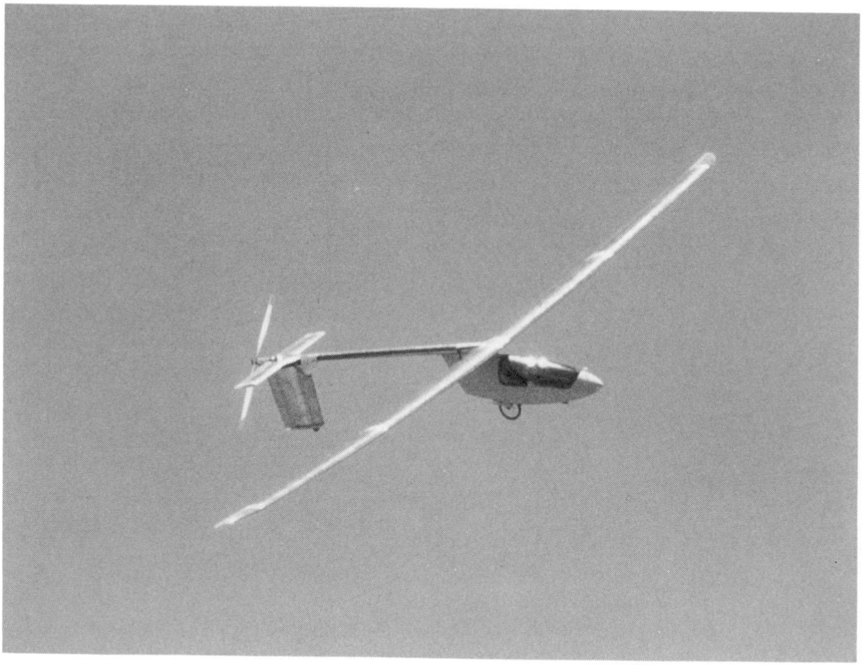

FIGURE 12.12 Solar air plane.

cell submodule. Calculators are usually used in the light, so a backup battery is not necessary. Figure 12.15b shows a circuit diagram for a wristwatch equipped with a solar cell. Wristwatches must operate even in darkness, so a backup secondary battery is also included. A protection diode prevents current from flowing backwards from the secondary batteries into the solar cell, and a recharge control circuit prevents overcharging. Figure 12.15c shows the system

FIGURE 12.13 Solar airship.

FIGURE 12.14 Hybrid solar car.

diagram for a solar air conditioner. As the air conditioning unit itself is a conventional inverter type, the direct current from the solar cells can be used to advantage. The DC output from the solar cells passes through a DC/DC converter and straight into the inverter. The AC from the commercial outlet passes through a diode rectifier and into the inverter unit, where it is combined with the current from the solar cells and sent on to the compressor, just as in standard systems.

3.4. Future Prospects for Thin-Film Solar Cell Modules

Uses for thin-film solar cell modules continue to spread. Beginning with watches, calculators, and other small general-use devices, they will eventually be used as small power sources for household appliances. From now on, if power-generating solar cells enter widespread, everyday use and are joined to electric power systems, a local electric network based on them will emerge. As this spreads throughout the country, connecting different local electric grids into a trunk network, it will become a country network. As it expands further, a global electric network will develop. In this way, we foresee a worldwide electric network based on solar cells, an energy system that we call the GENESIS Project (Fig. 12.16) (Kuwano, 1989), where GENESIS means Global Energy Network Equipped with Solar cells and International Superconductor grids. Energy demands in the year 2000 will be the equivalent of 14 billion tons of crude oil per year. To meet this, assuming that the conversion

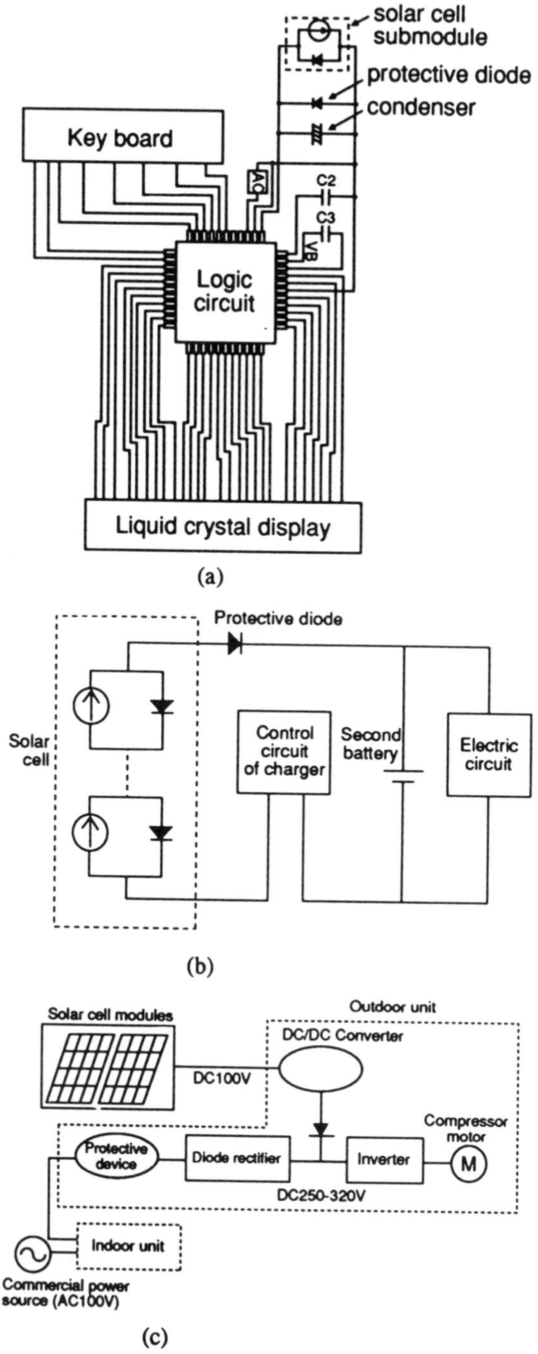

FIGURE 12.15 The circuit diagrams and system diagrams of applications of solar cell submodules. **(a)** calculator, **(b)** wristwatch, and **(c)** solar air conditioner.

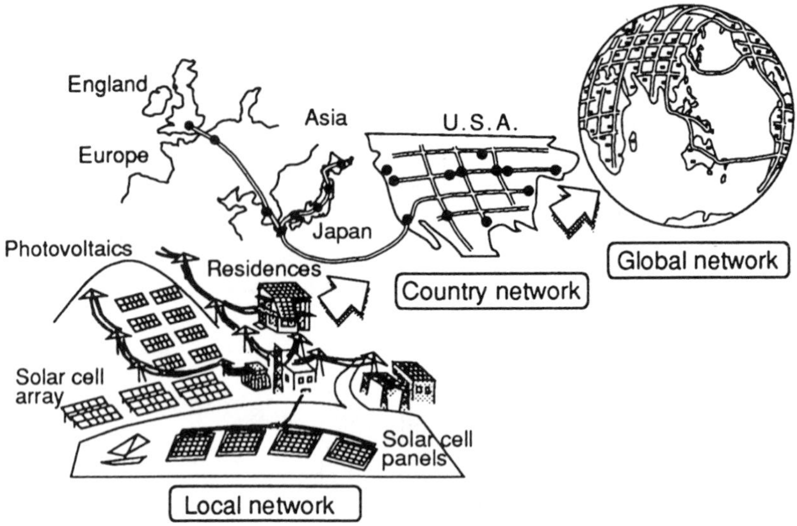

FIGURE 12.16 Step toward GENESIS (Global Energy Network Equipped with Solar cells and International Superconductor Grids) project.

efficiency of thin-film solar cells is 10%, 800 km square of thin-film solar cells would be needed. This is a feasible plan; barely 4% of the world's desert area would suffice. In the midst of the worsening energy crisis and environmental concerns, this plan must be put into effect for the sake of a prosperous twenty-first century.

4. SUMMARY

Thin-film photovoltaic cells have been developing step by step. The problem from here on is to improve performance of thin-film solar cell modules and to lower the cost still further, as well as to develop new applications.

Solar cells gain much attention as a clean energy source, and in this chapter we discussed thin-film solar cell modules, which have a wider range of uses and possibility of a low cost solar cell. We addressed their distinctive features, recent trends, latest applications, and prospects for the future. World environmental problems are in the spotlight these days, and solar cells show great promise as a new source of clean energy. Their performance continues to improve, and there is a steady progression of new products using solar cells. Thin-film solar cells should now play an important role alongside other PV cells.

ACKNOWLEDGMENTS

The work discussed in this chapter is supported in part by NEDO under the Sunshine Project of MITI, Japan.

REFERENCES

Carlson, D.E., and Wronski, C.R. (1976), *Appl. Phys. Lett.* **28**, 671–673.

Cunningham, D., and Barnett, A. (1990), *21st IEEE Photovoltaic Specialists Conf. Record*, IEEE, New York, pp. 307–310.

Duchemin, S., Bougnot, J., El Ghzizal, A., and Belghit, K. (1989), *9th E.C. Photovoltaic Solar Energy Conf. Proc.*, pp. 476–479.

Hamakawa, Y. (1992), *6th Int. Photovoltaic Solar Energy Conf. Proc.*, p. 3–10.

Ichikawa, Y., Fujikake, S., Yoshida, T., Hama, T., and Sakai, H. (1990), *21st IEEE Photovoltaic Specialists Conf. Record*, IEEE, New York, pp. 1475–1480.

Izu, M., and Ovshinsky, S.R. (1983), *Int. Soc. Optical Engineering Proc.*, SPIE, Washington, DC, **407**, pp. 42–46.

Kishi, Y., Inoue, H., Murata, K., Tanaka, H., Kouzuma, S., Morizane, M., Fukuda, Y., Nishiwaki, H., Nakano, K., Takeoka, A., Ohnishi, M., and Kuwano, Y. (1990), *5th Int. Photovoltaic Solar Energy Conf. Technical Digest*, Kyoto, pp. 645–648.

Kuwano, Y. (1987), *AIP Conf. Proc.* **157**, 126–133.

Kuwano, Y. (1989), *4th Int. Photovoltaic Solar Energy Conf. Proc.*, pp. 557–564.

Kuwano, Y., Imai, T., Ohnishi, M., and Nakano, S., (1980), *14th IEEE Photovoltaic Specialists Conf. Record*, IEEE, New York, pp. 1408–1409.

Kuwano, Y., and Ohnishi, M. (1981), *9th Int. Conf. Amorphous Liquid Semicond. Proc.*, pp. C4-1155–C4-1164.

Kuwano, Y., Ohnishi, M., Tsuda, S., Nakashima, Y., and Nakamura, N. (1982), *Jpn. J. Appl. Phys.* **21**, 413–417.

Marfaing, Y. (1979), *2nd E.C. Photovoltaic Solar Energy Conf. Proc.*, pp. 287–290.

Matsumoto, H., Kuribayashi, K., Uda, H., Kamatsu, Y., Nakano, A., and Ikegami, S. (1984), *Solar Cells* **11**, 367–373.

Matsuoka, T., Yagi, H., Waki, Y., Honma, K., Sakai, S., Ohnishi, M., Kawata, H., Nakano, S., and Kuwano, Y. (1990), *Solar Cells* **29**, 361–368.

Matsuyama, T., Wakisaka, K., Kameda, M., Tanaka, M., Matsuoka, T., Tsuda, S., Nakano, S., Kishi, Y., and Kuwano, Y. (1990), *Jpn. J. Appl. Phys.* **29**, 2327–2331.

Matsuyama, T., Sasaki, M., Tanaka, M., Wakisaka, K., Tsuda, S., Nakano, S., Kishi, Y., and Kuwano, Y. (1992), *6th Int. Photovoltaic Solar Energy Conf. Technical Digest*, pp. 753–759.

Nakamura, N., Nishikuni, M., Watanabe, K., Tsuda, S., Nakano, S., Ohnishi, M., Kishi, Y., and Kuwano, Y. (1985), *Int. Conf. Solar Wind Energy Applications Proc.*, pp. B17–B22.

Nakano, S., Matsuoka, T., Kiyama, S., Kawata, H., Nakamura, N., Nakashima, Y., Tsuda, S., Nishiwaki, H., Ohnishi, M., Nagaoka, I., and Kuwano, Y. (1986), *Jpn. J. Appl. Phys.* **25**, 1936–1943.

Ohnishi, M., Shibuya, H., Okuda, N., Kishi, Y., Matsuoka, T., Tsuda, S., Nakano, S., Kuwano, Y., Ohara, S., Kiyama, S., Hosokawa, H., Hirano, Y., Kawata, H. (1989a), *4th Int. Photovoltaic. Solar Energy Conf. Proc.*, pp. 633–638.

Ohnishi, M., Shibuya, H., Okuda, N., Inoue, H., Kishi, Y., Kiyama, S., Kawata, H., Okamoto, S., Tsuda, S., Nakano, S., and Kuwano, Y. (1990), *21st IEEE Photovoltaic Specialists Conf. Record*, IEEE, New York, 1394–1399.

Ohnishi, M., Waki, M., and Kuwano, Y. (1989b), *Optoelectronics* **4**, 195–204.

Schmitt, J.P.M. (1989), *Thin Solid Films*, **174**, 193–202.

Spear, W.E., and LeComber, P.G. (1975), *Solid State Commun.* **17**, 1193–1196.

Sakai, H., Maruyama, M., Yoshida, T., Ichikawa, Y., Kamiyama, M., Ichimura, T., and Uchida, Y. (1985), *17th IEEE Photovoltaic Specialists Conf. Record*, IEEE, New York, pp. 76–81.

Sakai, A., Nakata, M., Shirai, H., Namikawa, T., Hanna, J., and Shimizu, I. (1990), *5th Int. Photovoltaic Solar Energy Conf. Technical Digest*, pp. 697–700.

Tsuda, S., Nakamura, N., Nishikuni, M., Watanabe, K., Takahama, T., Hishikawa, Y., Ohnishi, M., Kishi, Y., Nakano, S., and Kuwano, Y. (1985), *J. Non-Cryst. Solids* **77 78**, 1465–1468.

Ullal, H.S., Stone, J.L., Zweibel, K., Surek, T., and Mitchell, R.L. (1992), *6th Int. Photovoltaic Solar Energy Conf. Technical Digest*, pp. 81–87.

Vervaet, A., Burgelman, M., Clemmick, I., and Capon, J. (1989), *9th E.C. Photovoltaic Solar Energy Conf. Proc.*, pp. 480–483.

Wagner, S., Shay, J.L., Migliorato, P., and Kasper, H.M. (1974), *Appl. Phys. Lett.* **25**, 434–435.

CHAPTER THIRTEEN

Silicon and Gallium Arsenide, Single Junction Space Modules

ROBERT E. NEFF, Space Systems/Loral, Palo Alto, CA 94303

1. BACKGROUND

Since the launching of Vanguard I in 1957, solar cells have been used to provide either primary or back-up power to a majority of military and commercial spacecraft. The growth in solar cell maturity over the years has been documented in Chapter 4. During that time frame, the methodologies used for design and the technologies available for assembling solar cells into functioning electrical circuits have also matured. Initially, small solar cells (1.0 × 2.0 cm) were soldered by hand, laying the rear side (p contact) of one cell onto the top ohmic bar (n contact) of the next series cell, thus providing the necessary voltage for a particular circuit. These small circuits were then bonded by hand to relatively heavy aluminum substrates and wired to a connector. Since that time, the methods for assembly have greatly matured to the point where large area solar cells with wrap through contacts (8.0 × 8.0 cm, used on the Space Station) are being automatically assembled in series and parallel configuration and bonded to extremely light, flexible kapton blankets that have conductor traces laminated between the layers of the kapton blanket. This chapter provides a brief overview of the many parameters that must be considered in designing both a solar array module for a particular space mission and flight-qualified assembly techniques that can be used to optimize different mission parameters such as subsystem mass and end-of-life (EOL)

Solar Cells and Their Applications, Edited by Larry D. Partain.
ISBN 0-471-57420-1

solar array capabilities. While beyond the scope of this Chapter, these "subsystem" design concepts must be integrated into the overall spacecraft "system" level design to provide a completely optimized spacecraft design. Additional information regarding system level design is given in Chapter 17.

2. COMPONENT DESIGN CONSIDERATIONS

Each particular discipline (electrical, mechanical, thermal, and so forth) has its own set of problems that must be solved in order for an overall spacecraft design to be optimized. The importance of open communications with these different subsystems early in the design of a solar array module cannot be over-stressed. Electrical design considerations, discussed below, require input from several different disciplines.

2.1. Choice of Cell Type

The optimized choice of a solar cell type is driven by many different factors, including expected on-orbit radiation fluence, as well as mass, power, operational, and orbital sun pointing requirements. The details of space-qualified solar cell construction have been provided in earlier chapters. However, to summarize briefly, single junction silicon solar cells typically have antireflective coatings on the surface of the cell to optimize absorption of the usable solar spectrum to the junction, and an aluminized back surface reflector (BSR) that aids in controlling operating temperatures. The cells can be small (2×4 cm) or large (8.0×8.0 cm), thick (200 μm) or thin (50 μm), or have a back surface field (BSF) applied by diffusion from aluminum paste or ion implanted boron. Silicon solar cells (with space flight heritage) have efficiencies ranging from 12.5% to 14.8%, depending on their construction, doping levels, and thickness of the cell, when tested at standard conditions (see below for definition). Gallium arsenide devices are typically manufactured with either the liquid phase epitaxy (LPE) or metal organic chemical vapor deposition (MOCVD) processes. In the case of the latter, the gallium arsenide junction can be placed on germanium substrates (GaAs/Ge) that are inherently less brittle than gallium substrates. These cells have efficiencies ranging from 18.0% to 19.5%. At the component level, however, these cells weigh more and cost more than conventional silicon solar cells. (Saga et al. [1993], however, have recently published data on the development of a textured-surface, space-grade silicon solar cell that has efficiencies as high as 18.3%. These cells, when compared with gallium arsenide, can be made thinner and have better radiometric properties, resulting in lower operating temperatures on orbit. These cells may compete very well with gallium arsenide cells in the future.) It is the responsibility of the design engineer to understand the affects of each of the above parameters, identify all of the requirements (power, mass, attitude control, thermal, cost, and so forth) and pick a candidate that best meets all mission criteria. This is never an easy task. The program office wants the cells to be inexpensive, the mechanical group wants the cells to be the lightest possible, the thermal group wants the best radiometric properties, and the power system wants the highest efficiency cells possible. These dichotomies are never totally resolved, and concessions by all subsystems are usually required.

Solar cells are typically electrically graded at 28°C while under one sun, air-mass zero (AM0) illumination. One sun intensity is considered to be the normal solar irradiance at one astronomical unit, approximately 1.496×10^{11} m from the sun. The solar intensity at this distance is 1,353 W/m^2. AMO indicates the spectral distribution of light in space. The output at the maximum power point is used to determine the efficiency of the solar cell. At this point in the design, however, it is not possible to determine the best solar cell candidate for a given mission profile.

Solar array modules are typically designed to meet all mission requirements at EOL, after the solar cells have been exposed to all on-orbit degradation. Currently, this is typically 10–15 years for geosynchronous earth orbit (GEO) type communications or weather satellites or can be as long at 30 years for the Space Station solar array, which will be in a low earth orbit (LEO). The two biggest contributing factors to degraded solar cell capabilities are the radiation environment and the radiometric properties of the solar cells and substrates on which they are mounted. The radiation environment, including electrons, protons, and solar flare events degrade both the current and voltage characteristics of solar cells. A stack density analysis is typically performed to determine the amount of protection provided by the coverglass and substrate material. Details on the failure mechanism are given in Chapter 17. Also, Tada et al. (1982) and Anspaugh (1989) provide additional information on ways of identifying radiation degradation factors. The radiometric properties, solar absorptance and hemispherical emittance, and the cell's degraded efficiency define the expected solar cell operational temperatures at EOL. As the expected operating temperature increases, the ability of the cell to provide power decreases.

Another point that needs to be clearly understood is how the solar cells are intended to be used. Silicon solar cells typically have a very constant (flat) current characteristic when a reverse bias voltage is applied. A cell's avalanche voltage can be greater than 30 V for silicon cells and are normally not damaged or degraded if the reverse potential is less than 20 V. This is not necessarily the case with gallium arsenide solar cells, which can have nonlinear reverse bias characteristics and can become permanently damaged if they are reverse biased with potentials as low as 2–3 V. While most solar cells are intended to be forward biased and used to "create power" when illuminated, there are some cases where solar cells are intended to be used in a reverse-biased condition where, due to the nature of their reverse bias characteristics, they control the amount of current flowing to a particular user load. Once these operational and EOL solar cell capabilities are defined, a candidate cell, meeting other mission requirements (cost, mass, and size), can be chosen.

2.2 Coverglass Protection—Minimizing Radiation Degradation

Once a candidate solar cell has been chosen, a thin piece of glass is bonded over the surface of the solar cell to reduce the degradation affects of the radiation environment, increase the cell's infrared emittance, and filter out ultraviolet radiation. This glass is typically 75–150 μm thick, but can be as thick 500–700 μm, depending on the expected radiation environment. The choice of a thicker coverglass to reduce degradation versus using thinner (lighter) coverglass and adding more cells to meet power is part of the optimization trade that needs to take place early in a module design. In addition to helping protect the cell junction from the radiation environment, the coverglass also acts as an ultraviolet (UV) filter, either by reflection or absorption, preventing the solar spectrum below (typically) 0.35 μm from reaching the solar cell junction. Solar cells do not respond well to wavelengths below this level, and the affect of light with these wavelengths reaching the solar cell junction is simply to increase the solar cell operating temperature, thus reducing efficiency. The coverglass helps to reduce this effect. Also, since the glass is a high emittance material, the cell operating temperature is reduced, thus increasing the cell's operational efficiency. Several different types of glass have been developed to match the coefficient of thermal expansion (CTE) of different solar cell materials. The proper choice of a coverglass material, in addition to protecting the cell junction, minimizes the bimetallic affect of the solar cell stack, once it has been attached to the solar cell. This reduces stresses on the solar cell assemblies when they are exposed to the thermal environment of space. Pilkington, Inc., and Optical Coating Laboratories, Inc., provide much of the space glass used for U.S. and European space solar array modules, while Asahi Glass provides covers for several Japanese satellite programs.

The coverglass is typically bonded to the solar cell using a thin layer (25–50 μm) of optically transparent adhesive. Dow Corning's DC-93-500 has been the candidate adhesive for many years, but other companies such as McGahn-Nusil and Wacker-Chemie have similar products. The process of attaching the coverglass to the solar cell can be very labor intensive and adds mass to the system, and the adhesive tends to darken slightly when exposed to the UV environment (even with the use of UV filters), reducing the transmission of the usable light spectrum to the solar cell junction. As a result, efforts have been made to qualify Teflon bonding (White and Jones, 1991) and adhesive-free application using electrostatic bonding (Koch and White, 1990). Regardless of the method used, a coverglass must be sized to cover the entire active area of the solar cell in order to optimize on-orbit capabilities. Even small areas of exposed active junction can result in large on-orbit electrical degradations (Jet Propulsion Laboratory, 1976).

2.3. Interconnection Material and Design

To allow for easy assembly of individual solar cells into series-parallel configurations, an interconnect is typically attached to the top surface (N contact for silicon and P contact for gallium arsenide) of the solar cell. This is attached using either welding or soldering techniques (discussed below) and is attached before the coverglass is bonded to the top surface of the cell. The choices of material and configuration are usually dependent on mission requirements, the type of solar cell used, the cell aspect ratio (length/width), and the type of substrate to which the solar cells are bonded.

Numerous papers have been presented in the literature (e.g., Neff et al., 1985; Arnold et al., 1985) detailing design materials and techniques that provide for a highly reliable configuration that takes into account electrical and mechanical characteristics, as well as improving reliability by accounting for the typical "crack plain" of a solar cell wafer material. Thin silver, 12–25 μm thick, die cut or chemically etched and formed with either an in-plane or out-of-plane stress relief loop, is often used. It is nonmagnetic and easily weldable. Silver-plated kovar is another material that is often used. This material has good mechanical properties and has a good CTE match to silicon. It does, however, have magnetic properties and may not be acceptable when an extremely clean magnetic design is required. Silver, while having extremely good electrical characteristics, is, however, easily degraded when exposed to an atomic oxygen environment.

Engineers must be aware of the intended orbit and/or the amount of time spent in the atomic oxygen environment when considering an interconnect design. The Long Duration Exposure Facility (LDEF), retrieved in 1990 after spending 69 months in LEO, carried numerous experiments that quantified the affects of exposure to atomic oxygen, as well as other environments such as space debris and micrometeoroids. Many types of materials were included in the experiment packages, including solar cell interconnects. Detailed findings are presented by Levine (1991).

Some solar cells have eliminated the need for a dedicated interconnect by providing either "wrap around" or "wrap through" interconnects. This basically involves plating around an edge or through a hole in the solar cell to position both the positive and negative contacts on the backside of the cell. Space Station cells currently are using a "wrap through" contact design. With this design, the "interconnect traces" are attached to the solar cells at a different point in the module construction sequence. Once the choices of solar cells, coverglass, and interconnect configuration have been made, these pieces are assembled into a compact unit called a Cover Integrated Cell (CIC), which becomes the electrical building block for module construction.

3. DESIGNING CIRCUIT CONFIGURATION

Circuit configuration will be determined by the required system bus voltage and the number of discrete solar cell circuits desired. For typical GEO communications spacecraft using a fixed bus voltage power system design, the required number of series cells required to meet the bus voltage requirements is dictated by the worst case degradation to the maximum power voltage point resulting from radiation degradation, increased operating temperature, voltage drops resulting from harness resistance losses, and diode voltage drops. The module must be designed to provide power at the user interface, typically inside of the spacecraft not necessarily at the connector on the solar array module. From a reliability point of view, the goal is to have the array operating at the maximum power point, during the hottest season (typically vernal equinox for earth-orbiting GEO satellites, but could be winter solstice, depending on the ability of the solar array to track the sun) after degradation by all on-orbit environments (radiation, thermal cycling, micrometeoroids, and so forth). Equation (1) provides a first order approximation for determining the series string length for a fixed voltage system.

$$\text{No. of series cells} = \frac{(\text{Bus voltage} + \text{diode voltage drop})}{[V_{P_{max}} \cdot D - (T_{op} - 28) \cdot dV/dT]} \tag{1}$$

where $V_{P_{max}}$ is maximum power voltage of a single cell at 28°C, D is environmental degradation factor (radiation, temperature cycling, and so forth), T_{op} is expected hottest on-orbit operating temperature, and dV/dT is change in $V_{P_{max}}$ as a function of operating temperature.

As the solar array cools, usually due to reduced solar view angles and/or increased distances from the sun, the operating point on the current–voltage curve for the module will move back toward the constant current portion of the curve. Once this initial series string length requirment is defined, the number of cells in parallel required to meet the total current requirements of the bus can be determined. The current output from a group of cells is directly proportional to the amount of sunlight reaching the cell junction. In the case of some LEO designs, solar cells with "gridded" contacts on both sides of the cell are bonded to flexible, transparent substrates. In these cases, the affects of reflected light, (albedo) from the earth, passing through the cell from the rear side, must also be accounted for. This results from the characteristic of silicon being transparent to the infrared spectrum. These parallel cells are typically configured into small circuits to allow for control and regulation of the array output and to improve system reliability. If the array is designed with a small number of circuits, the loss of a circuit due to a short (e.g., a micrometeoroid impact) or cracked cell could result in a large percentage loss of power-generating capability. As a result, array designers usually include a relatively large number of smaller sized, diode isolated circuits to minimize power loss should a short occur. Equation (2) provides a first-order approximation for determining the number of parallel cells required to meet bus current requirements.

$$\text{No. of parallel cells} = \frac{(\text{Bus load/bus voltage})}{[I_{op} \cdot D + (T_{op} - 28) \cdot dI/dT] \cdot S} \tag{2}$$

where I_{op} is single cell current capability at the operating point, at 28°C, one sun (it may be necessary to guess at this value for the first iteration), D is environmental degradation factor (this includes all effects that may degrade the cell's ability to produce current or reduce the transmission of light to the cell junction, including radiation, adhesive darkening due to UV, micrometeoroid impacts, and current mismatch), T_{op} is expected operating temperature at lowest solar intensity, dI/dT is change in I_{op} as a function of operating temperature, and S is solar intensity factor taking into account worst case solar distance ($1/R^2$) and off-pointing (cosine) effects.

The values provided by Eqs. (1) and (2) above for series and parallel cell requirements will give initial values to use in an array layout. However, defining the final values to be used in an optimized design is an iterative approach. Harness resistance and increased operating temperatures also have the affect of "softening" the knee of the solar cell's I–V curve (thus affecting the values of $V_{P_{max}}$ and I_{op}), and these second order affects need to be accounted for in a final design. These affects can be modeled with a series resistance where the resulting ΔV voltage drop is defined as $I \cdot R$ for a constant R and an I at a particular point on the circuit's current–voltage (I–V) curve. Determining the proper modeling values to use for knee softening resistance can be determined in the laboratory by measuring the affect on the shape of the knee of the I–V curve at different operating temperatures. Values for dI/dT and dV/dT can also be determined in this way. Simple computer programs can be written, using either a programming language or a spread sheet format, to help with the iterations and optimization of a design.

Table 13.1 provides typical reference values for a number of different design parameters for both silicon and gallium arsenide solar cells. The ranges of values given for the silicon cells result from the many different construction options available for silicon. These characteristics affect operating temperatures and degradation values. Caution must be used when referencing these values. No single parameter or characteristic can be used to define a "best" cell type for a particular solar array design. The design engineer must be very cautious when drawing conclusions from any one parameter. For example, silicon cells without a BSF typically degrade less in a radiation environment than cells with a BSF. However, they also have significantly lower output capabilities. Gallium arsenide solar cells have much higher conversion efficiencies than silicon, but also have poorer radiometric properties compared with silicon, thus resulting in increased operating temperatures, which reduces the perceived power increase. The density of the substrate material used for gallium arsenide cells is also about 2.3 times denser than silicon, resulting in increased mass at the component level. Before a final cell selection can be made, the effects of all design parameters for a particular mission must be taken into account.

Figure 13.1 shows typical beginning and end of life solar array capability for a 5,000 W solar array. Note the EOL equinox operating point near the knee of the I–V curve. Once on the voltage side of the knee, small changes in intensity or temperatures can cause major losses in power capability. As a result, defining an accurate series string length requirement is critical.

TABLE 13.1 Typical Solar Cells and Coverglass Parameters[a]

Temperature coefficients				
dI/dT (I_{sc}, mA/cm^2/°C)	BOL		EOL[b]	
Silicon	0.02–0.03		0.07–0.09	
GaAs/Ge	0.020		0.025	
dV/dT (V_{oc}, mV/°C)				
Silicon	−2.0 to −2.2		−2.0 to −2.2	
GaAs/Ge	−1.8 to −1.9		−1.9 to −2.3	
Material density (g/cm^3)	Silicon		GaAs/Ge	Coverglass
	2.40		5.46	2.20–2.62
Radiometric properties[c]	Silicon		GaAs/Ge	
Solar absorptance (nonoperating)	0.75–0.76		0.88–0.91	
Emittance (hemispherical)	0.81–0.82		0.81–0.82	
Electrical characteristics[d]	Silicon		GaAs/Ge	
I_{sc}(mA/cm^2)	36–42		30	
V_{oc}(mV)	545–605		1,000	
P_{max}(mW/cm^2)	17–20[e]		25–26	
Radiation degradation[f]	Silicon		GaAs/Ge	
	$1 \times 10^{+14}$	$1 \times 10^{+15}$	$1 \times 10^{+14}$	$1 \times 10^{+15}$
I_{sc}	0.95–0.99	0.82–0.94	0.89	0.80
P_{max}	0.86–0.99	0.72–0.90	0.91	0.74
V_{oc}	0.92–0.99	0.85–0.97	0.97	0.92

[a]This table provides ranges of values for the parameters listed. Values may vary from vendor to vendor, and specific values to be used in a detailed should be obtained from vendor data or laboratory measurements.
[b]End-of-life values after exposure to $1 \times 10^{+15}$, 1 MeV electrons.
[c]These characteristics can vary as a function of coverglass type and the coatings used. These values are typical for a cell covered with a Pilkington CMX coverglass.
[d]Values represent typical capabilities when tested at 28°C under one sun, AMO equivalent illumination. Range of values for silicon result from cell construction, thickness, and vendor. Gallium arsenide values are typical for an 18.5% efficient cell.
[e]See Saga et al.(1993) for details of a silicon design having the ability to provide 24 mW/cm^2 at AMO, one sun conditions.
[f]Relative degradation after exposure to the 1 MeV electron fluence shown. Range of values for silicon result from cell constructive, doping levels, thickness, and vendor. A low degradation value does not necessarily mean more on-orbit power.

For spacecraft bus designs that provide capability for "maximum power tracking," or for missions that have orbits that constantly have the array moving off the sun, the design techniques above may not apply. Elliptical or LEO missions, where the solar array must provide charge current to batteries during much of the orbit, require a higher level system design to guarantee that energy balance is maintained during each orbit. The methods above apply, conceptually, to these designs. The details of solar array system optimization, however, for these type of missions are beyond the scope of this chapter. (The

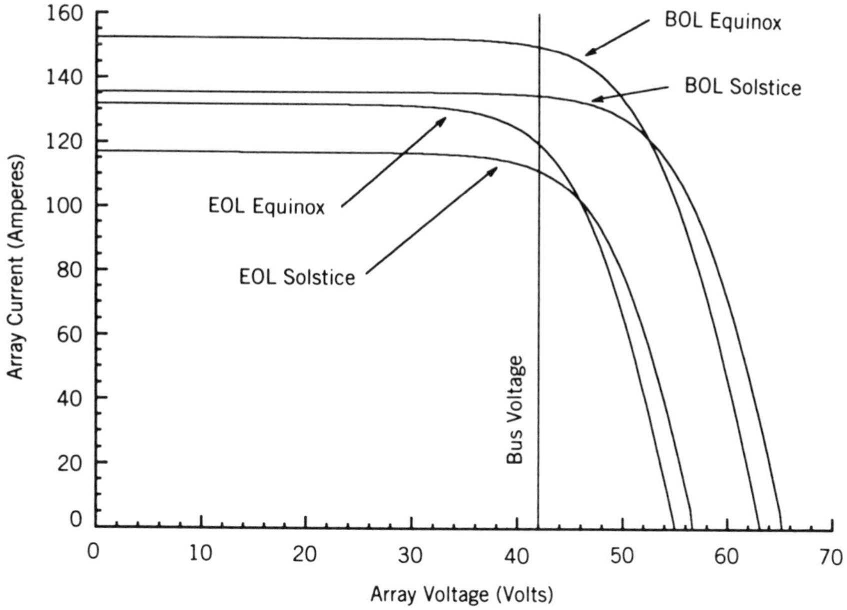

FIGURE 13.1 Typical solar array electrical characteristics.

IEEE sponsors an annual forum, the Intersociety Energy Conversion Engineering Conference [IECEC], that typically presents, as part of the conference, system level design concepts. Proceedings from this forum are recommended as a reference for further details on overall system level design concepts and optimization.)

Once the number of parallel and series solar cells for a solar array module design are defined, it becomes necessary to accommodate the mechanical constraints and for the design engineer or the solar cell vendor to optimize the layout on the panels. Whether using flexible blankets or honeycomb panels to support the solar cell circuits, the dimensions of the panels are typically driven by launch vehicle shroud clearance requirements. Once panel dimensions are defined, the optimization of a panel's packing factor takes place. The *packing factor* is a number less than one and is obtained by dividing the total area of all the solar cells on a particular panel by the total surface area of the panel. Spaces between the solar cells and areas around the panel edges used for circuit terminations, wires, diodes, or connectors account for the remaining available area. Providing symmetry in module layout simplifies manufacturing, makes cancellation of magnetic moments easier, and simplifies wire harness requirements.

The size of the solar cell required to allow for symmetrical assembly needs to be reviewed by the solar cell vendor to determine if the chosen size provides for optimized wafer usage by the vendor. Silicon cells are typically fabricated on wafers with diameters of 100 or 125 mm. Choosing a poor cell dimension in order to provide a good packing factor for a particular panel dimension could result in virtually doubling the number of wafers required by the cell vendor, thus greatly increasing costs. At this point, the dimensions of the substrate should be reviewed as it might be possible to change only slightly a particular dimension to allow for wafer usage optimization at the cell vendor. In general, the largest solar cell possible that both optimizes wafer usage and allows for a symmetrical layout should be used in a design. This minimizes the number of piece parts that must be fabricated and optimizes the panel packing factor. At this point, the final cell dimensions should be reviewed, along with the interconnect geometry, to verify that the typical crack plain of the solar cell wafer material has been accounted for in the interconnect design. This can minimize the risk that a cracked or damaged cell will current limit a circuit's output capabilities. With proper choices of cell and panel dimensions, it is possible to achieve packing factors with values approaching 95%. Optimizing the packing factor is one of the key methods to achieving increased power to weight ratios, whether one uses silicon or gallium arsenide solar cells. For small-sized panels, it may be necessary to reduce the size of the solar cell (increase the number of individual components) in order to optimize wafer usage and fully propagate the panel to achieve a high packing factor.

Once a symmetrical design is established, magnetic moments are minimized, since the current loops of adjacent modules can be oriented opposite each other and both the positive and negative wires carrying current can be routed next to each other, or as twisted pairs, virtually eliminating problems.

4. THE QUESTION OF DIODES—ARE THEY NEEDED?

The need for diodes in a modular system must also be defined as part of the initial design process. Diodes provide two different functions in a solar cell circuit. When connected in series with the circuit, they provide isolation between the other circuits and, depending on the system's electrical design, the spacecraft battery. These diodes are essential from a reliability point of view. Should one circuit be shorted to ground (e.g., from a micrometeoroid impact), the series diode provides isolation and prevents the other circuits from being grounded. They also prevent the battery from discharging through the individual strings during nonsunlit portions of any orbit or season. The series diode can be easily placed anywhere in the circuit harness and as a result can often be protected from the extreme thermal environment that the solar cells are expected to encounter.

The second function diodes provide is in a shunting mode. The criticality of including shunt diodes in a modular design is very much mission and cell type dependent. The degrading effect of either a cracked solar cell or a shadowed portion of a cell or circuit on the output of a circuit is well documented (Jet Propulsion Laboratory, 1976). Shunt diodes are usually placed in parallel with a small series string of solar cells for one of two reasons, either to allow strings to operate in a damaged or shadowed environment or to prevent permanent damage to solar cells when they are shadowed. Solar cells can become reverse biased, either from damage or from a shadow on one or more of the cells. When this occurs, the output of the string will normally be current limited, proportional to the area of the cell being shadowed or damaged. Under such conditions, the diode will allow the full current capability from the remaining series cells to be shunted around the affected cells. The circuit voltage characteristics are reduced, of course, by the number of cells being shunted and the voltage drop of the diode itself.

The second reason for implementing shunt diodes is to prevent permanent damage to the affected cells. The reverse bias characteristic of solar cells are difficult to maintain consistently and control during the manufacturing process. Silicon cells are typically very forgiving when they are reverse biased and will not be damaged when operated in a reverse biased mode of -15 to -20 V. As indicated earlier, this characteristic is one of the factors that allows silicon solar cells to be used as "current controls" rather than current generators. Gallium arsenide cells, on the other hand, can easily become permanently damaged when they are reverse biased by as little as -2 to -3 V. The use of shunt diodes in a gallium arsenide design may be imperative if shadow environments are expected to occur. Details of the Space Station shunt diode methodology are provided by Woike et al. (1990).

Implementing shunting diodes in a layout impacts the design, manufacturing, and assembly of a module. Connection points must be made available in the layout that may degrade the packing factor. In addition, the diodes will be exposed to extreme thermal and radiation environments that may impact their ability to function properly on orbit. Guaranteeing that the diodes will not fail, and thus cause a cascading failure to the solar cells they are intended to protect, is difficult. As a result, the impact of shadows (from antennas or other spacecraft appendages) on the modules needs to be cleary defined early in the design. Once defined, the risks associated with or without implementing shunt diodes into a design can be quantified.

5. FABRICATION AND TESTING OF MODULES

Individual companies have established different ways of fabricating and testing space modules. While the particular processes and techniques may be unique and proprietary to a given company, general requirements and methodology still apply. The individual CICs assembled above must be connected into discrete series-parallel configurations, either by soldering or welding. These are then attached, typically with a thin bondline of adhesive, to a substrate, either flexible or solid. Once assembled, the modules can be tested using natural sunlight, steady-state artifical illumination, or, more typically, pulsed illumination. The latter two cases usually try and match the space spectral environment, defined as air-mass zero (AM0), in order best to characterize the cell's response to the different wavelengths of space illumination. During the manufacturing sequence, process controls must be established to guarantee the integrity of the product. The purpose of this section is to present an overview of the different assembly techniques and testing methodology that can be used during and after the assembly to verify that the solar array circuits meet all requirements.

5.1. The Qualification Coupon

Regardless of the techniques used to fabricate a space module, they should be qualified to verify that the end product will be able to survive the space environment over the designed mission life. A test article or coupon should be designed such that all solar panel components and processes are represented. This includes CIC assembly, circuit assembly, module bonding, and harness routing and support. In addition, it is important to qualify a repair procedure that can be used if a cell panel gets damaged at some point in time during ground assembly and test. This repair procedure should include dielectric and substrate repair scenarios as well as CIC, interconnect, and wiring repairs. Once final processes and materials are defined, the coupon is fabricated and then tested. The primary test at the coupon level usually consists of an ambient pressure thermal cycle test over the expected on-orbit temperature extremes, followed by visual (infrared if possible) inspection and electrical testing to determine if degradation has occurred. Depending on the mission, LEO or GEO, this may involve several hundred to many thousand temperature cycles. Other test requirements, such as acoustic or sine vibration testing, are typically performed on a full-sized flight module or panel.

5.2. Finding Qualified Components

The temperature extremes on orbit are typically much more severe than any MIL SPEC requirements or levels to which vendors are willing to certify their products. This typically includes adhesives, wiring, connectors, diodes, and materials used in the assembly of the substrate itself. As a result, it becomes necessary to verify, through test, that the components and assembly methods can survive the rigors of the space environment. The use of "flight heritage" components and the qualification coupon discussed above serve to provide a high level of probability that a given design, once built, will be able to function properly on orbit.

5.3. Welding vs. Soldering

It is generally felt that welding provides a reliable, lower mass method for assembling solar cells into modules. Parallel gap welding techniques have long been used to assemble silicon arrays, and LaRoche (1988) presents details of its use on gallium arsenide solar cells. However, soldering techniques have been improved over time and have a long history of flight use. Initially, the rear side of a solar cell was completely covered with a layer of solder that resulted in a heavy cell. Since that time, the use of "zone soldering" or of solder "preforms" has greatly negated the impact of added mass. Also, in some respects, soldering is easier to perform, easier to inspect (visual filets), and easier to use when making repairs. The main concern with soldering is it's ability to survive over the mission life, since it has a tendency to crystallize after exposure to many deep thermal cycles. For GEO missions, where there are only typically a few thousand thermal cycles expected, this is probably not a problem. However, for LEO missions, the number of expected cycles can be orders of magnitude higher (albeit at less severe temperature limits). As a result, the effects of stress, fatigue, and crystallization need to be considered when using solder. In addition, since the solar cell contacts are typically silver plated, a solder with a silver component is recommended to eliminate concerns over silver leaching at the joint. Again, the qualification coupon is the vehicle that will determine if welding or soldering provides a more reliable, and perhaps simpler, assembly method.

5.4. Module Bonding to Substrates

Whether the modules are to be bonded to a flexible or a solid substrate, the bond line adhesive provides two functions. First, it provides mechanical integrity to allow the modules to survive the launch environment and on-orbit thermal stresses, and it helps protect the cells from damage during assembly and test. Second, the adhesive provides a thermal path to the substrate, resulting in reduced operating temperatures. To provide optimum characteristics, the bond line should be thin and uniformly distributed over the rear of the cell. Adhesives, or adhesive tapes, specifically designed to provide a low thermal resistance path, as well as good structural bond strength and low outgassing properties, are typically used in this application. Dow Corning, McGahn-Nusil, Wacker-Chemie, and General Electric, among others, have products meeting these requirements.

6. DESIGNING FOR RELIABILITY AND PROCESS CONTROL

Due to the nature of the space environment and the unlikelihood that a spacecraft, once launched, will be able to be repaired if damaged, it is imperative that design reliability and process controls be implemented to optimize the probability of meeting all mission objectives. Space does not permit going into the methodologies of providing high-reliability, low-risk process controls. However, Ott (1975) and O'Connor (1981) provide detailed information on process control and reliability design, respectively.

7. TESTING AND INSPECTION

Once a solar array module has been assembled, it needs to be tested and inspected. Aside from the obvious electrical testing to quantify circuit characteristics, thermal cycle testing, in vacuum, is typically performed. During this test, the electrical isolation between the cells and the substrate is typically monitored to verify that a short to substrate ground does not occur. This test usually consists of only a few cycles and is intended to verify workmanship and to pull out weak components (infant mortality). If necessary, repairs are made using the processes qualified on the test coupon described above. It is not uncommon to have several solar cells break during this test. If the reliability levels of the module are to be maintained, these cells must be found and replaced. Visual inspection with the naked eye or with low magnification can only identify obvious damage. Cracks in the silicon or gallium arsenide wafer material are very difficult to see and in many cases may not be noticed electrically since the grid lines on the surface of the cell may still be intact at this point. Both silicon and gallium arsenide, however, have the unique characteristic of being transparent to long wavelength light ($\lambda > 1\ \mu m$). To aid in the inspection of panels, techniques using an infrared camera have been developed that can easily locate cracked cells. Details of one such test facility are provided by Jennings et al. (1993).

8. SUMMARY

The purpose of this chapter was to identify some of the concerns and parameters that must be considered in designing a space solar cell module that not only will meet the requirements at the panel level but also will provide a design that interfaces well with other subsystems and optimizes the overall spacecraft design. Space does not permit discussing all concerns that might arise for a particular set of mission requirements or parameters, and it has been the author's experience that no two missions are exactly alike. As a result, "off-the-shelf" solar array designs typically do not result in the level of optimization required for meeting overall system level requirements. The reader is directed to the numerous references in this chapter for further information on solving mission-specific or unique design problems.

REFERENCES

Anspaugh, B.E. (1989), *Solar Cell Radiation Handbook*, Jet Propulsion Laboratory Publication SP 82-69 Addendum 1, California Institute of Technology, Pasadena.

Arnold, J.H., Mardesich, N., and Smith, B.S. (1985), *18th IEEE Photovoltaic Specialists Conf. Record*, IEEE, New York, pp. 663–668.

Jennings, C.M., Neff, R.E., and Pollard, H.E. (1993), *Proc. ASME Int. Solar Energy Conf.*, pp. 245–250.

Jet Propulsion Laboratory (1976), *Solar Cell Array Design Handbook*, Publication SP 43-38, California Institute of Technology, Pasadena.

Koch, J.W., and White, P.A. (1990), *21st IEEE Photovoltaic Specialists Conf. Record*, IEEE, New York, pp. 1294–1299.

LaRoche, G.J. (1988), *20th IEEE Photovoltaic Specialists Conf. Record*, IEEE, New York, pp. 974–978.

Levine, A.S., Ed. (1991), *LDEF—69 Months in Space*, First Post-Retrieval Symposium, NASA Conference Publication 3134.

Neff, R.E., Boron, W.R., and Pollard, H.E. (1985), *18th IEEE Photovoltaic Specialsts Conf. Record*, IEEE, New York, pp. 629–633.

O'Connor, P.D. (1984), *Practical Reliability Engineering*, John Wiley & Sons, New York, pp. 109–172.

Ott, E.R. (1975), in *Process Quality Control*, J. Robinson and L. Gardiner, Eds., McGraw-Hill, New York, pp. 34–84.

Tada, H.Y., Carter, J.R., Anspaugh, B.E., and Downing, R.G. (1982), *Solar Cell Radiation Handbook*, Jet Propulsion Laboratory Publication SP 82-69, California Institute of Technology, Pasadena.

Washio, H., Katsu, T., Tonomura, Y., Hisamatsu, T., Saga, T., Matsutani, T., Suzuki, A., Yamamoto, Y., and Matsuda, S. (1993), *23rd IEEE Photovoltaic Specialists Conf. Record*, IEEE, New York, pp. 1347–1351.

White, P.A., and Jones, D.E. (1991), *22nd IEEE Photovoltaic Specialists Conf. Record*, IEEE, New York, pp. 1508–1511.

Woike, T.W., Stotlar, S.C., and Woods, L. (1990), *21st IEEE Photovoltaic Specialists Conf. Record*, IEEE, New York, pp. 1288–1293.

CHAPTER FOURTEEN

Concentrator Modules Using Multijunction Cells

LEWIS M. FRAAS, JX Crystals, Inc., Issaquah, WA 98027

1. INTRODUCTION

Recently, large numbers of multijunction concentrator solar cells were reproducibly fabricated with energy conversion efficiencies over 30%. This has been accomplished using a light-sensitive gallium arsenide (GaAs) cell stacked on top of a gallium antimonide (GaSb) infrared-sensitive booster cell, as shown in Figure 14.1. In this stack, a conventional GaAs cell has been modified by providing a transparent metal grid on its backside to allow infrared (IR) radiation to pass on through to a second GaSb cell. The GaSb chip is sensitive to IR out to 1.8 μm. The details of fabrication of this GaAs/GaSb cell stack are described in Chapter 6. The most significant point for this chapter is that these cells are now available in sufficient quantity to begin the development of high-performance concentrator modules.

Why do we want to make higher performance solar electric power modules? This chapter will answer this question in some detail. Briefly, for space, the advantages are more power, less mass, less radiation damage, and lower cost. For terrestrial use, the advantage is a lower electric power cost. Lower electric power cost follows when more electric power is produced for a given investment in sunlight collector area, e.g., lenses, wiring, and tracking.

A priori, 30% efficient GaAs/GaSb cell stacks can be used in space or terrestrial concentrator modules using point- or line-focus optics. However, in practice, specific design constraints for space dictate a concentration ratio of approximately 50 suns, which allows both point- and line-focus options,

Solar Cells and Their Applications, Edited by Larry D. Partain.
ISBN 0-471-57420-1

FIGURE 14.1 Thirty percent efficient GaAs/GaSb cell stack in TAB frame.

whereas for terrestrial applications, economics dictates a higher concentration ratio, near 1,000 suns, which implies a point-focus design.

The next section of this chapter deals with the performance characteristics of the GaAs/GaSb cell stack for both space and terrestrial applications. In the third section, a two terminal voltage-matched cell group consisting of three cell stacks is then described along with the performance characteristics of this triplet building block. The fourth section then describes both point- and line-focus space modules. The fifth section describes terrestrial modules.

2. THE GaAs/GaSb TANDEM CELL STACK

Figure 14.1 shows a GaAs/GaSb tandem cell stack in a tape-automated bond (TAB) frame designed for use in a space point-focus concentrator module operating at 50 suns. The cells shown are 6.5×7.5 mm rectangular cells, 0.5 mm thick. These cells are fabricated as high-performance GaAs and GaSb single crystal chips. The off-center circular active area diameter on each chip is 5.5 mm. The two chips are rotated 90° with respect to each other and glued together with a transparent silicone adhesive. Electric contacts are made to this cell stack through three metal foil beam-lead sets in a 35 mm tape frame. The upper and lower beam-lead sets contact the top and bottom of the GaAs chip, and the third beam-lead set contacts the top of the GaSb chip. A tandem cell stack can be pretested and qualified in this tape frame by probing the two pads connected to the beam-lead sets for the GaAs cell or by probing to the third pad and the back of the GaSb cell for testing the IR cell.

The GaAs/GaSb cell stack shown in Figure 14.1 is the first cell stack designed with a specific concentrator module in mind. As noted, this cell stack was designed for a space point-focus concentrator module operating at 50 suns. a previous proof-of-concept design used a 3.5×5 mm rectangular cell with an active area diameter of 2.5 mm. As noted later in this chapter, different cell sizes and mask design will be appropriate for space line-focus and high-concentration terrestrial point-focus designs. However, all of these various GaAs/GaSb cell stacks will resemble the stack in Figure 14.1 in appearance in that off-center rectangular chips will be rotated 90° with respect to each other and the beam-lead sets will leave the cell stack in three different directions.

Because of the above cell dimension variations, this section will describe cell-stack performance in terms of the best cell-stack performance observed to date. Before beginning this description, however, the reader should be aware of some general trends. First, cell efficiencies quoted for space (air-mass zero [AM0]) are generally somewhat lower than for terrestrial conditions because of a broader AM0 spectrum. Second, cell efficiencies generally are higher for optimized cells operating at higher concentration ratios. These two trends combine to yield significantly higher efficiencies for a given cell tested at 1,000 terrestrial suns than for the same cell tested at 50 space suns.

Specifically for a GaAs cell, the best reported terrestrial efficiency at 500 suns is 28%. The same cell efficiency at 100 suns AM0 is 24%. Similarly for a GaSb cell behind GaAs, the best reported terrestrial efficiency at 500 suns is 9% while the efficiency for the same cell at 100 suns AM0 is 7%. Another point that should be made is that the typical cell produced in large quantity for module development work has an efficiency somewhat lower than the best cell efficiency. Based on the author's experience, this means that while the best GaAs 100 sun AM0 efficiency is 24%, a more typical GaAs cell efficiency is 22%. Similarly for GaSb behind GaAs at 100 suns AM0, the best is 7%, but 6% is more typical. The GaAs/GaSb cell-stack four terminal performance is obtained simply by adding these component cell numbers. Thus, $24\% + 7\% = 31\%$ represents a best stack efficiency at 100 suns AM0, while $22\% + 6\% = 28\%$ represents a more typical AM0 stack performance.

Summing the best terrestrial efficiencies for the component cells tested at 500 suns gives $28\% + 9\% = 37\%$. This is a spectacular number that deserves more careful consideration. For example, how does the GaAs/GaSb cell-stack performance vary from day to day and place to place here on earth? Since the sun's spectrum in space is constant, it is relatively straightforward to arrive at a single energy conversion efficiency for a given device. However, the sun's spectrum varies here on earth with humidity and turbidity among other things; so arriving at a single energy rating is problematic. Standard terrestrial direct normal and global reference spectra and standard reporting conditions have been adopted by the photovoltaic community. However, these conditions do not accurately reflect the actual operating conditions that a utility scale solar electric power plant might encounter. It is now becoming clear that utility solar electric power plants will not be located at sites with average amounts of sunlight but instead will be located at selected sites with well above average amounts of sunlight. For this reason, it is interesting to investigate tandem cell efficiencies for the sunny southwestern United States. Another reason for not using the standard AM1.5 spectrum is that it is not possible to determine sensitivities to changes in the sun's spectrum with just a single spectrum.

Researchers have previously examined the effect of solar spectral variations on the performance of AlGaAs/silicon tandem solar cells (King and Siegel, 1984). J.M. Gee at Sandia National Laboratories has written a computer program that calculates the performance of tandem solar cells for a specific location over a day and over a year. Inputs include location (latitude, longitude, and elevation) and component cell data (spectral response and equivalent circuit parameters). Also required are atmospheric data (precipitable water, turbidity, ozone, and so forth) as a function of day number. The direct-normal solar spectral irradiance is calculated as a function of day number and of time of day using a solar spectral modeling code (SPECTRL2) developed at NREL (Bird and Riordan, 1986) and a sun position algorithm. The calculated solar spectral irradiance is convolved with the spectral response of the component cells, and the tandem cell performance is calculated with a simple equivalent circuit model for the cell. Solar spectra and tandem cell performance are calculated every half hour after the sun reaches an elevation of 15°, and energy production from the tandem cell is summed over the day. Atmospheric data for Alburquerque are taken from a previous study that had extracted the data from various atmospheric data bases (King and Siegel, 1984).

The GaAs and GaSb component cell data described by Fraas and Avery (1990) were used to calculate the tandem gallium cell performance for various days through spring, summer, fall, and winter in Albuquerque. The results are shown in Tables 14.1 and 14.2. In Table 14.1, the day number in the year is shown in the left column and the precipitable water is shown in the right column. The cell currents, efficiencies at 25°C, and average direct light flux are shown in the intermediate columns. The turbidity also varies throughout the year (King and Siegel, 1984) but is not explicitly shown. It is interesting to note that in the winter (days 15 and 349), when the precipitable water is low, the GaAs cell efficiency is low at 26.6% while the GaSb cell efficiency is high at 9.1%. Later in the summer (day 227), when the precipitable water is high, the GaAs cell efficiency goes up to 27.6% while the GaSb cell efficiency goes down to 8.2%. Nevertheless the sum efficiency is nearly constant, changing only slightly from 35.7% to 35.8%. A GaAs/GaSb cell-stack efficiency of 35.7% is still a very spectacular number! Factoring in a lens optical efficiency of 90% suggests that one might hope for a module efficiency of 32% at 25°C.

TABLE 14.1 Tandem Stack Energy Conversion Efficiency for Albuquerque, NM (25°C)[a]

Day No.	I_{sc} (Top) $\left(\frac{mA}{cm^2}\right)$	I_{sc} (Bottom) $\left(\frac{mA}{cm^2}\right)$	Efficiency (Top) (%)	Efficiency (Bottom) (%)	Efficiency (Total) (%)	Light Flux $\left(\frac{mW}{cm^2}\right)$	Water Vapor (cm)
15	23.9	21.5	26.6	9.1	35.7	88	0.62
45	24.3	21.9	26.4	9.0	35.5	90	0.58
74	24.6	21.8	26.4	8.9	35.3	91	0.63
105	24.2	21.0	26.6	8.8	35.3	89	0.83
135	23.7	20.2	26.8	8.6	35.4	87	1.09
166	23.7	19.6	27.0	8.5	35.5	86	1.45
196	23.3	18.6	27.5	8.3	35.8	83	2.33
227	24.5	19.0	27.6	8.2	35.8	87	2.36
258	25.3	20.1	27.4	8.3	35.7	90	1.61
288	25.3	20.9	27.2	8.5	35.7	91	1.61
319	24.6	21.5	26.8	8.9	35.7	90	0.69
349	24.0	21.5	26.7	9.1	35.7	88	0.62

[a]Annual average total efficiency = 35.6%.

TABLE 14.2 Tandem Gallium Concentrator Cell Performance in Albuquerque, NM, at Normal Operating Temperature (NOT)[a]

Day of Year	GaAs Cell Eff. (65°C) (%)	GaSb Cell Eff. (55°C) (%)	Tandem Cell Eff. (NOT) (%)
15	25.0	7.6	32.6
45	24.8	7.5	32.3
74	24.8	7.4	32.2
105	25.0	7.3	32.3
135	25.2	7.1	32.3
166	25.4	7.0	32.4
196	25.9	6.8	32.7
227	26.0	6.7	32.7
258	25.8	6.8	32.6
288	25.6	7.0	32.6
319	25.2	7.4	32.6
349	25.1	7.6	32.7

[a]Annual average tandem cell efficiency = 32.5%. Annual average efficiency with 90% lens = 29.2%. Annual module energy production = 0.7 MW-h/m^2.

It is also very important to determine the tandem gallium stack performance at operating temperatures. The variation of both component cell efficiencies with temperature has been measured (Avery et al., 1990; and Boettcher et al., 1981). The stacked component cell operating temperatures have been described elsewhere as well (Gee and Chiang, 1990). It is noteworthy that the GaAs cell runs hotter at 65°C than the GaSb cell, which runs at 55°C (assuming 500 suns concentration). This is both because the GaAs cell receives more energy and because it is on top so that the heat flows down through the GaSb cell to the heat spreader. Since the GaAs cell in a tandem cell assembly is mounted on top of the GaSb cell, with an electrically isolating infrared-transparent bond, a 10°C temperature drop occurs across the interlayer between the top and bottom cells. The component cell and sum efficiencies at normal operating temperatures (NOT) are summarized in Table 14.2. The annual average stack efficiency drops to 32.5% at NOT. Assuming a lens efficiency of 90%, we project a module efficiency of 29.2% at 500 suns and an annual module energy production of 0.7 MW-h/m^2.

3. THE VOLTAGE-MATCHED TRIPLET CELL STACK GROUP

The cell stack shown in Figure 14.1 is a four terminal device possessing plus and minus terminals for each of the two cells. Conventional single junction cells are two terminal devices. It is therefore necessary to define a two terminal cell stack group. Since the V_{max} of GaSb is approximately one-third the V_{max} of GaAs, the two terminal group will be a voltage-matched triplet.

Figure 14.2 shows how three stacked GaAs/GaSb cell pairs can be interconnected with a flexible copper ribbon circuit into a triplet: a basic voltage-matched, three cell stack, two terminal unit. The ribbon circuit is patterned and bonded to the cells in such a way that the top GaAs cells are connected three in parallel while the bottom GaSb cells are connected three in series. The patterned traces finally join the cell strings in parallel to form a two-terminal, voltage-matched network. The triplet pattern is modular at this level and can be extended in series or parallel as if it were a single solar cell.

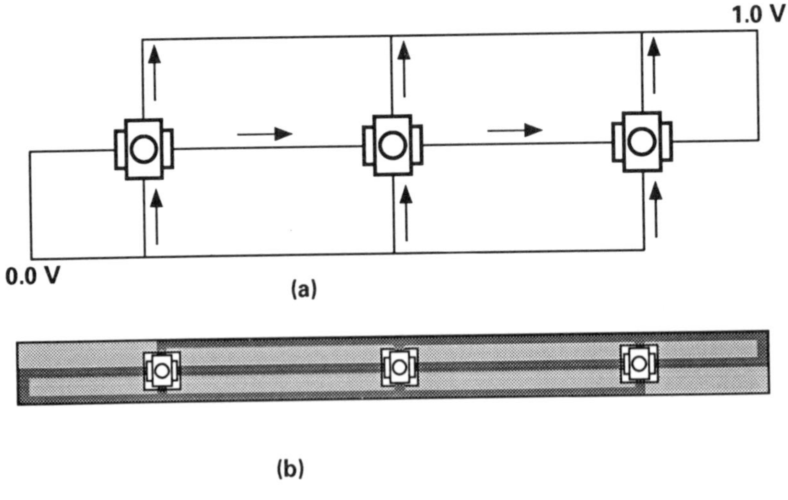

FIGURE 14.2 Triplet tandem cell circuit. **(a)** Circuit schematic. **(b)** Flex circuit ribbon.

The voltage match configuration is insensitive to current mismatch in the separate cell circuits. Peak power tracking is governed primarily by the max power voltage. Figure 14.3a shows V_{max} versus temperature for GaAs and a three-cell GaSb string for the case of 100 AM0 suns concentration. The crossover point where the two circuits are perfectly matched is at 90°C. At other temperatures the circuits will be slightly mismatched. Figure 14.3b charts

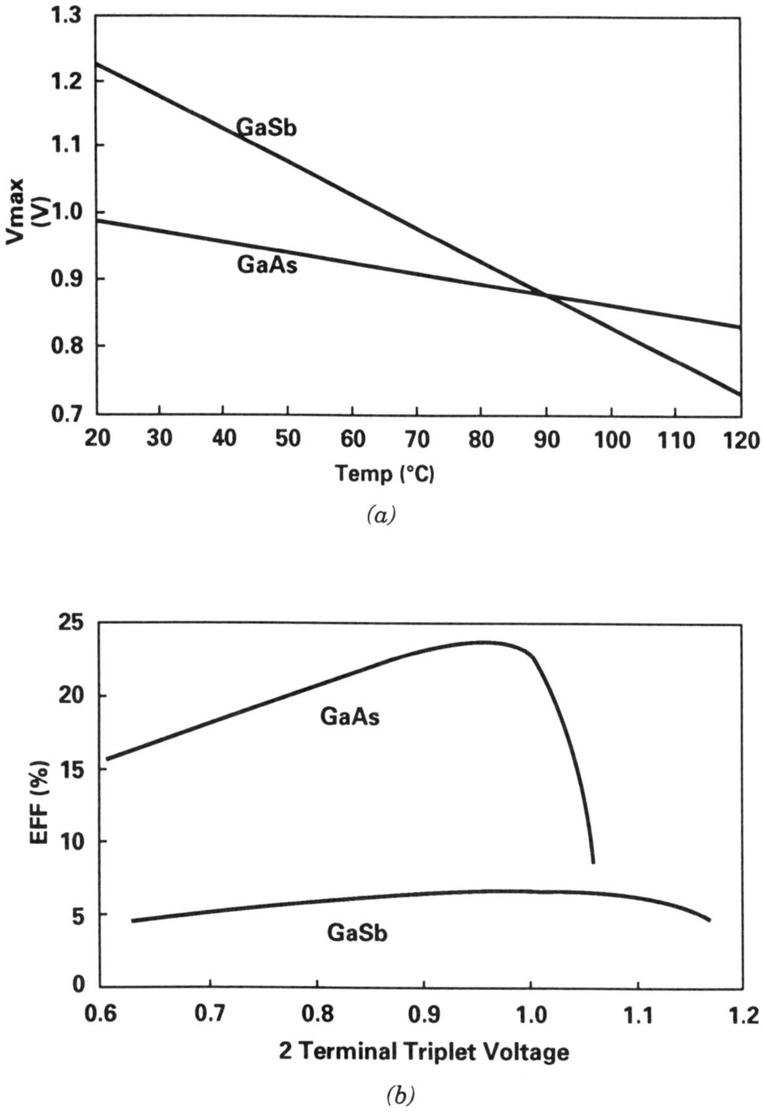

FIGURE 14.3 **(a)** Maximum power voltage matching for 3 GaSb:1 GaAs (100 suns, AM0). **(b)** Triplet circuit efficiency versus two terminal voltage at 60°C and 100 suns, AM0.

the efficiency of each circuit in a triplet as a function of the two terminal voltage, at 60°C and 100 AM0 suns. Note the sharp peak in the high-efficiency GaAs circuit in contrast to the flatter profile of the lower efficiency GaSb. If the two terminal voltage shifts away from the GaAs peak, there is a rapid fall off in combined efficiency while the GaSb contribution remains relatively constant. Therefore, the maximum power point of the triplet will approximately track the GaAs circuit. Table 14.3 lists projected efficiencies for a temperature range of 20°C–110°C and 100 AM0 suns of a GaAs circuit operating at it's optimum power point, a GaSb circuit at it's optimum power point, a GaSb circuit tracking the GaAs power point, and the combined efficiency for GaAs tracking. This provides a comparison between ideal four terminal performance and the two terminal case. In an expected operating range of 60°–100°C there is only a 0.1% efficiency loss in the GaSb circuit from tracking off its maximum power point.

The data presented in Tables 14.1 to 14.3 and in figure 14.3 represent circuit performance projections based on the best four terminal cell stack performance measurements (Fraas and Avery, 1990). Subsequently, live triplets have been fabricated and tested (Fraas et al., 1992b; Kuryla et al., 1992). Figure 14.4 shows a photograph of one such triplet.

The lens used in these units is the ENTECH minidome Fresnel lens described by Piszczor et al. (1990). This silicone RTV lens was designed as a lightweight lens for space concentrator modules and is used as-molded without a protective glass dome or protective coating. Each square lens has an aperature area of 13.8 cm^2. Typical optical efficiencies for these lenses range between 87% and 90%. This lens is designed to produce a 3 mm diameter spot on our cell. For a cell active area diameter of 5.5 mm, this provides an unilluminated guard band that leads to a tracking tolerance design point of

TABLE 14.3 Triplet Performance Versus Temperature

Temp. (°C)	GaAs Eff. (@ Pm_{GaAs}) (%)	GaSb Eff. (@ Pm_{GaSb}) (%)	GaSb Eff. (@ Pm_{GaAs}) (%)	Total Eff. (@ Pm_{GaAs}) (%)
20	24.9	8.4	7.4	32.2
30	24.5	8.0	7.3	31.8
40	24.2	7.6	7.2	31.4
50	23.8	7.3	7.0	30.8
60	23.4	6.9	6.8	30.2
70	23.0	6.5	6.5	29.5
80	22.6	6.2	6.2	28.8
90	22.2	5.8	5.8	28.0
100	21.8	5.4	5.3	27.1
110	21.4	4.9	4.7	26.1

FIGURE 14.4 Two terminal triplet test module incorporating ENTECH minidome Fresnel lenses and voltage-matched GaAs/GaSb tandem cell stacks.

$\pm 2°$. This lens produces a concentration ratio of 56 suns (lens area/cell active area).

Most of our module measurements were obtained using a flash simulator system, shown schematically in Figure 14.5. The application of a flash simulator for cell circuit testing, lens testing, and module testing has been described by Fraas et al. (1992a). Both high-performance single lens and triplet mini-modules have been fabricated and tested. Table 14.4 summarizes the performances of these items (including lens losses) for both space and terrestrial illumination conditions. The AM0 data shown in Table 14.4 were obtained using our flash simulator. The flash lamp power level was first set to obtain the

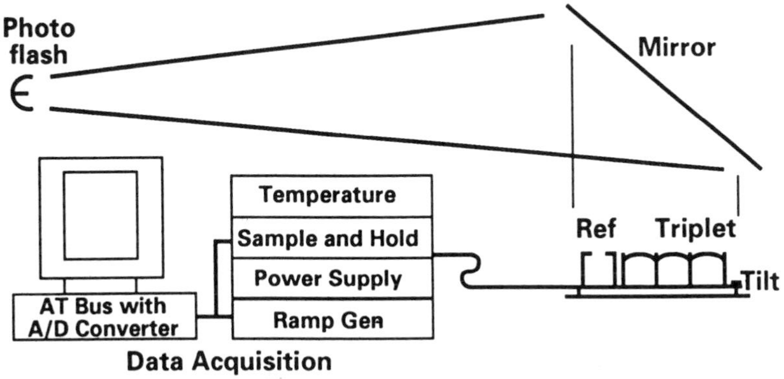

FIGURE 14.5 Flash test station for measuring concentrator module performance.

TABLE 14.4 Lensed Standard Assembly Performance[a]

	AM0 (22°C) Flash Simulated	Terrestrial: Actual Outdoor (50°C)		Terrestrial: Flash Simulated (22°C)
Single stack				
GaAs				
Efficiency	20.0%	21.0%	DNI = 827 W/m^2	21.9%
Fill factor	0.842	0.838		0.851
V_{oc}	1.144	1.097		1.143
I_{sc} (mA)	390	260		256
J_{sc} (mA/cm^2)	28.28	22.80		22.47
GaSb				
Efficiency	5.2%	4.6%	DNI = 788 W/m^2	5.3%
Fill factor	0.733	0.700		0.757
V_{oc}	0.452	0.418		0.450
I_{sc} (mA)	295	171		169
J_{sc} (mA/cm^2)	21.39	15.74		15.58
Combined				
Efficiency	25.2%	25.6%		27.2%
Triplet				
Efficiency	25.0%	25.2%	DNI = 829 W/m^2	26.4%
Fill factor	0.842	0.836		0.852
V_{oc}	1.160	1.106		1.151
I_{sc} (mA)	1438	935		922
J_{sc} (mA/cm^2)	34.77	27.26		26.90

[a]Single lens area = 13.79 cm^2; triplet lens area = 41.37 cm^2. Terrestrial J_{sc} are normalized to 1 kW/m^2 insolation. AM0 insolation = 1,360 W/m^2.

correct color ratio for the tandem stacked GaAs and GaSb cells. Then the lamp distance was adjusted to obtain the correct one sun intensity using GaAs and GaSb standard cells that were calibrated to NASA Lear Jet flight standards. The minimodule power output is the maximum product of the measured I–V data pairs, e.g., 1.4 W for the triplet under AM0. The power input is simply the illumination intensity times the total lens area, or 136 mW/cm^2 × 13.8 cm^2 × 3 = 5.6 W for the triplet under AM0. The triplet AM0 efficiency is then 1.4/5.6, or 25%.

The terrestrial performance data in Table 14.4 at actual operating temperature were obtained by direct outdoor measurements of the minimodules while tracking the sun. In this case the module I–V curve (power) was measured while the illumination level is measured with a standard Epply normal incidence pyroheliometer. The operating temperature is measured by a thermocouple located between the cell stack ceramic heat spreader and the anodized aluminum housing. For the triplet, the resultant outdoor energy conversion efficiency was 25.2% at 50°C. This is a spectacular module level efficiency! (The Department of Energy's goal for module efficiency under the Reagan and Bush Administrations was 25% for the year 2020 [Ramels, 1992]).

The flash tester was used to determine the V_{oc} and fill factor on the minimodules at room temperature using the measured terrestrial currents corrected to 22°C. The corrected currents were calculated from previously measured current versus temperature data. This procedure gives a temperature-corrected triplet efficiency of 26.4%. These triplet minimodule performance results (field operating without cooling and at room temperature) represent the highest prototype module efficiencies measured to date. It should also be noted that these terrestrial values are obtained from a unit with components designed to operate at 50 suns in space. Properly designed GaAs and GaSb cells can produce even higher performance at higher concentration ratios (500–1,000 suns). Based on our most recent triplet measurements, a terrestrial unit designed to operate at these elevated concentration levels would be expected to exhibit conversion efficiencies over 27% under actual operating conditions.

In addition to absolute performance measurements on fabricated triplets, measurement data have also been collected on performance versus temperature, effects of lens shading, and module reverse bias characteristics. The triplet efficiency temperature coefficient was measured and found to be −0.0507/°C (Fraas et al., 1992a), which is in agreement with the data calculated in Table 14.3 (−1% per 20°C).

The lens shading and triplet reverse bias measurements have proven to be quite interesting. Figure 14.6 shows I versus V curves for a triplet minimodule under forward and reverse bias in the light and dark and when one lens is shaded. These curves show that a GaSb infrared-sensitive cell also doubles as a bypass diode, providing shading protection for the GaAs cells.

Tunneling across a lower band gap is the simple physical reason why the GaSb cell allows uniform revese current flow at low voltages, thereby acting as a bypass diode. The curves in Figure 14.6 can be understood as follows. Note that a back-biased GaSb cell will allow approximately 600 mA of short circuit current to pass through it at slightly less than 0.6 V and that 600 mA is the bypass current required for a single shaded lens. Since three GaSb cells are parallel with each GaAs cell, the maximum voltage that can be applied to any given GaAs cell is 3 × 0.6 or 1.8 V. Now, when the triplet under test is illuminated and operated at its maximum power point, it generates 3 × the GaAs cell current of 350 mA plus the GaSb cell string current of 250 mA for a sum current of 1.3 A. If the whole triplet is shaded in a large module, the rest of the module will reverse bias the shaded triplet back to 2 V so that the 1.3 A can again flow through the whole circuit. What about the single shaded lens curve? In this case, shading drops out a GaAs cell (350 mA) and a GaSb cell (250 mA), hence the loss of 600 mA at the triplet max power point. However,

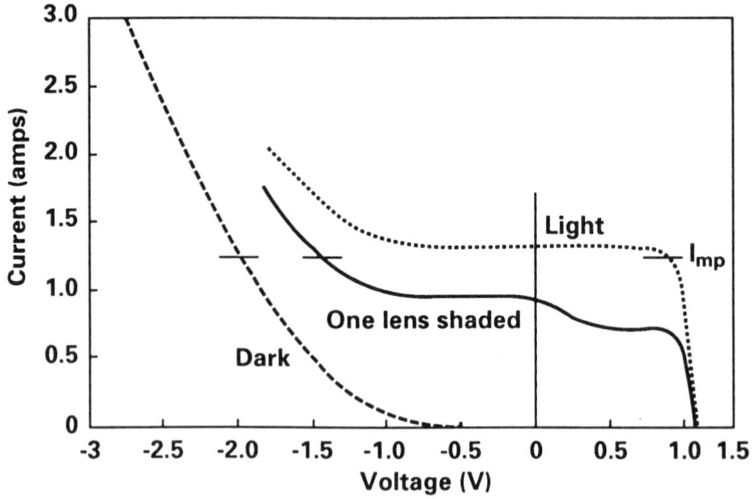

FIGURE 14.6 Measured forward and reverse characteristics of GaAs/GaSb tandem cell triplet.

as the rest of the module begins to back bias this cell pair and the voltage drops by 0.6 V from the max power point, the shaded GaSb cell begins to pass current. This explains the observed step. Thus near the 0 V point, the current loss from shading is only that from the shaded GaAs cell or 350 mA. This current cannot be recovered until the other two GaSb cells are reverse biased and break down under negative bias near −1.5 V.

This section began with measurements on four terminal cell stacks and predictions on two terminal triplet performance. These predictions were found to be valid by measurements on actual fabricated triplets (Fraas et al., 1992a).

4. SPACE CONCENTRATOR MODULE

4.1. Point-Focus Module Design

NASA Lewis and ENTECH have been developing a high-efficiency, lightweight space photovoltaic concentrator array (Piszczor et al., 1991). The goal is to develop an array with a power density of 300 W/m^2 and a mass-specific power of 100 W/kg. This power density goal translates to an array efficiency at operating temperature of 22%, a value at least two times higher than that obtainable with a conventional silicon flat-plate array.

From 1985 to 1990, ENTECH and NASA Lewis developed the concept of a lightweight point-focus domed Fresnel lens to allow the use of small concentrator solar cells in space. The use of lenses reduced the size of GaAs and other more exotic solar cell materials, bringing them into a reasonable cost range. This lightweight lens is fabricated as a thin silicone RTV dome with Fresnel grooves on its inside surface. It is domed rather than flat for two reasons. First, a domed shape provides for higher efficiency since it bends light rays by roughly equal amounts at both its top and bottom surfaces. Second, a dome shape allows for a smaller f/number. This means that the focal length and hence panel thickness can be smaller. Stowed panel volume is important in the launch vehicle. The concentration ratio is really set by panel thickness for reasonable cell sizes.

In 1989, Fraas and Avery (1990) demonstrated a 2.5 mm diameter tandem GaAs/GaSb concentrator cell with a 30.5% efficiency. In 1990, the 30.5% AM0 cell efficiency was verified by NASA in test flights. In late 1990, the Fraas team at Boeing began working with NASA and ENTECH to build a series of prototype solar modules for space flight. They built larger 5.5 mm diameter cells and lightweight structures and began to integrate these items with the ENTECH domed Fresnel lens. Figure 14.7 shows a small 12 lens concentrator module fabricated at Boeing and delivered in July 1991 to NASA and the Air Force for a Photovoltaic Array Space Power (PASP) flight experiment. This PASP module served as a vehicle for environmental testing done in collaboration with NASA and was launched by the US Air Force in the summer of 1994 on a test mission to prove the viability of the technology. Although this

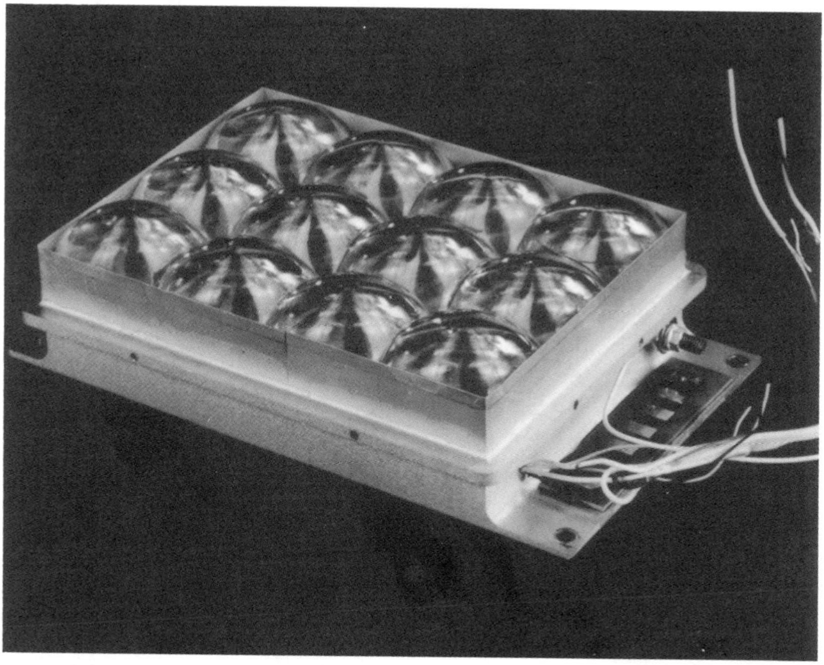

FIGURE 14.7 Photograph of 12 lens concentrator module fabricated by Boeing for photovoltaic array space power plus diagnostics (PASP PLUS) flight experiment.

module has achieved several of our initial goals, it is still only a test module, not a production module. This flight test module is significantly smaller than a production module, and it was assembled largely by hand.

The recent focus of our efforts has been to demonstrate that concentrator power modules using tandem cell stacks can be easily assembled with automated equipment to produce high-power performance at low cost. Figure 14.8 shows a drawing of our proposed production module assembly concept. For clarity, it shows a representative 4 × 6 lens/cell array. A production module would contain a larger number of array elements. The key assembly concepts shown are (1) a prefabricated lens parquet, (2) cell stacks interconnected through a prefabricated flex circuit mounted on the front surface of the panel back plane, and (3) pretested cell stacks excised from TAB tape and inserted onto the flex circuit cell sites.

The operations involved in the cell stack to module assembly are excising from the TAB site, beam-lead forming, pick and place, die bonding, and outer lead bonding. As of this writing, we have not yet implemented these operations for production. However, these operations are nearly identical to standard automated package assembly operations in use today in the semiconductor industry involving TAB. Once a cell stack is inner lead bonded (Chapter 6) and tested in a TAB frame, it is no different than any other single chip in a TAB frame.

The point-focus module work has now proceeded far enough so that it can be more generally evaluated. The high area related power density can now be

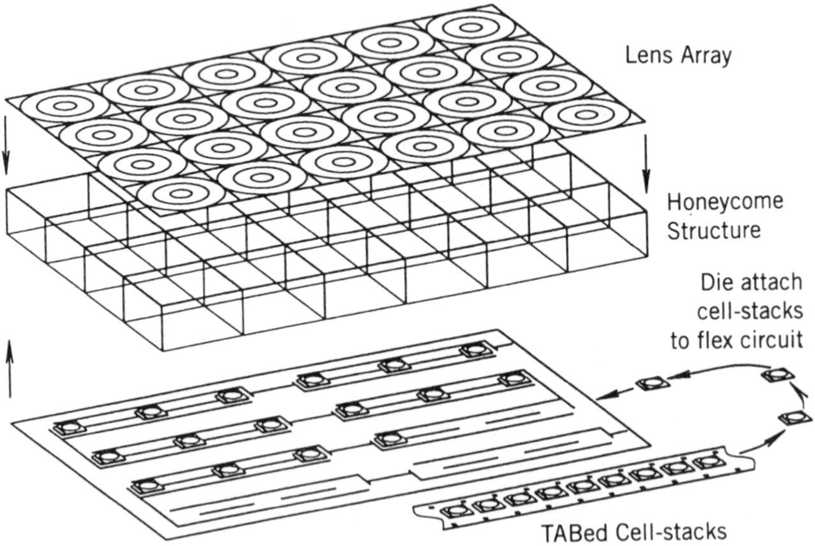

FIGURE 14.8 A manufacturable tandem cell concentrator module, consisting of a lens array, a honeycomb housing, and an array of TABed cell stacks bonded to a flex circuit on the module back plane.

considered to be a proven fact. Similarly, the high mass specific power is a proven fact. Additional advantages have now also been found. For example, concentrator modules now promise very high radiation resistance. This follows because the small area cells can be shielded with thick cover slides without the addition of significant weight.

Furthermore, the thick coverslide can be shaped as an optical secondary to provide a larger tracking band as shown in Figure 14.9. By shaping the thick coverslide as a cone, all the light entering the mouth of the cone will end up hitting the cell active area either by direct transmission through the cone or by deflection at the cone wall to vacuum interface. The diameter of the mouth of the cone in Figure 14.9 is 8.5 mm so that it captures all rays striking the array within a tracking angle of $\pm 4°$.

Thus, the original point-focus domed lens module concept has met three attractive objectives: high area related power density, high mass specific power, and high radiation resistance. Unfortunately, the low cost objective has proved illusive. The domed lens is difficult to fabricate in a form that allows passivation against UV and atomic oxygen. The fundamental problem with the domed lens is that the Fresnel grooves are reentrant on the inner surface so that the lens must be stretched to be removed from its mold. This is possible with silicone RTV but then the silicone lens has to be protected against the environment. The original idea was to laminate the lens into a glass dome superstrate. However, forming, trimming, and laminating the glass superstrate to the silicone lens has proven difficult. Boeing has recently developed a thin glassy coating for the silicone lens in order to avoid the glass superstrate. This coating has been used on the lenses in the PASP module. It is hoped that this

FIGURE 14.9 Shaped coverslide for improved pointing tolerance.

coating will not craze during handling and thermal cycling, degrading the lens optical efficiency over time. ENTECH has recently proposed the use of a linear Fresnel lens to resolve these manufacturing problems. As discussed in the next section, the line-focus approach may have other advantages as well.

4.2. Line-Focus Concentrator Module Design

While the domed lens array with GaAs/GaSb cell stacks offers the highest efficiency of any present space solar array technology, there are many missions that are not best served by the domed lens array approach. For example, for many small "light-sat" applications, array cost and weight are much more important to the mission planners than area-related power density. As another example, some missions can provide only single axis sun-tracking for the solar array. The domed point-focus lens array requires two axis sun-tracking and will not work for such missions. However, line-focus concentrators will work with single axis sun-tracking. This tracking advantage was one of the reasons that the Department of Defense (DOD) selected the line-focus reflector concentrator (SLATS) for the SUPER project (Allen, 1991).

For those missions that are not best served by the domed lens array (e.g., cost-driven or single axis array tracking), ENTECH has recently proposed a simple, very-low-cost, straightforward manufacturing approach for making the line-focus Fresnel lens shown in Figure 14.10. This lens is arched rather than domed. The 3 mil thick glass superstrate can be formed as an arch and then bent straight (without stretching). The nonreentrant Fresnel grooves can then be molded onto this flat glass superstrate using a flat mold. After molding, the glass lens is then allowed to relax back to its arched shape. Another important feature for manufacturing is that secondary molds can be fabricated from a first primary mold. This is not the case for the domed lens mole.

Given this line-focus lens, JX Crystals has recently proposed the line-focus GaAs/GaSb tandem cell receiver shown in Figure 14.11. As can be seen, the

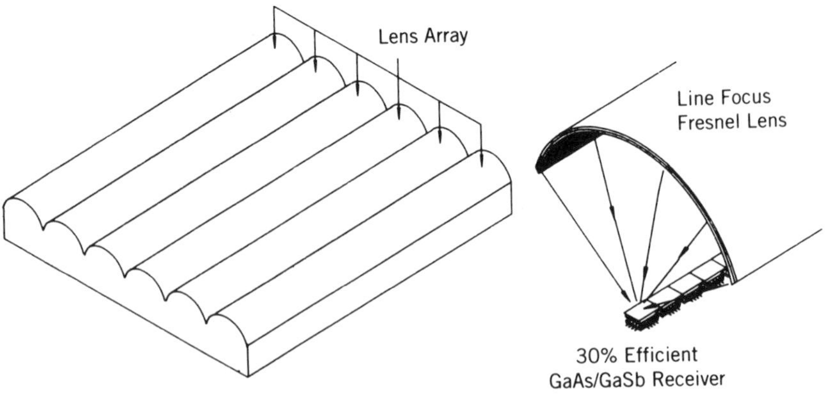

FIGURE 14.10 Space line-focus concentrator power module.

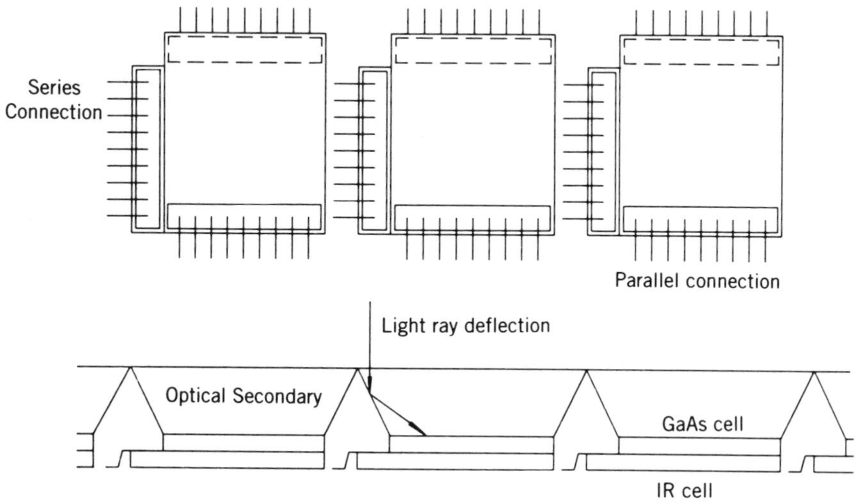

FIGURE 14.11 Line-focus tandem cell string.

addition of a truncated inverted pyramidal optical secondary adhesive bonded to the top of the GaAs cell immediately leads to a simple tandem cell string. A three-dimensional rendition of this concept is shown in Figure 14.12. As shown in Figure 14.11b, light rays on the line-focus that would otherwise hit the IR cell contact area are deflected by total internal reflection inside the glass optical secondary so that they now hit active cell area. Thus, the optical

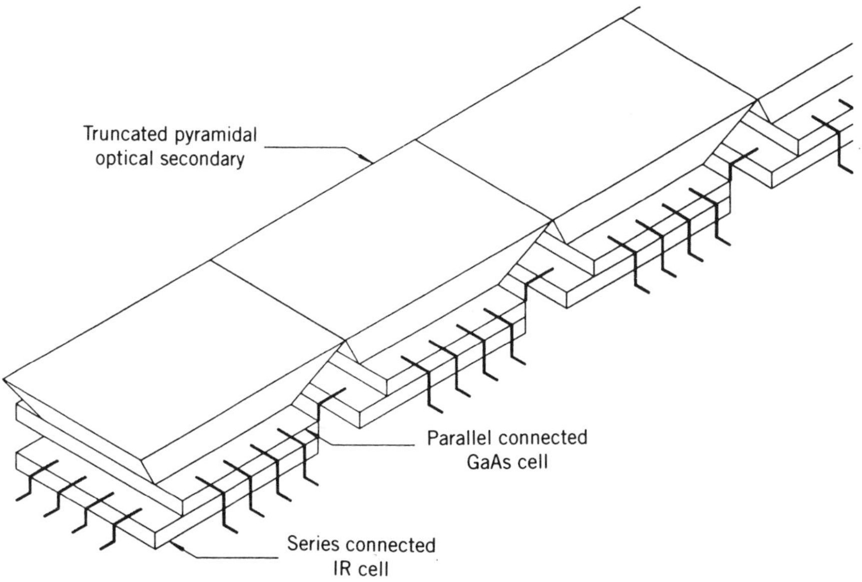

FIGURE 14.12 Three-dimensional view of a line-focus tandem cell string.

secondary provides room for the tandem cell interconnects, more specifically, the series interconnections for the IR cells.

The optical secondary also increases the concentration ratio. Suppose, for example, that the lens width is 5 cm. Further suppose a rectangular cell with dimensions of 4 × 5 mm and a rectangular active area of 4 × 3 mm. Finally, suppose that the length of a secondary element is 7 mm. Then the concentration ratio will be $(0.7 \times 5)/(0.3 \times 0.4) = 30$ suns.

The tandem cell string in Figures 14.11 and 14.12 evolves nicely from the point-focus module but it has some subtle advantages. For one thing, a line-focus string will build voltage more rapidly and the distance current flows between cells is considerably shorter. A second advantage may occur in the receiver assembly operation in that the pick and place is one-dimensionsal instead of two and travel distances are shorter.

The work on space concentrator modules has been interesting. It has been motivated by the expectation that space power modules can cost more than terrestrial power modules, thereby allowing the required development costs to be supported by high value sales. It has also been frustrating in that the costs of environmental testing are very high. Despite the high potential for major gains in power system performance, the space user community has been very conservative. Nervertheless, the NASA Space Station needs more power with less drag for low earth orbit and the DOD needs more radiation-resistant arrays for high radiation orbits. Finally, geosynchronous satellites can use more powerful, smaller, lighter, and cheaper arrays.

5. TERRESTRIAL CONCENTRATOR MODULE

Unfortunately, no experimental work has as yet been funded for adapting the tandem GaAs/GaSb cell stack to terrestrial applications. However, work on GaAs concentrator modules operating at 1,000 suns has been sponsored by EPRI and performed by Varian (see Chapter 11 and Kuryla et al., 1991). This work can be adapted to the tandem GaAs/GaSb cell roughly as shown in Figure 14.13. This artist's conception shows how cell stacks might be mounted on heat spreaders and interconnected. An optical secondary would undoubtedly also be used but is not shown as it would obscure the view of the cell stack.

Based on the Varian work one might imagine a 11 × 11 cm flat Fresnel lens focusing light down to a 4.5 mm × 5.5 mm rectangular cell with a 4 mm diameter active cell area. Since panel thickness is not a concern for terrestrial applications, a flat lens is preferred to a domed lens for lower cost. For this lens size, a square meter module will have a 9 × 9 lens array requiring 81 cell receivers. Coincidentally, from Chapter 6, a 4.5 × 4.5 cm wafer will produce 82 chips. This wafer area is about the same as a 2 inch diametric disc. Reflect for a minute! This is amazing! One 2 inch diameter wafer will provide all the cells needed for a square meter module. This illustrates the advantage of sunlight concentration.

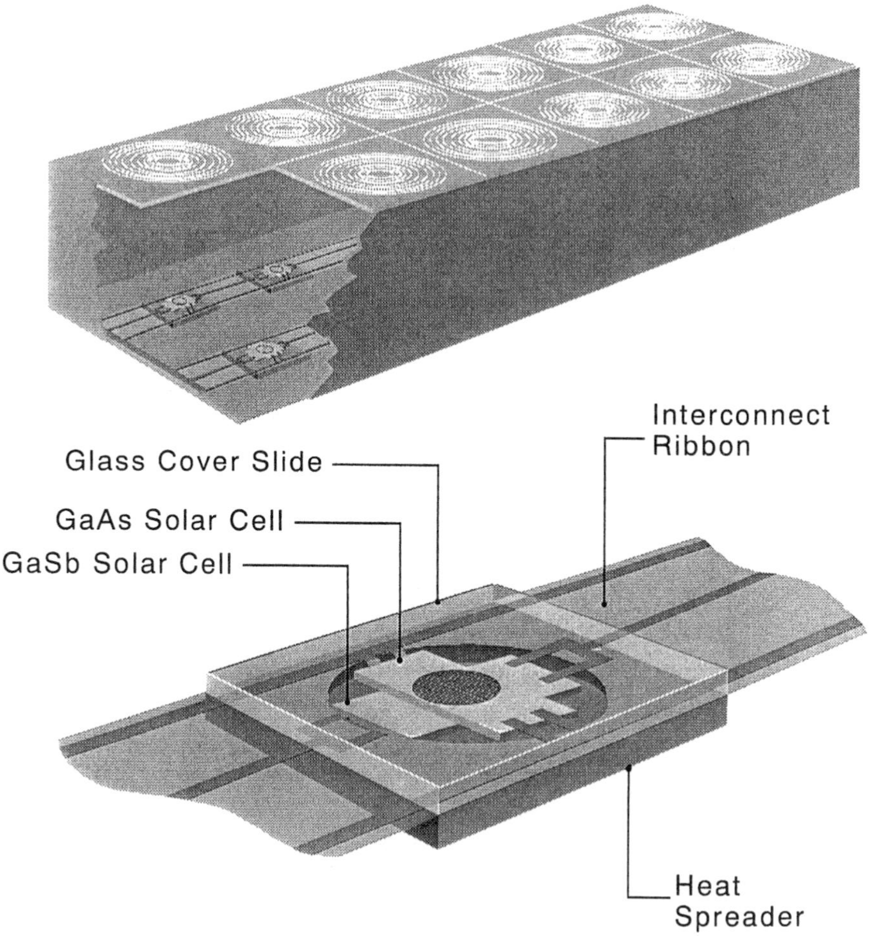

FIGURE 14.13 Terrestrial concentrator module and tandem cell stack concepts.

Varian did a module cost analysis and an energy cost analysis for their 1,000 sun GaAs module (Partain, personal communication, see Chapter 5). This analysis arrived at a module cost of $350/m^2. Approximately $150/m^2 (optical area) was allocated to the cell receiver. At 1,000 suns, the $150/m^2 translates to $15/cm^2 of cell active area or $1.88 per GaAs cell receiver, including the optical secondary.

A cross-check of the Varian assumptions can be provided by reference to Table 14.2. Recall that a 29% efficient module using GaAs/GaSb cells at NOT will produce 0.7 MW-h/m^2 of electric power per year in Albuquerque. At 10 cents/kW-h, this implies a revenue of $70/m^2 per year. If we assume a module cost of $350/m^2, the module pay back time will be 5 years. In 10 years, the investors will double their money for an annual rate of return of 7.5% without tax incentives. Since the electric power is produced free of pollution, tax incentives may be justified.

How important is module efficiency? Table 14.5 shows a calculation of the allowed module and cell costs for three different module efficiencies associated with the three possible concentrator cell types: Si, GaAs, and GaAs/GaSb. The first two rows show the assumed module efficiencies and the resultant module peak power ratings. The third row shows the allowed module cost resulting from the assumption of $\$350/m^2$ for the GaAs/GaSb case or a module cost of $\$1.40/W_{Peak}$. The fourth row simply notes that the lens and sheet metal costs are roughly independent of cell type. The bottom line result (fifth row) is that the allowed cell costs increases dramatically with module efficiency. Note that as the module efficiency went up from 22% for the GaAs cell case to 29% for the GaAs/GaSb cell case, the allowed cell receiver cost nearly doubled from $\$116/m^2$ to $\$200/m^2$. Specifically for the GaAs module, as footnote b indicates, the allowed GaAs cell cost is \$80 for the square meter module, implying that the allowed processed wafer cost can be as high as \$80 (\$1.00 per chip). This calculation assumed a cost for optical secondaries of $\$36/m^2$ (lens area). Given this result, footnote c indicates that a GaSb wafer cost of \$80 (\$1.00 per chip) is then allowed for the GaAs/GaSb module. As indicated in Chapter 6, these processed wafer cost targets for GaAs and GaSb are quite achievable for high volume cell production. In other words, the higher efficiency makes two chips in a tandem cell stack affordable. These cost projections assume volume production of at least 10 MW/year.

Although anyone would admit that these numbers are uncertain without substantially more work, they are calculated honestly and are quite exciting. The important thing to remember is that high efficiency leverages all the balance of system costs such that reasonable balance of system costs then

TABLE 14.5 Allowed Module and Cell Cost Dependence on Module Efficiency

	Si	GaAs	GaAs/GaSb
Module efficiency	17%	22%	29%
Module peak power at 860 W/m^2	145 W/m^2	190 W/m^2	250 W/m^2
Allowed module cost at $\$1.40/W_p$	$\$205/m^2$	$\$266/m^2$	$\$350/m^2$
Lens + sheet metal + wiring costs	$\$150/m^2$	$\$150/m^2$	$\$150/m^2$
Allowed cell receiver cost[a]	$\$55/m^2$	$\$116/m^2$[b]	$\$200/m^2$[c]

[a]All areas in m^2 are lens collector area.
[b]Assuming an optical secondary (OS) cost of $\$36/m^2$, this allows a GaAs cell cost of $\$80/m^2$.
[c]Assuming an OS cost of $\$36/m^2$ and a GaAs cell cost of $\$84/m^2$, this allows a GaSb cell cost of $\$80/m^2$.

become affordable. Suppose, for example, that all of the cost targets in Table 14.5 for all three cell technologies are achievable, which module would be preferred? Obviously, the GaAs/GaSb higher power module is preferred because more power can be produced for a given tracker and field installation cost. In other words, the balance of system costs will be lower for the most efficient technology.

6. CONCLUSIONS

Substantial progress has been made with tandem concentrator cells over the last 3 years. Recently, 30% space and 35% terrestrial cell stacks have been fabricated in research labs in sufficient quantity to begin concentrator module development. Concentrator minimodules using these cells have now been fabricated and tested with efficiencies over 25% at operating temperatures. This performance level is twice that available from single crystal silicon modules at operating temperature and four times the performance levels achievable from noncrystalline thin film options. Environmental testing of these advanced modules for space applications is beginning. Although considerably more developmental work is required before these modules become commercially available, the significant performance advantages will lead to major cost savings at the system level for space applications. Preliminary cost analysis also indicates that high efficiency terrestrial modules may produce economical utility electric power in the sunny southwestern United States.

REFERENCES

Allen, D.M. (1991), *22nd IEEE Photovoltaic Specialists Conf. Record*, IEEE, New York, pp. 20–22.

Avery, J.E., Fraas, L.M., Sundarem, V.S., Mansoori, N., Yerkes, J.W., Brinker, D.J., Curtis, H.B., and O'Neill, M.J. (1990), *21st IEEE Photovoltaic Specialists Conf. Record*, IEEE, New York, pp. 1277–1281.

Bird, R.E., and Riordan, C.J. (1986), *Climate Appl. Meteorol.* **25**, 87.

Boettcher, R.J., Borden, P.G., and Gregory, P.E. (1981), *IEEE Electronic Devices Lett.* **EDL-2**, 88–89.

Fraas, L.M., and Avery, J.E. (1990), *Optoelectronics Devices and Technol.* **5**, 297–310.

Fraas, L.M., Avery, J.E., Stanley, D.K., Yerkes, J.W., Piszczor, M.F., Flood, D.J., and O'Neill, M.J. (1992a), *ASME Solar Engineering* **92**, 825–830.

Fraas, L.M., Kuryla, M., Bigger, J.E. (1992b), *AIP-268, UA/NASA SERC Lunar Materials Technology Symposium*, p. II-9.

Gee, J.M., and Chiang, C.J. (1990), *21st IEEE Photovoltaic Specialists Conf. Record*, IEEE, New York, 41–46.

King, D.L., and Siegel, R.B., (1984), *17th IEEE Photovoltaic Specialists Conf. Record*, IEEE, New York, pp. 944–951.

Kuryla, M.S., Ladle-Ristow, M., Partain, L.D., and Bigger, J.E. (1991), *22nd IEEE Photovoltaic Specialists Conf. Record*, IEEE, New York, pp. 506–511.

Piszczor, M.F., Swartz, C.K., O'Neill, M.J., McDaniel, A.J., and Fraas, L.M., (1990), *21st IEEE Photovoltaic Specialists Conf. Record*, IEEE, New York, pp. 1271–1276.

Piszczor, M.F., Brinker, D.J., Flood, D.J., Avery, J.E., Fraas, L.M., Fairbanks, E.S., Yerkes, J.W., and O'Neill, M.J., (1991), *22nd IEEE Photovoltaic Specialists Conf. Record*, IEEE, New York, pp. 1485–1490.

PART THREE

SYSTEM TECHNOLOGIES

CHAPTER FIFTEEN

Terrestrial Off-Grid Photovoltaic Systems

CHRISTINA JENNINGS, Pacific Gas and Electric Company, Department of Research and Development, San Ramon, CA 94583

1. INTRODUCTION

The photovoltaic (PV) industry typically reports technology advances based on cells or modules, not systems. Laboratory cell efficiency records and present or potential module prices are commonly cited. It is important to realize that these subsystem advances, while valuable to monitor trends, are usually not directly or immediately applicable to PV systems. This chapter provides information on the applications, PV technologies, performance, reliability, price, and design considerations of terrestrial off-grid PV systems.

2. SYSTEM DEFINITION

An off-grid PV system is independent from the utility grid and consists of at least one PV module, a support structure, a battery bank (if storage is desired), an inverter (if AC loads are involved), control and safety equipment, and wiring. Multiple PV modules wired together constitute a PV array; the non-PV components are typically referred to as the balance of system (BOS). The support structure can be in a fixed position or have moving parts for the array to track the sun throughout the day. A hybrid system combines the PV system with a backup generator. Figure 15.1 indicates the direction of electricity flow in an off-grid PV system.

Solar Cells and Their Applications, Edited by Larry D. Partain.
ISBN 0-471-57420-1

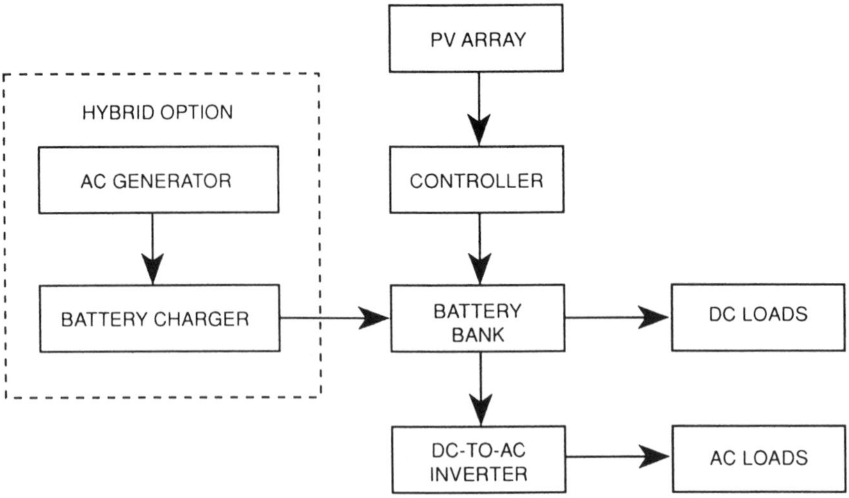

FIGURE 15.1 Block diagram of an off-grid PV system.

3. APPLICATIONS

Although megawatt-scale PV power plants are not yet competitive with conventional power plant technologies, PV is entering interim markets while approaching this goal. Steady growth in PV sales is almost entirely attributable to the terrestrial off-grid market (Fig. 15.2). The off-grid PV market consists of habitation (e.g., residential power, water pumping), industrial (e.g., communications, cathodic protection), and consumer indoor (e.g., calculators, watches).

Typical off-grid PV systems have PV capacities ranging from a few watts to several kilowatts, involve battery storage and fixed support structures, and are cost effective compared with other options considered, such as utility grid extensions, batteries, and engine generators. Since the price of stepping down grid voltage to power a small load such as a warning light may well exceed that of a PV system, cost-effective PV systems are not necessarily remote from the utility grid.

The electric utility industry is targeted as the ultimate end user of terrestrial PV. Although grid-connected PV is not yet economic, electric utilities have been using PV for an increasing number of off-grid applications (Table 15.1). Pacific Gas & Electric (PG&E) alone has over 1,000 cost-effective, in-house, off-grid PV installations, some in use since 1970s (Jennings, 1990). Cost-effective PV use by utilities will most likely follow PG&E's Diffusion Model of PV utility applications (Fig. 15.3) (Iannucci and Shugar, 1991). As PV system prices decrease, cost-effective PV use by utilities is anticipated to expand from today's off-grid applications to grid support,

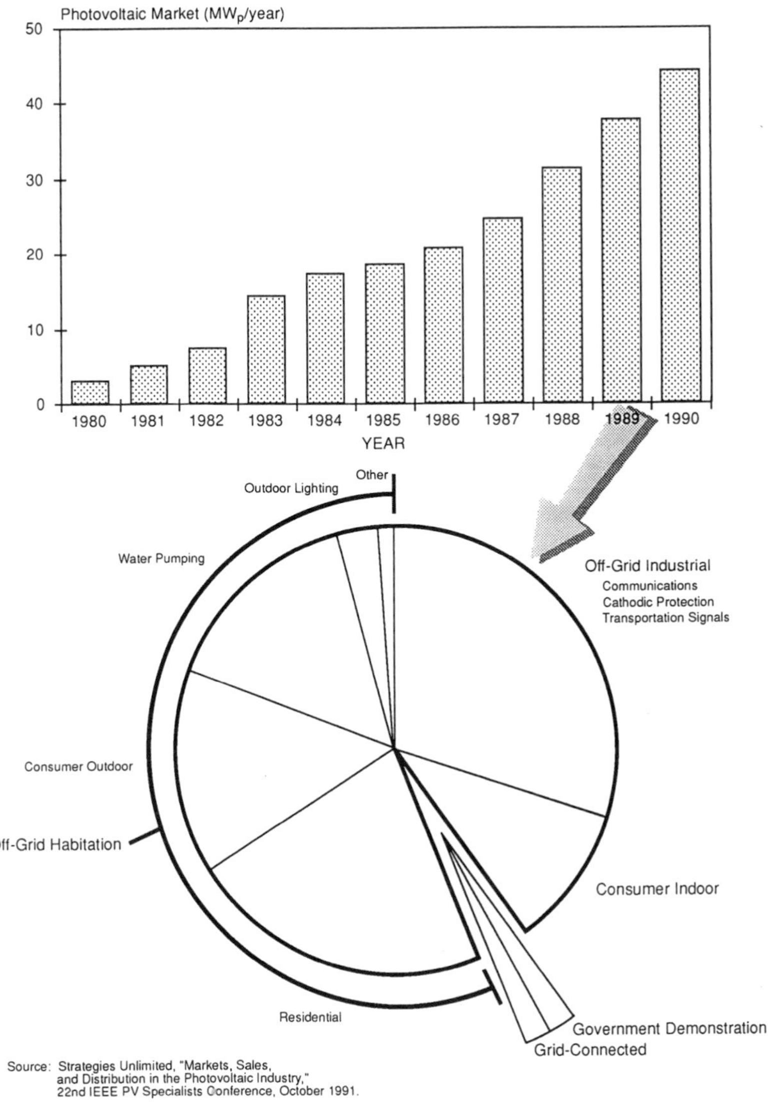

FIGURE 15.2 Terrestrial photovoltaic market. Source: Strategies Unlimited (1991).

villages and islands, and customer-sited applications. Continued PV system cost reductions, prompted by the economics of mass production through an increasing market share in at least one of these three imminent applications, should position PV to enter the peaking power market. PV must become economical and viable in the peaking market before the technology will be considered as a bulk power option.

TABLE 15.1 Cost-Effective PV Use by U.S. Utility Companies.

Utility Operations Area/ PV Application	Alabama Power	Arizona Public Service	Bonneville Power Administration	Boston Edison	Carbon Power & Light	Duke Power	Florida Power Corporation	Georgia Power	Gulf Power	Houston Lighting & Power	Indiana Michigan Power	K.C. Electric Association	Lower Colorado River Authority	New York Public Power Authority	Niobrara Electric Association	Northwest Rural Public Power District	Pacific Gas and Electric	Public Service of Colorado	Public Service of New Hampshire	San Isabel Electric Services	Savannah Electric and Power	Southern California Edison	Southwestern Electric Power	West Texas Utilities
Transmission and distribution																								
Tower obstruction beacons	X		X							X	X						X				X			
Sectionalizing switches		X						X									X							X
Cathodic protection							X																	
Power plants and facilities																								
Plant warning sirens		X		X													X							
Navigation aids							X						X											
Backup generators																	X							
Cathodic protection	X							X	X															
Automatic gate openers																	X							
Lighting																	X						X	
Communications																								
Microwave repeaters		X	X														X	X	X					
Remote metering																								
Emergency call boxes																								
Environmental monitoring																								
Weather stations																	X							
Water quality monitors						X																		
River level gauges	X																X							
Water temperature monitors																	X							
Cloud seeding																	X							
Insolation monitors														X			X							
Residential/commercial customer service																								
Remote residences																								
Water pumping and control					X						X	X			X	X				X				
Lighting					X															X		X		
Gas pipeline systems																								
Flow meters																	X	X						
Valve actuators																	X							
Cathodic protection																	X							
Gas flow computers																	X							
SCADA remote terminals																	X	X						

Source: Moore, T., "On-Site Utility Applications for Photovoltaics," EPRI Journal, March 1991.

4. PV TECHNOLOGIES

There are several competing PV technologies. Single crystal silicon, polycrystalline silicon (poly-Si), and hydrogenated amorphous silicon (a-Si:H) are commercially available. Ribbon silicon, gallium arsenide (GaAs), and thin-film technologies such as copper indium diselenide (CIS) and cadmium telluride

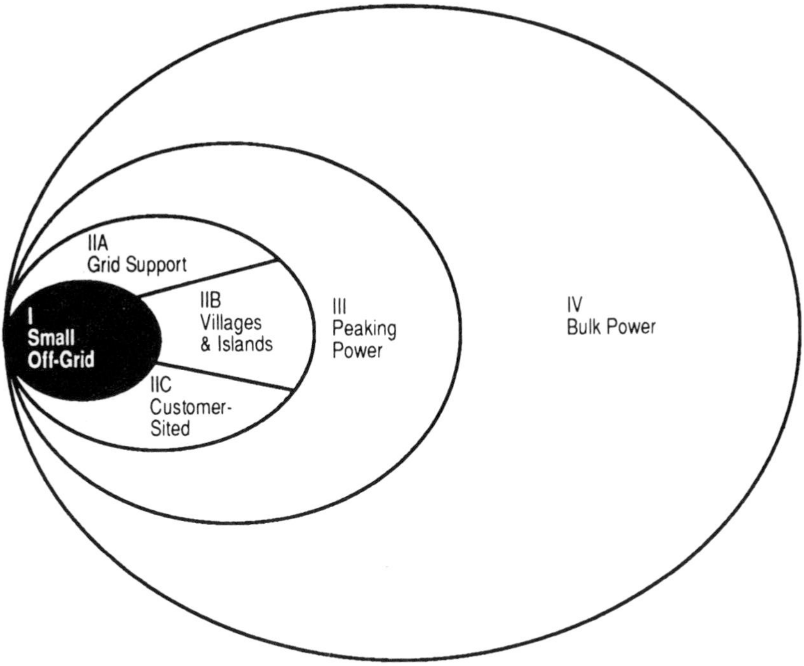

FIGURE 15.3 Diffusion model of PV utility applications.

(CdTe) are in various stages of precommercial development. Research continues in all PV technologies to improve performance and decrease costs.

There is no definite PV technology winner. At this time, single crystal and polycrystalline silicon flat-plate PV modules are the preferred technologies if (relatively) high efficiency, proven reliability, (relatively) low price, and availability are key criteria. However, the potential for low costs may be more promising for thin-film PV technologies than for conventional crystalline silicon, since thin films require much less PV material and may be more conducive to large-scale manufacturing techniques. Also, high efficiencies can be obtained by concentrating sunlight, although associated system price and operation and maintenance (O&M) costs may reduce the attractiveness of this option. Thin films, concentrators, and other emerging PV technologies have yet to approach their promising potentials. Single crystal and polycrystalline silicon flat-plate PV will probably remain the preferred technologies in the 1990s, with other PV technologies becoming attractive after 2000.

5. PERFORMANCE

PV cell, module, and system efficiencies have increased steadily (Fig. 15.4). Today's commercially available systems operate up to about 12% DC effi-

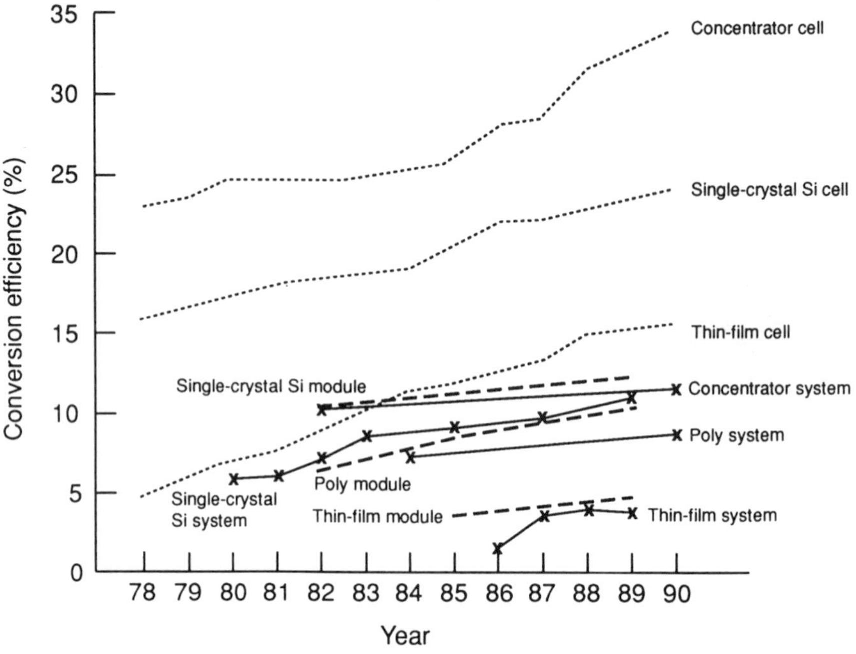

Sources: [1] U.S. Department of Energy, "Photovoltaics Program Plan FY1991-FY1995," October 1991.
[2] Jennings, C., and C. Whitaker, "PV Module Performance Outdoors at PG&E," 21st IEEE PV Specialists Conference, May 1990.
[3] Durand, S., and D. Bowling, "Field Experience with Photovoltaic Systems: Ten-Year Assessment," EPRI Report TR-102138, March 1993.
[4] Rosenthal, A.L., et al., "Photovoltaic System Performance Assessment for 1988," EPRI Report GS-6696, January 1990.
[5] Inglis, D.J., et al., "Photovoltaic Field Test Performance Assessment," EPRI Report AP-3244, September 1983.
[6] Wenger, H.J., et al., "Decline of the Carrisa Plains PV Power Plant: The Impact of Concentrating Sunlight on Flat Plates," 22nd IEEE PV Specialists Conference, October 1991.
[7] PG&E, "1991 PVUSA Progress Report," PG&E Report 007.5-92.6, 1992.

FIGURE 15.4 Laboratory PV cell,[1] commercial module,[2] and system[3–7] DC efficiency trends.

ciency, depending primarily on the PV technology. The time lag between laboratory cell efficiency records and equivalent performance by commercially available product is evident. Many years may pass before a laboratory cell efficiency record is achieved by a corresponding commercial module or system.

Since PV performance characteristics vary from cell to cell, cell mismatch causes module and system efficiencies to decrease slightly as the number of cells and modules in a system increases. Cell and module efficiency records and ratings are typically based on a specific, favorable set of weather conditions (known as *standard test conditions*), which are rarely duplicated outdoors. System efficiency results typically reflect real-world operating conditions.

Comparing PV technologies by efficiency alone is insufficient, since the highest efficiency concentrator technology tends to the most expensive to manufacture and operate.

6. RELIABILITY

Operation and maintenance costs are dependent on the PV application. The PV modules themselves are typically the most reliable system component. Backup generators, batteries, inverters, trackers, and controllers typically increase system O&M requirements. Site remoteness impacts O&M travel time and costs. (System design is also impacted; trackers are not typically used, and PV and battery capacities may be increased for a remote, unattended site.)

Minimal O&M cost data exist for PV systems. O&M costs for utility-scale, grid-connected systems can be very low: $0.004/kW-h for a fixed system and $0.008/kW-h for a tracking system (Lynette and Conover, 1989). Successive generations of grid-connected PV systems have exhibited decreased O&M costs. PV operation is silent, involves no fuel besides sunlight, and has no emissions and few, if any, moving parts—characteristics that support extrapolation of the limited O&M cost data to off-grid systems.

PV lifetime data are also limited. Although accelerated lifetime tests indicate that modules representing commercially available PV technologies should have a useful life of at least 20–30 years, the technology has not been fielded long enough to assess lifetime based on long-term, real-world exposure data. Module manufacturers provide warranties of up to 20 years. A 20 year commercial module life is a conservative estimate at this time.

7. PURCHASE PRICE

PV prices exhibit promising downward trends with time and system size (Fig. 15.5). PV system prices are logically higher than module prices, since modules are one of several components constituting a system (Fig. 15.6). System price reflects the sum of individual component costs and the value added by PV system integrators. To indicate price reductions associated with economies of scale, selected on-grid system prices are included in Figure 15.5 (although on-grid and off-grid systems are different in that on-grid systems involve a utility-grade inverter and no batteries or backup generator). As indicated by the steady market growth shown in Figure 15.2, PV is cost effective for many off-grid applications, even at around $20/W$_p$ for a typical off-grid system.

PV system price must be reduced to promote broader market diffusion of PV. Continued and significant price reductions are needed for PV to penetrate the grid-connected market. Reducing BOS cost is an increasing concern as PV manufacturers optimize production lines and reduce module costs.

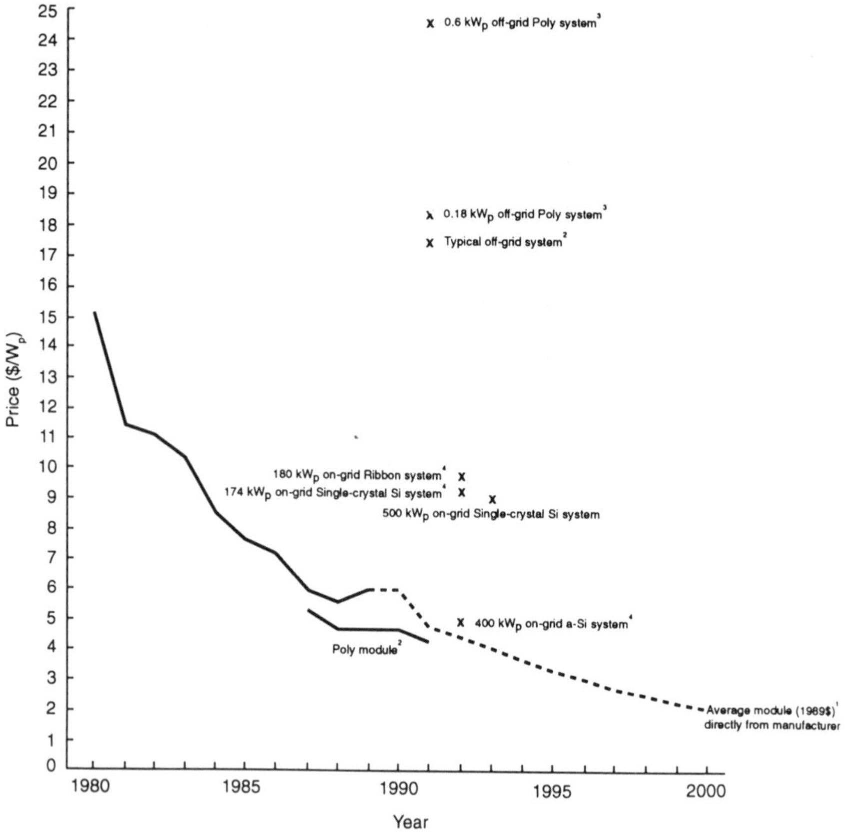

Sources: [1] Strategies Unlimited, "Five - Year Market Forecast 1989 - 1994," Report M-32, February 1990.
[2] Personal communication with John Berdner (Photocomm), 1991.
[3] Maycock, P., "PV News," December 1991.
[4] Candelario, T., et al., "PVUSA - Performance, Experience, and Cost," 22nd IEEE PV Specialists Conference, October 1991.

FIGURE 15.5 PV module and installed system prices.

8. DESIGN CONSIDERATIONS

Many resources exist to guide off-grid PV system design. PV system vendors (Solar Energy Industries Association, 1994) provide system design services as part of marketing their products and have the most design experience. System design workshops are held by the Florida Solar Energy Center, Siemens Solar Industries, Solar Technology Institute, Sandia National Laboratories, and others. Design manuals are available (Florida Solar Energy Center, 1991; Mrohs, 1990; Risser et al., 1990; Wiles, 1991). Although the detailed steps to design an off-grid PV system are beyond the scope of this text, key design considerations include the following:

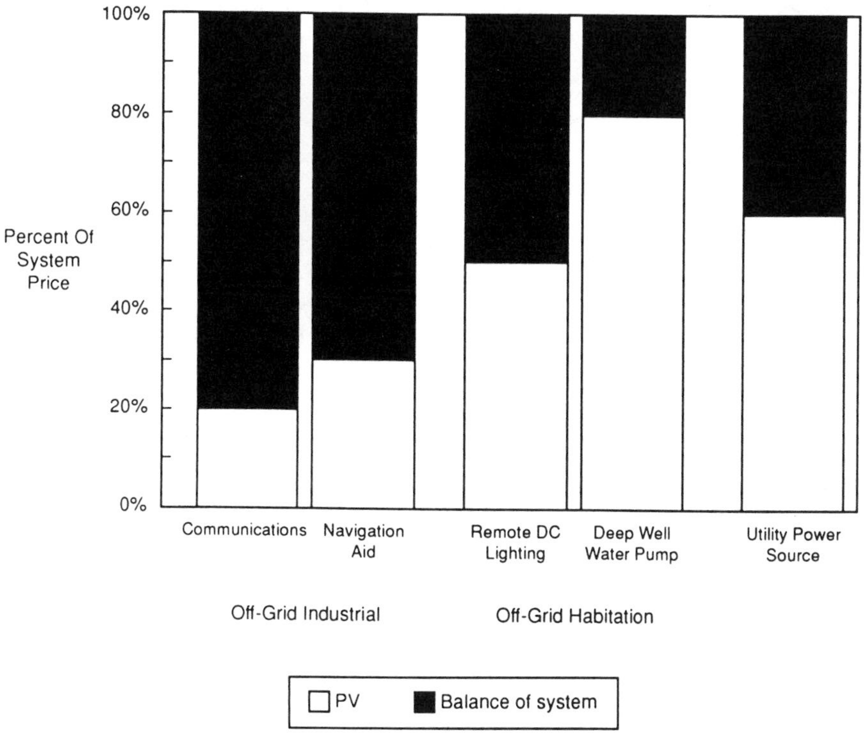

FIGURE 15.6 Breakdown of PV system prices. Source: Strategies Unlimited (1991).

- PV system design optimization requires knowledge or accurate estimation of the site-specific solar resource, the present and future DC and AC load usage patterns, the system availability requirements, and the O&M conditions. The designer typically determines if site-specific characteristics (e.g., terrain, site and system access, weather patterns, shading, seismic) will impact the design.
- Increasing PV system size to supply inefficient loads or wasteful habits is more costly than increasing energy efficiency.
- Simplifying the PV system design to the extent possible without sacrificing safety will maximize reliability and minimize technical support requirements.

9. UTILITY PERSPECTIVE

Favorable trends in PV performance, reliability, and price are increasing the number of applications in which PV is the preferred option. These off-grid PV systems foster awareness and acceptance of the technology. PV diffusion into grid-connected markets, promoted by continued PV system performance improvements, price reductions, and technology acceptance, is inevitable.

REFERENCES

Florida Solar Energy Center (1991), *Photovoltaic System Design*, Report FSEC-GP-31-86, FSEC, Cape Canaveral, FL.

Iannucci, J.J., and Shugar, D.S. (1991), *22nd IEEE Photovoltaic Specialists Conf. Record*, IEEE, New York, pp. 566–573.

Jennings, C. (1990), *21st IEEE Photovoltaic Specialists Conf. Record*, IEEE, New York, pp. 914–918.

Lynette, R., and Conover, K. (1989), *Photovoltaic Operation and Maintenance Evaluation*, Report GS-6625, EPRI, Palo Alto, CA.

Mrohs, M.F. (1990), *Photovoltaic Technology and System Design*, Siemens Solar Industries, Camarillo, CA.

Risser, V.V., et al. (1990), *Stand-Alone Photovoltaic Systems—A Handbook of Recommended Design Practices*, Report SAND87-7023, Sandia National Laboratories, Albuquerque, NM.

Solar Energy Industries Association (1994), *Solar Industry J.*, Vol. 5, Issue 1, pp. 54–76.

Strategies Unlimited (1991), *22nd IEEE Photovoltaic Specialists Conf. Tutorial*, IEEE, New York.

Wiles, J.C. (1991), *Photovoltaic Power Systems and the National Electric Code, Suggested Practices*, Southwest Technology Development Institute, Las Cruces, NM.

CHAPTER SIXTEEN

Terrestrial, Grid-Connected Systems: Field Experience, Status, Needs, and Prospects

F.R. GOODMAN, JR, J.C. SCHAEFER, and E.A. DEMEO, Electric Power Research Institute, Palo Alto, CA 94303

1. INTRODUCTION

The use of photovoltaics (PV) in cost-effective, off-grid applications has been demonstrated by thousands of reliable remote systems. Chapter 15 summarizes some of this experience. Two key features characterize these cost-effective remote installations: There is no utility or other power source nearby, and the power needed is relatively modest—usually less than 1 kW. In most cases batteries are necessary for continuous operation.

Looking to the future, the adoption of PV for large-scale or bulk power generation depends on PV's performance, cost, and value and on institutional factors such as corporate support. Of these, PV's performance on a large scale can now be estimated with considerable certainty, based on data from dozens of test installations.

However, no hard evidence exists that costs will decline enough for large-scale PV to compete with conventional power sources like coal. Although module costs have declined in the last decade, they still promise to be the major cost component for PV systems and must decline further if PV is to be competitive for bulk power generation. But studies (e.g., Stolte, 1992) do offer a basis for optimism. New materials, new production processes, and economies of volume production will all assist in lowering the cost of PV modules

Solar Cells and Their Applications, Edited by Larry D. Partain.
ISBN 0-471-57420-1

(DeMeo et al., 1990), and so will alliances among manufacturers, users, and research and development (R&D) organizations.

This chapter provides background for assessing how PV will reach cost-effectiveness. The experience of operational PV power plants, upon which PV's performance and requirements for operations and maintenance can now be predicted with considerable certainty, is presented, along with a brief history of operational PV plants. Next, a brief analysis of the economics of PV for grid-connected applications is developed, followed by discussion of conditions that PV must meet if it is to grow from its current niche applications alone to include bulk power generation by and for utility companies and their customers.

2. PV POWER PLANT SITING

An important factor in the cost of PV electricity is the insolation available to the PV power plant array; the more insolation available at a site, the more energy the plant generates and the lower is the cost per kilowatt-hour (kW-h), all other things being equal. Therefore sunny sites offer cost advantages over cloudy sites. The measurements upon which such siting choices depend are discussed in Chapter 20 (Emslie and Dollard, 1988; Knapp et al., 1980, 1982; Quinlan, 1981; Solar Energy Research Institute, 1983).

As an example of siting considerations, it will be noted below that the PV plant near Hesperia, California, performed better than any of the others, in part because of its high level of total insolation. Measured two axis tracking global insolation data from Hesperia for 1988 and 1989, two of Hesperia's best output years, show an average of 2,990 kW-h/m^2-year, but other sites may be even better. Calculations based on typical meteorological year (TMY) data from Daggett, California (about 60 miles north of Hesperia and the site of Solar One, a solar thermal power plant) show 3,390 kW-h/m^2-year. Similarly, Bryce Canyon data show 3,430 kW-h/m^2-year; Albuquerque and El Paso data show 3,470 kW-h/m^2-year, and Las Vegas data show 3,540 kW-h/m^2-year (Menicucci and Fernandez, 1986). This suggests that better performance and lower cost are possible at those other sites.

3. THE OPERATIONAL EXPERIENCE WITH PV POWER PLANTS

An important step in the commercialization of any product, including photovoltaics, is operational field testing, which demonstrates how various designs operate and provides credible performance data for economic and planning purposes. Since 1979 the U.S. Department of Energy (DOE), the U.S. Department of Defense Sandia National Laboratories, the National Renewable Energy Laboratory (NREL), the Electric Power Research Institute (EPRI), the Jet Propulsion Laboratory, various state agencies, photovoltaic manufacturers, systems developers, and electric utility companies have participated in a variety of such tests. At first their cooperation was informal, but since 1987 it has become more structured. PVUSA, a joint project involving all of the institutions named or categorized above, is now the most structured of these field test activities.

In addition to ongoing operation of PV systems as small power plants, field testing has generally included two other complementary activities: analysis of performance data and periodic, on-site systems tests to determine component reliability. Although most attention has focused on a few large systems that are judged most likely to represent future power plant designs, it is important to note that those whose results are reported extensively form only a portion of the more than 200 line-connected systems, totaling more than 10 MW of capacity, known to have operated in the United States in the late 1980s (Smith, 1989).

3.1. Modules in PV Power Plants

The history of line-connected flat-plate PV plants can be traced through four generations of PV modules. Early installations, from 1979 through about 1982, used experimental modules that suffered high failure rates. Most of the first-generation modules installed in systems at Mt. Laguna, California, and Natural Bridges, Utah, failed after several years of operation, mainly because of encapsulant deterioration.

To encourage the design and production of more reliable modules, the DOE and the Jet Propulsion Laboratory initiated in the early 1980s a system of so-called block purchases of PV modules, which were tested against increasingly stringent standards. Systems with second-generation modules built under a major DOE program and utilizing block III and IV modules showed much better reliability. In fact, most of those modules are still operating, even if not at their original installation sites. Most of the early plants, however, suffered from troublesome balance-of-system (BOS) equipment, which includes everything except the modules.

Most crystalline and polycrystalline silicon modules since the mid-1980s have met the block V standards. Plants with this third-generation of modules built after 1983 by utility companies, the DOE, and the ARCO Solar Corporation established a trend that is still apparent in the 1990s: The modules are generally the most reliable components in the plants. It was with these third-generation, block V modules that grid-connected PV power demonstrated its technical feasibility for the commercial market. The plants listed in Table 16.1 were designed not just to test whether the technology would work but also to assess their ability to function as commercial power plants. In most cases they achieved this goal, and one of them, the Lugo plant near Hesperia, California, set the standard that still exists for PV power plant performance.

PVUSA installations, whose fourth-generation modules were tested to an even stricter standard than block V, have generally demonstrated high reliability but have also illuminated ongoing BOS problems, which will be described below.

TABLE 16.1 Photovoltaic Power Plants

Plant	Nameplate Rating (kW)	Measured Rating (kW)	Module Type[a]	Tracking	Year in Service
Phoenix–Sky Harbor	225	173	Conc.	Two axis	1982
Dallas–Ft. Worth Airport	27	23	Conc., active cooling	Two axis	1982
Hesperia Lugo 1 & 2	1,000	932	Cryst.	Two axis	1982
SMUD PV-1	1,000	932	Cryst.	One axis	1984
Carrisa Plains 1–10	6,500	5,120[b]	Cryst.	Two axis	1984
SMUD PV-2	1,000	875	Cryst. and Poly.	One axis	1986
Austin PV-300	326	258	Cryst.	One axis	1986
Detroit	4	4	Amor.	Fixed	1987
Orlando–Solar Progress	15	13	Amor.	Fixed	1988

[a]Conc., concentrator silicon; Cryst, crystalline silicon; Poly., polycrystalline silicon; Amor., hydrogenated amorphous silicon.
[b]Calculated from observed outputs and PCU efficiencies.
SOURCES: DeMeo et al. (1990) and Mayorga and Hostetler (1986).

3.2. Power Ratings

Because of the wide range of irradiance, temperature, and wind speed conditions experienced by PV systems, the most credible power ratings are statistical estimates for specific conditions, based on linear regressions from a large number of observed data points. Such a linear regression for an entire 250 kW PV system is illustrated in Figure 16.1; each sloped line represents the linear regression against temperature for the irradiance level indicated. These data are from one sun, single crystal silicon modules from Siemens Solar Industries that were mounted on a one axis, passive tracking system in Austin, Texas. The rating for 1,000 W/m^2 is established as the intersection of that line and the vertical line at 20°C, just over 250 kW. The statistical process is the same for both modules and systems.

A number of organizations have tested individual PV modules on a long-term basis to ascertain how well they perform and to understand any problems that arise. One important finding from these tests and the calculations described in Figure 16.1 is that only a small fraction of the modules exceeded even 90% of the power ratings designated by their manufacturers (Lepley, 1990). Only one module of dozens in the Pacific Gas and Electric Company (PG&E) tests exceeded the manufacturer's rating (Jennings and Whitaker, 1990).

Performance data from a number of the line-connected systems have been recorded with computer-based data acquisition systems so that hourly values

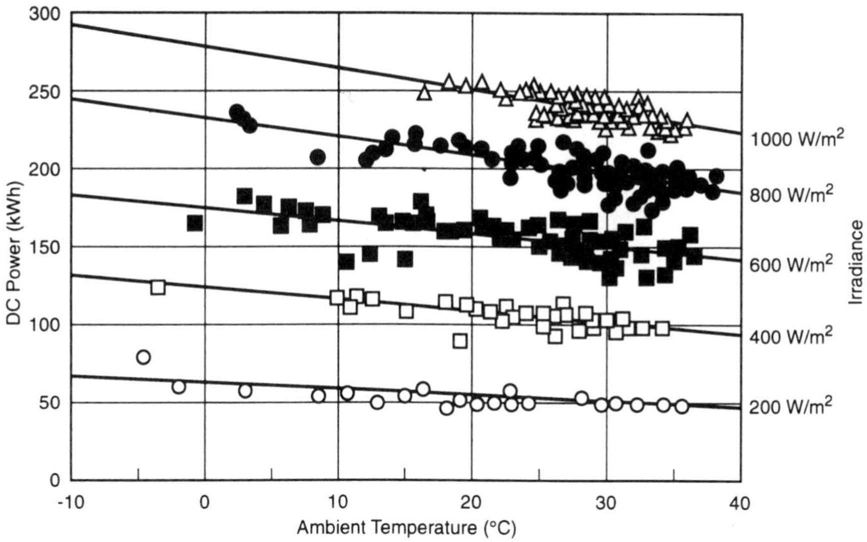

FIGURE 16.1 Linear regression showing determination of power rating. SOURCE: Rosenthal (1991).

are available for DC and AC power output, insolation, ambient temperature, module temperature, and in some cases wind speed (Southwest Technology Development Institute, 1990). From these data emerges the conclusion that until 1992 none of the line-conected PV systems met the kilowatt rating its supplier anticipated. Table 16.1 illustrates the discrepancy between measured and manufacturers' power ratings.

Part of this discrepancy arises from the PV industry's transition from one heavily engaged in research to one whose products are expected to meet commercial standards. Several rating standards are in use for power output and efficiency. Standard Test Conditions (STC) are widely used in the industry for cell research and for module performance testing in an indoor laboratory setting (IEEE, 1986). STC values are 1,000 W/m^2 (or 850 W/m^2 for concentrators), AM1.5 spectrum (described in Chapter 20 and Appendix A), and 25°C cell temperature. For indoor flash testing in which cell or module temperatures remain at indoor ambient temperature, this is a natural rating to use. However, under true sunshine conditions the cells in flat-plate modules will operate at temperatures at least 20°C higher than ambient, and their efficiencies decline as predicted by an equation. As a result, the STC-based nameplate ratings always exceed the measured values. Other system losses are due to mismatch among the modules, temperature discrepancies, and losses in the DC interconnections.

In an effort to make system ratings correspond more closely with true outdoor conditions, PVUSA adopted formally a standard under which system performance levels are measured at 1,000 W/m^2 (850 W/m^2 for concentrators), 20°C ambient, and 1 m/s wind speed. This rating has been designated the PTC

(PVUSA Test Conditions) and is the standard for the measured power ratings shown in Table 16.1.

To encourage the delivery of PV systems with more accurate ratings, the City of Austin (Texas), for its PV300 system, and PVUSA for all its systems specified that output would be measured under PTC and that suppliers would then be paid accordingly. Only one of these systems exceeded its manufacturer's rating.

Utility companies point out that a temperature of 20°C ambient does not truly reflect conditions when they often most need electrical generating capacity, so another rating standard has been proposed, prescribing 40°C ambient temperature (Firor et al., 1990). While this as-yet-unnamed rating standard will tend to make PV power rating appear even lower than its predecessors, it does represent better the actual conditions under which utilities in the desert Southwest experience their greatest need for electric power (Lepley, 1990).

3.3. Measured Energy Production

Performance measurements at operating PV power plants have provided data from which PV economics can be determined. In general these results suggest that PV technology can be expected in the future to operate reliably and with predictable output, except that of course in some parts of the country sunny days are difficult to predict. A few other problem areas remain as well.

Table 16.2 summarizes observed capacity factor measures from representative PV power plants. These systems range from the 1982 concentrator installation at the Phoenix, Arizona, Sky Harbor Airport (Eckert, 1987) to the 1988 Orlando, Florida, Solar Progress installation with amorphous silicon modules (Atmaram, 1993).

TABLE 16.2 Capacity Factors From Operating PV Power Plants

Plant	1984	1985	1986	1987	1988	1989	1990	1991
Phoenix–Sky Harbor	10	22	22	29[a]	OS[b]			
Hesperia Lugo 1 & 2	33	35	24	29	33	36	OS	
SMUD PV-1	NA	23	25	7	0	13		
Carrisa Plains 1–10	29	30	29	25	24	21		
SMUD PV-2			23[a]	23	19	9		
Austin PV-300				22[a]	25	23		
Detroit				14[a]	15	14	14[c]	12[c]
Orlando–Solar Progress					20[d]	16	18	18

[a]Six months of data.
[b]OS, out of service.
[c]Private communication.
[d]134 days of data.
SOURCE: DeMeo et al. (1990).

Capacity factor (CF) measures how well a power plant attains its possible or potential production. One that operates at its rated capacity for 8,760 hours in a year demonstrates a 100% capacity factor. CF is calculated mathematically as

$$CF = \frac{\text{Annual energy produced by the plant in kW-h}}{(\text{Empirically determined kilowatt rating}) \times (8{,}760\,\text{h})} \tag{1}$$

A solar power plant's capacity factor depends on the insolation at a given site, the plant's design (and particularly the type of tracking employed), and the reliability of the equipment. A sunny location in California's Mojave Desert such as Hesperia receives more insolation than one that is frequently cloudy in Detroit. Similarly, a plant with two axis tracking, like the one at Hesperia, captures the maximum amount of insolation and provides more energy and therefore a higher capacity factor than one with a fixed orientation.

It is important in the calculation of capacity factor that the plant's rating be consistently based on the same rating standard. For example, if PTC is used to establish the plant's rating, then that kilowatt rating should also be used for calculating capacity factor. All capacity factors reported here are based on PTC.

Table 16.2 reveals that although capacity factors vary from year to year as well as from plant to plant, values above 33% are possible at a good site with two axis tracking and a reliable plant. Hesperia illustrates this potential; a capacity factor of 36% was recorded for Hesperia in 1989.

Reliability is important because any time the plant is inoperable when the sun shines, the plant loses that energy. For example, the Sacramento Municipal Utility District (SMUD) PV-1 plant was out of service during part of 1987, 1988, and part of 1989 because of a cable failure (which had nothing to do with the PV portion of the plant) and resultant fire in its power conditioning unit (PCU). Capacity factors are low for those years. Similarly, the Hesperia capacity factors are low for 1986 and 1987 because the tracker gearboxes were replaced during that period. However, most of the plant continued to generate power even though some trackers were out of service.

The record of the Phoenix–Sky Harbor plant from 1983 to 1987 is noteworthy for two reasons. First, it exemplifies how a new and untested technology must be gradually improved with field testing. As one of the earlier and more complex installations, it experienced a number of problems in its original PCU, which had to be replaced, in the modules themselves, and in other BOS components (DC wiring, diodes, and trackers). The owner and operator, Arizona Public Service Company, steadfastly repaired, redesigned, and rebuilt parts as necessary to keep the plant running and to improve it. It achieved a capacity factor of 29% for the first 6 months of 1987, but was decommissioned later that year because the land was required for other purposes.

The second reason Phoenix–Sky Harbor is noteworthy is, as mentioned in Chapter 11, that it provides the only long-term operating history so far for point-focus concentrator systems (Eckert, 1987). Its 29% capacity factor if continued on an annual basis would have been exceeded only by that of the Hesperia plant. Therefore, given its first-generation design, Phoenix–Sky Harbor's 29% capacity factor suggests that at least 30% capacity factors are possible with well-designed and well-located concentrator systems.

The plants at Phoenix–Sky Harbor, Hesperia, and Carrisa Plains all utilized two axis tracking to capture the maximum insolation available. Point-focus concentrator systems, of course, must use two axis tracking. But two axis tracking systems are expensive and require some maintenance and repair, so a number of subsequent plants employed less complex, single axis tracking. The two PV plants at the Sacramento Municipal Utility District's (SMUD) Rancho Seco site have experienced few maintenance problems with their electrically driven, single axis trackers (Lynette and Conover, 1989).

Austin's PV-300, with its passive, single axis tracking, has experienced the lowest total maintenance costs of the large line-connected systems, roughly \$0.08/kW-h (Hoffner, 1989), although even the passive trackers have experienced some problems with seals. If daily inspections were excluded, maintenance costs would be even lower, at \$0.04/kW-h. Analysis concludes that \$0.05/kW-h is a reasonable projection for the future (Lynette and Conover, 1989).

Space PV applications have demonstrated high reliability, so the same was expected of terrestrial systems. Generally that expectation was met, with reliability improving markedly in more recent installations. For example, Austin's PV-300 has consistently demonstrated availabilities (the ratio of hours of operation to hours when insolation was adequate for the plant to operate) of 99%. Others such as Hesperia and Detroit have shown availabilities in excess of 95%.

3.4. Results from PVUSA

Recognizing the need for a consistent and structured nationwide PV system testing program, PG&E proposed the PVUSA program in 1986. Other participants include DOE, EPRI, Sandia National Laboratories, NREL, the California Energy Commission, the Department of Defense, Central and South West Services, the City of Austin, New York State Energy Research and Development Authority, Niagara Mohawk Power Corporation, Public Service Company of Colorado, Sacramento Municipal Utilities District, Salt River Project, San Diego Gas and Electric Company, Maui Electric Company, the state of Hawaii, Virginia Electric and Power Company, and the Commonwealth of Virginia.

Two sets of technologies are being tested (Hester, 1990). Emerging Module Technologies (EMTs), which had been tested as modules (by NREL, Sandia, or other independent laboratories) but not at the system level, are installed in systems with 20 kW nominal power rating; and Utility Scale (US) systems, utilizing modules that had been previously tested in small-scale systems, are installed with ratings up to 400 kW. In the case of EMTs, suppliers provided only the modules and their installation, and PVUSA installed identical DECC Helionetic PCUs for each system. The US systems are turnkey, meaning that their suppliers provided everything, including the PCUs. The systems in operation as of 1992 are listed in Table 16.3, along with key installation and operational data. The main PVUSA test site is in Davis, California; others are located on the island of Maui, Hawaii; in Austin, Texas; and near Kerman, California.

PVUSA requires all modules to pass the NREL factory qualification test (DeBlasio et al., 1990) plus a field wet-resistance test to ensure that excessive leakage currents to ground do not exist. Several manufacturers' modules in both the EMT and US segments failed the field wet-resistance test, which suggests that factory quality control was inadequate. Modules were returned to the factory for appropriate repairs.

TABLE 16.3 PVUSA Systems

Completion Date	Supplier	Technology	Efficiency (Percent)[a]	Power (kW)[a]
Acceptance tests completed				
1/89	Siemens Solar	Microgridded crystalline silicon	11.1	18.7
6/89	Sovonics (United Solar)	Tandem junction amorphous silicon, at Davis	3.5	17.3
10/89	Sovonics (United Solar)	Tandem junction amorphous silicon, at Maui	3.7	18.5
12/89	Utility Power Group	Tandem junction amorphous silicon	3.3	15.7
10/90	Solarex	Bifacial polycrystalline silicon	8.6	15.7
3/91	Entech	22×, linear Fresnel, crystalline silicon	11.3	16.5
10/92	Integrated Power	Ribbon silicon (Mobil EFG), single axis active tracking, Omnion PCU, at Austin	9.7	19.4
12/92	Advanced Photovoltaic Systems	Amorphous silicon, fixed tilt, Advanced Photovoltaic Systems PCU	4.1	479
Acceptance tests not completed				
	AstroPower	Thin film polycrystalline silicon	7.4[b]	19.3[b]
	Photon Energy	Cadmium telluride thin film	5.3[b]	18.7[b]
	Siemens Solar	Crystalline silicon, single axis passive tracking, Dickerson PCU	10.3[b]	174.0[b]
	Integrated Power	Ribbon silicon (Mobil EFG), single axis active tracking, Omnion PCU	7.4[b]	180.0[b]

[a]PVUSA test conditions: 1,000 W/m^2 (850 direct normal for concentrators), 20°C ambient temperature, 1 m/s wind speed.

[b]Manufacturer's estimates, as systems are not tested.

Another difficulty for the PVUSA systems has been installation delays, ranging from 1 month to more than 2 years. All of the early US systems except that at Kerman were behind schedule as of 1993.

Performance measurements formed the basis for payment to all contractors, and were calculated on the basis of PTC standards discussed earlier. Only the Advanced Photovoltaic Systems (APS) installation has met its expected PTC rating and in fact exceeded it by about 20%; others have fallen short by varying amounts, although the Siemens Solar (crystalline silicon) EMT was within 4%.

Annual capacity factors for all Davis fixed flat-plate EMT systems are similar at 21%–22%; the ENTECH installation, a two axis tracking concentrator system, achieves higher values than the fixed systems in the summer months but lower values in the winter.

PVUSA analysis of operational data shows that temperature coefficients of PV system output vary not only from technology to technology but also from one irradiance level to another and that, in contrast to widely held beliefs, are not always negative (PVUSA Project Team, 1991; Whitaker et al., 1991). This finding supports previous evidence (Pratt and Burdick, 1988).

3.5. Module Problems in PV Arrays

Between 1983 and 1990, the Southwest Technology Development Institute (SWTDI) tested the flat-plate modules at 35 different PV power plants to identify module output failures and their causes. Some of these plants were grid connected and some were not. A total of 76 visits were made, with two or more visits to 21 of the sites. Altogether, 282,447 module tests were carried out on 102,263 modules. Noncontributing modules were identified by monitoring each bypass diode current while shading one or more modules in each parallel block (Durand and Bowling, 1993). A module was itself defined as failed when its lack of output could not be traced to connections or components external to the module. Less than 5% of the 10,769 failures to contribute were attributed to defective modules: the rest were due to defective interconnections or other balance of system components.

Two widely used approaches for wiring modules into larger arrays embody potentially troublesome failure mechanisms. In one approach, standard modules having junction boxes fitted with screw or crimp terminals are simply wired together in the factory or by field electricians. These connections sometimes loosen, apparently as a result of thermal cycling or mechanical vibration, and the increasing electrical resistance of the connections not only increases field losses but in some cases even causes fires. The other approach is to interconnect nonstandard modules (often termed *laminates* if they have no frames) by welding cell interconnect ribbons from the negative and positive ends of the module directly to bus structures in the panels. In several cases these ribbons have failed, as a result of differential thermal expansion or mechanical or thermal cycling.

Vandalism may also prove to be a problem. PV modules at sites in Sacramento (McClelland Air Force Base) and Phoenix (John Long Homes) have been shattered by rocks, although most such modules continued to operate after being struck.

Several important sites were not included in SWTDI's visits. The most important of these were Hesperia, Carrisa Plains, and the PVUSA systems. In general, the reliability experience at Hesperia and PVUSA has been similar to that of other grid-connected systems. One notable exception, however, has been ethylene vinyl acetate (EVA) browning at Carrisa Plains.

3.6. EVA Browning

The most serious problem discovered at Carrisa Plains was originally known as *brown cell*, and it affected every module in the plant. In 1987, after 3 years of operation, operators noted a distinct decline in output. This performance degradation followed observations of what appeared to be a browning or yellowing of the cells in the first nine mirrored 600 kW segments of the plant. Figure 16.2 shows a mirrored tracker from one of these segments; in effect, it is a 2× concentrator but constructed with more or less standard flat-plate modules. By 1990, the sixth year of operation, the electrical degradation had reached about 40% (Wenger et al., 1991). This kind of degradation is obviously unacceptable for a power plant that must operate for 20 years or more.

Detailed examination revealed that it was not the cells that were browning but the EVA used as an encapsulant between the cells and the front cover glass. More correctly, *brown cell* should be called *EVA browning*. The problem turns out to be deterioration of the EVA under high temperature and exposure to intense ultraviolet energy, plus degradation of the electrical interconnects within the module. It is not known conclusively as of 1994 to what extent the output degradation is due to acetic acid attacking the interconnects, hot spots in the modules, or optical loss in the EVA. Nor is it known whether the problem will eventually affect all EVA. or only some formulations or batches. The modules in the tenth segment at Carrisa Plains, which have no mirrors to serve as concentrators, have browned at a reduced rate; as of 1991 there was only slight evidence of reduced electrical performance (Wenger et al., 1991).

EVA browning affects modules in other systems as well, particularly at high-temperature sites. However, it is important to note that not all modules with EVA, even in high-temperature locations, show browning. For example, a single axis tracking system at Ridgecrest—one of the hottest and sunniest locations in California—showed no evidence of EVA browning after 5 years of operation.

Research on EVA degradation continues (Pern et al., 1991; Petersen and Wohlgemuth, 1991; Wohlgemuth and Petersen, 1992), and yet it is not known whether a different formulation of EVA or a completely different encapsulant will be used in the future. EVA browning is not fatal to flat-plate PV, however, because other encapsulants have been used successfully in the past and can be used again (Mon et al., 1990). Polyvinyl butyrate (PVB), for example, worked well in the record-setting plant at Hesperia, and those modules show neither encapsulant browning nor evidence of reduced output.

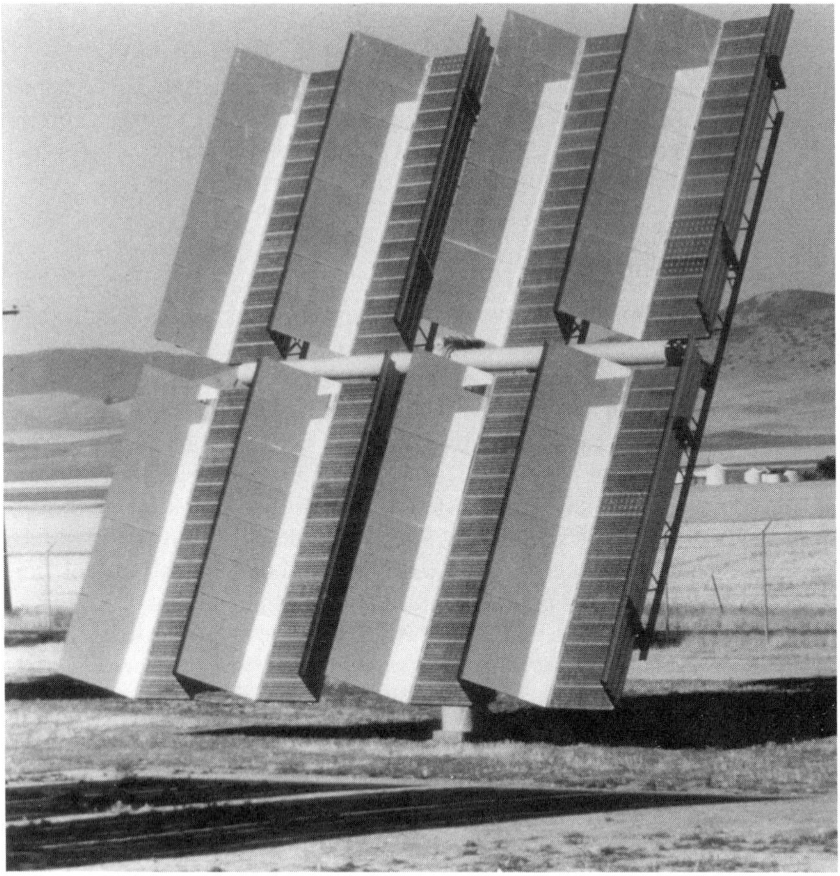

FIGURE 16.2 Mirrored tracker at Carrisa Plains.

3.7. Balance-of-System Failures

In addition to noncontributing modules discovered in SWTDI's tests, other array failures have become apparent during system operation, sometimes manifested by open circuits, short circuits, or even fires of varying severity.

Ground faults commonly initiate the fires, and several examples illustrate the difficulties that inadequate insulation can cause. One cool, damp morning at the Phoenix—Sky Harbor concentrator site, the kapton insulation between a blocking diode and a heat-sink failed in a string whose blocking diode had already failed as a short circuit. This permitted the full array current to flow backward through all cells from the bus to the ground fault, so that those cells and their modules burst into flame (Eckert, 1987). This situation might be viewed correctly as a multiple-contingency failure—water in the modules, insulation breakdown, and a failed blocking diode—which ought not to occur often. However, such multiple-contingency events do occur, so they must be considered.

Arizona Public Service installed fuses to protect against further such occurrences at Sky Harbor. However, no consensus exists on whether fuses should be used universally to protect against double contingencies, because the fuses and fuse holders themselves introduce one more set of components that can fail (Durand et al., 1990).

Ground faults also occurred at SMUD's PV-1 as a result of several component failures. For example, bypass diodes faulted to ground as a result of what was surely another unforeseen contingency: birds' nests behind the failed diodes, causing them to overheat (Collier and Key, 1988). Other ground faults were due to the installer's use of improperly specified underground cable for array interconnections. The cables at PV-1 failed, and all had to be replaced.

The most common array-related failures at Carrisa Plains were also caused by ground faults, which in turn were due to inadequate insulation between the module interconnect bus and the panel support structures. Operators were able to jumper out the affected panels with an almost imperceptible loss of output, and repairs could thus be delayed until enough failed panels were accumulated to justify the use of a crane.

Ground faults and other component failures in PV plants illustrate problems that can occur with the use of poorly conceived designs or underrated components. BOS failures may result from an attempt to save money, but such decisions generally turn out to be short sighted because they cost more to repair than they would have to build correctly in the first place.

On a broader philosophical note, designers must be aware that PV systems have to be reliable enough that they require very little service if they are to be cost effective on a large scale, because retrofitting thousands of small oversights is costly. The lesson that hindsight offers is twofold: there is no substitute for field testing, and in making design choices consideration should always be given both to previous field experience and to all possible failure modes.

3.8. Power Conditioning Units

Power conditioning units (PCUs) connect the photovoltaic array to the AC grid by convering the DC to AC, tracking the peak power operating point of the array, and providing proper startup and shutdown. They must be designed to minimize damage when electrical failures occur, and they should not disturb loads or other generators on the grid. From the perspective of reliable power generation, PCU outages are more serious than the module and interconnection problems noted above, because the whole PV array is disabled when the PCU is out of service.

The early reliability record for PCUs was poor; during the first decade of major PV system operations, more energy production was lost because of PCU problems than any other cause (Durand and Bowling, 1993), except possibly

EVA browning. In the order of their significance, problems observed include the following:

- PCUs shut down when they should not have
- They failed internally
- They failed to shut down when they should have, or they were not designed to provide adequate protection

The PCUs' tendency to shut down when they should not have was the most annoying problem for PV plant operators, who sometimes discovered upon visiting the plant (most have been unattended) that the PCU had shut down for no apparent reason. PCUs installed through the mid-1980s typically shut themselves down under conditions that were abnormal but not dangerous either to personnel or to equipment. For example, one common shutdown condition was an imbalance of more than a few milliamperes in positive and negative currents from a bipolar array or in a bipolar array's ground current. The theory was that an electric shock with more than 15 mAmp was potentially fatal to someone in contact with the high array voltage, so the PCU should shut down if it detected such an imbalance. Fortunately, nobody ever did make such a contact with the arrays. Unfortunately, a passing cloud can cause the same current imbalance, so that PCUs with that shutdown criterion tripped often. Another difficulty was PCU shutdown when the AC line voltages exceeded its limits, if even momentarily.

Since the mid-1980s, therefore, most modern PCU designs have either eliminated the sensitive shutdown functions or raised their thresholds. Another improvement reduced the outage durations by designing the PCU to attempt restart continually. The 99% availability of the PCU at Austin's PV-300 system is due in part to the fact that it is programmed to attempt restart continually rather than give up after a few tries.

In addition to the plants shutting down when they should not, utility companies are concerned that PCUs fail to shut down when they should, because of the potential for electrocution of linemen working on lines that they think are de-energized (Steitz, 1991). Controlled tests with the PCUs in PVUSA's EMT segments showed that islanding or run on (a PCU's continued operation in the absence of grid power) can last as long as 11 min under the rare circumstance with loads precisely equal to PCU output power (PVUSA Project Team, 1992). This finding corresponds with that of other investigators (Steitz, 1991). Therefore, until PCU designs demonstrate that they do not run on, utility companies will require PV power plants to provide a lockable disconnect switch to protect linemen against the unlikely event that a PV system might energize an isolated section of feeder.

A system's failure to shut down when it should can damage or destroy equipment. The most serious such event was the SMUD PV1 outage in 1987, in which the greatest damage resulted from the failure to protect against a line-to-line fault in the cables carrying array current into the inverter cabinet

(Collier and Key, 1988). More complete integration of appropriate protection functions, logically within the PCU, will eliminate such events in the future (Steitz, 1990).

Power quality tests have revealed that harmonic generation from the DECC Helionetic PCUs at PVUSA is within reasonable limits (PVUSA Project Team, 1991). Tests of other PCUs have generally reached similar conclusions PCU efficiencies have also been satisfactory, generally exceeding 95%. Similarly, with the use of self-commutating inverters rather than the line-commutating inverters built in the early 1980s, observed power factors have approached the desired value of 1.0 (Steitz, 1991).

Several challenges remain for PCU suppliers, however. One is to integrate PCU designs better with the PV power station requirements, particularly with regard to islanding and unnecessary shutdowns. Another challenge is cost; at $500 to $2,000 per kilowatt in 1992, PCU costs are too high to meet the requirements for cost-effective bulk power. PCU costs, part of what are termed *power-related BOS costs* in Section 4, will probably fall to below $100 per kilowatt (Stolte, 1992). One way to cut costs is to eliminate the 60 Hz transformer, although in doing so designers must ensure that even under failure conditions no DC is injected into the utility's grid. Finally, a PCU test site is needed, where manufacturers' designs can be tested to ensure their correct performance (Steitz, 1991).

3.8. Perspective on Field Experience

The casual reader might conclude—erroneously—that the foregoing description of problems encountered implies that PV systems are unreliable. Nothing could be further from the truth. Most of the grid-connected systems, and particularly those installed since 1986, have operated with few problems and with little attention. Problems that have occurred are almost exclusively in the BOS components and will continue to diminish as designers gain more experience.

4. COSTS AND THE ECONOMICS OF BULK PV POWER

In a bulk power context, the most important economic index for an electricity generating system is its levelized cost of energy (COE), usually expressed in dollars or cents per kilowatt-hour. Levelized over the life of the plant and based on a series of assumptions about the future, it is the revenue required to cover the sum of capital carrying charges and operating costs (DeMeo et al., 1992). For a power generation investment to be attractive to a utility or an independent power producer, its levelized COE must not exceed the value of that electricity as determined by the utility or other purchaser.

Levelized COE on an annual basis is composed of two major components: the annual equivalent cost related to the original investment in the plant plus the operating and maintenance expenditures necessary to maintain the plant in service. Mathematically:

$$\text{COE} = (\text{TCI} \times \text{FCR})/(\text{annual AC energy production}) + \text{O\&M} \quad (2)$$

$$= (\text{TCI} \times \text{FCR})/(\text{CF} \times 8760) + \text{O\&M} \quad (3)$$

$$= (\text{TCI} \times \text{FCR})/(\text{S} \times 0.8) + \text{O\&M} \quad (4)$$

where COE is the levilized cost of electricity (dollars per kilowatt hour); TCI is the total capital investment required to install the plant (dollars); FCR is the fixed charge rate or fraction that converts an initial investment into an annualized charge (it includes amortization of the original investment like a mortgage payment, plus insurance costs, property taxes, estimates of future inflation, risk premium, and other charges related to the investment [fraction]); O&M is the operating and maintenance expenditures (dollars per kilowatt-hour); CF is the capacity factor as defined in Eq. (1) and based on PTC (fraction); and S is total insolation (kilowatt-hours per square meter per year).

The factor 0.8 accounts for losses in conversion efficiency from sunlight to AC because of DC wiring losses, PCU losses, module mismatch, module heating above PTC temperatures, and AC losses in collection systems and transformers (DeMeo et al., 1992).

A PV power plant's constituent costs can be considered on the basis of module cost per square meter; BOS equipment cost per square meter for equipment that is a function of the plant's area, such as support structures, trackers, and electrical cables; and BOS equipment cost per kilowatt for equipment that is a function of the kilowatt rating of the plant, such as power conditioning, controls, and substation components like switches and transformers. Therefore the components of TCI can be expressed per PTC kilowatt as follows:

$$\text{TCI} = (1 + i)[(c_m + c_b)A + c_p] \quad (5)$$

where i is the fractional addition necessary for indirect costs like engineering, interest during construction, and permits; A is the module area in square meters per DC kilowatt, at PTC; c_m is the PV module cost per square meter; c_b is the area-related BOS cost per square meter; and c_p is the power-related BOS costs per kilowatt. Conveniently, the area required for one kilowatt of dc module output at PTC is the inverse of the module efficiency (η_m), so that

$$A = 1/\eta_m \quad (6)$$

Combining Eqs. (4), (5), and (6) with (3) yields a single equation for levelized cost of energy:

$$\text{COE} = (1 + i)[(c_m + c_b)/\eta_m + c_p](\text{FCR})/(0.8\,\text{S}) + \text{O\&M} \quad (7)$$

None of the variables in Eq. (7) can be projected with great certainty, so that ranges of values are employed to provide perspective. The approach adopted is to estimate baseline values—based on previous studies and listed in Table 16.4—and then to assign upper and lower limits to each variable.

Figure 16.3 illustrates the dependence of COE on key variable in Eq. (7), and the scale below it illustrates the upper and lower limits for each variable. Because they are not as significant in the calculation, indirect costs and O&M are not shown in Figure 16.3. The baseline values are reasonable estimates of future values, but they should be viewed more as illustrative than definitive. For example, although single crystal silicon modules with 11% efficiencies can now be purchased on the commercial market, they cost about $600 per square meter—compared to $400 per square meter estimated as the upper limit for module cost shown in Figure 16.3. Additionally, no commercial copper indium diselenide (CIS) modules are available on the market in 1993, so that both costs and efficiencies reflect only expert opinion about what will be available when the PV market grows enough to drive costs down.

Figure 16.3 portrays the COE resulting from the assumed baseline values—the single point at $0.11/kW-h—as well as from the values representing best and worst cases for each variable. For example, with all other variables being equal to their baseline values, a module efficiency of 18% will result in a COE of about $0.08/kW-h. Similarly, a fixed-charge rate of 16% instead of 11% will raise the COE to $0.17/kW-h. Changing only one variable at a time is somewhat unrealistic because some or all variables will probably differ from baseline values in the future. Still, Figure 16.3 reveals which of these variables offers the most leverage. The variable with the greatest effect is module cost; the relatively great distance between the upper right end of the module cost

TABLE 16.4 Baseline Values for Cost of Energy Calculations

Variable	Value	Comments[a]
Module cost	$133/m^2$	CIS modules, 100 MW/yr (adapted from Stolte, 1992, p. 8–33)
Efficiency	11.1%	CIS modules, 100 MW/yr (adapted from Stolte, 1992, p. 8–33)
Fixed charge rate	10.3%	Adapted from Stolte (1992, p. 8–16)
Capacity factor	26.5%	Fixed, Las Vegas (adapted from Menicucci and Fernandez, 1986, p. 29)
Area-related BOS cost	$28.60/m^2$	Adapted from Stolte (1992, p. 8–33)
Power-related BOS cost	$234/kW	Adapted from Stolte (1992, p. 8–33)
Indirect costs	0.225	Adapted from Stolte (1992, p. 8–33)
O&M	$0.005/kW-hr	DeMeo et al. (1990a), Lynette and Conover (1989)

[a]Stolte's numbers were modified to permit reference to array area and module performance rather than to AC output.

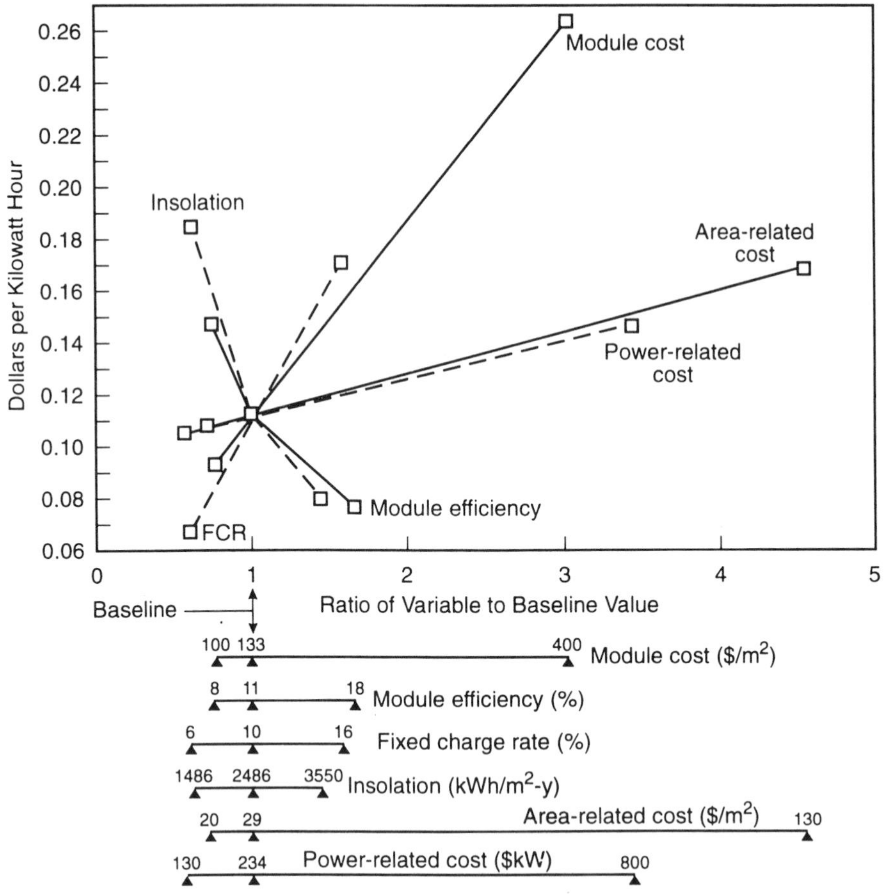

FIGURE 16.3 Flat-plate COE as a function of key variables.

line and the baseline point indicates how far module costs have to be reduced even to reach baseline conditions.

The baseline numbers reflect estimates for fixed flat-plate systems. Corresponding values for single axis tracking and two axis tracking systems would be somewhat different; capacity factors might rise as high as 38% with two axis tracking in a high-insolation site, but area-related costs would rise too. Changes in more than one variable cannot be estimated from Figure 16.3, but must be calculated from Eq. (7).

One useful approach in examining PV's future is to assume a target COE—a value that implies bulk power competitiveness in major segments of the utility industry—and then determine the cost and performance values that meet this target. The intersection of a horizontal line corresponding to such a cost target, say \$0.08/kW-h, and any of the single variable lines indicates, on its scale below the graph, the value which that variable alone must reach to yield a COE of \$0.08/kW-h. It is apparent that only efficiency improvement, higher insolation, and lower fixed charge rates alone can reach the \$0.08/kW-h target. However, because all variables are likely to vary from their baseline values, it is also useful to ask what combinations would reach such a target (DeMeo et al., 1992). Solving Eq. (7) for the cost variables in question will provide that information.

There is nothing sacred about the \$0.08/kW-h target, even for utilities. In some applications the target COE value will be lower and more difficult for PV to meet; in some cases it will be higher and therefore easier to meet. Electricity delivered to the grid is worth more during a utility's peak hours than in the middle of the night because utilities must provide adequate electrical generating capacity to supply the peak loads. Therefore one factor in evaluating PV is the correspondence by time of day of its power output with a utility's load. Not surprisingly, for summer-peaking utilities the correspondence is good (Hoff and Iannucci, 1990), particularly when there are heavy loads at midday such as with irrigation or commercial air conditioning.

PV is already meeting the COE requirements for many remote applications where any other source of power would cost more. In fact, a range of PV applications, from small remote systems that are economic even at high cost to bulk power generation that is feasible only at low cost, forms the path by which PV applications will broaden in the future (DeMeo et al., 1990; Iannucci and Shugar, 1991).

5. HIGH VALUE APPLICATIONS

The challenge for the PV industry now is to discover more diverse PV applications whose size can be larger and whose value per kilowatt-hour may be less than the remote installations described in Chapter 15. The place to look for such new applications is anywhere that electricity is already used or can be used. Two examples from the utility industry exemplify the range of such possibilities—one almost like a remote system and one almost like bulk power generation.

One early application of cost-effective PV is powering a growing population of electrically operated transmission line switches or sectionalizing switches. Even though a supply with thousands of volts may be nearby, it often turns out to be cheaper to power these low-energy loads with PV than it is to install stepdown transformers, insulators, fuses, and switches. Analysis shows that the smaller the load and the further it is from commercial power, the more likely it is for PV to be the most economic supply. For example, PV is more economical than conventional utility power for line extensions in excess of 200 feet and with loads of less than 10 kW-h per month (Bigger et al., 1991).

A much larger application was identified on PG&E's distribution system in California's Central Valley (Shugar, 1990). It illustrates how electricity can sometimes be worth more to a utility company than its average value, depending on when and where it is delivered. In this case, a PV system located on a Kerman Substation feeder is potentially cost-effective because

- The substation and feeder loads correspond closely with PV output on an hourly basis
- The PV system's presence will permit a delay in installing a new transformer at the substation plus other equipment
- Losses throughout the system, on both transmission and distribution levels, will be reduced with the PV system in place
- State and federal tax credits are assumed
- Customers' voltage levels will be held closer to standards because the PCU also provides reactive power
- The reliability of power delivery to the customer will be enhanced

The situation at Kerman Substation results from an unusual combination of circumstances that will not occur often (Strategies Unlimited, 1992). However, comparable circumstances will probably be found on other utilities, and developing them not only will provide a market for PV suppliers but also will save utility companies money. It should be noted that Kerman has been established as the site of a US PVUSA segment with 500 kW of PV capacity. This installation should allow actual quantification of the benefits assumed.

Another option is grid-connected PV installed on utility customers' residences (Goodman et al., 1993). The electricity provided is worth more to the customers, who would otherwise purchase it from utility companies at retail rates, than it is to the companies, who can only pay wholesale rates to independent generators. In order for this option to attain widespread success, the customers must be willing to undertake such projects, which involve both technical and economic issues. Hundreds of such systems are operating successfully. However, many electricity customers understand neither the economics of electricity sales and purchase transactions nor the operation of PV systems (Russell and Kern, 1991). Clearer explanations to customers as well as more attention to customers' concerns will be necessary. Unless a suitable PV service industry emerges to provide long-term support to PV owners, utility ownership may be preferable to customer ownership for most residential systems.

6. EXTERNAL COSTS AS A SOURCE OF VALUE

Consideration of external costs as well as internal costs in evaluating electricity supply options works to the economic advantage of electricity from PV or other pollution-free sources. Internal costs are those paid by electricity producers (and ultimately by their customers) and include fuel, amortization of investment, operation and maintenance, insurance, and taxes—generally cash outlays by the electricity producers. External costs are borne by someone other than the producers. Examples are the cost of acid-rain damage done to farmland or to a fishing resource and or adverse health effects from air pollution or chemical toxics. Quantification of these external costs is difficult at best and certainly controversial. Estimates of their magnitude range up to approximately that of internal costs, so that the sum of internal and external costs (the total *social costs* in economists' terms) may be twice the internal cost of production (Hohmeyer, 1988; Ottinger, 1990).

Added to the uncertainty about external costs is the way in which utility regulatory commissions will require the companies to include them (Ottinger, 1990; Wiel, 1992). Regardless of how that occurs, however, the values suggested as surrogates for external costs of conventional generation, a few cents per kilowatt-hour in 1993, are not yet large enough to make PV competitive with conventional electricity production.

7. INSTITUTIONAL ISSUES

The difficulty in finding and exploiting larger scale, grid-connected PV applications is not only a technical and economic challenge, but also an institutional one. With more than 100 years of history, utility companies are accustomed to planning on the basis of large, centralized generation sources. Decentralized or distributed generation sources (not only PV but also co-generation, storage, wind, fuel cells, and small hydro) involve a new and somewhat unfamiliar paradigm (Rueger and Manzoni, 1990). The Kerman application exemplifies the concept: a small power source close to loads offering utility companies more degrees of freedom in reducing costs and enhancing the concepts of sustainability, energy efficiency, modularity, and strategic power delivery.

An institutional challenge that faces all of the renewable energy industry is attaining closer cooperation with its major potential customers, which include electric companies. In general, renewable energy supply industry capabilities do not yet overlap with their needs. This is more the case with PV than it is with wind power and solar thermal trough technologies (Holl, 1989), which have achieved lower costs and far greater penetration in the bulk power market. Except for concentrators, the enabling technology for bulk PV power does not yet exist, and the gap is relatively quite large. Thus, the basis for establishing commercial cooperation in PV is in early, high value applications such as those noted above and in Chapter 15.

Meanwhile, PV must continue to serve ever-expanding niche markets, to attain further cost reductions, and to sustain focused R&D to achieve the cost and performance needed to break into energy-significant power markets. This will occur most readily through cooperative alliances of suppliers, users, and researchers (both public and private) whose eventual common goal is to thrust PV technology into the bulk power market.

Government and industry R&D organizations can facilitate the formation of nationwide (or even international) alliances by helping to bring together parties with common interests. They can also help by underwriting some of the financial risk that is always present when investing in an uncertain future. But it is essential that the industrial partners take on the majority of the risk and responsibility, so that they retain a strong commitment to success in the commercial market. Put in other terms, too much outside support may soften their focus and impede progress; it may even preclude commercial success. However, too little outside support could also preclude commercial success.

Relative to solar thermal generation at $3,000 per kilowatt or wind at $1,000 per kilowatt, it would appear at first glance that PV faces a bleak future if its modules alone now cost $5,000 per kilowatt. But appearances are deceiving and perhaps incorrect, because PV holds a much greater potential for cost reduction than do any of the other technologies. Moreover, PV offers siting flexibility that solar thermal technologies do not: It needs no cooling water, which is a limited quantity in desert areas where insolation is most abundant and where the most economic solar power plants will be built.

8. DESIGN AND INSTALLATION OF UTILITY SCALE PV PLANTS

The design and installation of bulk power PV power plants will not be a complex process. By the time they are built at large scale, the process will be relatively straightforward because the technology is so much simpler than any other used to generate utility-scale power.

The steps involved will be the following: selecting the site, determining electrical capacity, determining tracking type, designing support structures for soil and wind loads expected, obtaining permits (perhaps more quickly than for other generation options because PV is environmentally benign), ordering components from the factory, building access roads, installing the electrical collection system (similar to an electrical distribution system), installing and connecting components, and energizing the plant as each segment is completed. Designing and building the roads and foundations, and perhaps the support structures, will be the most complex steps because they are the only ones that will vary from one site to another. Based on ARCO Solar's experience at Carrisa Plains, PV plant construction is essentially a factory mass production process utilizing standardized components. Of interest to utility companies, which have sometimes taken more than 10 years installing power plants, PV can be installed quickly. For example, ARCO Solar energized the first segment of its own Carrisa Plains plant less than a year after the decision to proceed.

9. CONDITIONS NECESSARY FOR BULK PV POWER

Unquestionably, the most important condition PV must satisfy is cheaper and more efficient modules. The technological path by which this will come about is not yet clear. It appears unlikely that a major breakthrough will suddenly provide PV at a much lower cost than current levels, although such an event is not impossible. More likely the process will be evolutionary.

From a corporate perspective, the conditions for PV to attain success are not entirely technical or technological. As in any business, companies such as utilities will make the right decisions about PV when they understand the implications involved. Information sometimes is not complete in a rapidly changing field like PV, but history suggests that large, unsuccessful investment decisions have been made in the PV business without the benefit of information that was available. Keeping current on emerging technologies is good insurance against making the wrong choices; it also enables an organization to know what the right choices are before everyone else has already made them.

10. SUMMARY

The field experience from more than 10 MW in over 200 grid-connected PV systems generally confirms the promise offered a decade ago: reliable power production whenever the sun shines. Although some problems have arisen with BOS equipment, as can be expected with any new technology, none of them has proven insurmountable; most important, the modules themselves have generally proven to be reliable. There is no technical barrier to the wide-scale adoption of PV as a bulk power source. Moreover, PV offers to utility companies a number of demonstrated advantages: modularity, simplicity, the potential for rapid installation, and virtually no adverse environmental impact.

PV's only barrier to wide-scale adoption for bulk power is the high cost of modules, and, with the exception of concentrators, no module technology has been demonstrated that will overcome this barrier. Studies do suggest that some cost-effective applications may be found for residential use and local grid support; if that is the case, they may provide what for the current PV industry is a relatively large market.

Several improvements on the part of the PV industry will help but cannot by themselves make PV acceptable as a bulk power source. These are more credible output ratings; on-time equipment delivery; and improved designs for BOS equipment like power conditioning units, diodes, fuses, switches, and intermodule wiring—all of which need to be as reliable as the PV modules.

Unless a major breakthrough reduces module costs, PV's prospects for bulk power generation will lag behind those of other, less expensive, renewable energy sources such as wind and solar heat. PV concentrators are an exception in that judicious engineering could result in more rapid cost reduction without a breakthrough. PV commercialization could be accelerated if a more complete economic treatment, which ascribes value to PV for *all* avoided costs associated with other options, was adopted by the public, by regulatory agencies, and by utility companies. In the absence of such treatment, PV markets for commercial grid connection will continue to grow only as fast as may be justifiable by PV cost reductions.

REFERENCES

Atmaram, G. (1993), *Performance and Reliability of Solar Progress Photovoltaic Plant, 1990–1991*, Report TR-102168, EPRI, Palo Alto.

Bigger, J.E., Kern, E.C., and Russell, M.C. (1991), *22nd IEEE Photovoltaic Specialists Conf. Record,* IEEE, New York, pp. 486–491.

Collier, D.E., and Key T.S. (1988), *20th IEEE Photovoltaic Specialists Conf. Record,* IEEE, New York, pp. 1035–1042.

DeBlasio, R. Mrig, L., and Waddington, D. (1990), *Interim Qualification Tests and Procedures for Terrestrial Photovoltaic Thin-Film Flat-Plate Modules*, Report SERI/TR-213-3624, NREL (formerly SERI), Golden, CO, pp. 4–13.

DeMeo, E.A., Goodman, F.R., Jr., Peterson, T.M., and Schaefer, J.C. (1990), *21st IEEE Photovoltaic Specialists Conf. Record,* IEEE, New York, pp. 16–23.

DeMeo, E.A., Weinberg, C.J., and Tassiou, R. (1992), *Proc. IEA/ENEL Exec. Conf. Photovoltaic Systems Electric Utility Applications,* OECD/IEA, Paris.

Durand, S., and Bowling, D. (1993), *Field Experience With Photovoltaic Systems: Ten-Year Assessment*, Report TR-102138, EPRI, Palo Alto.

Durand, S.J., Bowling, D.R., and Risser, V.V. (1990), *21st IEEE Photovoltaic Specialists Conf. Record,* IEEE, New York, pp. 909–913.

Eckert, P. (1987), *Lessons Learned and Issues Raised at Sky Harbor—A Utility-Interactive Concentrator Photovoltaic Project, 1987 Am. Soc. Mech. Eng. Solar Energy Conf.*, Honolulu.

Emslie, W.A., and Dollard, C.J. (1988), *Photovoltaic Resource Assessment in Colorado,* Report AP-5883, EPRI, Palo Alto.

Firor, K., Whitaker, C.M., and Jennings, C. (1990), *21st IEEE Photovoltaic Specialists Conf. Record,* IEEE, New York, pp. 932–936.

Goodman, F.R. Jr., DeMeo, E.A., and Zavadil, R.M. (1994), *Solar Energy Materials and Solar Cells* **35**, 375–386.

Hester, S.L., Townsend, T.U., Clements, W.T., and Stolte, W.J. (1990), *21st IEEE Photovoltaic Specialists Conf. Record,* IEEE, New York, pp. 937–943.

Hoff, T., and Iannucci, J.J. (1990), *21st IEEE Photovoltaic Specialists Conf. Record,* IEEE, New York, pp. 892–897.

Hoffner, J. (1989), *Proc. 1989 Conf. Am. Solar Energy Soc. Solar 89*, ASES, Denver.

Hohmeyer, O. (1988), *Social Costs of Energy Consumption*, Springer-Verlag, New York.

Holl, R. (1989), *Status of Solar-Thermal Electric Technology*, Report GS-6573, EPRI, Palo Alto.

IEEE (1986), *Recommended Criteria for Terrestrial Photovoltaic Power Systems*, ANSI/IEEE Std. 928–1986.

Iannucci J.J. and Shugar, D.S. (1991), *22nd IEEE Photovoltaic Specialists Conf. Record,* IEEE, New York, pp. 566–573.

Jennings, C., and Whitaker, C. (1990), *21st IEEE Photovoltaic Specialists Conf. Record,* IEEE, New York, pp. 1023–1029.

Knapp, C.L., and Stoffel, T.L. (1982), Direct Normal Solar Radiation Data Manual, NREL (formerly SERI), Golden, CO.

Knapp, C.L., Stoffel, T.L., and Whitaker, S.D. (1980), *Insolation Data Manual*, NREL (formerly SERI), Golden, CO.

Lepley, T. (1990), *21st IEEE Photovoltaic Specialists Conf. Record,* IEEE, New York, pp. 903–908.

Lynette, R., and Conover, K. (1989), *Photovoltaic Operation and Maintenance Evaluation*, Report GS-6625, EPRI, Palo Alto.

Mayorga, H., and Hostetler, R. (1986), *Photovoltaic Field Test Performance Assessment: Technology Status Report Number 4*, Report AP-4466, EPRI, Palo Alto.

Menicucci, D.F., and Fernandez, J.P. (1986), *Estimates of Available Solar Radiation and Photovoltaic Energy Production for Various Tilted and Tracking Surfaces Throughout the U.S. Based on PVFORM, a Computerized Performance Model*, Sandia National Laboratories, Albuquerque, NM.

Mon, G., Gonzalez, C., Willis, P., Jetter., E. Sugimura, R., and Ross, R. Jr. (1990), *21st IEEE Photovoltaic Specialists Conf. Record,* IEEE, New York, pp. 1043–1050.

Ottinger, R.L. (1990), *Environmental Costs of Electricity*, Oceana, New York.

Pern, F.J., Czanderna, A.W., Emery, K.A., and Dhere, R.G. (1991), *22nd IEEE Photovoltaic Specialists Conf. Record,* IEEE, New York, pp. 557–561.

Petersen, R.C., and Wohlgemuth, J.H. (1991), *22nd IEEE Photovoltaic Specialists Conf. Record,* IEEE, New York, pp. 562–565.

Pratt, R.G., and Burdick, J. (1988), *20th IEEE Photovoltaic Specialists Conf. Record,* IEEE, New York, pp. 1272–1277.

PVUSA Project Team (1991), *1989–1990 PVUSA Progress Report*, Pacific Gas and Electric Company, San Ramon, CA.

PVUSA Project Team (1992), *1991 PVUSA Progress Report*, Pacific Gas and Electric Company, San Ramon, CA.

Quinlan, F.T. (1981), *Typical Meteorological Year User's Manual*, National Climatic Data Center, Asheville, NC.

Rosenthal, A.L. (1991), *Photovoltaic System Performance Assessment for 1989*, Report GS-7286, EPRI, Palo Alto.

Rueger, G.M., and Manzoni, G. (1992), *Proc. IEA/ENEL Exec. Conf. Photovoltaic Systems Electric Utility Applications*, OECD/IEA, Paris.

Russell, M.C., and Kern, E.C. (1991), *Experiences and Lessons Learned With Residential Photovoltaic Systems*, Report GS-7227, EPRI, Palo Alto.

Shugar, D.S. (1990), *21st IEEE Photovoltaic Specialists Conf. Record,* IEEE, New York, pp. 836–843.

Smith, K. (1989), *Survey of U.S. Line-Connected Photovoltaic Systems*, Report GS-6306, EPRI, Palo Alto.

Solar Energy Research Institute (1983), *Solar Radiation Directory*, NREL (formerly SERI), Golden, CO.

Southwest Technology Development Institute (1990), *Data Acquisition for Photovoltaic Power Plants*, Report GS-7082, EPRI, Palo Alto.

Steitz, P. (1991), *Photovoltaic Power Conditioning: Status and Needs*, Report GS-7230, EPRI, Palo Alto.

Stolte, W.J. (1992), *Engineering and Economic Evaluation of Central-Station Photovoltaic Power Plants*, Report TR-101255, EPRI, Palo Alto.

Strategies Unlimited (1992), *Distributed PV Applications in the Utility Grid*, Report PM-36, Mountain View, CA.

Wenger, H.J., Schaefer, J., Rosenthal, A., Hammond, B., and Schlueter, L. (1991), *22nd IEEE Photovoltaic Specialists Conf. Record,* IEEE, New York, pp. 586–592.

Wiel, S. (1991), *Elec. J.* **Nov. 1991**, pp. 46–54.

Whitaker, C.M., Townsend, T.U., Wenger, H.J., Iliceto, A., Chimento, G., and Paletta, F. (1991), *22nd IEEE Photovoltaic Specialists Conf. Record,* IEEE, New York, pp. 608–613.

Wohlgemuth, J., and Petersen, R. (1992), *Proc. Photovoltaic Performance Reliability Workshop*, NREL (formerly SERI), Golden, CO, pp. 313–326.

CHAPTER SEVENTEEN

Silicon and Gallium Arsenide Space Systems

DENNIS J. FLOOD, NASA Lewis Research Center, Cleveland, OH 44135

1. INTRODUCTION

The first space solar array was carried aloft on Vanguard I on March 17, 1958. Vanguard I was the second satellite successfully orbited by the fledgling U.S. civilian space program, which had been initiated following the launch of Sputnik I by the U.S.S.R. on October 4, 1957. Sputnik I was powered by electrochemical batteries and worked for 21 days. The array on Vanguard I consisted of six photovoltaic panels mounted to the outer surface of the satellite and produced approximately 1 W of power for over 6 years. The cells on that first array were single crystal silicon, each 2×0.5 cm in size and about 10% efficient at 28°C. Since then, solar arrays have been the primary source of power generation for earth-orbiting satellites in the United States, Europe, and Japan.

Space solar arrays and satellite power systems have grown in size and complexity since 1958. The largest satellite solar power system flown by the United States thus far was on Skylab I, which was launched into low earth orbit on May 14, 1973. There were two separate solar arrays on Skylab I: the Apollo Telescope Mount (ATM) array and the Orbital Workshop (OWS) array. The ATM array was comprised of four deployable wings that produced a total of 10 kW of power in orbit at its beginning of life. The OWS array consisted of two deployable wings intended to produce a total of 12 kW at its beginning of life in orbit. An inadvertent partial deployment of the OWS tore off one wing during ascent to orbit. The remaining OWS wing and the entire

Solar Cells and Their Applications, Edited by Larry D. Partain.
ISBN 0-471-57420-1

ATM array operated completely successfully for the duration of the Skylab I mission.

Silicon solar cells have been the predominant cell type used on space arrays for over 30 years. There is now an extensive variety of silicon cell designs, sizes, and configurations that are available from commercial suppliers. It became necessary to develop this large "catalog" of cells because of the wide variations of mission requirements and operating environments that are usually encountered by mission planners. This often results in a unique set of requirements for optimizing each mission in terms of its capability, lifetime, or some other aspect of satellite performance. This uniqueness means that solar array (and hence power system) performance cannot be optimized by using a single silicon solar cell design for all missions. Prior chapters in this volume have given an indication of the wide range of silicon space solar cell types and their operating characteristics.

Gaining further improvements in solar array performance (i.e., higher efficiency, lower mass, longer life) now essentially requires the use of solar cells made from new materials. Each new cell type will enable space power system optimization for certain combinations of mission requirements. As is the case with silicon cells, one cell type will not work for all missions. The result is that silicon solar cells will not necessarily be replaced on all future missions; instead, the "new catalog" will have to expand to include the additional cell types that will become available over the next several years. There are many such cells in various stages of laboratory development (Flood, 1991, 1992). However, the only nonsilicon cells commercially available for space application at present are made of gallium arsenide. We shall discuss their use on space solar arrays in more detail later in the chapter.

The solar array is one subsystem of a full satellite power system. In this chapter we outline the basic elements of a complete space power system and briefly discuss some of the fundamental concerns involved when determining its size with regard to power level and mass. Most of the discussion, however, will focus on the solar array and will explore those factors that will result in an optimum array design for a given mission. We conclude the chapter with an elementary, but illustrative example of a satellite solar array trade study that demonstrates the impact new cell technology can have on array performance and satellite capability.

2. SPACE SOLAR POWER SYSTEMS

2.1. Power System Description

Figure 17.1 provides a schematic diagram of the fundamental aspects of a photovoltaic space power system. Each function is handled by a separate subsystem. Array output power is maximized by the orientation subsystem, subsequently gathered by the power collection subsystem, and routed in

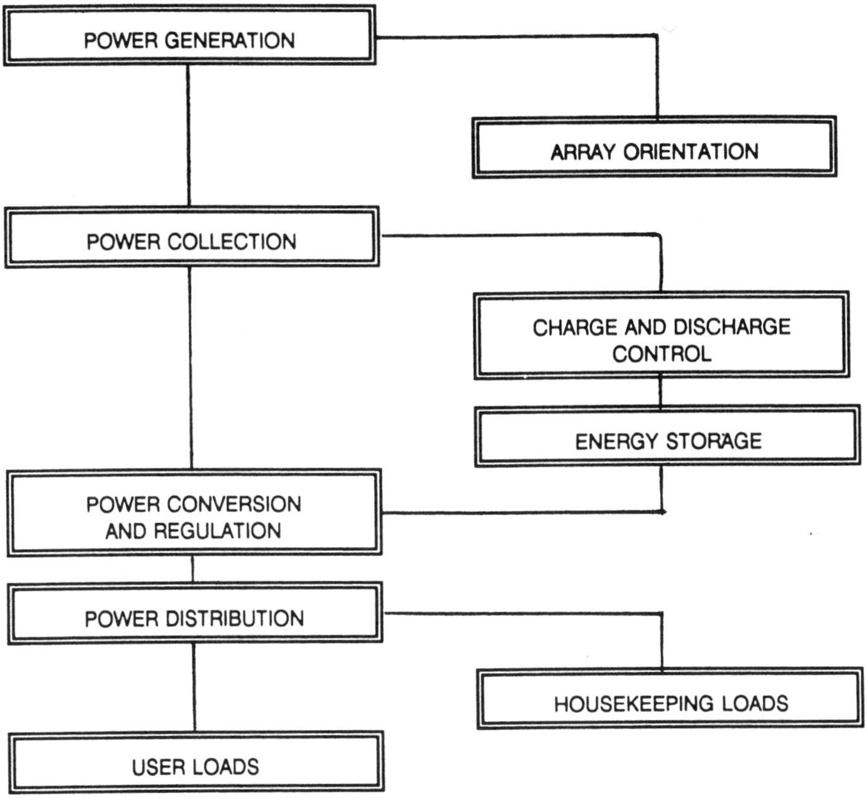

FIGURE 17.1 Schematic diagram of a generic space solar power system.

varying parts to the storage subsystem and the conversion and regulation subsystem. The power distribution subsystem takes the regulated power from either the array or the storage subsystem and distributes it to the user and housekeeping loads. The latter are those needed to operate the array orientation drive motors or needed for satellite subsystem status and monitoring circuits, operating cooling equipment, and so forth. It is readily apparent that power system design and development require a multidisciplinary approach and that there are complex interactions between subsystems. The specific details of the power system designs can vary considerably and are not pursued here. Those readers wishing an extensive bibliography of papers covering essentially every aspect of space power system design and development should consult the series of proceedings from the Intersociety Energy Conversion and Engineering Conferences, especially from the mid-1970s to the present.

There are two figures of merit that are used to measure the performance of a space solar array, as well as the entire power system: power per unit mass in watts per kilogram (W/kg) and power per unit area in watts per meter squared (W/m^2). These are referred to simply as *specific power* and *area power density*, respectively. (The inverses of these quantities are also often used and are known as *specific mass* (in kg/kW) and *specific area* in kW/m^2). Typical values for state-of-the-art (SOA) space solar arrays, using silicon solar cells mounted on rigid panels, are 30–40 W/kg and 90–110 W/m^2 at the start of the mission, or beginning of life (BOL). The end-of-life (EOL) values for any given array are dependent on mission time and location. Chief among the factors affecting the ratio of EOL to BOL is solar cell radiation damage caused by charged particle bombardment from electrons and protons trapped in the earth's geomagnetic field or from solar flares. This is followed by mechanical and electrical degradation of cells, interconnections and array components from other environmental effects, such as plasma interactions and thermal cycling. Elimination, or at least substantial mitigation, of such effects is at the heart of all space photovoltaic device and system research and development efforts.

2.2. Space Mission Requirements and Power System Technology Drivers

Future space missions will encompass a broad range of energy requirements, from several hundred watts to multihundred kilowatts, and have operational lifetimes ranging from less than 5 years to well beyond 20 years. Table 17.1 lists some important mission subsets, a qualitative assessment of their associated power requirements, and the key attributes a power system must have to be suitable for use within each mission subset. The key attributes for each mission subset have been listed in relative priority order, based on general mission concepts. The actual prioritization for any mission will, of course, depend critically on its final design and the result of mission and system trade-off studies.

TABLE 17.1 Mission Subsets

Mission Subset	Power Level	System Attributes
Unmanned near earth (LEO, HEO, GEO) and unmanned planetary	Low to intermediate	Low mass, long life
Space Station	High	Minimum area, low mass, low cost, long life
GEO Platform	Intermediate	Long life, low mass
Lunar Base, manned planetary	Intermediate to high	Low mass, portability, long life
Electric Propulsion Orbit Transfer	High	Reusability, minimum area, low mass

SOURCE: Flood (1989).

It is clear from Table 17.1, however, that low mass and long lifetime are important power system attributes in virtually all potential space missions. Power system cost and size have greater or lesser importance depending on the mission objectives and operational environment (low earth orbit, geosynchronous orbit, planetary surface, interplanetary, deep space, and so forth). For example, low total cross-sectional area is a critical attribute for the space station because of the effects of residual atmospheric drag (Baraona, 1990). The array, as the largest physical feature on the station, has a major effect on total life cycle cost because of the expense of constantly reboosting the station back to its desired altitude. Such considerations do not apply to a mission to the lunar surface, for example. In the latter case, although total area may be important because of other factors such as ease of construction and deployment, it will not be a primary driver in selecting a particular technology for the mission. Specific power and resistance to proton radiation damage from solar flares will certainly be among the more important factors for selecting a lunar surface solar array technology.

The desired power system attributes for each of the mission subsets can be transcribed into attributes at the subsystem and component levels, where they can serve as guides for developing new technology for future missions. At the array level, the associated attributes are low mass, high strength, environmental durability and compatibility, and, very often, minimum stowage volume. At the cell level, the system attributes of low mass, minimum area, and long life transcribe into high cell efficiency and radiation tolerance, with cell mass and cost usually having relatively lesser importance. The principal reason is that the solar cells are themselves a relatively lesser fraction of the total mass and cost of a full space power system, while their efficiency and usable lifetime are major determinants of the balance-of-system (BOS) mass and cost. Cell efficiency determines array area, which in turn determines array mass. Array lifetime is mission specific and is loosely defined to be length of time the array operates before its output power has fallen to a level below that needed to operate the satellite (or surface system) reliably. In general, array lifetimes are determined by the rates at which solar cell electrical output degrades, assuming that the mechanical aspects of the cell and array have been properly engineered to withstand thermal cycling, vibration, and other operational and environmental effects. The chief cause of electrical degradation is bombardment of the cell by constituents of the natural charged particle radiation environment. The ability of a solar cell to operate while, or after having been, subjected to such bombardment is a measure of its radiation resistance. The extent to which a solar cell is radiation resistant depends on many factors: the material from which it is made (i.e., silicon, gallium arsenide, indium phosphide), its actual device structure, and its ability to anneal (or be annealed) as the damage occurs. We shall investigate how cell efficiency and radiation tolerance affect the cost and performance of space solar arrays in a later section.

2.3. Space Power System Design

Design considerations vary depending on whether the satellite is intended for low earth orbit and has low power requirements (Dakermanji et al., 1991) or high power requirements (Winslow et al., 1989; Patil et al., 1990; Tam et al., 1991). Designs will differ for high earth orbit and geosynchronous satellites (Lovgren et al., 1989; Malachesky et al., 1991; Winter and Teofilo, 1989). Discussions of generic power systems and solar array considerations may be found elsewhere (Moser, 1990; Kenny et al., 1990; Bercaw and Cull, 1991; and Slifer, 1989).

One of the most significant areas where orbit location (LEO, HEO, GEO) affects the power system design is the storage subsystem and its interaction with the array subsystem. Battery charge and discharge cycles differ extensively, depending on the sunlight-to-eclipse ratio of the orbit. As a result, much of the attention on power system design is devoted to managing the storage subsystem. The storage subsystem must be designed to provide all the satellite power requirements during eclipse and to be fully recharged during the sunlight portion of the orbit. In low inclination, low earth orbits, the sunlight-to-eclipse ratio is about 2 to 1: approximately 1 h in sunlight followed by approximately 30 min in eclipse. Battery life is greatly affected by the depth of discharge and the rate of charge, while battery usefulness is measured in terms of energy storage capacity, in watt-hours/kg. The extent to which they can be discharged, and the rate with which they can be recharged, without causing irreversible damage to the electrodes must also be included. Requiring a smaller number of batteries to provide the eclipse power means that they will have to undergo a deep depth of discharge and then be fully recharged in less than 1 h. Increasing the number of batteries will reduce the depth of discharge required, but will greatly increase the total mass of the storage subsystem. This trade off must be made carefully, since the storage subsystem mass is often the largest part of the total system mass. A simple example can illustrate the situation.

At present, NiCd batteries are the ones most often used for space applications. They are commercially available with energy densities in the range of 10–20 W-h/kg. For a "typical" LEO eclipse time of 30 min, a complete discharge (i.e., 100% depth of discharge [DOD]) over the 30 min results in a battery-specific power of 20–40 W/kg. This figure will, of course, be made lower by the mass of the rest of the storage subsystem components (regulators, thermal control elements, wiring, and so forth). It will also be made lower by using a lower DOD, a necessary requirement to preserve battery function. At 20% DOD, commonly used for space NiCd batteries to guarantee a 5 year life in a low earth orbit (Hord, 1985), the specific power ranges from 4 to 8 W/kg. If the satellite requires 2 kW of power to function, the battery mass will range from 67 kg to 100 kg at 100% DOD and from 250 kg to 500 kg at 20% DOD.

This should be contrasted with the mass of the solar array. Estimating the array mass must take into account the battery charging requirements along with the load requirements. The energy consumed from the batteries by the loads during the eclipse in our example is 1 kW-h, regardless of the battery DOD, and must be replaced by the array during 1 h of sunlight operation. Hence, without allowing for any losses during battery charging or in the power management and distribution circuits, the array must be able to generate at least 3 kW of power. (Typically, inefficiencies in the battery charging and load distribution circuits require the array to have about twice the power generating capability needed to operate everything.) Current rigid panel solar arrays on U.S. satellites have specific powers in the range of 15–30 W/kg (Flood et al., 1989). The array mass in this example ranges from 85 to 120 kg, which is one-third or less of the battery mass at 20% DOD.

The situation is somewhat different for GEO applications. A LEO satellite will complete nearly 6,000 orbital revolutions every year. A satellite in a circular geosynchronous equatorial orbit will complete 365 orbits per year, but will undergo only about 90 eclipses (Rauschenback, 1980). The maximum eclipse time is about 1 h, but since so few battery cycles are required a much greater DOD may be used. In addition, the charging may take place over a 23 hour period, which will help to preserve battery life. The storage subsystem mass may thus actually be lower than that required for LEO, depending on the DOD allowed. Although the array need not be oversized by as large a factor as for LEO, it must now operate in a charged particle radiation environment (primarily electrons trapped in the earth's geomagnetic field) which will degrade its output significantly over time. Arrays in low inclination LEO will experience very little radiation damage degradation, since the density of charged particles that can be trapped there is very low (Tada et al., 1982). The array must be oversized at the beginning of the mission by an amount at least equal to that expected to be lost because of radiation damage. As mentioned, this important aspect of power system design will be discussed later.

3. SPACE SOLAR CELL PERFORMANCE VERIFICATION

3.1. The Space Solar Spectrum

Proper design of a space photovoltaic power systems depends critically on the ability to predict accurately the performance of its component solar cells in the space environment. Given the inconvenience and cost of actually measuring solar cell performance in space, the usual procedure is to perform such measurements in the laboratory using a xenon arc lamp as a solar simulator. Figure 17.2 contains a plot of the solar spectrum outside the earth's atmosphere, known as the air-mass zero (AM0) spectrum (given in Appendix B). Figure 17.2 also contains a plot of the terrestrial solar spectrum taken at solar noon (air-mass 1 [AM1.5G], see Appendix A), under prescribed conditions of atmospheric clarity and moisture content. The differences between the

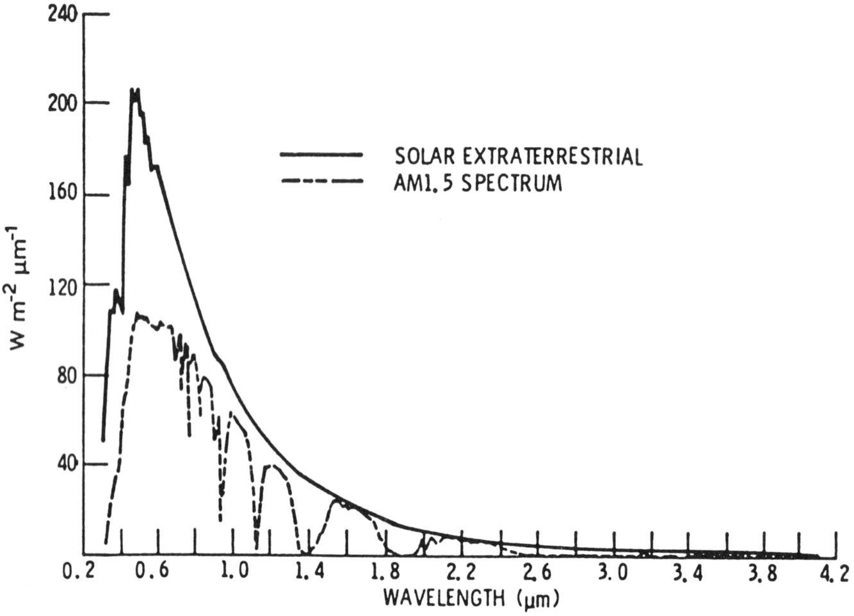

FIGURE 17.2 Solar spectral irradiance.

terrestrial spectrum and the AM0 spectrum are too great and too variable, because of uncontrollable meteorological conditions, to allow use of the former in predicting space solar cell performance with confidence. The match between the xenon arc lamp spectral intensity distribution and that of the AM0 spectrum is much better. Nonetheless, considerable care must still be exercised when using laboratory solar simulator measurement results to predict AM0 performance values.

3.2. Laboratory Measurement Techniques

The first step in the procedure is to obtain a standard cell, the total output of which has actually been measured in the AM0 spectrum. The output of this standard cell is then used as a reference against which to adjust the intensity of the solar simulator. The proper simulator intensity is reached when the reference cell has the same output as obtained in the AM0 spectrum. Before the measurement of the unknown cell can proceed, however, it is also necessary to determine the spectral response of both the standard cell and the unknown cell to make certain that they are closely matched. The spectral response of a solar cell is its output under uniform illumination as a function of the wavelengths that span the AM0 spectrum. The matched spectral response is required because the xenon arc lamp and AM0 spectra have important differences, most notably in the longer wavelength region where the typical solar simulator spectrum has several large spikes. Differences also exist in the short wavelength region, where the AM0 spectrum is somewhat more intense than that of the simulator. If a reference cell that has a very strong spectral response at the shorter wavelengths and a relatively weaker response at the longer wavelengths is used to set the simulator output, the resulting spectrum will be "red rich," with an artificially high intensity in the long wavelength region. If the cell to be tested has the opposite relative values of spectral response, i.e., strong in the red and weaker in the blue, it will appear to have a higher output than it should. The difference can amount to several percent, depending on the degree of spectral response mismatch between the cells. The opposite condition can also occur, giving lower values for the unknown cell as well.

The net result of the above is that it becomes necessary to maintain a large library of AM0 calibrated solar cells of all sorts, with as many different types of spectral response as can be obtained. The problem is compounded because the radiation damage caused by charged particle bombardment affects the spectral response of solar cells in many different ways, causing it to vary differently from cell to cell, depending on details of the cell structure and material (doping density, layer thickness, and so forth). Ensuring accurate performance prediction for any particular cell type as a function of charged particle radiation fluence thus requires the availability of radiation-damaged, spectral response–matched standard cells in the library for each different cell material and structure as are likely to be encountered.

3.3. AM0 Calibration Techniques

The need for large numbers of standard cells cannot be met by actual space flight calibration measurements, because of both the limited access to and the cost of even the simplest space missions. An alternative is to make the required measurements at very high altitudes, where the atmospheric attenuation is a very small correction and thus the results are unaffected by water vapor or other meteorological conditions. The measurements are made using either a high altitude balloon (Anspaugh et al., 1992) or a high altitude aircraft (Brinker, 1993). Because the aircraft measurements are made at altitudes below 50,000 feet, a small correction must be made for the light absorbed by the ozone in the residual atmosphere above that altitude. The full I–V characteristics of the cells can, however, be measured under precisely controlled temperature conditions. On the other hand, while altitudes approaching 100,000 feet can be reached for the balloon measurements, essentially eliminating the need for atmospheric corrections, measurements are restricted only to a determination of the cell's output near its short circuit operating condition. In addition, the measurements are not made under controlled temperature conditions. The temperature at which the measurements are made is determined, however, and a correction made to the data to provide cell output current at the standard reporting condition of 25°C. Both techniques give results that are in essential agreement with each other, and both have been used to build up the library of standard cells that NASA uses to make accurate laboratory measurements.

4. SPACE SOLAR ARRAY TECHNOLOGY

4.1. Rigid Arrays

Rigid solar arrays presently in use are essentially of two types: a body mounted design (i.e., the cells are attached directly to the body of the spacecraft) and a deployable panel design, such as the OWS array on Skylab. Deployable rigid panel arrays are comprised of one of more panels (hinged together if there are more than one) and can reach power levels of 10–20 kWs. The specific power is typically in the range of 25–35 W/kg at BOL (Bennett et al., 1990), although early versions were often less than 10 W/kg (Rauschenback, 1980). The panels on most U.S. satellites that have been launched since the early 1970s are made of a sandwich of two face sheets bonded to an aluminum honeycomb core. The face sheets are either of thin aluminum or glass/epoxy material, and the panel total thickness is typically 1 cm or less. If required, an insulating sheet is bonded to the panel surface on which the solar cells are to be placed to provide for electrical insulation between the cells and the panel.

The cost of satellite design, development, and space qualification is such that once a satellite/array configuration has been successfully flown, the satellite manufacturer becomes reluctant to make drastic changes in it. Consequently, every effort is made to meet the need for additional power for a particular mission by making only small changes in the design or size of the solar array, or in the overall array configuration (such things as array aspect ratio and total deployed area). In many cases, the area available for solar cell deployment may be restricted by the size of the spacecraft, as it is on single axis stabilized vehicles. The mechanical problems associated with deploying large, rigid panel arrays and the subsequent need for managing a large moment of inertia on orbit also restrict the size of that type of array and limit the amount of power available for future missions. Once all the small, relatively low cost changes to the basic satellite design have been incorporated, the only option for increasing array power is to use advanced, higher efficiency solar cells. Until that point is reached, however, the cost of using advanced solar cells compared with the cost of silicon cells is a major consideration; the increased cell cost is constantly traded against the cost of making any changes in satellite design. New cells will simply not be used unless it is cost effective to do so, or the mission cannot be accomplished in any way without using them. In the latter case cost might not be an issue at all.

4.2. Flexible, Lightweight Planar Arrays

An alternative to rigid deployable arrays has been developed and flown that uses thin, flexible panels in one of two configurations: (1) flat panels hinged together so they fold in an accordion-style arrangement during stowage and (2) a continuous, roll-up version that is stowed around a drum inside a cylindrical container (Rauschenback, 1980). The Hubble Telescope uses a roll-up array, for example, and the Space Station Freedom will use a fold-up array. Specific powers for these designs range from about 25 W/kg for early roll-up versions to 66 W/kg for the fold-up design; the latter had been developed to supply power for an electric propulsion drive on a satellite intended for a comet chasing mission (Rauschenback, 1980). The mission was never flown, and the so-called SEP (Solar Electric Propulsion) array was never used, although it was tested in space on a 1984 shuttle flight. The Space Station Freedom array is a derivative of that array, but has a lower specific power because of changes required in the array stowage container to make its operation compatible with the safety requirements of a manned mission. A higher specific power version of the SEP array has since been developed by NASA, the Advanced Photovoltaic Solar Array, or APSA, for short (Kurland, and Stella, 1989). The specific power at BOL exceeds 130 W/kg compared with a typical value of 35 W/kg for rigid arrays and is twice that for the SEP array. The area power density is essentially the same, since it is a function primarily of cell efficiency, although array design can affect it as well (such things as packing fraction of the solar cells, so-called dead area required for other purposes such as wiring and hinge attachment, and so forth). A schematic diagram of the array is shown in Figure 17.3 (Stella and Flood, 1989).

Two key features of this array design are (1) for about 1 kW sizes and above, about half the total mass is taken up by the storage container, the mast, and the deployment system (e.g., motors); and (2) the stowage volume is very

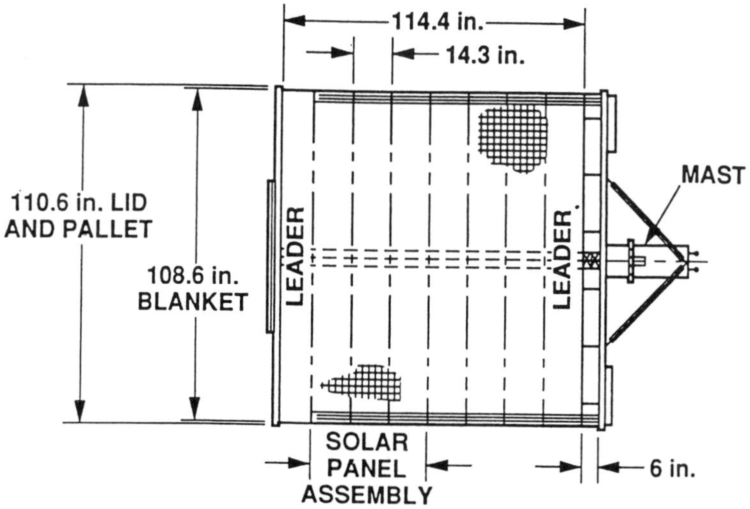

FIGURE 17.3 Schematic diagram of the NASA/JPL 1 kW prototype advanced photovoltaic solar array. SOURCE: Stella and Kurland (1989).

compact. In general, the masses of the stowage container and deployment mechanisms do not change significantly as the array size increases, while that of the blanket and mast increase in a nearly linear fashion to about 75 kW. The result is that array-specific power increases dramatically as array output power grows from a few hundred watts to about 1 kW and is then approximately constant to the 75 kW level. It is this feature that makes this array technology attractive for those future missions requiring both increased power and reduced mass. The specific power eventually saturates, however, and may even fall as power levels grow beyond the above range. As with rigid arrays, the use of advanced solar cells eventually becomes the only way to improve array performance. The issue again becomes one of cell cost and mass, since the array design permits growth in array size without any redesign or development cost.

5. THE SPACE ENVIRONMENT

5.1. Space Radiation

Before entering into a discussion of array cost and performance trade offs, it is important to understand the factors that affect array performance in the first place. One of the most important of these is solar cell degradation caused by charged particle bombardment from the naturally occurring space radiation environment. Figure 17.4 depicts the region of space in the general vicinity of the earth known as the *magnetosphere*. The radiation environment near the earth consists primarily of electrons and protons trapped in the geomagnetic field, along with transitory corpuscular radiation (atoms, ions, and protons) associated with solar flares. There is a lesser contribution from galactic cosmic rays. The earth's (approximately) dipole magnetic field is responsible for the radiation belts near the earth, trapping and holding the charged particles for long periods of time, thereby forming a low-density plasma in an inhomogeneous field. The plasma is in a dynamic state, with particles entering and escaping at all times. Solar flares are the source of particles entering the plasma. Most of the solar flare protons and high-energy electrons simply pass through the magnetosphere, while some with lower energies are trapped in the dipole field. Escape from the magnetosphere can also occur along the magnetotail.

Figure 17.5 shows the distribution of charged particle types in near earth space, for altitudes of one earth radius and higher. The densities and energies of the trapped protons make them the predominant cause of solar cell degradation below about four earth radii, with a small contribution from the inner zone electrons. The situation is reversed at geosynchronous altitude, where the effects of trapped protons may essentially be ignored and the damage is caused by the trapped, or outer zone, electrons. Energies of the particles trapped in the two regions range to a few MeV, while solar flare protons will have energies up to several hundred MeV. Measurements of solar array

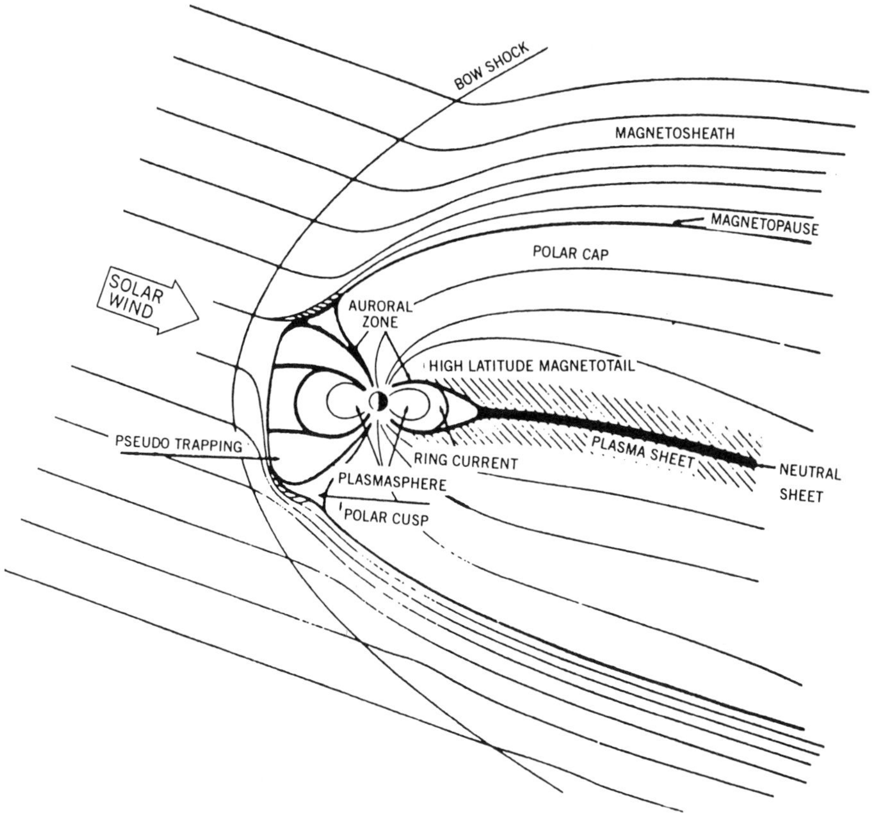

FIGURE 17.4 The earth's magnetosphere. SOURCE: Tada et al. (1982).

degradation in geosynchronous orbit, where at least some geomagnetic shielding occurs, have shown that solar flare protons account for about the same amount of power loss over a full solar cycle (11 years) as the trapped electrons (Weinberg, personal communication). The only difference is that the trapped electrons result in a steady, cumulative loss of power, while the solar flare protons cause a sudden, step-like drop in output. A single solar flare event may last up to a few days, but the total fluence absorbed by the solar cell can be sizeable. For more information specifically about the near-earth radiation environment, including computer codes to model it, see Strassinopolous and Raymond (1988) and Sawyer and Vette (1976).

5.2. Other Environmental Effects

Other factors that affect array performance are its electrical and mechanical interactions with the local space environment in which it is to operate. The effects of such interactions are strongly dependent on orbit altitude, and to

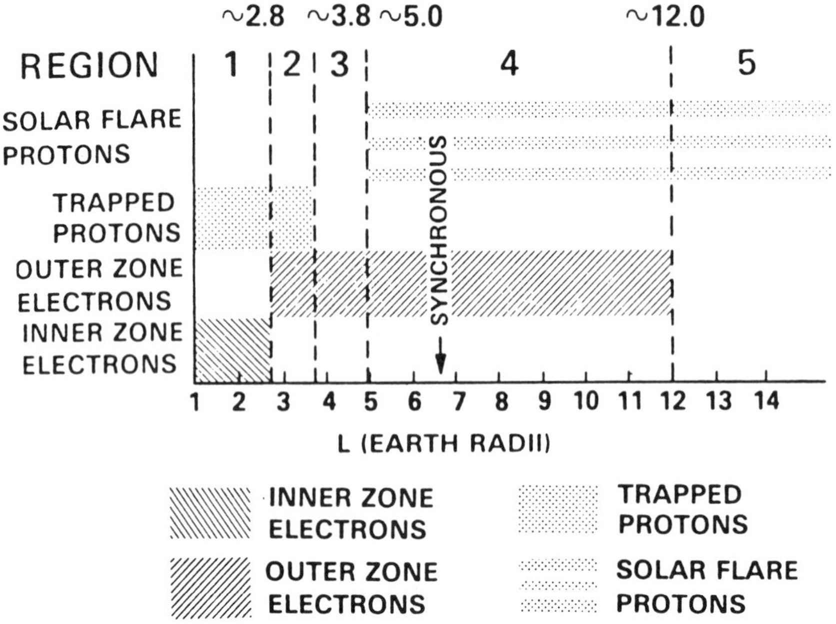

FIGURE 17.5 Charged particle distribution in the magnetosphere. SOURCE: Stassinopoulos and Raymond (1988).

some extent on orbit inclination. They include potential mechanical failure from thermal cycling as the array moves into and out of the solar eclipse each orbit (Scheiman and Smith, 1991); damage caused by micrometeroid bombardment and attack by atomic oxygen (Rutledge et al., 1991); array power drain from current leakage into the space plasma (Ferguson, 1991); arcing caused by electrostatic charge build-up on insulating surfaces (Fredericks and Scarf, 1973); contamination from other spacecraft subsystems (Roux and McKay, 1984; Marvin et al., 1988); and interactions with the earth's magnetic field (Atzei and Capart, 1972; Rauschenback, 1980). All of these effects may be mitigated or eliminated by appropriate engineering design and, since they are essentially independent of solar cell type, will not be pursued further here.

6. RADIATION EFFECTS ON SPACE SOLAR CELLS AND ARRAYS

6.1. Basic Theory of Solar Cell Operation

The preceding brief discussion serves to illustrate the need to understand radiation damage effects in solar cells caused by both electrons and protons. Although a cell may encounter primarily only one or the other of the two particles on any given mission, the same cells should be usable in both environments. Some very elliptical orbits may even result in significant exposure to both electrons and protons during the same mission. In this section the basic theory of solar cell operation and the effect on cell performance of radiation-induced defects are reviewed, and the concept of damage equivalence is discussed. It concludes with a brief summary of the present status of knowledge for a variety of solar cell types.

A simple p n junction solar cell can be thought of as a diode operating in parallel with a constant current source as shown in Figure 17.6. For completeness, the circuit also includes the internal series and shunt resistances, R_s and R_{sh}, associated with an actual device. The relationship between the current and voltage at the output terminals of the solar cell is (Sze, 1981)

$$I = I_s\{\exp[(q/kT)(V - IR_s)] - 1\} - I_L + (V + IR_s)/R_{sh} \qquad (1)$$

where I_s is the diode saturation current, and I_L is the light-generated current in the solar cell. The diode saturation current is given by

$$I_s = Aq[(D_p p_{no}/L_p) + (D_n n_{po}/L_n)] \qquad (2)$$

q is the value of the electric charge, D_p is the diffusion coefficient for holes in the n region, D_n is the diffusion coefficient for electrons in the p region, p_{no} is the equilibrium concentration of holes in the n region, n_{po} is the equilibrium concentration of electrons in the p region, A is the total area, and L_p and L_n are the diffusion lengths of holes and electrons, respectively.

It is well documented that shunt resistances as low as 100 Ω have little effect on the output of a solar cell. Hence the last term in Eq. (1) may be neglected. Under this approximation, the output power of the device is simply

$$P = IV = I\{[kT/q]\ln[(I + I_L)/I_s + 1] - IR_s\} \qquad (3)$$

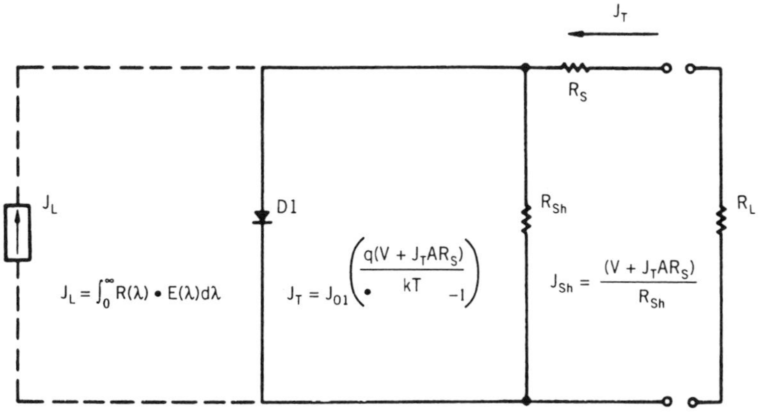

FIGURE 17.6 Circuit representation of a solar cell.

Under short circuit conditions, $V = 0$ and Eq. (1) becomes

$$I_{sc} = I_s\{\exp[(q/kT)(-I_{sc}R_s)] - 1\} - I_L \tag{4}$$

Under one sun conditions, I_{sc} is essentially equal to I_L for low values of R_s, and Eq. (4) becomes

$$\ln[2I_{sc}/I_s + 1] = qI_{sc}R_s/kT \tag{5}$$

The preferred structure for a space silicon solar cell is a thin ($\leqslant 0.5\,\mu$m) n-type emitter on a thick ($\geqslant 200\,\mu$m) p-type base. In such a cell, the base component of short circuit current predominates for all but the shortest wavelengths of the incoming solar illumination. For the case of minority-carrier electrons in a p-type material,

$$I_s = qAD_n n_{po}/L_n \tag{6}$$

where A is the cell area. In addition, $I_{sc} \gg I_s$, and Eq. (5) becomes

$$\ln(1/I_s) = qI_{sc}R_s/kT + \ln 2I_{sc} \tag{7}$$

from which

$$I_{sc} = (kT/qR_s)[\ln(L_n/qD_n n_{po}) + \ln(2I_{sc})] \tag{8}$$

Equation (8) will be useful when we consider the effects of radiation damage on the cell. As we shall see then, the short circuit current is a slowly varying function of the particle fluence, and to a reasonable approximation Eq. (8) may be rewritten in the form

$$I_{sc} = a\ln(L_n) + b \tag{9}$$

where a and b are constants.

From Eq. (1) the open circuit voltage is simply

$$V_{oc} = (kT/q)\ln[I_{sc}/I_s + 1] \tag{10}$$

where I_L has again been approximated by I_{sc}, an approximation valid for the cells of concern here. Again, it is a straightforward matter to show that, using Eq. (9),

$$V_{oc} = (kT/q)\{[b + a\ln(L_n)]/AqD_n n_{po}/L_n\} \tag{11}$$

Equation (11) gives the dependence of open circuit voltage on diffusion length and shows it to be a somewhat more complex relationship than between the short circuit current and diffusion length.

The maximum power out of the solar cell under illumination is given by

$$P_{max} = I_m V_m \tag{12}$$

where I_m and V_m are the current and voltage at the maximum power point. The ratio of current and voltage at the maximum power point to I_{sc} and V_{oc} is known as the fill factor, FF:

$$FF = I_m V_m / I_{sc} V_{oc} \tag{13}$$

from which

$$P_{max} = FF(I_{sc} V_{oc}) \tag{14}$$

The efficiency (η) of the cell is simply the ratio of maximum power out to the total power in, or

$$\eta = P_{max}/P_{in} = FF(I_{sc} V_{oc})/P_{in} \tag{15}$$

Equation (14) or (15) may be used to predict the performance of a solar cell as a function of diffusion length. As we shall see in the following paragraphs, radiation damage affects primarily the diffusion length of the cell; the above equations should therefore predict cell behavior as a function of radiation damage, provided the fill factor is relatively unaffected by it.

It is well known that the lifetimes of the excess electron and hole minority carriers in a solar cell, under low-level injection conditions, are inversely dependent on the density of recombination centers present in the cell for each carrier type. Bombardment of the cell by high-energy electrons and protons increases the number of recombination centers by creating defects in the solar cell material. To a first approximation it can be assumed that the density of additional recombination centers at any instant during the exposure is proportional to the total accumulated fluence. The minority-carrier lifetime (τ) is then given by

$$1/\tau = 1/\tau_o + c\phi \tag{16}$$

where t_o is the initial lifetime, ϕ is the fluence (number of particles/cm^2), and c is a constant for a given material and particle type. Since the diffusion length is defined as

$$L = (D\tau)^{1/2} \tag{17}$$

Eq. (16) may be rewritten as

$$(1/L)^2 = (1/L_o)^2 + K\phi \tag{18}$$

where L_o is the initial diffusion length, and $K = c/D$ is the diffusion length damage coefficient. In general, K will be different for different types of particle radiation and can vary to a greater or lesser degree as a function of the doping density and type of dopant in the semiconductor. It will vary as a function of the energy of the incoming particles. If the diffusion coefficient varies only slowly with accumulated fluence, K can be considered to be approximately constant for a given energy and type of particle radiation. The same can be said if D is a slowly varying function of doping concentration as well. The assumptions work reasonably well in silicon solar cells, so that for a given dopant and incoming particle energy it is adequate to specify one damage coefficient for electrons and a separate coefficient for protons for the entire range of doping concentrations of normal interest for solar cell fabrication. There are as yet insufficient data to determine whether the same assumptions are reasonable for other than silicon solar cells, such as those made from the III-V compounds. It is common practice to make the assumptions, however, and to use the results to help compare and understand the radiation resistance of even those materials.

Inserting Eq. (18) into Eq. (9) yields the following expression:

$$I_{sc} = b - (a/2)\ln(K\phi + 1/L_o^2) \tag{19}$$

which predicts that the short circuit current is expected to vary as the natural logarithm of the total fluence. Similarly, the expression for the diode saturation current in Eq. (6) becomes

$$I_s = qD_n n_{po} A[K\phi + (1/L_o)^2]^{1/2} \tag{20}$$

and combining Eqs. (20), (19), and (10) gives

$$V_{oc} = (kT/q)\ln\{[b - (a/2)\ln(K\phi + 1/L_o^2)]/qD_n n_{po}[K\phi + (1/L_o)^2]^{1/2}\} \tag{21}$$

for the open circuit voltage. This more complex relationship can be simplified by noting that the logarithm of the fluence, which appears in the numerator, is a more slowly varying function that the fluence itself, so that the variation of V_{oc} as a function of fluence can be written approximately as

$$V_{oc} = \text{const.} - (kT/2q)\ln\{qD_n n_{po}[K\phi + (1/L_o)^2]\} \quad (22)$$

Combining Eqs. (19) and (22) with Eq. (15) results in an expression of the form

$$P_{max} = \text{const} - g\ln[K\phi + (1/L_o)^2] \quad (23)$$

where g is (to a good approximation) a constant, and we have neglected the square of the logarithm compared with the first power.

Although the preceding equations for I_{sc}, V_{oc}, and P_{max} will not necessarily allow precise numerical prediction of the rate at which any given space solar cell output will decrease as a function of radiation fluence (because of all the limiting assumptions that were made, e.g., p-base cell with thin emitter), they do provide at least a good qualitative description of the effects of radiation damage in such devices. The expressions are also useful in introducing the concept of damage equivalence, by which the different amounts of damage caused by a given particle type (electron or proton) at different energies can be related to the damage caused by the same particle at a single energy. Equation (23) can be written in the form

$$P_{max} = P_{max\,o} - \text{const}\ \ln(1 + \phi/\phi_c) \quad (24)$$

where ϕ_c is an arbitrarily chosen constant referred to as the critical fluence. The critical fluence is simply the value of fluence at a given particle energy that will degrade the solar cell output to some chosen level. It will vary with the energy of the particles and with the doping density of the material. Once the variation of ϕ_c with energy is known for a given dopant, however, it is then possible to determine cell degradation at one energy and, with the use of ϕ_c, determine the total fluence required at all other energies to cause the same amount of degradation. Figure 17.7 shows the variation of the electron critical fluence for various base doping densities in n/p silicon solar cells. In this case, the critical fluence has been defined to be that for which the output has been reduced to 75% of its pre-irradiation value. Plots such as this one are not available for the III–V compound or thin-film solar cells.

A major restriction on the use of the concept of damage equivalence is that the electron or proton energies must be high enough to cause damage uniformly throughout the solar cell. That will be the case for all trapped electrons, but, for protons with energies low enough to be stopped completely, in the cell, the use of a simple damage equivalence, where the amount of damage that occurs at any arbitrarily chosen energy may be used as the reference, will no longer be valid. A modified form of the concept may be used that relates the damage at all proton energies to that which occurs at the energy for which the relative damage is greatest in actual space environments.

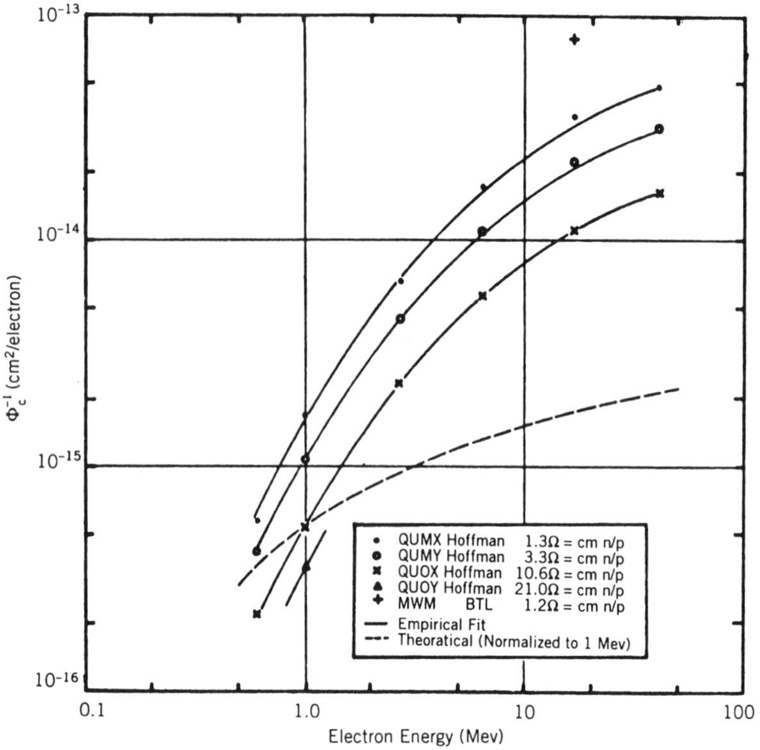

FIGURE 17.7 Energy dependence of the electron critical fluence for n/p silicon solar cells. SOURCE: Downing et al. (1964).

For silicon cells, the damage caused by 10 MeV protons is used for that purpose (a more detailed discussion can be found in Tada et al., 1982). In general, the same equations as derived above for I_{sc}, V_{oc}, and P_{max} can be used, but the constants, determined from empirically fitting the data to the form of the equations, are somewhat greater than for 1 MeV electrons in the same material. For this reason, the concept of equivalence between electron and proton radiation damage must be used with care, since it is basically an approximation. The relationship between the critical fluence for 1 MeV electrons and 10 MeV protons has been found to be (Tada et al., 1982)

$$\phi_c(1\,\mathrm{MeV\,e^-}) = 3{,}000\,\phi_c(10\,\mathrm{MeV\,p^+}) \tag{25}$$

for n/p silicon cells, essentially independent of doping density.

The numerical factor has been found to be 1,000 for the same relationship in GaAs cells (Anspaugh and Downing, 1984) and 650 for InP cells (Takamoto et al., 1990). The latter have junctions fabricated using the closed tube diffusion technique. Data of this sort are not yet available on InP cells made using epitaxial growth techniques, nor, for that matter, on most of the advanced cell types currently under development.

6.2. Experimental Results on Cells

The preceding brief discussion illustrates the need for more extensive studies of the effects of radiation damage in solar cells, particularly, as we shall see below, for other than single crystal silicon cells. From the viewpoint of atomic physics the situation is quite clear. There are two fundamental interactions between the incoming electrons or protons and the atomic constituents of the semiconductor material: inelastic interactions with the bound atomic electrons in the solid, or elastic collisions with the atoms of the solid. The first interaction, which results in ionization, is transient in nature and constitutes the primary energy loss mechanism for the incoming particles; the second can result in atomic displacement in the solid, provided the incoming particle energies are above a threshold value that is characteristic for each semiconductor type. It is the second interaction that affects the performance of solar cells, since the defects created by atomic displacement may become recombination centers for the light-induced minority carriers, reducing their lifetime and lowering cell output.

While much can be done to characterize the performance of solar cells as a function of radiation fluence, so that reliable performance predictions may be made for array design purposes, such as approach is of little help in attempting to improve their radiation resistance. As was shown above, the cell parameter most important in determining the output of a solar cell as a function of radiation damage is the diffusion length. Attempts to improve cell performance must start by developing an understanding of the nature of the defects that are created by atomic displacement and determining which of them have the greatest impact on the cell diffusion length. Then it is necessary to introduce ways either to reduce the rate at which defects are introduced into the cell material or to anneal them. The ideal annealing technique would be one that does so on a continuous basis at normal operating temperature, while power is being generated by the cell during irradiation. Acceptable, but not as desirable from the viewpoint of spacecraft operations, is the ability to anneal the damage with simple techniques and at the modest temperatures that are compatible with most array structural materials. Typically $\leqslant 100°C$ is desirable.

It has been shown in n/p silicon space solar cells that there are three defects that have the largest impact on minority-carrier diffusion length. The most important one is a boron–oxygen interstitial complex (Weinberg et al., 1984a, b). Its creation starts when a mobile interstitial silicon ion, created by the primary radiation damage event, is captured by a boron substitutional ion, subsequently replacing the boron and ejecting it into an interstitial site (Watkins, 1974). The boron then forms a complex with an interstitial oxygen ion, and the result is a stable recombination center. (The oxygen is essentially always present at low densities as an unwanted impurity in silicon.) The B_i–O_i defect can be reduced by thermal annealing at temperatures above 200°C, too high to be of practical use for restoring power on a space solar array. Even worse, the defect changes to a boron–oxygen–vacancy complex, which does not anneal until temperatures above 400°C are reached. Keeping the unwanted oxygen content of the silicon below $5 \times 10^{15}\,cm^{-3}$ has been found to reduce the extent of the radiation-induced performance degradation at all 1 MeV electron fluences normally used in laboratory tests (up to $10^{15}\,cm^{-2}$) (Weinberg and Swartz, 1980). Furthermore, introduction of low levels of lithium, which is an n-type dopant in silicon, into the p base of the cell has been found to reduce significantly the amount of power loss over the same range of fluences and to allow substantial annealing of the cells at 100°C (Weinberg, 1984a, b). The accepted explanation is that the lithium selectively combines with some of the oxygen ions and prevents them from entering into the reaction that forms the recombination centers described above (Weinberg et al., 1984a, b). Of the other two major radiation-induced defects in p-type silicon that are known to affect cell performance, one is the divacancy while the other is most likely a vacancy–oxygen–carbon complex (Weinberg et al., 1984a, b). (Carbon is also a common unwanted impurity in silicon.) Again, the primary event is the creation of silicon interstitials by the incoming electrons. Although the defect structure becomes much more complex with the addition of lithium to the silicon lattice, it appears that it may also combine with single vacancies and divacancies to some extent, and inhibit the formation of the corresponding recombination centers as well (Weinberg et al., 1984a, b).

Silicon space cells appear to be satisfactorily understood (Flood and Brandhorst, 1987). That is not at all the situation for cells made from any of the III–V, thin-film, or II–VI materials. In many cases there are very few measurements of I_{sc}, V_{oc}, or P_{max} as a function of fluence, whether for 1 MeV electrons or protons of any energy. Data are beginning to accumulate sufficiently for GaAs cells, however, to be able to determine damage coefficients for both electrons and protons for a few of the more commonly used dopants and establish the validity of using the concept of damage equivalence to predict actual on-orbit cell performance. The defect structure in the compound semiconductors is considerably more complex than in silicon and considerably harder to understand, with the result that only limited progress has been made to date. A recent review by Weinberg (1990) provides an excellent summary of the situation and illustrates the complexity of issues surrounding the subject.

Detailed studies of the radiation-induced defects in high-quality epitaxially grown InP cells have only recently begun and are not discussed here. A recent review, again by Weinberg (1991), indicates that a great deal of work is still required to identify conclusively the types of defects that are formed. Extensive work, however, has shown that electron radiation damage is readily annealed thermally at low temperatures (approx. 100°C), and by forward bias injection of carriers (Yamaguchi et al., 1986; Sugo and Yamaguchi, 1989; Sugo et al., 1990). The results suggest that continuous annealing of the damage as it occurs, even in the severe environment of the Van Allen radiation belts, may be a distinct possibility. Clearly, a great deal of additional work is required before we have a complete understanding of the basic reasons for the superior radiation resistance of this material. Its full development into a radiation hard, flight ready device may well be enabling technology for spacecraft using solar electric propulsion to traverse the Van Allen radiation belts. There should also be a major impact on the life of satellites in geosynchronous orbits. We shall examine this briefly in Section 7.

6.3. Array Flight Results

A great deal of actual spaceflight experience has been accumulated with regard to the radiation damage of silicon solar cell arrays, although data are not often in the open literature. The effects of several recent solar flare events on several communication satellites have been dramatic, however, and data from them has been reported (Gelb and Goldhammer, 1991; Goldhammer, 1990; Murray et al., 1991). Several solar flares occurred in quick succession in 1989, on August 12, September 29, October 19, and December 1. In some cases satellites exhibited degradation only after the October 19 solar flare, while in other cases degradation was observed after each of the four flare events listed. Such a variation in array performance is not easily explained. The October 19 event was the most intense, causing degradations ranging from 2% to 5% of total array output. For those satellites that exhibited degradation after any of the other events, the observed degradation was less than 1% each time. The degradation is a strong function of cell type, coverglass thickness, and array design. Body mounted arrays will protect the cells from backside irradiation more than a deployed rigid panel array. Similarly, the steady-state degradation expected from the trapped electrons has the same sort of dependence and is on the order of 2%–4% per year for most satellites. Although NASA has developed models for predicting solar flare proton fluences (Sawyer and Vette, 1976), the statistical nature of their occurrence makes accurate prediction of array degradation somewhat imprecise. A rough rule of thumb for estimating solar array power after several years in GEO (generally after about 60%–70% of the total period of a solar cycle) is that the degradation caused by solar flare protons amounts to approximately 50% of the total. The actual degradation at any instant may differ from this significantly, depending on the activity of the cycle and the number of flares actually containing significant proton fluences.

7. SOLAR ARRAY TRADE OFFS

Spacecraft designers and mission planners must have some incentive to use new technology to accomplish their goals. If the technology lowers mission cost but increases the risk of mission failure, in whole or in part, beyond some predetermined range, the technology will not be used. On the other hand, if the risks are acceptable and mission capability is significantly enhanced by using the new technology, slightly higher cost may become completely acceptable. The use of GaAs solar cells on currently available space solar arrays falls primarily in the latter category, as the examples that follow show.

Datum and Billets (1991) have compared the cost and weight impacts of using GaAs solar cells that have been manufactured on germanium substrates (Chu et al., 1991; Chen et al., 1991) and Si solar cells on a rigid panel array intended for use in LEO. The mission duration and orbit inclination are such that a significant accumulation of electron and proton radiation fluences will occur. At the operating temperatures expected (60°C for Si, 87°C for GaAs), the GaAs/Ge to Si BOL power ratio is 1.38. At EOL, which occurs after exposure to a total equivalent fluence of 1.44×10^{14} MeV electrons/cm^2, the power ratio is 1.5. They show that, even with the higher manufacturing cost of GaAs/Ge cells, presently over five times higher than for Si (Applied Solar Energy Corp., personal communication), the panel cost ratio, based on present assembly costs, is only 1.17.

An even more striking result is obtained when comparing panel weights. Starting with a cell weight ratio of 4.8 (the GaAs/Ge cell used in the comparison, although only about 2.2 times more dense, is thicker than the baseline Si cell used), the panel actually becomes 16% lighter, at the same EOL power level, when GaAs/Ge cells are used instead of Si cells. The reduction in panel weight can be traded against the slightly higher panel cost per watt for savings in launch costs or against the value of adding more payload capability to the satellite. The impact of the higher area power density that results from using GaAs/Ge cells can be seen in other aspects of satellite operations as well. A smaller array area reduces the propellant requirement for momentum management and drag make-up, which means that mission life can be expected at no mass penalty and without redesign of the propellant tankage.

Bailey et al. (1991) have examined the impact of InP cells on one of the Earth Observing Satellites (EOS) originally intended for polar orbit deployment. The flight array proposed in this case, e.g., was a small version of the Space Station Freedom array, i.e., a flexible fold-up array. The solar cells considered as baseline by the mission planners are 8 × 8 cm, 0.2 mm thick silicon with a 0.125 mm thick coverglass on each one for radiation protection. The orbit chosen for this platform is at 98.2° inclination and at 705 km altitude. The total power level required after 5 years in orbit is 12.2 kW, with 9.76 kW remaining at 7.5 years.

Because InP solar cells are still under development, cost comparisons were not possible in this study. There is, however, sufficient performance data, including results from an extended spaceflight test (Brinker and Weinberg, 1990) to make a size comparison at EOL. The total fluence expected after 5 years on orbit is such that, even with a 0.125 mm thick coverglass, the silicon cell output will degrade by 18%. The InP cell, on the other hand, with only a 0.05 mm thick cover, will degrade by 0.3% in the same period of time. Assuming 20% BOL efficiency for the InP cells, the total reduction in array size will exceed 50% and can have a significant impact on mission lifetime and cost. Two factors should be noted in this example: Compared with the situation with the rigid panel, cell weight is a larger fraction of the weight of a flexible, fold-up panel, and far less radiation protection is provided for the backside of the solar cell. On the APSA, for example, a 0.1 mm thick cell accounts for about 52% of the panel weight. On the rigid panel in the example above, the cells account for less than 10% of the panel weight, which is itself about 50% of the total array weight (Kurland and Stella, 1989). Using a heavier cell compared with Si cells has a much greater impact on the weight of a fold-up array than on the rigid panel array. In spite of that, advanced cells still have a major performance impact. The cost remains an open issue, but expectations are that, once in production, InP/Si cells should be no more expensive than GaAs/Ge cells, so that costs at the panel level should behave comparably to the example given.

8. CONCLUSION

The design, development, fabrication, and operation of space photovoltaic space power systems is a complex, multidisciplinary set of tasks involving several subsystems and their interactions. Although the solar array is but one of the many subsystems in the power system, its performance establishes the basic overall performance of the total system.

Significant improvements in array performance have been made since the first one was launched in 1958 aboard the Vanguard I satellite. Silicon solar cells have been the only space solar cells available for most of the 34 years since then. However, GaAs/Ge cells are now commercially available and are beginning to replace them on a cost-effective basis for missions where radiation damage will be significant. We have reviewed the basic mechanisms by which solar cells degrade after charged particle (electron and proton) irradiation and have seen how the greater radiation resistance of GaAs/Ge cells compared with Si cells may actually result in lower overall mission cost. Newer cell types, such as InP, which are still in laboratory development, hold promise for even greater radiation resistance and more gains in satellite power system lifetime.

REFERENCES

Anspaugh, B.E. (1991), *Results of the NASA/JPl Balloon Flight Solar Cell Calibration Program*, JPL Publication 91-36, JPL, Pasadena, CA.

Anspaugh, B.E., and Downing, R.G. (1984), *Radiation Effects in Silicon and Gallium Arsenide Solar Cells Using Isotropic and Normally Incident Radiation*, JPL Publication 85-61, JPL, Pasadena, CA.

Atzei, A., and Capart, J. (1972), *9th IEEE Photovoltaic Specialists Conf. Record*, IEEE, New York, pp. 206–216.

Bailey, S., Weinberg, I., and Flood, D. (1991), *Proc. 26th Intersoc. Energy Conversion Engineering Conf.*, American Nuclear Society, Illinois, pp. 250–255.

Banks, B., Rutledge, S.K., de Groh, K.K., Auer, B., Mirtich, M., Gebauer, L., Hill, C.M., and Lebed, R.F. (1991), *22nd IEEE Photovoltaic Specialists Conf. Record*, IEEE, New York, pp. 1433–1439.

Baraona, C.R. (1990), *21st IEEE Photovoltaic Specialists Conf. Record*, IEEE, New York, pp. 30–35.

Bercaw, R.W., and Cull, R.C. (1991), *26th Intersoc. Energy Conversion Conf.*, American Nuclear Society, Illinois, pp. 332–339.

Brinker, D.J. (1993), *NASA Solar Cell Calibration Techniques*, NASA Technical Memorandum (in press).

Brinker, D.J., and Weinberg, I. (1990), *21st IEEE Photovoltaic Specialists Conf. Record*, IEEE, New York, pp. 1167–1171.

Chen, J.C., Ristow, M.L., Cubbage, J.J., and Werthen, J.G. (1991), *22nd IEEE Photovoltaic Specialists Conf. Record*, IEEE, New York, pp. 133–136.

Chu, C., Iles, P., Yoo, H., Reed, B., and Krogen, J. (1991), *22nd IEEE Photovoltaic Specialists Conf. Record*, IEEE, New York, pp. 1512–1517.

Dakermanji, G., Carlsson, U., Culver, H., Rodreguez, E.G., Ahmad, A., and Jagielski, J. (1991), *25th Intersoc. Energy Conversion Conf.*, pp. 45–54.

Datum, G.C., and Billets, S.A. (1991), *22nd IEEE Photovoltaic Specialists Conf. Record*, IEEE, New York, pp. 1422–1428.

Ferguson, D.C. (1991), *Proceedings of a Conference on Space Photovoltaic Research and Technology*, NASA Conference Publication 3121, Lewis Research Center, Cleveland, OH, pp. 47.1–47.4.

Flood, D.J. (1989) *Acta Astronautica* **19**, 805–812.

Flood, D.J. (1991), *Proc. Eur. Space Power Conf.*, ESA SP-320, pp. 531–536.

Flood, D.J. (1992), *NASA Programs in Space Photovoltaics*, NASA Technical Memorandum 105428.

Flood, D.J., and Brandhorst, H.W. (1987), in *Current Topics in Photovoltaics*, J.D. Meakin and T.J. Coutts, Eds., Vol. 2, Academic Press, New York, pp. 143–202.

Flood, D.J., Piszcor, M., Stella, P., and Bennett, G. (1989), *Proc. Eur. Space Power Conf.*, ESA SP-294.

Fredericks, R.W., and Scarf, F.L. (1973), in *Photon and Particle Interactions with Surfaces in Space*, R.J.L. Grard, Ed., Reidel, Boston, pp. 227–308.

Gelb, S.W., and Goldhammer, L.S. (1991), *22nd IEEE Photovoltaic Specialists Conf. Record*, IEEE, New York, pp. 1420–1433.

Goldhammer, L.J. (1990), *21st IEEE Photovoltaic Specialists Conf. Record*, IEEE, New York, pp. 1241–1248.

Hord, R.M. (1985), *Handbook of Space Technology Status and Projections*, CRC Press, Boca Raton, FL, pp. 56–61.

Kennedy, B.H., Cull, R.C., and Kankam, M.D. (1990), *25th Intersoc. Energy Conversion Conf.*, pp. 484–489.

Kurland, R.M., and Stella, P.M. (1989), *Proc. 24th Intersoc. Energy Conversion Conf.*, pp. 829–834.

Lovgren, J.G., Teofilo, V.L., and Murdock, R.L. (1989), *Proc. 24th Intersoc. Energy Conversion Conf.*, pp. 13–17.

Lu, C. (1990), *Proc. 25th Intersoc. Energy Conversion Conf.*, pp. 1–6.

Malachesky, P.A., Simburger, E.J., and Zwibel, H.S. (1991), *22nd IEEE Photovoltaic Specialists Conf. Record*, IEEE, New York, pp. 1572–1575.

Marvin, D.C., Hwang, W.C., and Simburger, E.J. (1988), *20th IEEE Photovoltaic Specialists Conf. Record*, IEEE, New York, pp. 913–917.

Mills, M.W., and Kurland, R.M. (1988), *Proc. Conf. Space Photovoltaic Res. Technol.*, NASA Conference Publication 3030, Lewis Research Center, Cleveland, OH, pp. 122–137.

Moser, R.L. (1990), *Proc. 25th Intersoc. Energy Conversion Conf.*, pp. 66–71.

Murray, J.F., Neff, R.E., and Pollard, H.E. (1991), *22nd IEEE Photovoltaic Specialists Conf. Record*, IEEE, New York, pp. 1540–1543.

Patil, A.R., Kim, S.J., Cho, B.H., and Lee, F.C. (1990), *Proc. 25th Intersoc. Energy Conversion Conf.*, pp. 96–103.

Rauschenback, H.S. (1980), *Solar Cell Array Handbook*, Van Nostrand Reinhold, New York, pp. 440–445.

Roux, J.A., and McKay, T.D., Eds. (1984), *Spacecraft Contamination: Sources and Prevention, Progress in Astronautics and Aeronautics*, Vol. 91, American Institute of Aeronautics and Astronautics, New York.

Rutledge, S., Olle, R., and Cooper, J. (1991), *22nd IEEE Photovoltaic Specialists Conf. Record*, IEEE, New York, pp. 1544–1547.

Sawyer, D.M., and Vette, J.T. (1976), *Report NSSDC*, National Space Science Council Data Center, Greenbelt, MD.

Scheiman, D.A., and Smith, B.K. (1991), *Proc. Conf. Space Photovoltaic Res. Technol.*, NASA Conference Publication 3121, Lewis Research Center, Cleveland, OH, pp. 36.1–36.9.

Slifer, L.W., Jr. (1989), *Proc. Conf. Space Photovoltaic Res. Technol.*, NASA Conference Publication 3107, Lewis Research Center, Cleveland, OH, pp. 204–217.

Stassinopolous, E.C., and Raymond, J.P. (1988), *IEEE Photovoltaic Specialists Conf. Record*, IEEE, New York, pp. 1423–1442.

Stella, P.M., and Flood, D.J. (1989), *NASA Technical Memorandum 103284*, Lewis Research Center, Cleveland, OH.

Stella, P.M., and Kurland, R.M. (1989), *Proc. Conf. Space Photovoltaic Res. Technol.*, NASA Conference Publication 3107, Lewis Research Center, Cleveland, OH, pp. 423–432.

Sugo, M., Takamashi, Y., Al Jassim, M.M., and Yamaguchi, M. (1990), *J. Appl. Phys.* **68**, 540–547.

Sugo, M., and Yamaguchi, M. (1989), *Appl. Phys. Lett.* **54**, 1754–1756.

Sze, S.M. (1991), *Physics of Semiconductor Devices*, John Wiley & Sons, New York.

Tada, H.Y., Carter, J.R. Jr., Anspaugh, B.E., and Downing, R.G. (1982), *Solar Cell Radiation Handbook*, 3rd Ed., JPL Publication 82-69, National Aeronautics and Space Administration, New York, pp. 5.1–5.24.

Takamoto, T., Onazaki, H., Takamura, H. Ura, M., Ohmori, M., Yamaguchi, M., Ikegami, S., Arai, H., Hayashi, T., Ushirokawa, A., Takahashi, K., Kubata, M., and Ohnishi, A. (1990), *Proc. 5th Photovoltaic Sci. Eng. Conf.*, Kyoto, Japan, pp. 547–552.

Tam, K., Yang, L., and Dravid, N. (1991), *Proc. 25th Intersoc. Energy Conversion Conf.*, pp. 212–217.

Watkins, G.D. (1974), in *Lattice Defects in Semiconductors*, F.A. Huntley, Ed., Inst. Phys. Conference Publication 23.

Weinberg, I. (1990), in *Current Topics in Photovoltaics*, T.J. Coutts and J.D. Meakin, Eds., Vol. 3, Academic Press, New York, pp. 87–133.

Weinberg, I. (1991), *Solar Cells* **33**, 331–348.

Weinberg, I., Mehta, S., and Schwartz, C.K. (1984a), *Proc. 4th Eur. Conf. Photogenerators Space*, ESA SP-210, Nordvijk, Netherlands, pp. 131–136.

Weinberg, I., Mehta, S., and Swartz, C. (1984b), *Appl. Phys. Lett.* **44**, 1071–1075.

Weinberg, I., and Swartz, C.K. (1980), *15th IEEE Photovoltaic Specialists Conf. Record*, IEEE, New York, pp. 490–495.

Winslow, C., Bilger, K., and Baraona, C. (1988), *Proc. 24th Intersoc. Energy Conversion Conf.* pp. 283–287.

Winter, C.P., and Teofilo, V.L. (1989), *Proc. 24th Intersoc. Energy Conversion Conf.*, pp. 19–24.

Yamaguchi, M., Yamamoto, M., Uchida, N., and Uemura, C. (1986), *Solar Cells* **19**, 85–90.

PART FOUR

OTHER ASPECTS

CHAPTER EIGHTEEN

An Overview of Environmental, Health, and Safety Issues in the Photovoltaic Industry

PAUL D. MOSKOWITZ, Brookhaven National Laboratory, Upton, NY 11973

1. INTRODUCTION

A large variety of material and processing options are used to fabricate photovoltaic modules. Traditionally, the potential viability of these options would be evaluated solely by such factors as module efficiency, reliability, and production cost. In the past decade, however, many governmental organizations, industry groups, and private citizens have been placing greater emphasis on the need to ensure that new energy-producing technologies will not endanger environment, health, or safety (EH&S). To meet this objective, the photovoltaic industry must aggressively manage the EH&S issues relevant to the specific material and processing options under investigation or in commercial operation. Proactive management can take various forms, but is best accomplished through the development of detailed and systematic analyses of EH&S issues associated with all aspects of the product design process to minimize the potential impacts from manufacturing, use, and disposal of the spent modules. Fundamental to this process is a comprehensive understanding of the hazards and risks presented by material and processing options used by the photovoltaic industry and the associated control strategies needed to ensure that EH&S risks are minimized throughout all stages of the module's

Solar Cells and Their Applications, Edited by Larry D. Partain.
ISBN 0-471-57420-1

product life. This chapter summarizes these hazards and risks and the controls that will help the photovoltaic industry continue to grow in an environmentally conscious way.

2. GENERAL METHOD FOR CHARACTERIZING HAZARDS AND RISKS

In the evaluation of the EH&S issues related to the production and use of photovoltaic modules, a basic framework is first needed. In this context, it is important to clarify the differences between the terms *hazard* and *risk*. *Hazard* is a measure of the damage caused to life or property given an exposure to a physical or biologically important stress. *Risk* is formally defined as the product of hazard and event probability. Thus, for example, a material may be very hazardous because of its inherent biological toxicity; however, the risk to health from this material may be very small because exposures rarely occur.

In Figure 18.1, a framework for characterizing hazards and risks for different materials and processing options is presented. To implement this framework, basic information on the material and process must first be collected. Based on this information, hazards to health or environment are then identified, but only in qualitative terms. Next, regulatory controls or constraints associated with these materials or hazards are reviewed. This information is then used to guide the preparation of a more detailed risk analysis.

A risk analysis can range from a simple qualitative description of the risk to a full-blown quantitative probabilistic risk assessment like those conducted for nuclear power plants (e.g., Moskowitz et al., 1989a). The effort to produce these analyses can range from minutes for the first type of approach to many man-years for the latter alternative. In practice, the level of effort required for any risk analysis increases as the risk increases; the larger the risk the more comprehensive the analysis should be. Preparation for the risk assessment revolves around four basic steps: Identification of Accidents; Quantification of Event Probability; Quantification of Event Consequences; and Risk Estimation. Risks may be deemed to be acceptable using various criteria, for example, a marginal excess lifetime risk to the individual of dying from cancer is $<10^{-5}$. If the risks are acceptable, then further technology development can be pursued. If the risks are unacceptable, then the system must be modified by incorporating more controls or, if all else fails, abandoned.

Although the precise evaluation of these risk estimates in theory seems very simple, their estimation and interpretation is in fact quite complex. Two issues that complicate this process greatly are uncertainty and fear. Uncertainty is important because knowledge is imperfect. Thus, any estimate prepared has inherent uncertainty due, for example, to simplifying assumptions made or inadequate scientific knowledge. If the uncertainty is not explored, reviewers of these types of analyses often ask how far off can the answer be. If the answer can differ by more than one or two orders of magnitude and the risks are still below the "threshold" of interest, then it is presumably safe to proceed. If,

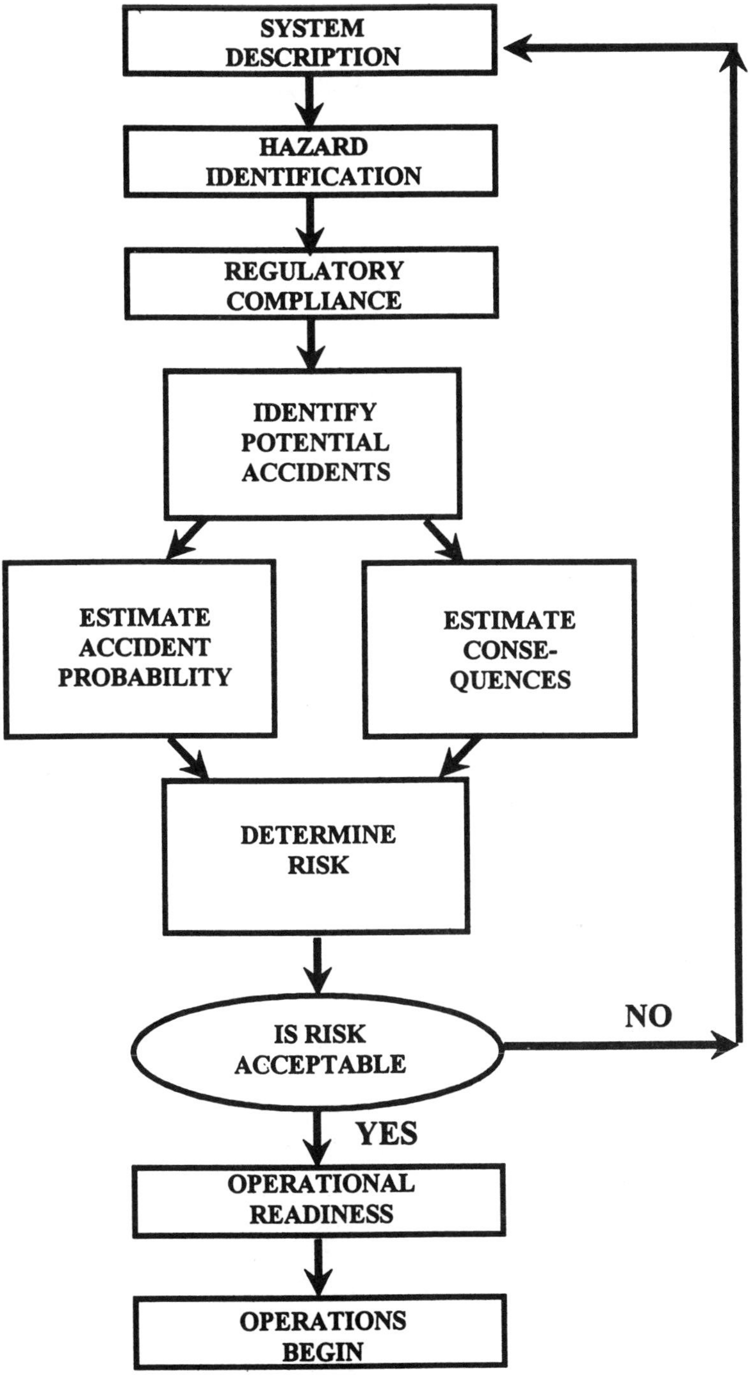

FIGURE 18.1 Framework for assessing hazards and risks and photovoltaic material and processing options.

however, the upper bound estimate infringes into the threshold level, then the response may be unsafe.

A second complicating issue is that of fear or perception of risk. As various studies have shown (e.g., Covello, 1987), the general public's perception of risk is often vastly different than the actual observed risk levels. In general, people tend to fear more those risks that are dread (e.g., cancer from radiation) and involuntary or personally uncontrollable. In an oxymoronic way, people may protest activities that realistically present very small risks, while at the same time have lifestyles (e.g., smoking, drinking, or driving without seat belts) that can present very large risks. Although we can argue simply that we are unwilling to tolerate any risk imposed from new technologies, the underlying measure of merit should actually be based on the balance between risks versus benefits. Still, the calculation of these risks and benefits and boundaries to which they apply will continue to be arguable and be colored by the eyes of the viewer.

3. SYSTEM DESCRIPTIONS

3.1. Manufacturing Options

Photovoltaic modules can be made from many materials and formed in a variety of designs. In general, the fabrication of a module generally involves a series of discrete production processes, with each process requiring several operations. In Table 18.1, representative photovoltaic process options are identified (Moskowitz et al., 1986). Note that the exact specification of a process sequence depends on the technology and module designs, and does not necessarily require every process step or follow the order shown.

Characterization of the hazards and risks presented by these options depends on the collection of a baseline data set including information on manufacturing system characteristics (e.g., deposition efficiency and rate); module characteristics (e.g., electric conversion efficiency); feedstock requirements (e.g., kg/MWp); and environmental residuals. This is a time-consuming, but necessary step in the evaluation process. Table 18.2 gives a representative system description for a process used to manufacture a hydrogenated amorphous Si (a-Si: H) module (Fthenakis et al., 1984). Clearly, the type and degree of hazard present will vary greatly among different technologies. Large differences can also exist within the same material grouping due to physical variations (e.g., temperature, pressure, feedstock) in the manufacturing process (e.g., electrodeposition vs. co-evaporation vs. metalorganic chemical vapor deposition) and even the feedstock (e.g., use of H_2Se vs. elemental Se).

3.2. Application Options

Like manufacturing, the markets or applications for photovoltaic modules vary from very small-scale consumer products (several watts) to large-scale utility

TABLE 18.1 Process Steps and Options in Fabricating Photovoltaic Cells

Step	Option	Technical Status
Substrate preparation	Detergent wash	Commercialized
	Methanol wash	Commercialized
	Acid etch	Commercialized
	Alkali etch	Commercialized
	Plasma etch	Commercialized
Metalization	Chemical vapor deposition (CVD)	Commercialized
	Evaporation	Commercialized
	Electroplating	Commercialized
	Sputtering	Commercialized
Transparent contact deposition	Chemical vapor deposition (CVD)	Commercialized
	Evaporation	Commercialized
	Chemical spraying	Commercialized
	Sputtering	Commercialized
Junction formation	Chemical vapor deposition (CVD)	Commercialized
	Glow discharge	Commercialized
	Evaporation	Commercialized
	Reactive sputtering	Commercialized
	Ion implantation	Commercialized
	Diffusion	Commercialized
	Metal organic chemical vapor deposition (MOCVD)	Commercialized
	Close-spaced vapor transport (CSVT)	R&D
	Molecular beam epitaxy (MBE)	R&D
Passivation	Anodic oxidation	R&D
	Heat treatment	R&D
	Chemisorption	R&D
Scribing	Mechanical	Commercialized
	Laser	Commercialized
Antireflective	Chemical vapor deposition (CVD)	Commercialized
	Evaporation	Commercialized
Cell interconnection	Soldering	Commercialized
Encapsulation	Vacuum bonding	Commercialized
Module testing	Simulated sunlight	Commercialized

arrays (100s kW_p to MW_p). From an EH&S perspective the key differentiating features among these applications are related to the rated electrical capacity of the module/array and the numbers of modules/arrays installed. The first is important because of the potential for electrical shock hazards, while the second is important from the perspective of potentially hazardous materials (e.g., As and Se) present in the modules themselves.

TABLE 18.2 Sample Process Characterization

ACTIVITY:	Hydrogenated Amorphous Silicon (a-Si:H) Thin-Film Deposition
PROCESS:	Chemical Vapor Deposition (CVD) from Higher Silanes
METHOD:	Thermal Atmospheric Pressure CVD
PRESENT STATUS:	Commercial

Process Description

The silanes are decomposed, and an a-Si:H film is formed on heated substrate by application of heat alone

System Characteristics

Total flow rate (1 pm)	2.1
Pressure (atm)	1
Substrate temperature (°C)	380–450
Deposition rate (nm/min)	10–60
Deposition efficiency	50%

Solar Cell Characteristics

Electric conversion efficiency	5% at AM1
Structure	p/i/n
Layer thickness	p = 10 nm, i = 500 nm, n = 20 nm

Feedstock Requirements (kg per 10 MW_p annual cell production)

Silane	610
Diborane	1.35
Phosphine	2.10

Environmental Residuals and Control Technologies

Air Pollutants	Quantity Released (kg/10 MW_p)	Standards	Control Technology	Control Cost ($/10 MW_p)
SiH_4	396	N.A.	N_2 dilution and caustic water scrubber	Capital = $18,000; operation and maintenance = $15,000 per year
B_2H_6	0.9	N.A.	Reaction with $KMnO_4$	
PH_3	1.36	N.A.	Reaction with $KMnO_4$	

Water Pollutants	None
Solid Wastes	None

Occupational Health and Safety Issues

Hazard	Source	Standards (TLV)	Comments
PH_3	Feedstock, exhaust	0.3 ppm	Highly poisonous, flammable
B_2H_6	Feedstock, exhaust	0.1 ppm	Highly poisonous, flammable
SiH_4	Feedstock, exhaust	0.5 ppm	Flammable, explosive
Non-ionizing radiation	Heater	—	—
a-Si	Reactor, ducts	10 mg/m^3	Dust inhalation suspected harmful

4. HAZARDS PRESENT IN THE PHOTOVOLTAIC INDUSTRY

4.1. Manufacturing

In a module manufacturing facility, hazards to occupational health can arise from physical (e.g., repetitive motions, electromagnetic radiation, laser) or chemical/biological (i.e., hazardous chemicals) agents. In Table 18.3 the principal types of hazards associated with different material and product types are highlighted by Moskowitz et al. (1988, 1989) and others (e.g., Coutts et al., 1987, 1988). In general, the hazards present in this industry are primarily related to materials being used (see Table 18.4). Therefore, in the discussions presented below, special attention will be given to several materials of principal interest.

4.1.1. Silane. Silane (SiH_4) is the principal compound used in the fabrication of a-Si : H photovoltaic modules. SiH_4 and related byproduct gases can under specific conditions autoignite and cause a fire or explosion (Moskowitz et al., 1989c; Fthenakis and Moskowitz, 1990). The temperature at which the gas spontaneously ignites depends on many factors, including concentration, leakage flow rate, air flow conditions, ignition delay, and environmental effects (e.g., expansion volume, pressure, and oxygen content). In addition to the possibility of autoignition, SiH_4 may also be ignited by electric sparks in reactor systems or if a leak develops in the workspace that contains several electric components, including high-voltage RF or DC power supplies. Another interesting source of ignition is static electricity. In addition, SiH_4 and other flammable gases when released into space may carry enough electric charge to initiate a flame. Serious accidents have also occurred with liquid and solid

TABLE 18.3 Overview of Hazards by Material and Module Types

Material	Model Type	Types of Potential Hazards
Si	Crystalline	HF acid burns
	Polycrystalline	HF acid burns
	Amorphous	SiH_4 fires/explosions
CdTe	Polycrystalline	Cd chronic toxicity, carcinogenicity, module disposal
CIS	Polycrystalline	Se acute/chronic toxicity, module disposal
GaAs	Single Crystal	As chronic toxicity, module disposal
	Polycrystalline	As acute/chronic toxicity, carcinogenicity, module disposal

TABLE 18.4 Typical Hazardous Production Materials Used in Photovoltaic Module Manufacture

Hazardous Production Material Name	DOT Hazard Classification[a]	OSHA Permissible Exposure Level	OSHA Immediately Dangerous to Life or Health
Arsenic	Poison	—[b]	100 mg/m^3
Arsine	Highly toxic gas	—[b]	6 ppm
Cadmium	Poison	—[b]	9 mg/m^3
Diborane	Flammable gas	0.1 ppm	40 ppm
Diethyl Silane	Flammable liquid		
Diethyl Zinc	Pyrophoric liquid		
Dimethyl Zinc	Spontaneously combustible		
Hydrochloric Acid	Corrosive material	5 ppm	100 ppm
Hydrofluoric Acid	Corrosive material	3 ppm	30 ppm
Hydrogen	Flammable gas	0.05 ppm	2 ppm
Hydrogen Selenide	Highly toxic	10 ppm	300 ppm
Hydrogen Sulfide	Flammable gas		
Indium	Not regulated		
Methane	Flammable gas		
Molybdenum Hexafluoride	Toxic and corrosive gas		
Oxygen	Gaseous oxidizer		
Phosphine	Highly toxic and pyrophoric gas	0.3 ppm	200 ppm
Phosphorus Oxychloride	Corrosive material		
Selenium	Poison	0.2 mg/m^3	
Silane	Pyrophoric gas		
Silicon Tetrafluoride	Toxic and corrosive gas		
Tellurium		0.1 mg/m^3	
Tertiarybutylarsine	Pyrophoric and highly toxic liquid		
Tertiarybutylphosphine	Pyrophoric liquid		
Tetramethyltin	Pyrophoric and highly toxic liquid		
Trimenthylindium	Pyrophoric solid		
Trimethylaluminum	Pyrophoric liquid		
Trimethylgallium	Pyrophoric liquid		
Tungsten Hexafluoride	Toxic and corrosive gas		

[a]DOT, Department of Transportation.
[b]Carcinogenic; reduce exposure to lowest feasible concentration.

TABLE 18.5 Silane Accident Scenario Analysis

		Distance (m)		
Scenario No.	Assumptions	1 psi	2 psi	5 psi
1	16 kg release	50	30	23
2	130 l/min for 30 s	5	2.5	1
3	2 l/min for 30 s	1.2	0.6	0.2

silanes in vacuum systems. In the event of an explosion, the potential damage would depend on many factors, including the distance to all types of targets. In Table 18.5, estimates of potential energy fields for three representative silane-release scenarios and three critical values are given. The 1 psi values corresponds to a community safe threshold level; the 2 and 5 psi levels define limits for damages to pressurized and low-pressure equipment. As shown, the potential area impacted by the catastrophic release of an entire 16 kg cylinder could be very large and could extend beyond the boundaries of a production facility. Recently presented data (Shiban, 1992) suggest that this type of event could only occur if the SiH_4 were stored within a confined space (e.g., gas cabinet or shed). Under unconfined conditions similar to those found in an outdoor bunker, the experimenters could not produce an explosion. This strongly supports the concept of outdoor bunkering versus indoor gas cabinet storage of SiH_4 to reduce the risk presented by this material.

4.1.2. Cadmium Compounds. Cadmium (Cd) is the principal material used in the fabrication of CdTe modules and is a secondary component in CIS modules. Details on the hazards presented by Cd in photovoltaic module manufacture have been reviewed by Moskowitz et al. (1990) and by Moskowitz (1992). In theory, the toxicology and hazards from different Cd compounds (e.g., CdCl, CdS, CdTe, CdO) will differ due to physical (e.g., solubility) and pharmocokinetic (e.g., absorption across the gastrointestinal tract) factors. However, due to the general lack of compound-specific information, the toxicology summaries presented below are for all inorganic Cd compounds. The principal effects of continued exposure to low levels of Cd are on the kidneys, lungs, and bones. In the body, Cd accumulates in the renal cortex, where it is effectively retained. The renal cortex is generally accepted to be the critical organ for Cd accumulation, and proteinuria (abnormal excretion of low molecular weight proteins) to be one of the classic effects of Cd poisoning. The evolution of measured renal proteinuria to the more severe renal dysfunction is usually slow. The latent period before the clinical observation of excessive proteinuria depends on the intensity of exposure, but in general exceeds 10–20 years of exposure. Chronic exposure to excessive Cd levels may also cause bone disease characterized by softening, bending, and reduction in bone size. Other possible effects include hypertension, nausea, and nasal irritation. In some workers exposed to Cd, loss of sense of smell has also been observed. Various types of lung disturbances, including emphysema, obstructive lung disease, pulmonary fibrosis, and lung cancer, have been found in workers chronically exposed to Cd dust and fume. The U.S. Environmental Protection Agency (EPA, 1984a, b) has concluded that inhaled inorganic Cd compounds are probable human carcinogens.

The Occupational Safety and Health Administration (OSHA, 1990) has recently issued a directive to alert employers and employees of the inadequacy of OSHA's own exposure standards and of the interim need to reduce exposures in accordance with the American Conference of Governmental and Industrial Hygienists (ACGIH) standards for Cd dust and salts of $50\,\mu g/m^3$ for an 8 hour TWA (vs. the OSHA standard of $200\,\mu g/m^3$ and for Cd oxide fume) and use $50\,\mu g/m^3$ as a 15 min ceiling (vs. the OSHA standard of $300\,\mu g/m^3$). OSHA has recently been challenged about the appropriateness of classifying all Cd compounds as carinogenic because of compound-specific differences in solubility, bioavailability, and so forth (OSHA, 1991). OSHA's position has recently been upheld by the courts.

4.1.3. Selenium Compounds. Selenium (Se), in the form of a pure element or as a hydride gas (H_2Se), is used in the production of CIS modules. Selenium is an essential trace element. Nevertheless, little is known about the detailed pharmacology and toxicology of inorganic selenium compounds at elevated exposure levels. Possible health effects of elevated Se exposures include selenosis, anticarcinogenicity, and dental caries (Moskowitz et al., 1985; Moskowitz and Fthenakis, 1990b).

With respect to acute toxicity, it is well known that H_2Se and SeO_2 are highly toxic; the OSHA IDLH concentrations for these materials are 2 ppm and $100\,mg/m^3$, respectively. Accidental releases resulting in exposures in excess of these concentrations can cause severe irritation of the respiratory tract, which may be followed by pulmonary edema and possible death.

Potential health effects associated with chronic long-term exposures are based principally on data collected from animal studies. In these studies, a range of effects have been reported, including alkali disease and blind staggers syndrome in hoofed animals. Concentrations of Se in the diet that lead to toxicity depend strongly on the chemical form of the element. Elemental Se, for example, is virtually insoluble in water and has little significant toxicity. Selenites (SeO_3) and selenates (SeO_4) are soluble and present greater risks. In general, chronic toxicity in animals appears to occur when the daily food intake of selenium exceeds 0.5 ppm. This concentration is equivalent to a human dose of between 3,500 and 7,000 μg Se per day.

Debate continues about whether Se is a carinogen or an anticarcinogen. The principal argument for carcinogenicity is based on a single report published more than 45 years ago and has not been verified by more recent studies. In contrast, more recent animal and human epidemiological evidence suggests that Se has anticarcinogenic properties.

4.1.4. Arsenic Compounds. Arsine (AsH_3) gas is used in the production of thin-film GaAs modules (Lee and Moskowitz, 1985). Acute arsenic exposure results in the hemolysis of blood erythrocytes. As AsH_3 breaks down the hemoglobin, the blood's capacity for carrying oxygen is diminished. Symptoms of AsH_3 poisoning usually develop within a few hours of exposure. Headache, dizziness, nausea, and weakness occur early, followed by kidney failure and, in the most severe cases, death. Since the discovery of AsH_3 in 1775, over 450 cases of AsH_3 poisoning have been reported, including several deaths arising from exposure to AsH_3 gas from the handling and treatment of compressed gas cylinders and AsH_3-contaminated process equipment (Moscovsky et al., 1992). Inorganic arsenic compounds, including AsH_3, are thought to be human carcinogens. Although OSHA has not yet formally recognized arsenic as a human carcinogen, the National Institute for Occupational Safety and Health (NIOSH, 1987) has issued an alert indicating that gallium arsenide compounds should be treated as carcinogens.

4.1.5. Phosphine. Phosphine (PH_3) is used as a feedstock in the production of ZnP and InP modules and as a dopant in the production of other types of modules (Fthenakis and Moskowitz, 1987). Pure PH_3 is considered to be odorless up to 200 ppm. Although classified as pyrophoric, extremely pure PH_3 will not ignite spontaneously under 300°F. This is based on laboratory and field studies (Moscovsky et al., 1992; Hazards Research Corporation, 1988). In the absence of in-air detonation, the most significant hazards presented by PH_3 are due to its toxicity. Acute phosphine exposure can result in pulmonary edema, hemorrhage, cardiac failure, and brain damage. Chronic exposures result in bronchitis, neurolgic symptoms, and liver dysfunction.

4.2. Application

Worker or public exposure to energized DC circuits could conceivably arise when a contractor or homeowner is installing, maintaining, or removing a rooftop photovoltaic system (Moskowitz et al., 1983). Physiological and anatomical effects of DC electric shocks, important in evaluating health and safety risks, include perception of electric current flow and release thresholds. The main factors affecting electric shock severity are type of circuit (AC or DC), circuit voltage, impedance of the human body, size of current flows through tissues, current pathway through the body, and contact duration. In Table 18.6, selected effects of DC and AC electric currents on humans are summarized. In general, low-voltage DC is not considered to be as harmful as the corresponding AC because it does not produce strong tetanic contraction of muscle.

TABLE 18.6 Selected Effects of Electric Currents on Humans

	DC (mA)		60 Hz AC (mA)	
Effect	Men	Women	Men	Women
Slight sensation	1	0.6	0.4	0.3
Perception threshold, median	5.2	3.5	1.1	0.7
Let-go threshold, median	—	—	16	10
Release threshold, median	76	51	—	—
Ventricular fibrillation, 3 s shock	500	500	675	675

5. TYPES OF RISKS

Identification of all risks from all materials, technologies, and applications is a large undertaking. This section illustrates important classes of risks that may be present in the photovoltaic industry. These examples are drawn from more detailed analyses referenced through the text.

5.1. Acute Occupational Risks

Accidental release of toxic gases in research and development or commercial-scale manufacturing facilities can present acute risks to workers (Moskowitz et al., 1986, 1989b). These can arise from a variety of situations (e.g., mishandling, equipment failure) and for a range of materials used in these manufacturing facilities. The dangers to health in these instances can be due to physical (e.g., SiH_4-based explosion) or biological (e.g., inhalation of AsH_3 gas resulting in hemolysis) stresses. Another source of acute occupational exposure is from high-voltage electrical equipment. In the semiconductor industry, electrocution is one of the largest sources of fatalities.

Crandall et al. (1992) have recently examined risks to occupational (and public) health associated with the handling of toxic gases at the National Renewable Energy Laboratory in Colorado. Their findings are contained within a Safety Analysis Report (SAR) that required the authors to characterize quantitatively the risks associated with storage, distribution, use, and disposal of toxic gases. Their report provides detailed descriptions of accident-initiating events, event probabilities, and consequences associated with 30 different accidents; an example drawn from that effort is presented below.

5.1.1. Accident Description. This hypothetical accident is associated with the leaking of H_2Se, AsH_3, or PH_3 from a gas cylinder in a gas cabinet. The leak is postulated to occur immediately after a cylinder change operation. The cause of the leak is identified to be due to the cross-threading of valve CGA fitting by the cylinder change-out team and to their failure to helium leak check the system.

5.1.2. Event Probability. In preparing the SAR, gas vendors were contacted and asked about the observed rate of return of gas cylinders due to cross-threading. A compressed gas vendor indicated that they had processed more than 500 cylinders per month for the last 5 years, using stainless steel CGA connections, and have not had any cylinders returned that had clearly been cross-threaded. One cylinder was returned with some thread damage, but the company spokesman was not sure of how the threads were damaged: "[Cross-threading] is just not a problem of any consequence in this industry, especially for stainless steel 350 CGA connectors." Based on these empirical data, Crandall et al. (1992) concluded that this is not likely to occur in the 3 year operating life cycle of the system, but is possible (defined by the authors to occur with an annual probability of occurrence of 10^{-4} to 10^{-2}). Crandall et al. further noted that for this accident to occur, not only must the valve be cross-threaded during cylinder change-out, but the operator must fail to helium leak check the connection. Due to the combined set of events that must occur, the overall probability of this event was further reduced to extremely remote (defined by the authors to occur with an annual probability of occurrence of 10^{-6} to 10^{-4}).

5.1.3. Consequences and Risks. The leakage rate in this transient depends on the pressure in the cylinder and the opening of the built-in flow restrictor. A fully opened valve on an H_2Se cylinder at 125–150 psi, equipped with a 7 mil flow restrictor, will release gas at a maximum flow rate of 1.15 slpm. Assuming leakage into the gas cabinet at a rate of one-tenth of the maximum flow due to the fact that the cylinder could be near empty and connected to the gas handling system, it was calculated that H_2Se concentrations of 20–40 ppm could build up within a purged cabinet (IDLH 2 ppm). A release of AsH_3 from a 205 psi cylinder, through a 6 mil orifice, would release gas at a maximum flow rate of 1.7 slpm; leakage of one-tenth of this mass would generate concentrations of 23–50 ppm in the cabinet (IDLH = 6 ppm). For PH_3 releases, under the same assumptions, the maximum flow is 6.9 slpm and the concentration would build to 100–200 ppm, which is at or below the IDLH level of 200 ppm. Even though the concentrations of AsH_3 and H_2Se exceed the IDLH, Crandall et al. (1992) concluded that the consequences would be only marginal because they would occur outside of the occupied work space. In the external environment, these materials would be further diluted by atmospheric dispersion processes and would therefore not present significant risk to public health.

5.2. Chronic Occupational Risks

Chronic occupational health risks can arise from workers being exposed to low levels (subacute) of toxic materials over long periods of time. Of the large variety of materials present in the photovoltaic industry, those that are potential human carcinogens are of greatest interest (Moskowitz and Zweibel, 1992). This includes inorganic arsenic and cadmium compounds. In production facilities, workers may be routinely or accidentally exposed to hazardous compounds through the air they breathe, as well as by ingestion from hand-to-mouth contact. Of these two pathways, inhalation is probably the most important because of the larger exposure potential and higher absorption efficiency of some metal compounds (e.g., Cd) through the lung than the gastrointestinal tract. In general, absorption through the skin is not recognized to be a large source of exposure to metal compounds. The physical state in which a compound is used and/or released to the external environment is another determinant of risk. Processing options, for example, in which Cd compounds are used or produced in the form of fine fumes or particles, especially those $\leqslant 5\,\mu m$ diameter, which are readily inhalable and penetrate deep into the lung, present larger hazards to health. Similarly, processing options that use or produce volatile or soluble Cd compounds also must be more closely scrutinized. Fortunately, several facility managers where CdTe research is being conducted have reported in private discussions that existing exposures at these facilities are below regulatory levels and biomonitoring data (urinary Cd levels) and indicate that no significant adverse exposures have occurred.

5.3. Acute Public Health Risks

Accidental exposure to accidentally released toxic gases from module manufacturing facilities and to energized circuits are likely to be the most important source of acute public health damage. Risks to public health from accidentally released toxic gases are similar to those that could be experienced in the workplace. Because workers tend to be healthier than the general public, a given exposure to each group should present larger health risks to the public. However, because of the physical separation in time and space between a manufacturing facility and the general public, the overall hazard to the public is likely to be smaller than that of the worker. Nevertheless, protecting the public from accidentally released toxic gases is one of the most important issues facing module manufacturers.

5.3.1. Combustion-Related. Combustion-related processes may occur during accidental fires at residential, commercial, or utility sites or during routine combustion in an incinerator fueled by municipal solid wastes containing spent photovoltaic modules. Public health hazards associated with the release of As, Cd, and Se from related modules during a fire have been examined in some detail by Moskowitz and Fthenakis (1990a). Fire scenarios involving a residential array of $5\,kW_p$, a commercial rooftop array of $100\,kW_p$, and a $500\,kW_p$ ground-mounted substation were examined. Of these scenarios, the first two are the most likely since residential and commercial buildings are often involved in fire-related incidents; the third is highly improbable since it is difficult to conceptualize a mechanism leading to the complete combustion of a large ground-based array field. Nevertheless, as a "worst case" exercise, all three scenarios were evaluated by defining the source term, examining dispersion of emitted materials in the atmosphere, and estimating health risks associated with exposure to emitted compounds.

If photovoltaic material is released in the form of either fume or vapor, this release will last only a few minutes and, consequently, outdoor human exposures, if any, will be of about the same duration. Given the short duration of the predicted exposures, the OSHA/NIOSH IDLH level (referring to 30 min exposures) is best used as a measure of hazard for comparison with the calculated ground-level ambient concentrations.

A representative result from the modeling exercises is presented in Figure 18.2 which compares estimated ground-level and IDLH concentrations. As shown, the concentrations decline rapidly as distance from the source increases. Similarly, smaller releases result in lower estimated concentrations across all distance classes. The analysis showed that the maximum concentrations of CIS vapors or fumes from residential fires are much lower than the IDLH concentrations at downwind distances of about 100 m, while the maximum estimated concentrations from a CdTe residential array fire results in concentrations approximately equal to the IDLH. Releases of CIS and CdTe from commercial systems under the conservative scenarios may result in hazardous

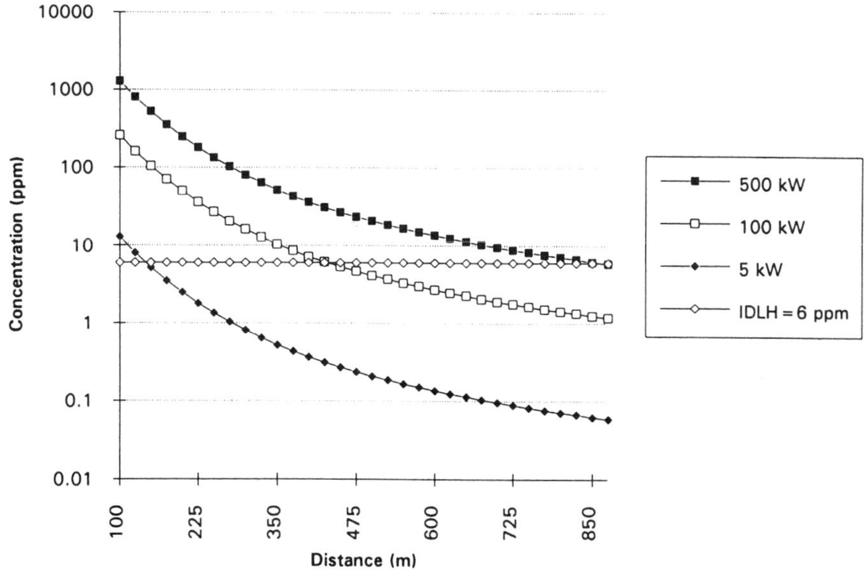

FIGURE 18.2 Estimated ground-level concentrations of As from a fire consuming GaAs modules of different sizes—100% release.

ground-level concentrations downwind from a fire at much larger distances. In the case of GaAs residential fires, exceedances occur at the lower bound IDLH equivalent of 6 ppm; the upper bound concentration limit of 150 ppm is not exceeded. Thus, fires involving both commercial and substation-level GaAs systems may present public health hazards under the most conservative release and dose–response conditions.

The probability of a fire in a residence and on a roof is small when compared with the incidence in a commercial building. Due to the larger concentrations possible in commercial fires, and the increased probability of fire in such a structure, the health risks from these applications may be larger and mitigation strategies may need to be developed for these events. Although a substation-level fire presents the greatest hazard due to the larger mass of material present, the probability of complete combustion of all arrays is likely to be very small. Hence, the risks from these applications are probably smaller than those of commercial operations.

5.3.2. Electrical. A homeowner who is installing, maintaining, or removing a rooftop photovoltaic system could be exposed to electric shock hazards. Although engineering safeguards are incorporated into all systems and literature is available on the same handling of these devices, it is not possible to make a system completely fault-free or to ensure that safeguards will never be circumvented.

Moskowitz et al. (1983) have examined the potential hazards to human health presented by the circuit voltage in a rooftop module. Results of their analysis are summarized in Table 18.7. In this analysis, they assumed that contact with the energized array was made at some point along the circuitry, after the stepwise addition or removal of the modules in a rooftop array. Their analysis suggests that the voltage generated by six modules connected in series is sufficient to cause ventricular fibrillation and possibly death under nominal conditions. In this analysis, however, the authors failed to evaluate event probability. Therefore, it is not possible to define a numeric value for this type of risk. Because of the internal safeguards, it should be small.

5.4. Chronic Public Health Risks

Public health may be affected principally from chronic exposure to compounds released to the environment as a byproduct of different manufacturing steps or as a waste from the uncontrolled disposal of spent photovoltaic modules at the end of their useful life. Large-scale disposal of spent photovoltaic devices will not occur until 20–30 years after their initial installation. In the decommissioning of these devices, the principal concern will be associated with the presence of heavy metals (e.g., Cd and Se) in the solid wastes (Moskowitz and Zweibel, 1992). In the case of disposal in a municipal waste stream, hazards to the public could arise from leaching of landfilled waste and mobilization of metal compounds into ground and surface waters and possibly terrestrial foodchains

TABLE 18.7 Physiological Effects Associated With Exposure to DC Energized Circuits

No. of Modules in Series	Open Circuit Voltage (V)	Total Body Impedance (Ω)	Current Passing Through Body (mA)	Physiological Effects[a]	
				Men	Women
1	17.6	500	35.2	A	A
2	35.2	500	70.4	B	B
3	52.8	500	106	B	B
4	70.4	500	141	B	B
5	88.0	500	176	C	C
6	106	500	211	C, D	C, D
7	123	500	246	C, D	C, D

[a]A, Shock is painful, but not fatal. Secondary injuries and minor contact burns possible. B, Shock is painful and could produce temporary respiratory arrest or cause unconsciousness; death is possible but highly unlikely. C, Death will probably occur for all "sensitive" individuals weighing less than 50 kg; secondary injuries and contact burns can occur in all members of this population. D, Death will probably occur for all "sensitive" individuals weighing less than 70 kg; secondary injuries and contact burns can occur in all members of this population.

or from combustion and mobilization of compounds to the atmosphere. Data collected from leachate plumes beneath landfills where municipal solid wastes and coal ash have been disposed suggest that the Cd present in these wastes can be leached, sometimes in significant quantities (Table 18.8). Of course, the actual leachability of waste will probably be less for two important reasons: First, the metals will be sandwiched between two layers of glass and reasonably isolated from the environment. Second, some of the metals (e.g., CdTe) are insoluble in water and would therefore exhibit only limited mobility in the soil environment found beneath landfills.

TABLE 18.8 Median Concentrations of Substances Found in Municipal Solid Waste Landfill Leachate Compared With U.S. Exposure Limits

Substance	Concentration (ppm)	U.S. Drinking Water Standard	EPA Leachate Concentration Limits
Arsenic	0.005–0.218	0.05	5.0
Cd	<0.0001–0.044	0.001	0.1
Copper	<0.005–24		
Lead	<0.0005–2.92	0.05	5.0
Selenium	0.0025–0.037	0.01	1.0

TABLE 18.9 Concentrations ($\mu g/Nm^3$) of Selected Trace Elements in Municipal Incinerator Stack Sases

	Type of Facility		
Substance	Mass Burn	Modular	RDF
Arsenic	0.452–233	6.09–119	19.1–160
Cd	6.22–500	20.9–942	33.7–373
Lead	25.1–15400	237–15500	973–9600

In the event of combustion in a municipal incinerator, current measurements of Cd and other elements in the stack gas are again useful to place this issue in proper perspective. In Table 18.9 concentrations of As and Cd in the stack gas effluent from municipal incinerators are given. Again, it can be seen that Cd and other trace elements are present in the gaseous waste stream, sometimes in significant quantities. The source of the Cd is principally, but not solely, from the large presence of Ni/Cd household batteries. The EPA (1987) has examined this issue and concluded that the cancer risk to the public presented by these emissions can pose significant potential risks. They estimate, for example, that the Cd emissions levels identified in Table 18.9 would result in an annual excess cancer incidence of 0.2 and a maximum individual risk of 10^{-4} to 10^{-6}. Regulatory controls are often required by the EPA when the annual excess cancer incidence exceeds 1 or the individual risk level to the public is greater than 10^{-6}. This risk estimate is close to the threshold where the risks would remain unregulated.

6. RISK MANAGEMENT STRATEGIES

6.1. Regulatory Oriented

In the United States, statutes and regulations protect health by controlling environmental releases of liquid, solid, and gaseous materials to the environment; by controlling work practices and worker exposures to toxic and hazardous materials; or by controlling building and fire codes where these operations may occur. Two critical sources of guidance are the Clean Air Act Amendments of 1990 and Article 80 of the Uniform Fire and Building Code.

6.1.1. Clean Air Act Amendments of 1990. In November 1990, the President signed into law the 1990 Clean Air Act Amendments. These Amendments contain some of the most detailed guidance ever included within environmental legislation passed by Congress. Two key provisions of this Act that have aroused great interest to operators/owners of facilities that generate hazardous air emissions or handle extremely hazardous chemicals are Sections 301 and 304.

In 1990, major revisions were made to the statutory and regulatory approaches for controlling toxic air pollution emissions from point sources. These changes were incorporated into the newly authorized Clean Air Act. In this Act, an aggressive strategy for controlling hazardous air pollutant emission standards for routine releases from stationary sources was adopted. More specifically, Title III lists 189 hazardous air pollutants (including As, Cd, and Se compounds) and directs the EPA to promulgate maximum achievable control technology (MACT) standards for industrial sources emitting these contaminants in quantities exceeding 10 tons per year. MACT standards may be achieved through process changes, installation of pollution controls, materials substitution, or operator training and certification. Should these controls fail to provide an ample margin of safety to public health, for example, a residual cancer risk exceeding one in 10,000 to the most exposed person, the EPA Administrator is required to develop more stringent emission limits. In developing MACT standards, the EPA Administrator may apply different emission standards to new and existing sources. The criteria used in deciding such standards would be based on special considerations of the cost and feasibility of control, energy impacts, and environmental factors. However, any new source must achieve the maximum degree of reduction in emissions that is deemed achievable for the best-controlled existing similar source. Thus, standards for new sources are likely to be more stringent than those for existing sources. Although it is highly unlikely that a photovoltaic manufacturing facility could emit these quantities, this program is still important because of the regulatory focus placed on several materials used in photovoltaic cell manufacture (Table 18.10).

In contrast with past environmental legislation, which focused principally on the roles and responsibilities of the EPA, Section 304 defines a joint role for both the EPA and OSHA in chemical accident prevention. More specifically, Section 304 directs the Secretary of Labor to promulgate, under authorities of the Occupational Safety and Health Act of 1970, chemical process safety standards to protect employees from hazards associated with accidental releases of extremely hazardous substances. The first set of regulations responding to this Section were promulgated by OSHA (1992) in February 1992. These regulations focus on the development of a comprehensive management

TABLE 18.10 Selected Hazardous Air Pollutants To Be Regulated Under the Clean Air Act; Title III Requirements

Antimony compounds	Glycol ether
Arsenic compounds	Lead compounds
Beryllium compounds	Mercury compounds
Cadmium compounds	Nickel compounds
Chromium compounds	Selenium compounds
Cobalt compounds	

program, according to which individual companies can develop a safety strategy tailored to their specific processes. This strategy should be based on integrating technologies, procedures, and management practices. The standard details the elements and performance measures so that each facility can develop the most appropriate management system to meet its particular needs. Representative items to be included in the development of such safety plans include, but are not limited to,

- Employee involvement in process safety management. The act states that employers are to consult with their employees and representatives regarding the employers' efforts to develop and implement a process safety and hazard assessment program.
- Compilation of process safety information. Complete and accurate written information concerning process chemicals, process technology, and process equipment is essential to an effective process safety management program.
- Process hazard analysis. A process hazard analysis (PHA) is one of the most important elements of the process safety management program. A PHA is an organized and systematic effort to identify and analyze the significance of potential hazards associated with the processing or handling of highly hazardous chemicals.
- Operating procedures and practices. Operating procedures describe tasks to be performed, data to be recorded, operating conditions to be maintained, samples to be collected and safety and health precautions to be taken.
- Employee training. All employees, including maintenance and contractor employees involved with highly hazardous chemicals, need to understand fully the safety and health hazards of the chemicals and processes they work with for protection of themselves, their fellow employees, and the citizens of nearby communities.
- Contractors. Employees who use contractors to perform work in and around processes that involve highly hazardous chemicals will need to establish a screening process so that they hire and use contractors who can accomplish the desired jobs without compromising the safety and health of employees at the facility.
- Pre-start-up safety. The initial start-up procedures and normal operating procedures need to be fully evaluated as part of the pre-start-up review of any new or modified equipment to ensure a safe transfer into the normal operating mode for meeting the process design parameters.
- Mechanical integrity. Employers will need to review their maintenance programs and schedules to see if there are areas where "breakdown" maintenance is used rather than an on-going mechanical integrity program. Equipment used to process, store, or handle highly hazardous chemicals needs to be designed, constructed, installed, and maintained to minimize the risk of releases of such chemicals.

- Nonroutine work authorizations. Nonroutine work that is conducted in process areas needs to be controlled by the employer in a consistent manner. A work authorization notice must have a procedure that describes the steps the maintenance supervisor, contractor, or other person needs to follow to obtain the necessary clearance to get the job started.
- Managing changes. Changes in equipment, operating procedures, raw materials, and so forth, must be actively managed.
- Investigation of incidents. Incident investigation is the process of identifying the underlying causes of incidents and implementing steps to prevent similar events from recurring. The intent of an incident investigation is for employers to learn from past experiences and thus avoid repeating mistakes.
- Emergency preparedness. Each employer must address what actions employees are to take when there is an unwanted release of highly hazardous chemicals.
- Compliance audits. Employers need to select a trained individual or assemble a trained team of people to audit the process safety management system and program. The audit should include an evaluation of the design and effectiveness of the process safety management system and a field inspection of the safety and health conditions and practices to verify that the employer's systems are effectively implemented.

The standard identifies more than 130 toxic and reactive chemicals and threshold quantities that trigger compliance with this Act. A partial listing of these chemicals and their threshold quantities is given in Table 18.11. These quantities are likely to exceed those present in large-scale photovoltaic module manufacturing facilities. Thus, module manufacturers will probably not be directly covered by this statute. Nevertheless, it is clear that the spirit, intent, and procedures outlined in these regulations should not be ignored.

TABLE 18.11 Partial Listing of Toxic and Reactive Highly Hazardous Chemicals That Present a Potential for a Catastrophic Event at or Above the Threshold Quantity

Chemical Name	Threshold Quantity (lb)
Arsine	100
Dichlorosilane	2,500
Hydrofluoric acid, anhydrous	1,000
Hydrogen fluoride	1,000
Hydrogen selenide	150
Hydrogen sulfide	1,500
Phosphorous oxychloride	1,000

6.1.2. Uniform Building and Fire Codes. The Uniform Building and Fire Codes (UBC/UFC) provide an excellent example of the influence a local community can have on the design and operation of a high-technology manufacturing facility. These codes are used by more than 1,200 jurisdictions in all parts of the United States; they are the exclusive codes used by almost all municipalities west of the Mississippi. They are of special interest because specific codes have been developed for the semiconductor industry. Article 80 of the UBC/UFC, described in detail by Goldberg and Fluer (1987), governs conditions hazardous to life and property from fire and explosion and provides guidance for the issuance of permits. Examples of such requirements include limits on the quantities of hazardous production materials present in any single facility, ventilation controls, fire separations, and use of sprinklers.

6.2. Industry Oriented

The photovoltaic industry has the most to gain by proactively managing its own risks. As noted elsewhere (Moskowitz and Fthenakis, 1990a), this can be best approched through the application of a hierarchical hazard management strategy which includes

1. Selection of inherently safe systems and materials. The most efficient strategy to reduce hazards is simply to choose technologies that do not require the use of large quantities of hazardous materials or use processing options that are inherently hazardous (e.g., high temperature and pressure).
2. Prevention of accident-initiating events. Once specific materials and systems have been selected, strategies to prevent accident-initiating events need to be implemented. Administrative and engineering options (e.g., remote storage, maintenance, inspection and testing, guidelines for system integrity, safeguards against process deviations) should be identified, evaluated, and implemented.
3. Prevention/reduction of releases. The quantity and type of materials released to the environment from a manufacturing facility depend on the production and control technology alternatives used as well as specific standards set by federal, state, and local regulatory agencies. Sources of potential accidental releases must be closely matched with appropriate control technologies.
4. Control/reduction of external releases. If an accident occurs and safety systems fail to contain a hazardous material, especially a gas release, then engineering controls must be relied on to reduce/minimize environmental releases. If the release is confined, it may be possible to divert the contaminated media to pollution control equipment. Releases to open or partially confined spaces, however, are much more difficult to control.
5. Prevention of exposures. Prevention or minimization of human exposures is needed if, in spite of the previous strategies, a chemical release

occurs. This may include remote siting of toxic materials storage facilities, exclusion zones adjacent to plant boundaries, and early-warning systems.

6. Prevention of consequences. If medical interdiction is required, knowledge and availability of medical procedures and supplies at nearby medical facilities can reduce the consequences of human exposures to hazardous materials.
7. Mitigation of consequences. If all else fails, industry in the aftermath of an incident may need to mitigate consequences by implementing relief and health surveillance programs and possibly by providing compensation to exposed individuals.

6.3. Public Oriented

In contrast to occupational health, where many active hazard control strategies are needed, the role of directed hazard management in the public sector is limited. In general, public health will be protected by the establishment and implementation of strategies to manage the release of byproducts to the environment from manufacturing facilities and from waste disposal operations. There are two instances, however, where the public may need to play an active role to protect their own health and safety.

The first situation can occur when the public could be exposed to combustion byproducts from a fire involving a metal (As, Cd, and Se) containing array fields larger than 100 kW_p. In this situation, the public should stay indoors for a short period of time with all windows closed until the transient cloud containing the dust and fume disperses. Procedures for educating and notifying the public (e.g., sirens) should be planned before the installation of such devices. Furthermore, practice exercises should be conducted routinely.

The second role of the public is to help ensure the proper disposal of metal-containing modules (especially CdTe) used in residential applications. Unfortunately, the institutional and technical mechanisms do not yet exist for the controlled disposal of these devices. When institutional and technical options for the proper disposal of these devices are established, the public should be encouraged to use them. Although hazards presented by the uncontrolled disposal of these devices by any single individual are relatively unimportant, they will be of larger concern to state and federal organizations where the inherent hazards associated with the large-scale disposal of modules from thousands of residences will be more carefully evaluated.

7. CONCLUSIONS

The photovoltaic industry must continue to work hard to maintain safe working environments and to produce environmentally acceptable devices. The options for maintaining these safe environments exist. The optimal allocation of limited resources, however, demands that the exact mix of options adopted be carefully matched to the material, processing, and applications of interest. A balance must also be achieved between measurable versus perceived risk. Failure to optimize in these selections may waste limited financial resources and leave health at risk.

REFERENCES

Coutts, T., Kazmerski, L.L., and Wagner, S. (1987), *Solar Cells* **19(3–4)**, 287.

Covello, V.T. (1987), *J. Environ. Psychol.*

Crandall, R.S., Nelson, B.P., Moskowitz, P.D., Fthenakis,V.M. (1992), *Safety Analysis Report for the Use of Hazardous Production Materials in Photovoltaic Applications at the National Renewable Energy Laboratory*, NERL/MP-451-4778A National Renewable Energy Laboratory, Golden, CO.

Fthenakis, V.M., Lee, J.C., and Moskowitz, P.D. (1984), *Amorphous Silicon and Gallium Arsenide Thin-Film Technologies for Photovoltaic Cell Production: An Identification of Potential Health and Safety Hazards 51768*, Brookhaven National Laboratory, Upton, NY.

Fthenakis, V.M., and Moskowitz, P.D. (1987), *Solar Cells* **22**, 303–317.

Fthenakis, V.M., and Moskowitz, P.D. (1990), *Solid State Tech.* **33**, 81–85.

Goldberg, A., and Fluer, L. (1987), *H-6 Design Guide to the Uniform Codes for High Tech Industries*, GRDA Publications, Mill Valley, CA.

Lee, J.C., and Moskowitz, P.D. (1985) *Solar Cells* **18**, 41–54.

Luft, W. (1988) *AIP Conference Proceedings 166—Photovoltaic Safety*, American Institute of Physics, New York.

Hazards Research Corporation (1988), *Fire and Explosion Properties of Phosphine*, 6568, Hazards Research Corp., Rockaway, NJ.

Moscovsky, J.A., Rainer, D., Asom, M.T., and Quinn, W.E. (1992), *Appl. Occup. Environ. Hyg.* **7**, 375–384.

Moskowitz, P.D. (1991), *Curr. Top. Photovoltaics* **4**, 134–181.

Moskowitz, P.D. (1992), *Environmental, Health and Safety Issues Related to the Production and Use of CdTe Photovoltaic Modules* (in press).

Moskowitz, P.D., Coveney, E.A., Rabinowitz, S., and Barancik, J.I. (1983), *Solar Cells* **9**, 1–10.

Moskowitz, P.D., and Fthenakis, V.M. (1990a), *Solar Cells* **29**, 63–71.

Moskowitz, P.D., and Fthenakis, V.M. (1990b), *Solar Cells* **30**, 89–99.

Moskowitz, P.D., and Fthenakis, V.M. (1990c), *Hazard Assessment Chemicals* **7**, 83–109.

Moskowitz, P.D., and Fthenakis, V.M. (1992), *Solar Cells* **29**, 63–71.

Moskowitz, P.D., Fthenakis, V.M., and Kalb, P.D. (1988), *Preventing and Controlling Accidental Gas Releases.* Photovoltaic Safety, AIP Conference Proceedings, American Institute of Physics, New York, pp. 161–174.

Moskowitz, P.D., Fthenakis, V.M., and Lee, J.C. (1985), *Potential Health and Safety Hazards Associated with the Production of Cadmium Telluride, Copper Indium Diselenide, and Zinc Phosphide Photovoltaic Cells, 51832*, Brookhaven National Laboratory, Upton, NY.

Moskowitz, P.D., Fthenakis, V.M., and Lee, J.C. (1989a), *Solar Cells* **27**, 149–158.

Moskowtz, P.D., Fthenakis, V.M., Morris, S.C., and Hamilton, L.D. (1989b), *Approaches for Identifying, Characterizing and Managing Risks from Accidentally Released Toxic Gases.* Hazard Assessment and Control Technology in Semiconductor Manufacturing Meeting, American Conference of Governmental Industrial Hygienists, Lewis Publishers, pp. 271–284.

Moskowitz, P.D., Fthenakis, V.M., Youngblood, R.W., and Mendez, S.R. (1989c), *Evaluating Risks Associated with the use of Silane*, Hazard Assessment and Control Technology in Semiconductor Manufacturing Meeting, American Conference of Governmental Industrial Hygienists, Lewis Publishers, pp. 97–114.

Moskowitz, P.D., Kalb, P.D., Lee, J.C., and Fthenakis, V.M. (1986), *An Environmental Source Book on the Photovoltaics Industry, 52052*, Brookhaven National Laboratory, Upton, NY.

Moskowitz, P.D., and Zweibel, K.M. (1992), *Recycling of Cadmium and Selenium from Photovoltaic Module Production and Use: A Workshop Report* (in press).

Moskowitz, P.D., Zweibel, K.M., and Fthenakis, V.M. (1990), *Health, Safety and Environmental Issues Relating to Cadmium Usage in Photovoltaic Energy Systems, 44946*, Brookhaven National Laboratories, Upton, NY, SERI/TR-211-3621, DE9000310.

National Institute for Occupational Safety and Health (1987), *NIOSH Alert-Request for Assistance in Reducing the Potential Risk of Developing Cancer From Exposure to Gallium Arsenide in the Microelectronics Industry*, U.S. Department of Health and Human Services, Cincinnati, OH.

Occupational Safety and Health Administration (1990), *Fed. Reg.* **55**, 4052–4147.

Occupational Safety and Health Administration (1991), *Fed. Reg.* **56**, 47348–47349.

Occupational Safety and Health Administration (1992), *Fed. Reg.* **57**, 6356–6417.

U.S. Environmental Protection Agency (1984a), *Health Effects Assessment for Cadmium*, EPA/540/1-86-038, Cincinnati, Ohio.

U.S. Environmental Protection Agency (1984b), *Updated Mutagenicity and Carcinogenicity Assessment of Cadmium—Review Draft*, EPA-600/8-83-025B, Washington, DC.

U.S. Environmental Protection Agency (1987), *Municipal Waste Combustion Study: Assessment of Health Risks Associated With Municipal Waste Combustion Emissions*, EPA/530-SE-87-02, Washington, DC.

CHAPTER NINETEEN

Photovoltaic Market Analysis

ROBERT HILL, Newcastle Photovoltaics Application Center, University of Northumbria, Newcastle-upon-Tyne NE1 8ST, England

1. INTRODUCTION

The photovoltaic (PV) effect was discovered in silicon at Bell Labs in 1954 at a time when they were urgently seeking a replacement for batteries to power their remote telephone lines. In those days of valve (tube) equipment, the need to recharge batteries was frequent and expensive. The first market for silicon photovoltaic cells was therefore to recharge these batteries in the Bell system. However in those early days, PV was even more expensive and the project was dropped.

PV continued to be developed for the emerging space program under the inspirational leadership of William Cherry, and since Vanguard I in 1958 almost all space satellites have been PV powered. Throughout the 1960s PV had a small but steady market and access to research and development (R&D) funds from government agencies far in excess of those that could have been allocated on a purely commercial basis. The very active R&D in silicon cells and in its challenger Cu_2S/CdS thin-film cells led to steady progress. Silicon cells were always bound to win this contest because their performance gained enormously from the investments in material preparation, device preparation, and theoretical understanding made by the electronics industry.

The activity in terrestrial applications was very small, although by the early 1970s the first company specializing in PV had been set up, when Joe Lindmeyer left Comsat to found Solarex. Following the oil shock of 1973, the National Science Foundation set up the Research Applied to National Needs program and a meeting was held in Cherry Hill in 1974 to consider PV as part

Solar Cells and Their Applications, Edited by Larry D. Partain.
ISBN 0-471-57420-1

of this program. The market was perceived as GWatt-sized central ground-based stations in the sun belt. The improvements over 1974 PV performance needed to meet the cost goals were enormous, but to the credit of all concerned, these were perceived, not as obstacles, but as challenges, and these programs set the goals and the ethos of the PV community that exist to this day.

The cost and performance goals are still those needed for large central ground-based stations, and the PV community and the PV industry are still dominated by technology and technologists. The industry is beginning to change and a very few companies are now market-led, but the change from technology-push to market-pull as the dominant force in the industry is essential if PV is to take its rightful place as one of the major technologies of the twenty-first century.

If the PV industry is to become market-led then it needs to understand its markets and feed market intelligence back into its investment and production processes. All companies must make predictions about future markets to inform their investment decisions. These are usually based on an analysis of past trends and a best guess at their extrapolation into the future. This can be done at the aggregate level but is most useful at the disaggregated levels, for each market segment. This provides no feedback to the company as to the needs of present or future markets or new market opportunities, so additional marketing effort is required to obtain this market intelligence. The techniques for doing this will depend on the nature of the market and on the type of customer. There is a wide variety of materials and technologies that the PV industry can use to meet the needs of its customers in different market segments, and it is useful to review these options and relate them to possible market opportunities.

2. PV TECHNOLOGY OPTIONS

There are two broad routes to achieving cost and performance goals. The first is to drive for high efficiency and then seek innovative materials processing techniques in large-scale production to bring down costs. The other route is to devise low-cost technologies and then work on the materials and device science and engineering to increase the efficiency. These routes are typified by the single crystal and the thin-film technologies, respectively. Both of these routes have been advocated for both space and terrestrial cells.

For space cells the requirement of maximum power/mass ratio to minimize launch penalties and maximum power/unit area to minimize drag in the residual atmosphere have favored the highest efficiency cells as discussed in Chapters 4 and 17. Single crystal silicon wafer cells have been the standard since the start of the space program, and these have generated such a long track record that the space industry has high confidence in their performance.

Solar cells in space must withstand the radiation environment, and the crucial parameter is the cell efficiency after radiation equivalent to $10^{15}\,cm^{-2}$

of 1 MeV electrons. Silicon degrades to around one-half to two-thirds of its initial efficiency after this dose, while GaAs-based cells degrade by about 20% and InP-based cells by about 5%. Thin-film cells do not stop most electrons and protons and so degrade very little. There is thus a strong interest in moving to the III–V materials, which have not only better radiation hardness but also a higher starting efficiency. The thin-film materials could be of interest for satellites in geostationary orbits where drag from residual atmosphere is negligible and the potentially high power/mass ratio could have significant launch benefits.

In terrestrial PV markets, the proven reliability of silicon wafer cells will allow this technology to retain a major market share throughout the 1990s. The various thin-film technologies discussed elsewhere in this book are likely to challenge, silicon wafer cells from about 1995 onward. It seems unlikely that modules using silicon wafer cells will be able to be sold at much less than $\$2/W_p$ even for bulk purchase from large-scale (100 M watt-peak [W_p] per annum [p.a.]) production plant. Within this decade we should see commercial silicon wafer cells with efficiencies around 20%, giving 18% modules. The thin-film materials are unlikely to match this performance. The copper-indium-diselenide (CIS) and CdTe technologies should be capable of module efficiencies around 14%, although the thin-film multicrystalline silicon-on-ceramic could be developed to give efficiencies close to those of silicon modules if integrally integrated module production becomes possible. Hydrogenated amorphous silicon (a-Si:H) must overcome its photodegradation problem if it is to have a future for power production, although it has advantages in other market sectors that will continue to provide market opportunities, particularly for double or triple junction cells.

The cost of a PV module to a customer is very dependent on the scale of the PV production and distribution systems and of course on the size of the customer's order. In small (1 MW_p p.a.) production the costs of capital and labor are predominant. In the thin-film technologies, vacuum processes are expensive to install and give high unit costs at small production rates (Ricaud, 1992). Unit costs fall considerably as production rate is scaled up to 100 MW_p p.a., by which time most of the benefits of scale have been achieved. The problem with investing in large production facilities is that the output must be sold, so the market must be capable of absorbing the output of the plant, and the company must be capable of distributing that output. The profitability of manufacture is quite critically dependent on the plant achieving its design throughput and on selling the product. The scale of manufacture can thus increase only in line with market size. Of course, market size will rise as the module price falls, so the choice of plant size for a new investment calls for a fine judgement on the market size and company market share some years in the future.

At the present time, there are only two commercial PV technologies, wafer silicon and a-Si:H. There is a continuing debate between crystalline and multicrystalline wafers, and there is concern for the future availability of low cost feedstock—the tops and tails of boules from the electronics industry—since the demand from the PV industry could outstrip the supply. Through the 1990s new PV technologies such as CIS, CdTe, thin-film multicrystalline-Si, and possibly the silicon sphere technology of Texas Instruments, will become commercial. These new materials offer both lower cost (potentially) and new market opportunities. The a-Si:H already supplies the demand for various sizes of plate for original-equipment-manufacturer (OEM) products. The advent of the new thin-film technologies will widen these markets considerably if production is organized so that plate can be cut to a range of sizes to satisfy the different needs within the OEM market.

It seems likely that there will be no single "PV technology," but a range of products with the different materials, each aimed at specific market niches, although the largest potential market, grid-connected power, is likely to see all the different technologies competing for market share.

3. MARKETING PHOTOVOLTAICS

Marketing of any product involves identifying potential customers, discovering their needs, and meeting those needs with the product. The "customers" in the different segments of the PV market are very different. For space solar cells the customer is usually an array manufacturer who is contracted to the satellite manufacturer who has the prime contract from the purchaser of the satellite. The cells chosen must meet the approval of all three. They are technically sophisticated with a paramount concern for reliability, and they are cautious about any product that does not have a track record. They are much happier with gradual developments than with radical innovations. For terrestrial cells, the professional market has similar customers among the utilities, telecommunications companies, and so forth, and these companies are used to making decisions on a life-cycle cost basis. All of these customers want to produce electricity from PV in a specific location with high reliability at the lowest cost. In most of the markets for PV the customer is not buying electricity, but is interested in some specific service: lighting, sound, water, and so forth. The cost and reliability of the whole service-providing system is the key concern and the cost is usually the up-front capital cost of the system, which they compare with the up-front cost of alternative ways of providing the service.

All potential customers, before making a purchase, ask themselves five key questions about a product:

(1) Does it do the job?
(2) Is it better/cheaper than alternatives?
(3) Can I afford it?
(4) Can I get it/maintain it?
(5) Does it fit in with the way I do things/my lifestyle/my self image?

The PV industry has spent many years demonstrating a positive answer to the first question in a wide variety of applications, and new applications for PV appear frequently. Similarly, positive answers to the second question have been shown in careful studies of many applications from distributed or embedded generation in utility grid systems to rural electrification in developing countries. It is important to be clear just what the competing alternatives are, since they are sometimes not other sources of electricity. For instance, the alternative to embedded generation is line up-grading, while the cost of alternatives to PV-powered marine buoys is mainly the labor cost of replacing gas cylinders. The full cost of providing the particular service must be included in the comparisons.

The third question confronts the issue of the up-front cost of PV. In the "First World", if we wish to make a capital purchase, we arrange a loan, and pay off the loan over time. The absence of such credit facilities is one of the major impediments to the use of PV systems in developing "Third World" countries, along with the very high interest rates charged on personal loans in these countries. The solution to this problem of converting the up-front capital cost into an affordable recurrent cost is the key to opening major markets in developing countries and should be a focus for international funds from the United Nations or World Bank.

Question 4 relates to the distribution and maintenance network that is essential if PV systems are to gain widespread public acceptance. Establishing such networks is expensive and time consuming. It calls for extensive and intensive training of local personnel and the organization of supply chains for systems and spares.

The final question is the one addressed by the usual advertising for, say, cars or clothes, but it is also a serious issue for PV. A utility engineer charged with maintaining a secure supply is not used to dealing with generators whose output is beyond their control, and they need much convincing that such variable supplies have a place amongst their generator technologies. Equally, a farmer in, say, Pakistan who practices flood irrigation has little use for a PV pump that could supply his daily water usage, but only at low rate and spread over all the daylight hours. The use of PV in both cases calls for a major change in their working practices, and they will consent to this only if they perceive other very clear benefits.

The collection of individuals, institutions, and companies who can or could answer the five questions in the affirmative constitutes the potential market for photovoltaics. Taking all the different market segments, and assuming that supply, maintenance, training, and financing structures were all in place, the potential total market probably includes just about every person, company, and institution in the world. We are far from that situation at present, although solar calculators have penetrated almost everywhere in the industrial countries. We must differentiate between real present markets where the five conditions are already met and potential future markets that could open if conditions change and investments are made in organization, training, and financial infrastructures.

The large and growing PV market in Mexico results from the decision of the Mexican Government to demand that the state electricity utility provide electricity to all areas of the country. The utility found that PV systems were the cheapest way of discharging this responsibility in rural areas and set up the structures needed for this task (Huacuz Villamar, 1992). In this case the alternative to PV was "no electricity," and the Mexican Government considered that the social and political cost of this alternative was too high. The political will to counteract the greenhouse effect is also causing governments to consider social and environmental costs, and this is changing the ground rules in economic comparisons of PV and conventional electricity generation. Political decisions such as these can transform potential markets into real markets.

In the real markets, the penetration of a product into the market is usually considered to follow an "S-shaped" curve. This is exemplified beautifully by the sales of solar calculators, shown in Figure 19.1. Only at the bottom end is this curve approximately exponential, so extrapolations of past trends on the basis of a constant percentage growth need to be made with care unless it is clear that the market is far from saturation. Apart from some types of consumer products, this seems to be the case in general for PV. The following sections on future PV markets are estimated on the basis of extrapolation of past trends, with due allowance for the effects of possible future political decisions.

4. SOLAR CELLS FOR SPACE SATELLITES

The environment to which space cells are subjected is very different from that for terrestrial cells, and the criteria to which they are optimized are also quite different. Space cells must withstand the large forces induced during launch and the large temperature and voltage cycles induced by moving into and out of the shadow of the earth at regular intervals. The major stress for space cells is, however, the radiation encountered in space, as discussed in Chapters 4, 13, 17, and 21.

The radiation environment in space consists mainly of electrons and protons with energies from a few keV upwards. The glass coverslips on the cells stop most of the low energy protons, but the electrons and protons that stop in the active region of the cells cause radiation damage and reduced efficiency. The density of electrons and protons varies with altitude and reaches a peak in the Van Allen belts, so most satellites orbit below or above these belts and transfer through them as rapidly as possible.

Solar cells made from III–V materials are much less sensitive to radiation damage than wafer silicon cells, while thin layers of material allow most electrons and protons to pass straight through and thus suffer little radiation damage.

The lifetime of a space satellite is, barring accidents, governed by the length of time it can maintain itself in its design orbit. There is a tenuous residual

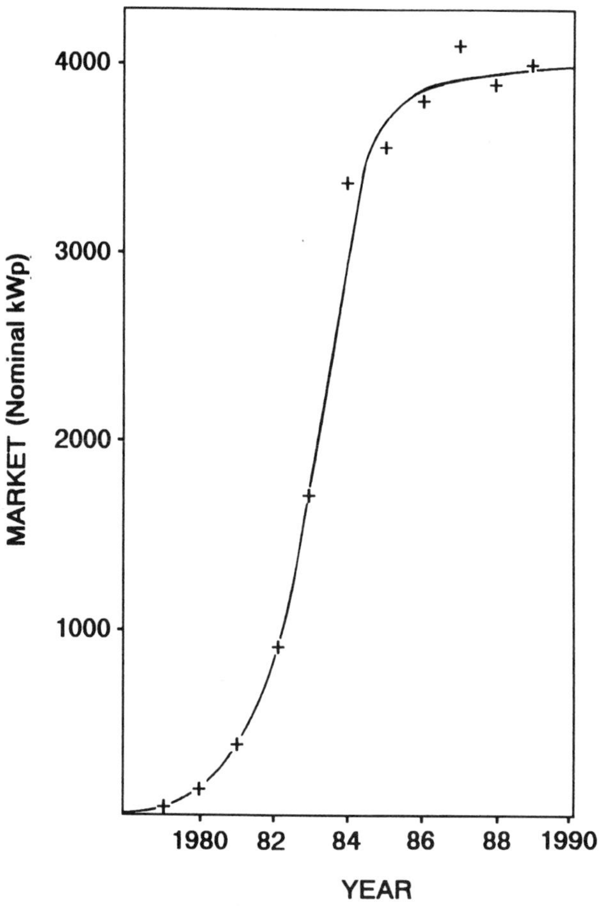

FIGURE 19.1 Sales of solar calculators between 1979 and 1989.

atmosphere extending far out into space. The density of this residual atmosphere decreases with altitude, but even at geostationary altitudes the drag, plus the gravitational effects of the moon, can shift the satellite off station. Satellites have small rocket motors to correct for these influences, and the lifetime depends on how long the rocket fuel lasts. The greater the perturbation of the orbit, the more often the rockets must be fired, so fuel lasts for a shorter time.

The solar array must power the satellite up to the end of its lifetime, so the array is sized according to the efficiency of the cells at the end of the satellite lifetime. To aid comparisons, there is a standard definition of lifetime in terms of radiation dose, and end-of-life (EOL) efficiency is defined as that after $10^{15}\,cm^{-2}$ of 1 MeV electrons. The cell efficiency at beginning-of-life (BOL) is higher than that at EOL, and EOL/BOL efficiency ratios vary from about 0.6 for silicon cells and about 0.8 for GaAs-based cells to about 0.95 for InP and thin film cells. Since BOL efficiencies for III–V cells are around 18%–20% compared with about 14% for silicon, the array size for a given EOL power is around 70% larger for silicon cells than for GaAs cells.

The cost of a complete fold-out solar array for a satellite is around \$0.25M m^{-2}. If this array contains silicon cells, these cells might account for \$10,000 or so of this cost and generate about 120 W m^{-2} at EOL. GaAs cells would cost around \$100,000 and generate about 200 W m^{-2} at EOL. A 2 kW_p array would thus cost about \$4.2M and have an area of 16.7 m^2 with silicon cells. With GaAs cells the area is reduced to 10 m^2 with an array cost of about \$2.4M and cell costs of \$1.0M. These figures are illustrative only, but they do show that the high area-related BOS costs of satellite arrays provide a very high incentive to achieve the highest possible EOL efficiency. With InP, which might cost at present \$800/W_p or \$200,000 m^{-2}, the area of a 2 kW_p array would be reduced to 8.2 m^2 at a cost of \$2.0M with cell costs of \$1.6M. The additional benefits of reduced launch mass and hence costs, and the extended lifetime resulting from the smaller array size, all enhance the pressure to drive for high EOL efficiency even at the expense of high cell costs.

These arguments on the importance of high EOL power per unit area apply particularly to satelites in low earth orbit (LEO), i.e., a few hundred kilometers altitude. For high earth orbit (HEO) and geostationary orbits (GEO), launch costs can be large and the power/mass ratio becomes an important parameter. The drag from the residual atmosphere is much reduced and gravitational effects on the orbit become important. If plans for solar power satellites ever became a reality, they could use roll-out arrays of thin cells on lightweight backing to provide the 10 GW_p or so of projected power.

There have been a number of comparative studies of the suitability of the different cell technologies for different space missions. Typical results of such studies are shown in Table 19.1 with higher numbers indicating better relative choices.

The data in Table 19.1 reflect the technical performance of the different types of cell in LEO and GEO orbits, but do not include factors such as cost, reliability, or handleability in array manufacture, which all influence the final choice. The necessary caution of satellite manufacturers and operators toward new technology is a major factor in the speed with which new materials are adopted. A long period of trial panels, small subarrays, and so forth must be undertaken to demonstrate reliability and confirm the predicted performance. GaAs cells have emerged satisfactorily from the trials and have been used as the primary power source in some satellites. They are now serious commercial contenders with silicon cells.

Since the beginning of the space age, the trend has been toward larger, heavier, and more complex satellites, with ever increasing power demands. Such satellites now have power demands averaging about 3 kW, but some plans for the space station show demands up to 100 kW. These large satellites are highly complex multifunctional systems. They take years to build and cost up to \$500M. Any failure can cause major setbacks in, for instance, telecommunications services and has very serious financial implications, beyond the loss of \$500M worth of satellite.

TABLE 19.1 Marks Out of 10 for Technical Performance of Different Types of Cell

	Orbits	
Cell Type	LEO	GEO
Si	7	9
GaAs	10	9
GaAs/Ge	10	7
InP	9	10
Thin film	5	8

As discussed in Chapter 21, there is growing interest in smaller satellites with specific functions that are quicker and cheaper to build and can often be launched as companions to a large satellite. These satellites require much less power, perhaps up to 500 W for a minisatellite and under 100 W for a microsatellite, so it becomes possible to mount the cells on the body of the satellite rather than on fold-out wings. Clearly each of these small satellites uses less solar cells than a large one, but the numbers likely to be launched will probably represent an increase in the total space cell market. Because the cells are body mounted and area is thus limited, the power per unit area is a crucial parameter, and the III–V materials have significant advantages over silicon.

The world market for space solar cells is around 40 kW p.a. for commercial applications and a similar or larger amount for government and military satellites. With the constant increase in telecommunications traffic and global networks and the increasing needs of remote sensing, the commercial market could double by the end of the 1990s. If some of the more ambitious global communications proposals such as Iridium become reality, then the market could be over 100 kW p.a. at the turn of the century.

It is difficult to predict the noncommercial, mainly military market in the present circumstances, but it seems unlikely for instance that the Russian space program will match that of the former U.S.S.R. The space cell manufacturing facilities in Russia are underutilized and are beginning to look for markets outside and could add to the competition through the 1990s. The military market in the NATO countries may not expand rapidly, but is unlikely to fall.

The future market size and character for space cells is determined by the market for space satellites. More and more regions want to have satellites dedicated to their communications needs and the actual and potential orders for large GEO satellites total around 50, most of them larger and more powerful than their predecessors. There is, however, an interesting development in the use of fleets of LEO satellites to provide telecommunications services at potentially lower cost than the GEO-sats. The lifetime of GEO-sats is increasing from about 8 to around 15 years, mainly from more accurate launch positioning saving on thruster fuel. The LEO fleets, 77 working satellites plus spares in orbit and on the ground in the case of Iridium, must be launched quickly to have the whole network in place, with around 10–15 launches a year of replacement satellites once the system is fully operational. Extending the lifetime of these LEO-sats reduces the frequency of replacement launches and hence significantly reduces the operational costs. The use of III–V materials for the cells on these satellites could thus be of great commercial benefit.

These developments in satellites markets influence not only the size of the space cell market but also the optimum material. For the large GEO commercial satellites, where total cost in orbit is a major consideration, thinned high-efficiency silicon cells are likely to be specified for some time to come. If the geostationary orbits become so crowded that Molniya orbits have to be used, then radiation hardness will be of crucial importance, and such satellites are likely to use InP cells. The smaller satellites in LEO are likely to turn increasingly to III–V materials, probably GaAs/Ge, and would be the main market for multijunction cells when these become commercially available.

The present domination of silicon cells in the \$200M p.a. world market is very likely to be eroded through the 1990s by GaAs and its alloys, as the market expands to around \$500M p.a. at the turn of the century. Beyond that time, the dominance of the III–V materials and alloys, probably on Si or Ge substrates, will be fully established, and GaAs may well be replaced by InP for orbits where radiation resistance is important.

5. TERRESTRIAL MARKETS FOR PV

The terrestrial markets is very diverse, using different products and often with different criteria for defining the "best" product. To analyze the market for PV it is essential to divide it into market segments that are relatively homogeneous in their characteristics. There are many ways of subdividing the PV market, none of them fully satisfactory, but a useful set of categories are consumer products, industrial markets, remote communities and grid-connected systems.

5.1. Consumer Products—PV for Families or Individuals

These products are bought by individual consumers from shops or mail-order catalogs. The marketing of PV by the PV companies is then to product manufacturers or retail chains. Sometimes PV companies will themselves manufacture consumer products, but this is rare, and it is likely that there will be increasing specialization as the market expands. PV consumer products are themselves quite diverse, characterized only by the fact that they are bought by individual consumers. We can identify three separate categories within products each with their own characteristic.

5.1.1. Domestic Power Supplies. This category includes PV systems for houses remote from the grid, holiday homes, caravans (mobile homes), boats, and so forth. The PV system provides power for lighting, appliances, and so forth and is sized to provide reliable power either all year round for remote houses or during leisure periods for the other applications. For the systems designer, this category is identical to the other remote applications, but in marketing strategy they are quite different. In this category the purchaser is the homeowner using his or her own money and supply of the product to the purchaser is via the retail distribution system.

5.1.2. Individual Power Supplies. In this category the consumer is buying a product in order to enjoy a specific service. The manufacturer of the product has incorporated PV because it makes the product cheaper, enhances its market appeal, or allows a novel product to be developed. Within this category come garden lights, fountains, home security systems, sun-roofs or fans for cars, battery chargers, and personal electronics.

The PV power is usually an a-Si:H silicon plate, cut to the product manufacturer's specification to fit onto or into the product. The concern is to generate the required voltage, usually between 3 and 12 V, in a restricted area, which gives the integrally interconnected plate a distinct advantage over the wiring up of individual cells. Current requirements are usually small, although the increased current density available from CIS or CdTe cells would be welcomed. CdTe would be particularly appropriate, as fewer cells would be needed to provide a given voltage.

The cost of electricity from PV is quite irrelevant in these applications, as the alternative is usually a small-sized battery from which electricity costs around $1,000/kW-h. The cost per piece is, however, quite critical to the economics of manufacturing the whole product, and the cost of cutting and packing can be a large fraction of this cost.

Marketing PV to the potential manufacturer is often by building demonstration units to prove the principle. The competition is between the different suppliers or sometimes with Hong Kong minimodule suppliers, and the winner is the supplier at the lowest cost, even though, if anyone bothered to do the sums, it comes out at $20–50/W_p$.

5.1.3. Indoor Applications. These applications include such things as calculators, watches and clocks, fans, toys, instruments, and novelties. Sizing PV to provide even tens of milliwatts can be difficult because of the small areas available and the low power in indoor lighting (even 1,000 lux is only about $3\,W\,m^{-2}$) (Pearsall and Hill, 1989). However, many of these applications call for microwatts at a few volts, and a few integrally interconnected cells each of a small area are sufficient. For larger intermittent power demand, the cells can charge a small battery or capacitor in, for instance, toys or fans, and of course this is also the case for clocks or watches, which must work throughout the hours of darkness.

There are two selling points in these applications: (1) "PV means never having to change a battery!" and (2) the possibility to develop novel products such as credit card calculators. The marketing is to the manufacturers of the products, often by demonstration, although so many of these products are now widely available that other manufacturers whose products might use PV have some confidence in the technology.

The consumer products market is often ignored or treated as outside the main concerns of the PV industry. There are two good reasons why this attitude is wrong. The first is that the consumer products can sell in millions and some do—100M p.a. for the solar calculator—and even though the power per product is small, the market as a whole is substantial, perhaps a quarter of PV output. The second reason is the influence that such products have on the public perception of PV. If it becomes an established part of their everyday lives through consumer products, they will be much more receptive to proposals that it could help meet the needs for bulk electrical power. It is good advertising for PV as a whole to the general public and also to opinion formers and politicians whose decisions are important to the future growth of the PV industry.

Consumer products are market-led in a purely commercial environment. The growth of the market is influenced by general economic factors such as the 1993 world recession, but it benefits from the strong entrepreneurial spirit of the consumer products business. Figure 19.2 shows estimates of the growth of this market to the end of the 1990s, based on an extrapolation of past trends with due allowance for market saturation effects.

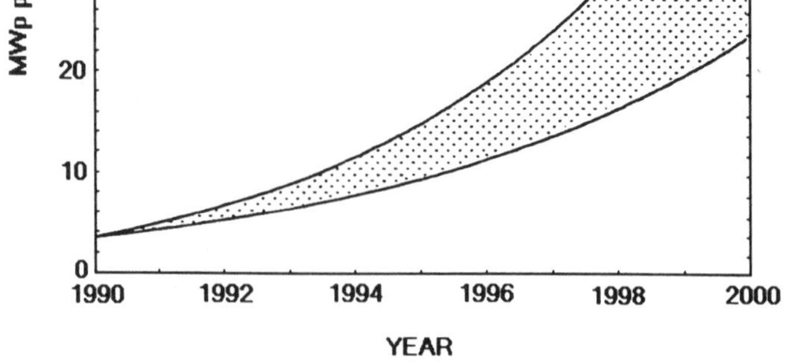

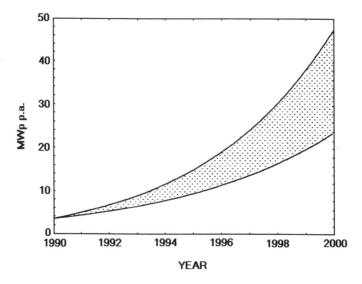

FIGURE 19.2 Projections for the sale of consumer products up to the year 2000.

5.2. Industrial Markets

The industrial markets, or Professional Systems as they are often called, are characterized by the sale of PV modules or systems to industry, which then uses PV for its own purposes. This could be for purposes within the purchasing company or the company could sell-on the PV system as a package with its own products.

The industrial market is not homogeneous, and there are three broad categories that need to be considered: communications, cathodic protection, and remote power. PV is cost-competitive in these markets because of the high costs associated with refuelling and maintaining internal combustion engines or in changing batteries or liquid gas cylinders. It is essential in the economic comparisons to consider the total lifetime capital and operational costs of PV and alternative systems.

5.2.1. Communications. The category covers all forms of communication, including relay and repeater stations for telecommunications, monitoring control and reporting stations, and warning and signal lights on land and sea. Selling to industry, particularly the technical manufacturing industry, calls for a different approach from the consumer products market. Companies in the communications industry buying a PV system to power, for instance, a relay station, are very concerned to ensure that the PV system is at least as reliable as their relay station, so that their commercial reputation is not adversely affected by problems with the PV. They will usually have, or quickly acquire, a technical expertise in PV systems so that dealings with the PV industry are on a technically literate level. The sizing of PV systems is also different in professional and consumer systems, because the acceptable Loss-of-Load Probability (LOLP) is quite different. A LOLP of 0.1%–1.0% (around 1 day down p.a.) is usually acceptable in consumer systems, while a LOLP of 0.001%–0.01% would usually be demanded in say the telecommunications field. This may entail the use of PV–diesel hybrid systems to keep costs down (Dichler, 1989). The demand for reliability by the communications markets is similar to, though not as extreme as, that of the space industry. However, in space the irradiance is almost constant, except when the satellite is in shadow so the solar input is predictable throughout the satellite lifetime. On earth, the solar input is randomly variable about mean daily and seasonal variations and is predictable, and only on a probabalistic basis, only if good solar data sets are available. The estimation of LOLP in areas where there is no prior experience or measurement is therefore very difficult, and PV systems are often oversized to compensate for the uncertainty or have too high a LOLP because the PV system has been undersized to maintain a competitive price. As solar data throughout the world improve and PV companies and PV user companies gain operational experience, the sizing for a given LOLP becomes more of a science and less of an art. The market can then develop in a more orderly manner, with user industries issuing realistic specifications and the PV industry responding with realistic sizing. Since reliability is so important, particularly in telecommunications, there is a strong preference among the PV-buying industry to stick with wafer silicon modules. Only when other materials have a proven track record in the field will they have the confidence to move away from the present industry standard.

5.2.2. Cathodic Protection. Metal structures require protection against corrosion. The simplest protection is a coat of paint, but where the potential for corrosion is high then cathodic protection is used. A voltage is applied to the structure to be protected, against a sacrificial anode, which is large enough to oppose the electrochemical potential difference of the two metals, so the anode corrodes instead of the structure. The voltages required are small and DC, but the currents required to protect large structures can be quite high. The application is thus ideal for PV in most circumstances. The sizing of the PV system is not as critical and LOLP values of 0.1%–1% or so are quite acceptable since corrosion is a slow process.

The alternatives to PV are mainly small diesels or perhaps thermoelectric generators on gas pipelines, using gas bled from the pipeline. In both cases maintenance costs can be very high, and in hazardous areas with flammable gases PV has obvious advantages. The cost comparisons for PV systems are the lifetime costs of the alternatives, and the cost of corrosion to the structures if cathodic protection is not used.

Replacement of PV modules in cathodic protection systems would often be expensive, and companies will tend to favor PV modules using wafer silicon since they have confidence in its performance and longevity. For the more accessible locations, however, the lower capital costs of thin-film modules could prove attractive and provide the initial opening into this market.

5.2.3. Remote Power. This market sector includes applications such as electrified fencing, perimeter security at large sites, intruder detection, and temporary lighting and other small power on building sites, military sites, farms, and other locations remote from utility supply. The customers for the PV systems are companies and military and other government authorities, and the systems may have to meet civil or military product specifications.

The selling points for PV are "power where you want it," the cost of running long power lines from a utility supply, particularly on a temporary basis, or the cost of maintenance and refuelling small engines. For military applications and in areas such as game reserves or national parks, the silent operation of PV compared with small engines is a significant factor.

The industrial market sector has been growing at about 25% p.a. overall for some years, with cathodic protection being the fastest growing area of this sector. An estimate of the future growth up to the turn of the century is shown in Figure 19.3.

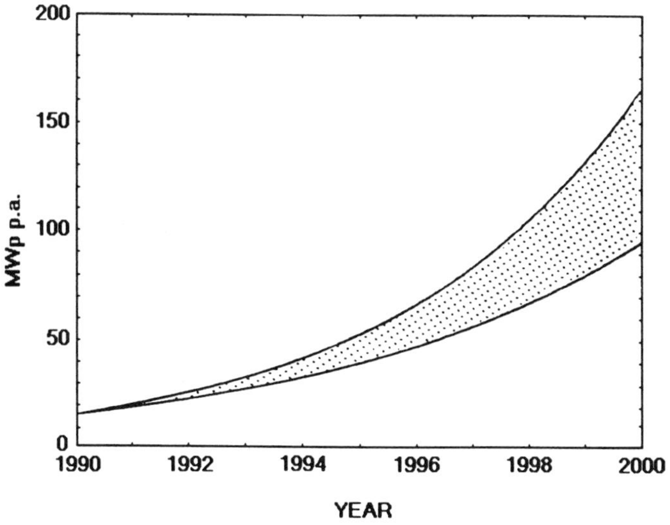

FIGURE 19.3 Projections for the sales of professional systems in industrial markets.

5.3. Electricity Services to Remote Communities

Photovoltaics provides electricity for remote communities in both the industrialized, (as discussed in Chapter 15) and the developing countries. The range of services supplied is very wide and includes water pumping and water treatment, village supplies for domestic and small industry use, medical uses, educational uses, and communications via telephone, television, and radio and battery charging for personal radios, and so forth.

Some of these applications overlap with the domestic consumer products segment discussed in Section 5.1.1, but in this case the equipment is purchased by a utility, government agency, or international agency for deployment in the community. The marketing and, in the Third World, the supply and maintenance implications are very different. In Australia, for instance, the utility is supplying Remote Area Power Supply (RAPS) systems to communities in the outback as a lower cost alternative to extending grid lines. The consumers are charged by the utility on the same basis as urban consumers, as a matter of social equity. In the industrialized countries the utility or other authority uses PV or PV–diesel hybrids such as the RAPS systems, as a good engineering solution to their problems when life-cycle cost comparisons show it to be the cheapest option. In developing countries the potential market is huge, but the problems of introducing PV on a scale appropriate to their needs are daunting.

The "developing countries" are a very inhomogeneous group. About a dozen countries have good technological competence and the capability of using international technologies for their own benefits. In this group, more accurately labeled as "newly industrializing countries," are China, India, Pakistan, Indonesia, the Phillipines, and Malaysia in Asia; Brazil, Argentina, and Mexico in Latin America; and possibly Kenya and Zimbabwe in Africa. These countries have an endogenous capability and can or could produce PV cells, modules, and systems using their own industry and the rate of application of PV is limited more by poor organization and lack of political will power than by technical or industrial factors. The great success of the Mexican PV program shows what can be done when the political will is there and the program well organized.

A further group of 30–40 countries have some indigenous capability, i.e., they have reasonable technical competence in some areas and the educational and institutional infrastructure to make some use of technologies from abroad. These countries need to import some, perhaps most, of the hardware, but can make some components for themselves. Their educational and institutional infrastructures may need external support to cope with the introduction of PV, but their local managerial and entrepreneurial capabilities can be developed.

The majority of countries, about 100 of them, do not have the technical or institutional capabilities to adopt new technologies for their own benefit without outside help. They will need to import the hardware and much of the technical and managerial skills needed to make use of PV and will need overseas aid throughout the 1990s.

The task of bringing all the nations of the world up to a high level of technical and managerial competence is rather beyond the capabilities of the PV industry, but it is possible to identify the capabilities required to promote the widespread application of PV and give appropriate support to those countries that are trying to develop the necessary infrastructure. There are four areas of infrastructure that need to be in place before a country can develop an endogeneous capability in PV. The educational infrastructure must provide an adequate output of people at each technical level with the capability of being trained in PV and giving training in PV. The institutional infrastructure must provide the managerial competence to organize industry and commerce and provide efficient planning, finance, and marketing networks. The industrial infrastructure must be able to provide either directly or by trade the inputs needed to manufacture PV cells modules and systems. Finally, the political infrastructure must provide consistent and helpful policies within which the PV industry can operate and allow free expression of user needs and the development of entrepreneurial skills.

With these infrastructures in place it is possible to develop the expertise in installation and maintenance, the supply and distribution networks and financial credit facilities needed at the point of scale that are a necessary, but not sufficient condition for the growth of the PV market. It is also necessary to have a PV "product champion" at the decision-maker level to ensure that the other processes of government coordinate with the growing use of PV, preferably with the active support of the energy utilities. In almost all developing countries, internal efforts at social and economic progress are very dependent on external factors, such as world trade and debt interest payments, levels of international and bilateral assistance, and global funds for energy investments. The international pressures resulting from environmental concerns and the emergence of PV product champions in some of the international funding agencies are positive factors in promoting PV.

The rural markets in developing countries include village electrification, power for medical centres and schools, communications, and domestic supplies for key personnel. These are usually funded by government agencies or utility companies and often with external assistance. Other markets include lighting for market stalls and small shopkeepers who could pay cash, and such surveys as have been done suggest that this market is just waiting to be opened up by the development of distribution networks.

The urban markets for PV in developing countries are often ignored but should be regarded as important. These markets include lighting for small shopkeepers who now use kerosine, domestic supplies, and toys for the rich urban elite (who are often the decision-makers) and utility supplies. Grid-connected systems will be discussed in the next section but all the arguments for distributed and embedded generation apply with much greater force in developing countries where demand is usually growing much faster than can be supplied by conventional fossil-fueled central stations with associated distribution grids.

The markets in the Remote Communities sector are heavily influenced by political and general economic factors. The ability of utilities or governments in industrialized countries to invest in new facilities for communities is influenced by the state of their balance sheet or economies, both of which are less than healthy in most nations at present (1993). As the national economies pick up, so should this market for PV.

The market growth in developing countries depends on the state of the world economy and crucially on any positive forces coming from the environmental concerns expressed at the Earth Summit in Rio de Janeiro in June 1992. The alleviation of some or all of the net outflow of $40 billion p.a. that the developing countries pay to the industrial countries in debt interest (net of all inflowing international aid) would allow major investments in their infrastructures and in rural development. The recognition by the World Bank and others that their previous policies of investing in large-scale energy systems has caused great environmental damage and been of dubious economic benefit to the host country might be reflected in reallocation of investments to more appropriate tecnnologies such as PV. The establishment of the GEF (Global Environmental Facility of the World Bank) and FINESSE (Financing Energy Services for Small-Scale Energy Users, see Chapter 22) are both positive factors in promoting the growth of the PV market.

In view of these political uncertainties, the projection for the future markets in this sector is also very uncertain, but the probable ranges of the market are shown in Figure 19.4.

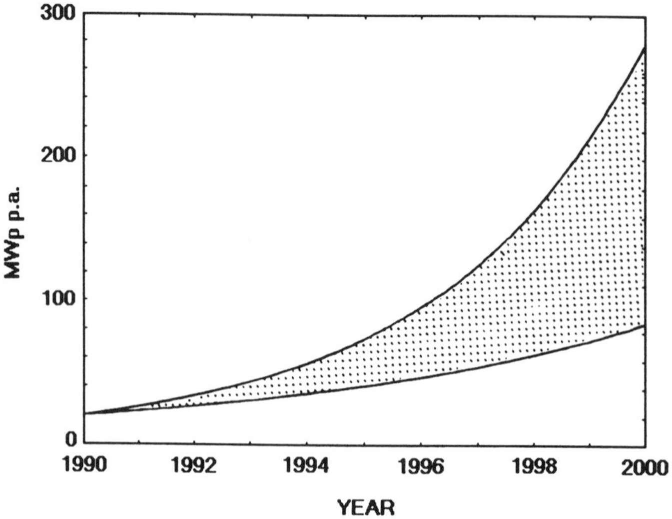

FIGURE 19.4 Projections for the sales to Remote Communities markets, mainly for developing countries.

15.4. Grid-Connected Systems

The large central station PV system feeding power into the utility grid was the vision of the Cherry Hill meeting in 1974, as mentioned earlier. It is still the "holy grail" sought by much of the PV industry, and Chapter 16 discusses the means whereby this market can be developed. There are two grid-connected markets that are much closer to being real market opportunities: embedded generation and distributed generation.

5.4.1. Embedded Generation. The pioneering work of Dan Shugar of Pacific Gas & Electric (PG&E) has shown that the use of embedded PV generation in a utility distribution network can be a cost-effective solution to the problems of overload or power quality at critical points in the network (Shugar, 1990; Ianucci and Shugar, 1991).

When a distribution network has a long line from the generating source, loads along that line can result in poor power quality at the far end of the line. This is reflected in voltage losses from line resistance and high harmonic content and poor power factors from nonresistive loads along the line. If these result in power quality below an acceptable standard then the line and its associated switch-gear and transformers would normally be up-graded. Also, as the loads along the line grow, there comes a time when some critical point on the distribution line/feeder network becomes overloaded, requiring up-grading of lines, transformers, and switch-gear.

Shugar has shown that on the PG&E network it could be cost-effective to install a PV plant feeding power to a substation at the critical point rather than undertake the expense of up-grading. Iannucci and Shugar (1991) conclude that embedded PV generation could cost twice as much as conventional central station generation (e.g., from fossil fuels) and still be cost-effective in carefully selected applications. They also show that with Federal and State solar tax credits it is already cost-effective in one case study that has been carried out. In this paper, Iannucci and Shugar (1991) allocate some benefits (avoided costs) to the environmental benefits of generating electricity from PV rather than the present energy mix of PG&E, to a level of about \$20/kW-year, i.e., about \$0.25/kW-h. There is an emerging consensus that the environmental costs of fossil or nuclear stations are in the region of 5–10 cents/kW-h, considerably larger than that assumed by Iannucci and Shugar and larger than the solar tax credits. If these external environmental costs were to be internalized through legislation or the State Regulatory Commissions, then embedded generation would be attractive at a much larger number of sites. As PV system costs fall through the 1990s this application will be increasingly attractive to a growing number of utilities, and the market for 100 kW_p to 1 MW_p embedded generation PV systems is likely to increase considerably.

The consideration that make embedded PV generation attractive for PG&E apply with much greater force in developing countries. The growth in demand for electricity in these countries consistently outstrips the supply capabilities, with consequent brown-outs and black-outs that disrupt their industrial, commercial, and personal activities. There have as yet been no formal studies on a distribution network of a utility in a developing country, but such studies should be undertaken as soon as possible to unlock a potentially very large market. The increased reliability of supply could bring considerable and quantifiable benefits to the economy of the host country, which should make it an attractive proposition for international aid.

5.4.2. Distributed Generation. *Distributed generation* is used here to denote the supply to individual buildings from PV attached to each building. The buildings are also connected to the grid and take electricity from the grid supply when demand exceeds PV output or feed power into the grid when PV output exceeds demand. It should be noted that the electricity price with which PV generators must compete is the retail price, as distributed generation avoids buying electricity from the utility. The price paid by the utility for electricity exported to the grid depends on governmental regulations in some countries or on negotiations between the building owners and the utility. In Switzerland buildings with distributed PV generation have meters that record only the net import of electricity from the grid, i.e., exports are credited with the same price per kW-h as imports. The best known program for installing distributed generation is the German 1750 Roof project, to install an average of about 3 kW_p on the roofs of 1750 houses. The Swiss program to install 200 MW_p by the turn of the twenty-first century also has a major fraction of installations on domestic rooftops. However, there is a considerable mismatch between domestic demand and PV output that could cause major problems for utility companies if PV becomes very widely used.

In higher latitudes the lower average sun angle, particularly in winter, makes the use of PV as a wall cladding very attractive. The vertical orientation reduces the large summer/winter output ratio, and when applied to commercial buildings there is a reasonably good match between the demand (mainly during working hours) and the PV output. The cost of the conventional claddings for commercial buildings varies from a low of \$100 m^{-2} to as much as \$1,000 m^{-2}. The PV modules are doing two jobs: (1) they act as a cladding for the building, for which \$100–\$1,000 m^{-2} can be charged; and (2) they generate electricity, and only the remainder of the PV system cost is charged against electricity generation. If the total PV system costs fell to \$1,000 m^{-2} and PV replaced a conventional cladding costing \$1,000 m^{-2}, then electricity would be a free byproduct of the PV cladding. Recent studies by the author of PV-clad commercial buildings in England suggest that a module price of \$2/W_p would allow electricity to be generated at a cost about equal to the present retail price, if the PV modules replaced mid-range cladding.

The great advantage of distributed generation from buildings is the very large area that could be covered without employing any additional land area. A typical commercial city center could generate an annual average of about 15 MW per km^2 of land area in the United Kingdom, more in sunnier climates, so the total resource is very large. The lower the latitude, the more beneficial it becomes to place PV on roofs rather than walls, because of the large sun angles, but areas of roofs in most town and city commercial areas are also large, giving the probabilities of a large distributed generating capacity.

This market is probably close to reality in some situations at the present time and will become very large early in the next century.

5.5. Central Ground-Based PV Stations

The large ground-based central PV stations are the most familiar of the grid-connected systems and are discussed in Chapter 16. Central stations such as the one at Carrissa Plains was operated for years with high reliability and a high degree of availability during daylight hours. Its well-known problems with elevated operating temperature and degradation of the ethylene vinyl acetate should not detract from its outstanding performance as a generating system, and such teething troubles are to be expected in the first few such systems. Many other countries are building or operating megawatt-scale central stations to gain operational experience of such systems within a utility generation and distribution network. Large central generating stations are familiar to utility engineers, and their desire to incorporate PV into their normal operational philosophy is quite understandable (see question 5 in Section 3). In California and other hot, dry areas, peak loads are typically due to air conditioners, and so the demand is in reasonable synchronization with PV output. The output from the PV plant can thus be sold as peaking power, with a value to the utility much higher than base-load power. The PV system does, however, have characteristics of base-load plant, high capital cost but low operating costs, so it must be operated so as to use all of the output it can give, with other, higher operating cost plants being taken off-line to accommodate the PV output.

As more PV plant is added to a utility system, the PV output will fill more and more of the load peak and will eventually begin competing with base-load generators. The PV plant is thus competing against different classes of conventional generating plants as the PV fraction of total capacity rises and the costs of electricity from PV must fall to compete with these different plants. The market for PV central stations will thus first be in selected areas for the top of the peak load, and as PV costs fall it will be able to compete for a larger share of daytime peak loads in a wider range of utilities.

Holiday resorts and other leisure facilities usually experience a large increase in electricity demand during the summer months, and a central PV plant could be a useful component of their generating capacity. It could even be a tourist attraction in itself.

Ultimately PV must address the problem that people have the "annoying" habit of wanting to switch on lights at night, when solar energy is not available. Some intermediate storable energy vector is required between the PV plant and the customer, and hydrogen seems at present to be the most likely candidate. It could be envisaged that the hot, dry deserts of the United States, Africa, Australia, India, and so forth could be the major energy suppliers to the world via hydrogen. These markets, however, are likely to come about well into the next century, although it does provide a technical route to the maintenance of electrical supplies with almost no CO_2 emissions and minimal environmental impact.

The rate at which the market for grid-connected systems will grow depends on the timescales over which PV generation can become cost-competitive with other electricity generators and on the willingness of utilities to incorporate PV into their generation mix once it is competitive. The time at which PV becomes cost-competitive depends on the cost of electricity from other sources as well as its cost from PV. It has been estimated that the U.S. taxpayer unwittingly provides $40 billion a year in subsidies to fossil and nuclear fuels, and this does not take into account the environmental costs of these fuels. The fossil fuel and nuclear industries do of course have very powerful "product champions" as well as many voters whose livelihoods depend on them. It will not be easy to ensure that the economic comparisons between PV and the conventional fuels are made on an equitable basis, but it should be part of the market strategy of the PV industry to ensure that this happens. The fact that the major PV companies also have heavy involvement in fossil fuel or nuclear industries does of course give conflicting interests, but others in the PV community could take the lead on these issues.

Except in areas where cloudless skies are the norm, utility engineers worry about incorporating a source that is variable between perhaps 100% and 10% of maximum output on timescales from seconds upward. PV plants are usually given a capacity factor of around 20% of name-plate output and cost comparisons performed on this basis. All of these data are based on solar measurements from one point, solar irradiance data are effectively the output of one solarimeter. However, the input to the grid from a large number of widely dispersed PV plants would have quite different characteristics. The average output from PV plants over geographically dispersed locations is likely to be far less variable and have timescales of its variation of minutes rather than seconds. It should thus be much easier for the utility grid to respond without dangerous transients, and the use of a large PV fraction of capacity should not cause instabilities in grid operation. The data on solar inputs averaged over a large geographical area, and the variability of this average do not seem to be available at present. These measurements should be made over some years to allow good statistical analysis as part of the marketing strategy for selling to utilities.

The market for grid-connected systems is very much influenced by political decisions. Government departments set targets for PV installations not on a commercial basis but as a means of stimulating the PV industry and also as a means of alleviating voters' demands on environmental issues. If all of the plans so far announced did actually come to fruition, there would be about $2\,GW_p$ of PV installed by the turn of the century, much of it grid-connected It might just happen. If the present growth patterns continue, with a 25% p.a. growth rate there would be a world market of $300\,MW_p$ p.a. by the end of the 1990s and around $2\,GW_p$ of PV installed if we include all consumer products. The split in the market segments would have a fairly small fraction of grid-connected systems on a purely commercial basis, certainly less than 10% of the world market. Any prediction of the grid-connected market is thus uncertain as it is so dependent on politically driven decisions. After the turn of the century, the grid-connected market will become a real commercial market, but of a size and with a growth rate that will initially at least reflect the political decisions of the 1990s. Figure 19.5 shows an estimate, with large uncertainties, of the possible growth of the grid-connected market. No split is given for embedded, distributed, or central generation, as this is so uncertain, although it is my belief that the main commercial market in the 1990s will be for embedded generation, with distributed generation rising strongly toward the end of the decade.

6. CONCLUSION

All of the different segments of the PV market are growing but at different rates, and the future mix of the PV industry is likely to be very different from the present one. As has been emphasized throughout this chapter, the two potentially enormous markets—developing countries and grid-connected systems—are very dependent for their growth on political decisions, and there is thus a large uncertainty on their size in the 1990s. They will eventually become

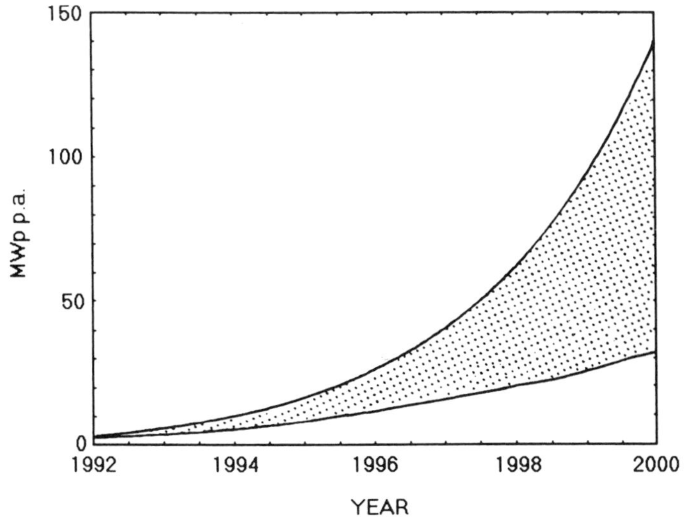

FIGURE 19.5 Projections for the sales of grid-connected systems.

real commercial markets independent of politics, but the rate of which this happens is outside the direct control of the PV industry and provides good targets for lobbying.

The consumer products sector takes a large fraction of PV output at present. The growth of this market depends on a constant stream of new PV-powered products emerging from the consumer products industry to add to the products whose markets are saturating. The entrepreneurial consumer industry is constantly looking for new products, and the PV industry could benefit from a closer liaison with this industry with more attention paid to its needs.

The industrial market has been the staple of the PV industry for many years and is still very far from saturation in almost all applications. It is still the case that many companies whose products could benefit from PV are unaware of PV technology. Part of the future marketing strategy for this sector should be an awareness-raising campaign, with articles on existing and potential PV applications in specialist industry journals in all sections that might become users of PV. There are large numbers of applications that the PV industry has not yet thought of, but appear as problems in some industries. If the specialists in that industry become aware of the potential of PV to solve their problems, another PV market can develop.

The Remote Communities market is driven by the need to provide electrical services to communities not cost-effectively served by grid extension. In countries such as Australia and Mexico, the utility companies take this commitment to customers seriously, and PV is seen as the best engineering solution to the meeting of these needs. In many developing countries the utility has been ignored when PV programs have been set up. As a consequence the utility company sees PV either as an irrelevance or as a threat. If the company can be involved in the PV programs, then their view can be changed and PV can become a means of expanding their role and their revenues. For some, perhaps many, utilities in developing countries, their terms of reference governing their operations make it difficult for them to engage in activities other than central generation and distribution (SADCC, 1991). This is a matter that could be brought to the attention of the governments controlling those utilities by the international agencies and points again to the need for a product champion at the decision-maker level to bring such changes about.

The strategy needed to ensure growth in the market for grid-connected systems has two aspects. The first is the need for demonstration projects to identify any problems and allow utilities to become familiar with the use of PV in their system. These demonstration projects are established for MW_p-sized ground-based central stations in many countries, and the German 1750 Roof program has made a start for distributed systems. Demonstrations of embedded generation and widespread demonstrations of PV-clad buildings beyond the small systems in Switzerland are also needed. There are a number of national plans for the development PV generating capacity. The most ambitious of these is the Solar 2000 program, discussed in Chapter 22. These plans call for this demonstration and market development and will play a crucial role in accelerating the penetration of PV into a true large-scale commercial market.

The second prong of this attack must be to establish the full costs of electricity generated by fossil and nuclear fuels and either to internalize these costs or to credit PV and other renewables with the avoidance of these costs. At present tax credits for renewables are seen as subsidies and not as equalizing payments. There is a need to move as soon as possible to an international consensus on the external costs of all energy technologies so that comparisons can be made on the basis of the full costs to society of the different technologies.

If all of the predictions for the different market segments present in this chapter actually come to pass, then the world PV output in the year 2000 will be in the range of 180–950 MW_p p.a. The upper limit presents an extreme challenge for the PV industry. Growth in output requires prior investment, and the rate of growth is limited not only by the scale of investment received but also by the time needed to master scaled-up production plant (Ricaud, 1992) and the need to source increasing volumes of materials as input to the production process (Bonda, 1992). From a purely financial point of view, the world PV industry could cope with expansion to production levels of 1 GW_p p.a. by 2000 (Hill and Day, 1991), but the problems identified by Ricaud and Bonda may prove more intractable than finding the capital investment. Taking the middle ranges of the projections given in this chapter, a world PV output of 300–500 MW_p p.a. would seem to be achievable on all criteria.

REFERENCES

Bonda, J. (1992), *Prospects for Photovoltaics—Commercialisation, Mass Production and Application for Development*, ATAS Issue 8, R. Hill, E. Omeljanovsky, and I. Chambouleyron Eds., United Nations, New York (in press).

Hill, R., and Day, J. (1991), *22nd IEEE Photovoltaic Specialists Conf. Record*, IEEE, New York.

Huacuz Villamar, J. (1992), *Prospects for Photovoltaics—Commercialisation, Mass Production and Application for Development*, ATAS Issue 8, R. Hill, E. Omeljanovsky, and I. Chambouleyron, Eds., United Nations, New York (in press).

Iannucci, J.J., and Shugar, D.S. (1991), *22nd Photovoltaics Specialists Conf. Record*, IEEE, New York.

Pearsall, N.M., and Hill, R. (1989), in *Applications of Photovoltaics*, R. Hill, Ed., Adam Hilger, Bristol.

Ricaud, A. (1992), *Prospects of Photovoltaics—Commercialisation, Mass Production and Application for Development*, ATAS Issue 8, R. Hill, E. Omeljanovsky, and I. Chambouleyron, Eds., United Nations, New York (in press).

SADCC (1991), *NRSE Technologies and the Electric Power Utilities*, SADCC Annual Technical Seminar, Swaziland, November 1991, Southern African Development Co-ordination Conference, Technical and Administrative Unit Luanda, Angola.

Shugar, D.S. (1990), *21st Photovoltaics Specialists Conf. Record*, IEEE, New York, pp. 836–843.

CHAPTER TWENTY

Solar Resource Characteristics

CAROL RIORDAN, National Renewable Energy Laboratory,
Golden, CO 80401

1. THE RELATIONSHIP BETWEEN SOLAR RADIATION AND PHOTOVOLTAIC SYSTEM PERFORMANCE

The power produced by a photovoltaic (PV) system is the product of incident solar radiation (S) and the PV system's energy conversion efficiency (η): $S \times \eta$. Much of the focus of PV research is on improving η to increase power output. How do we use our knowledge of S to improve PV system performance? Knowing the characteristics of S allows us to choose the locations with the best resources, evaluate design options (such as flat plates vs. concentrators or tracking vs. nontracking systems), avoid overdesign (costly PV systems), avoid underdesign (unreliable PV systems), and evaluate a potential match between the time-delivery of the resource and peak electricity loads. Thus, knowing the characteristics of S allows us to optimize the design (performance and cost) of PV systems and reliably predict the power produced by a PV system using $S \times \eta$.

Solar radiation resource characterization involves measurement and modeling techniques, measurement instrumentation, and multiyear data bases for simulating PV system performance so that we can optimize system design and reduce the uncertainties in the performance predictions. The U.S. Department of Energy's (DOE) Solar Radiation Resource Assessment Project at the National Renewable Energy Laboratory (NREL) is the national solar resource characterization program for the United States. The National Weather Service (NWS) of the National Oceanic and Atmospheric Administration (NOAA) operates the national solar radiation (SOLRAD) network, which provides

Solar Cells and Their Applications, Edited by Larry D. Partain.
ISBN 0-471-57420-1

some of the data for United States resource characterization. Regional and project-specific resource characterizations are performed by NREL and other national laboratories, universities, utilities, instrument manufacturers, climatologists, and others.

The remainder of this chapter describes the complex variability of solar radiation resources, which makes resource characterization a challenging task. This chapter also identifies the data products that are available as a result of ongoing resource characterization activities in the United States to improve our knowledge of resource characteristics.

2. SOLAR RADIATION VARIABILITY

Solar radiation resource variability is generally characterized in three dimensions: spatial, temporal, and spectral. *Spatial variability* is geographic variability due to different climates and latitude (sun angles and day length). For example, downwelling solar radiation at the earth's surface in the dry southwestern climates is greater than in the rainforest regions of the Pacific Northwest. *Temporal variability* refers to the variation of solar radiation resources over time: near instantaneous (e.g., 1 min), hourly, daily, monthly, seasonal, and inter-annual. For example, the diurnal (daily) profile of solar radiation is different for clear days compared with partly cloudy days. *Spectral variability* is the variation of the spectral content of solar radiation caused by changes in atmospheric constituents that scatter and absorb solar radiation in selected wavelengths. For example, after a volcanic eruption, such as Mt. St. Helens, El Chicon, or Mt. Pinatubo, the volcanic ash and sulfuric acid particles in the atmosphere cause more scattering and absorption of solar radiation, especially in the shorter wavelengths. This scattering and absorption results in a shift from blue to more yellow skies and enhanced red sunrises and sunsets.

Solar radiation resources are also characterized with respect to the direct-beam and diffuse components of total (global) solar radiation. Direct-beam radiation comes directly from the sun's disk and is called *direct-normal* when the receiving surface is normal (perpendicular) to the sun's rays. Diffuse solar radiation comes from the sky, excluding the sun's disk. Global solar radiation is the sum of the direct-normal multiplied by the cosine of the incidence angle and the diffuse (sky) radiation. For a horizontal surface, the relationship is

$$\text{Global horizontal} = \text{direct normal} \times \cos(z) + \text{diffuse horizontal}$$

where z is the solar zenith angle (the angle of the sun with respect to the vertical direction). For an arbitrarily oriented surface, the relationship is

Global (total) = direct normal × $\cos(i)$ + diffuse (in the field of view of the collector surface) + ground-reflected radiation (in the field of view of the collector surface)

where i is the angle of incidence of the direct-beam radiation (the angle at which the direct beam strikes a surface with respect to the normal to the surface).

The characterization of global solar radiation into direct and diffuse components is important because different types of PV systems are designed to use the available solar radiation differently. Concentrator systems track the sun and focus the direct-beam solar radiation, so direct-normal resources must be characterized for these applications. Concentrators with high concentration ratios have stringent tracking requirements so that the concentrators remain in focus. Strong gear boxes and rigid support structures prevent wind from deflecting the concentrators out of focus. Flat-plate collectors utilize both direct-beam and diffuse solar radiation, plus a ground-reflected component. The amount of plane-of-array incident solar radiation for a flat-plate collector depends, in part, on whether the PV arrays are fixed-tilt, single axis tracking, or two axis tracking.

Fixed-tilt collectors are tilted toward the equator at an optimum tilt angle to reduce the angle of incidence of the direct-beam irradiance, thereby increasing the resource availability. The annual energy production is not very sensitive to the tilt angle as long as it is within plus or minus 15° of the latitude. As a general rule, to optimize the performance in the winter, the collector can be tilted 15° greater than the latitude. To optimize performance in the summer, the collector can be tilted 15° less than the latitude. Adjusting the collector tilt throughout the year to maximize energy production is generally not of sufficient benefit to offset the labor costs and the increased cost due to a more complex collector structure. Tracking systems are designed to reduce the angle of incidence of the direct beam by tracking either the elevation or the elevation and the azimuth of the sun throughout each day.

For PV applications, one can either measure the plane-of-array incident solar radiation or estimate it from computer models that calculate incident solar radiation using direct-normal and diffuse solar radiation values, the position of the sun, and the orientation of the PV device (Reindl et al., 1990). Models are also available to convert global solar radiation values to direct-normal solar radiation (Maxwell, 1987; Perez et al., 1990a, b). In the absence of direct-normal and diffuse measurements, models are available to calculate solar radiation from meteorological data (e.g., cloud cover), if available, and the sun position (Maxwell, 1991).

In a practical sense, PV system designers must judge the economic and performance trade offs among concentrators and flat plates, tracking or nontracking, based on system costs versus the amount and value of PV power produced with each option. To help in these system analyses, Sandia National Laboratories, Albuquerque, developed a model called PVFORM. Using hourly solar radiation data, PVFORM calculates the incident solar radiation and PV power produced by an arbitrarily oriented PV device (Menicucci, 1986; Menicucci and Fernandez, 1986). Sandia also developed a program called SIZEPV (Chapman and Fernandez, 1989) as well as design manuals for stand-alone PV systems (Sandia National Laboratories, 1988; Chapman, 1988). Other commercial software products are available as well (see announcements in solar energy journals). Sources of data for these PV system simulations are described in Section 4.

Although not discussed in this chapter, PV system performance also depends on PV cell or module temperature, and coincident data on ambient temperature, wind speed, and solar radiation are desirable for system simulations (Nann and Emery, 1992). Ambient temperature and wind speed are used to calculate PV cell or module temperature. Increased temperature is generally detrimental to power output. For crystalline silicon cells, power decreases about 0.4% for each degree C rise in cell temperature.

2.1. Spatial Variability

The dominant factors causing spatial (geographic) variability of solar radiation are cloudiness and latitude (which determines sun angles and day length). Other factors are atmospheric turbidity (haziness), water vapor, and altitude. The atmospheric factors are related to climate and topography.

A commonly used index that represents the combination of atmospheric effects is a clearness index (also called *cloudiness index* because of the dominant effect), K_t. K_t is the ratio of extraterrestrial solar radiation (i.e., radiation at the top of the atmosphere) on a horizontal plane to solar radiation on a horizontal plane at the earth's surface. It represents the fraction of horizontal solar radiation transmitted through the atmosphere. Figure 20.1a–d shows the mean monthly K_t for 4 months of the year for various locations in the United States calculated from the SOLMET/ERSATZ data base (Solar Energy Research Institute, 1990). This figure gives a general picture of broad-scale spatial variability in atmospheric transmittance across the United States.

Sun angles affect the availability of solar radiation in two ways. First, when the sun is on the horizon (vs. overhead), its rays travel a longer distance in the atmosphere. The slant path relative to the vertical path is called the *relative air mass*. When the sun is directly overhead, the air mass is 1.0; when it is 60° from the vertical, the air mass is 2.0. Higher air mass values result in a reduction in incident solar radiation because of more scattering and absorption of solar radiation by atmospheric constituents such as aerosols. Second, the intensity (power density) of solar radiation decreases as the angle of incidence of the radiation increases.

General contour maps of solar radiation resources (see Fig. 20.2a and b) confirm that cloudier locations such as Seattle have lower resource values than sunnier locations such as Albuquerque. However, higher resolution data (which were used to generate these general contour maps), as well as regional data, must be used to discriminate resource differences between, say, western and eastern Washington or the mountain and plain regions of Colorado. Regional differences may be due to topography, smog, coastal fog, and so forth.

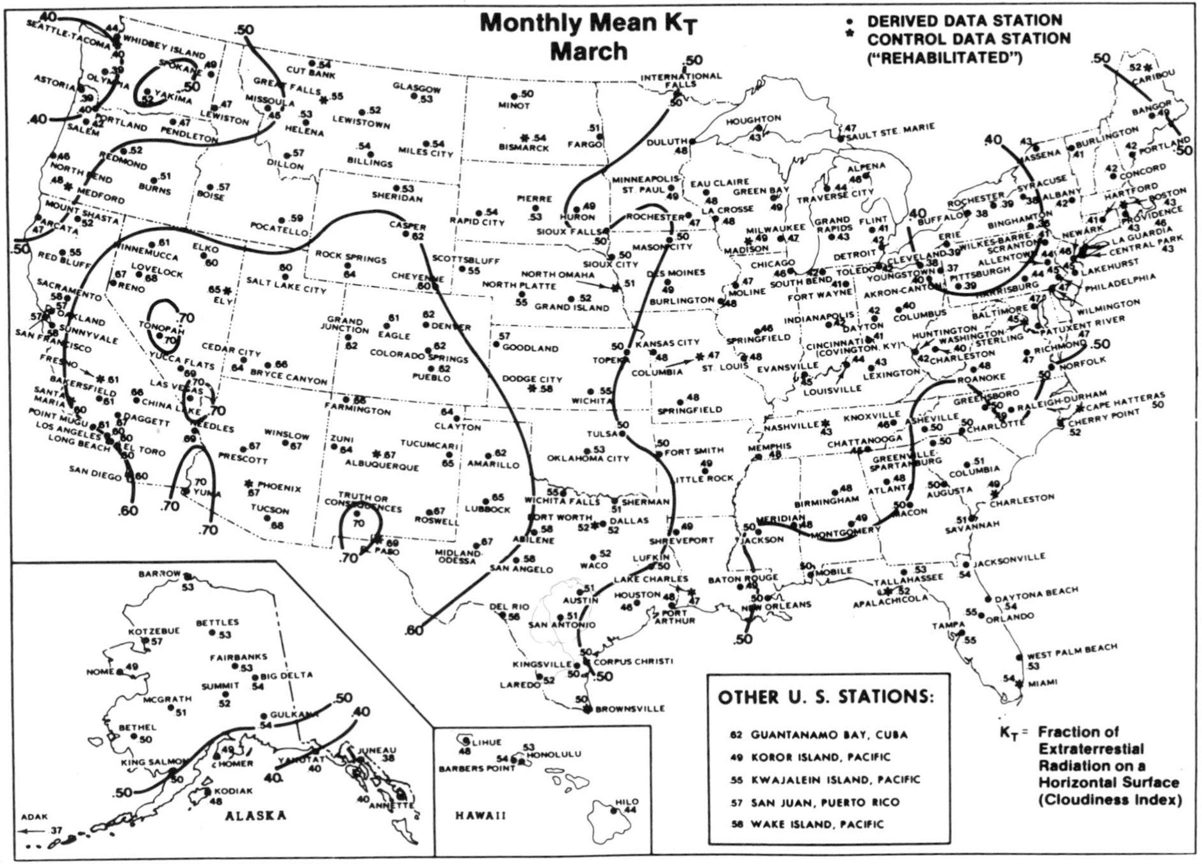

FIGURE 20.1a Mean monthly K_t for March. SOURCE: Solar Energy Research Institute (1990).

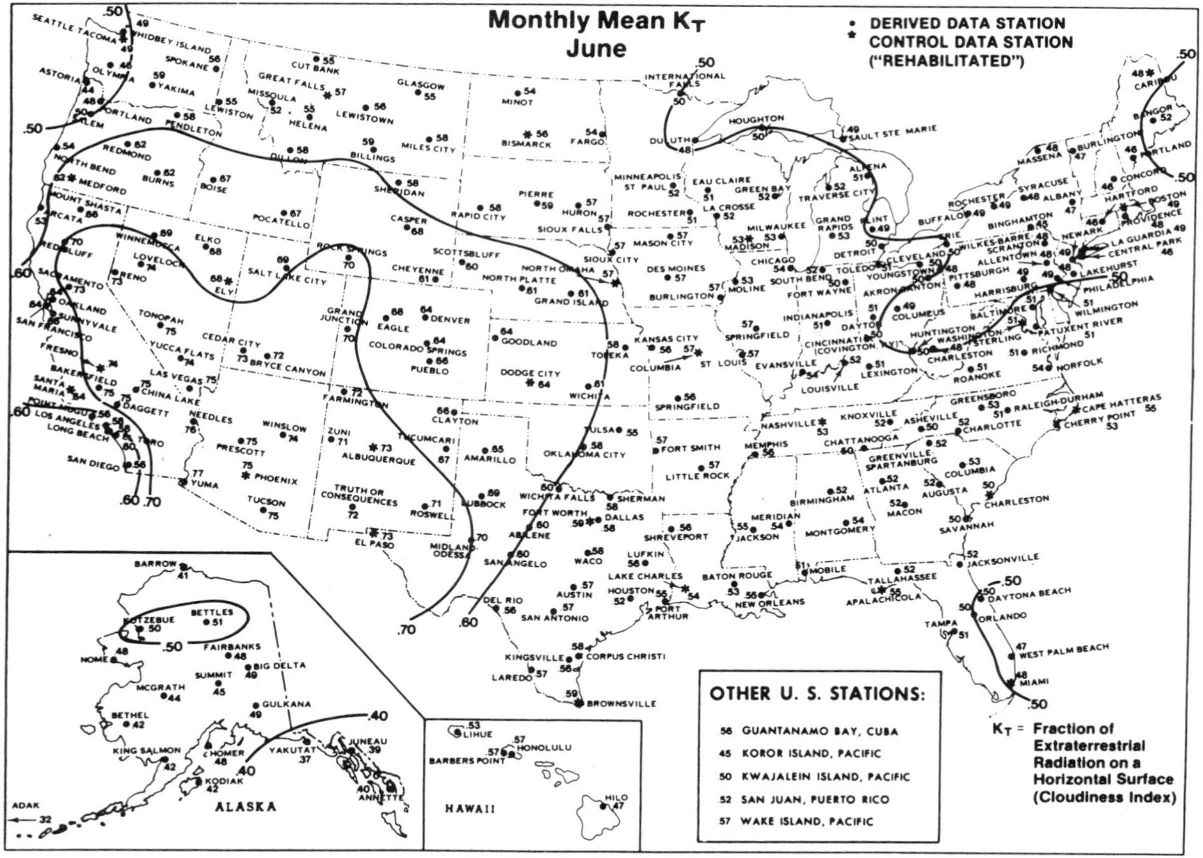

FIGURE 20.1b Mean monthly K_t for June. SOURCE: Solar Energy Research Institute (1990).

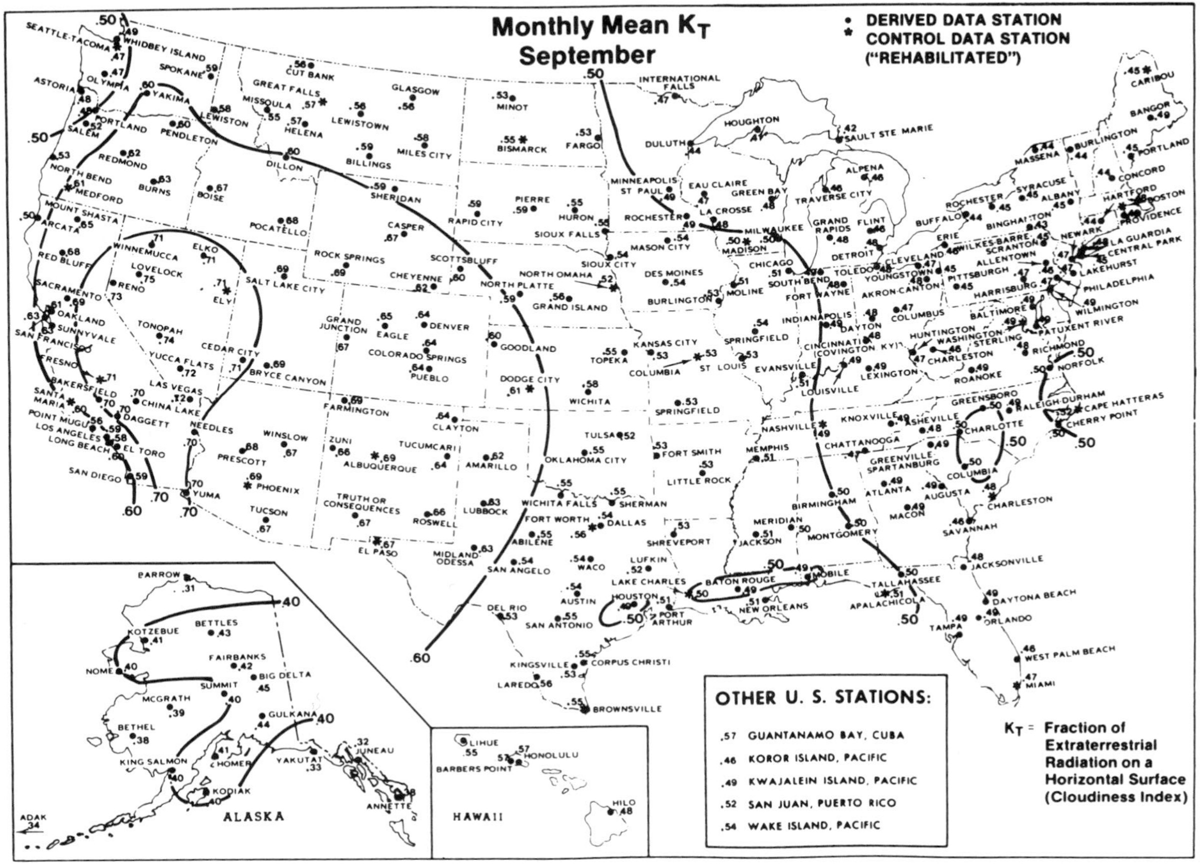

FIGURE 20.1c Mean monthly K_t for September. SOURCE: Solar Energy Research Institute (1990).

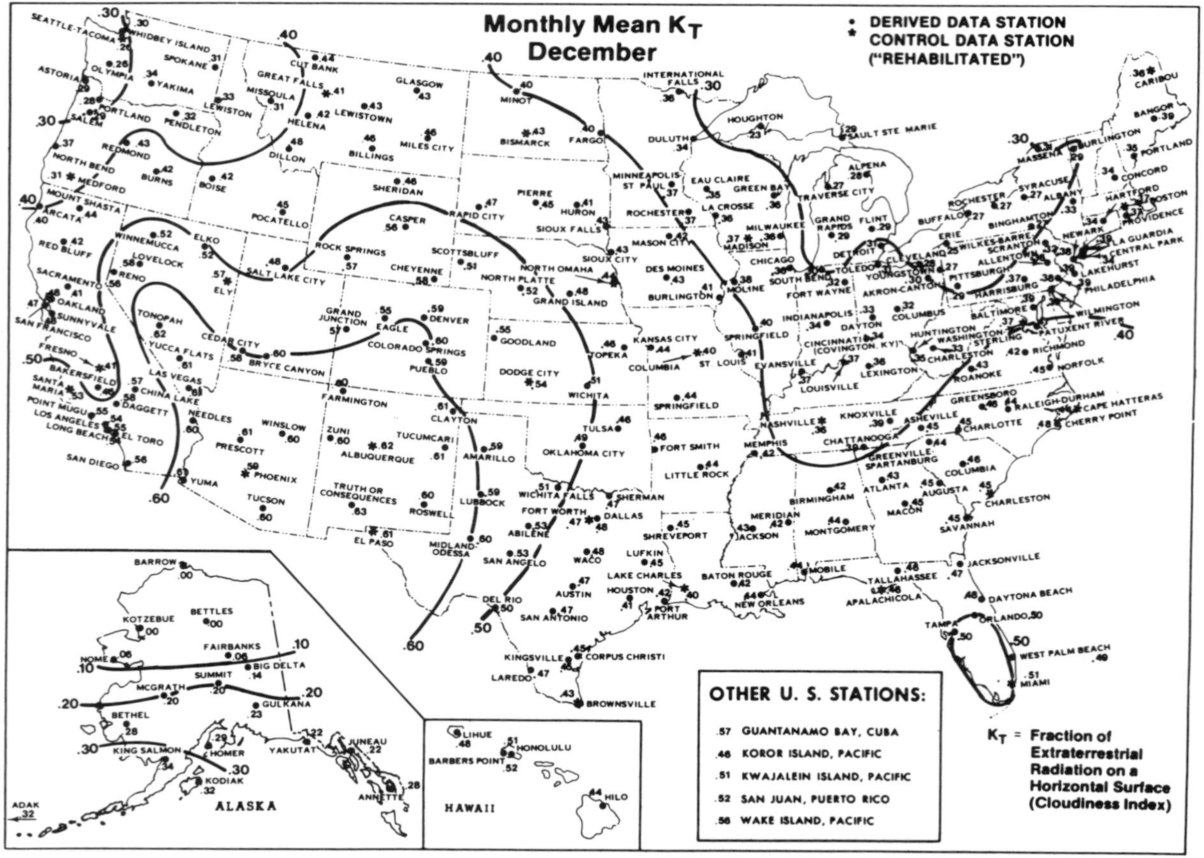

FIGURE 20.1d Mean monthly K_t for December. SOURCE: Solar Energy Research Institute (1990).

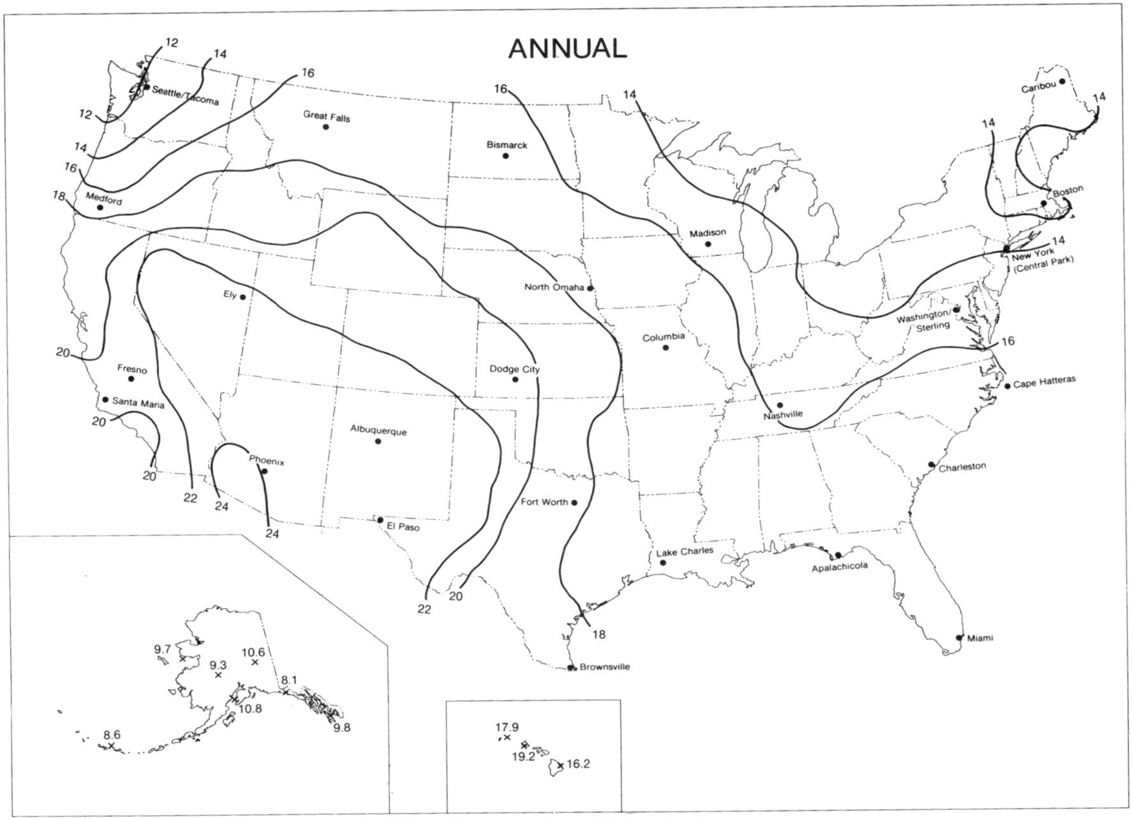

FIGURE 20.2a Annual average daily global solar radiation on a south facing surface, tilt = latitude, in MJ/m^2 ($1\ MJ/m^2 = 0.2778\ kW\text{-}h/m^2$). SOURCE: Solar Energy Research Institute (1981). (Also reproduced in Hulstrom, 1989.) Updated maps are being produced from the 1960–1990 National Solar Radiation Data Base at NREL.

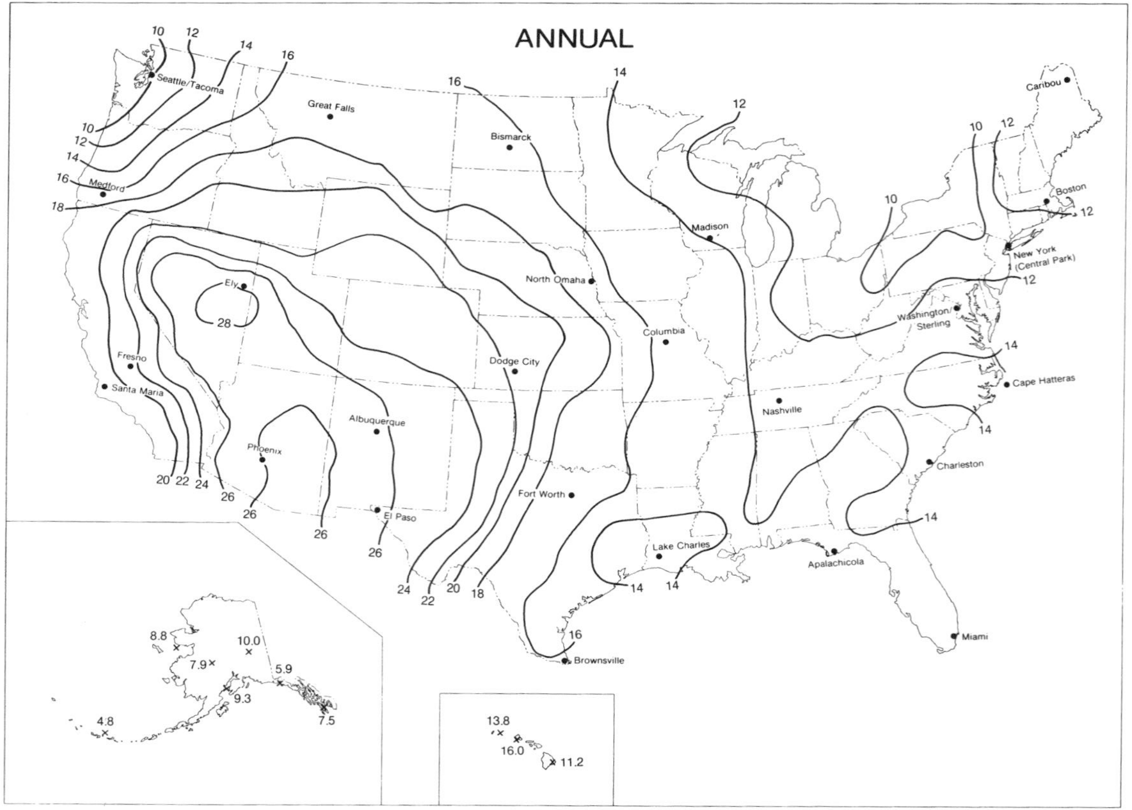

FIGURE 20.2b Annual average daily direct normal solar radiation in MJ/m^2 ($1\,MJ/m^2 = 0.2778\,kW\text{-}h/m^2$). SOURCE: Solar Energy Research Institute (1981). (Also reproduced in Hulstrom, 1989.) Updated maps are being produced from the 1960–1990 National Solar Radiation Data Base at NREL.

The significance of these spatial variations to PV system performance is generally understood by the PV community, but reliable data on the actual magnitude of the differences may be lacking for site-specific applications. Consequently, PV practioners must be more conservative in their design than ultimately desirable when they estimate solar radiation for a site using data from a nearby location.

2.2. Temporal Variability

Sun angles and atmospheric effects, predominantly cloudiness, also determine hourly, daily, monthly, and inter-annual variability of the solar resource at a particular site. For example, Figure 20.3 shows the diurnal variation of direct-normal solar radiation and global solar radiation on horizontal, tilted,

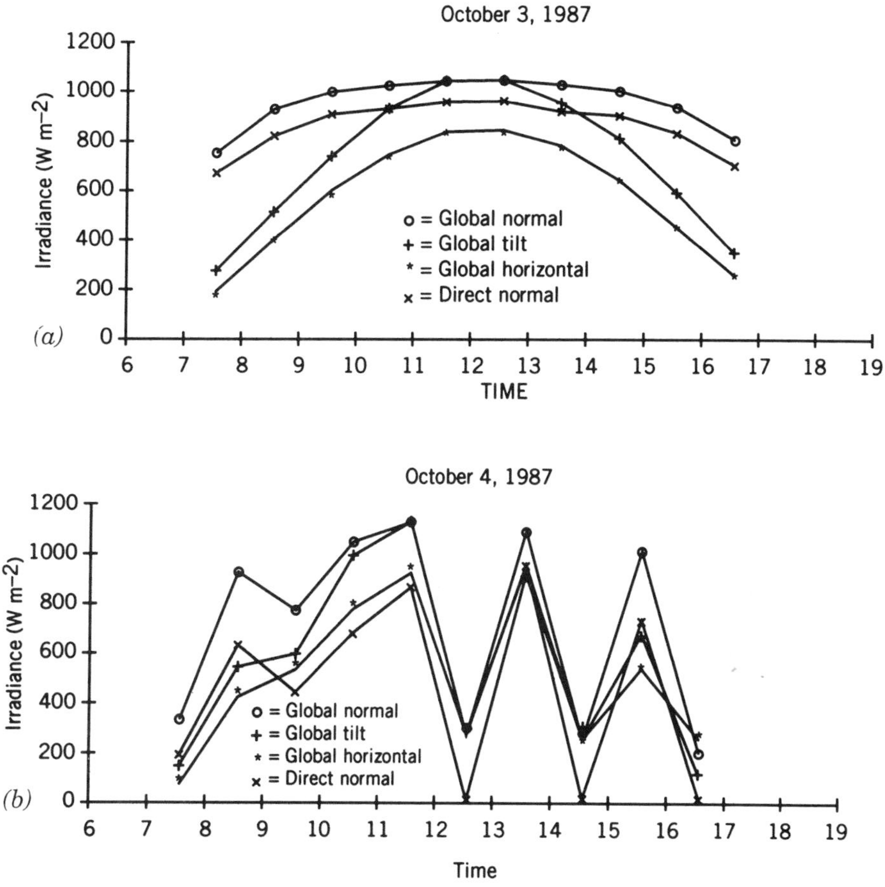

FIGURE 20.3 Examples of diurnal variation of direct-normal and global solar radiation for clear **(a)** and partly cloudy **(b)** days.

and sun-tracking surfaces for clear and partly cloudy days at Cape Canaveral, FL. Another example of temporal variability is given in Figure 20.4, which shows typical monthly values of global solar radiation for Washington, DC, for a 23 year period. Monthly values for individual years are superimposed to show the the inter-annual variability of the monthly values.

Temporal variability affects PV system performance evaluations in several important ways. First, it is important for electric utilities to understand the resource variability in order to integrate intermittent renewables into their plant operations (Parsons and Wan, 1992). Second, high value applications of PV include displacing high-cost power during peak load periods with PV generation when there is a match between the utility's peak load and the solar resource (e.g., a match between the sunniest part of the day and peak loads caused by air conditioning requirements) (Perez et al., 1991). Third, it is important to assess the "worst case" solar radiation conditions (such as persistence of cloudy skies) in order to design systems, including battery storage, that will reliably serve a critical application such as telecommunications or refrigeration of vaccines.

2.3. Spectral Variability

The extraterrestrial spectrum (shown in high resolution in Fig. 20.5) is modified by spectrally selective absorption and scattering by atmospheric constituents such as clouds, ozone, water vapor, aerosols, air molecules, and gases in the atmosphere. Examples of spectra measured at the earth's surface are shown in Figure 20.6. Some of the major absorption features are indicated. Spectral variability at the earth's surface that is significant for PV performance is primarily influenced by changes at atmospheric turbidity and water vapor and by changes in the sun's position (relative air mass) throughout the day and year (Riordan and Hulstrom, 1990).

Figure 20.7 shows variations in measured direct-normal and global spectra throughout a clear day in Cape Canaveral. As the path of the solar beam through the atmosphere increases (when the sun is closer to the horizon), more scattering and absorption of solar radiation occurs, with a larger effect on the shorter wavelengths. This shifts the direct-normal spectrum toward the longer wavelengths. Because some of the radiation that is scattered out of the direct beam reaches the Earth's surface as the diffuse component of global solar radiation, the shift to longer wavelengths is less dramatic for the global spectra.

Increased turbidity (higher amounts of aerosols, dust, and so forth) increases the scattering and absorption of solar radiation, especially at the shorter wavelengths. This also results in a shift of the spectrum toward longer wavelengths, with more of an effect on direct-normal spectra than on global spectra.

Increased water vapor increases the absorption of solar radiation in the water vapor absorption bands of the spectrum (e.g., 940 nm). Figure 20.8 shows spectra measured at Cape Canaveral, Fl, and Denver, CO, and illustrates

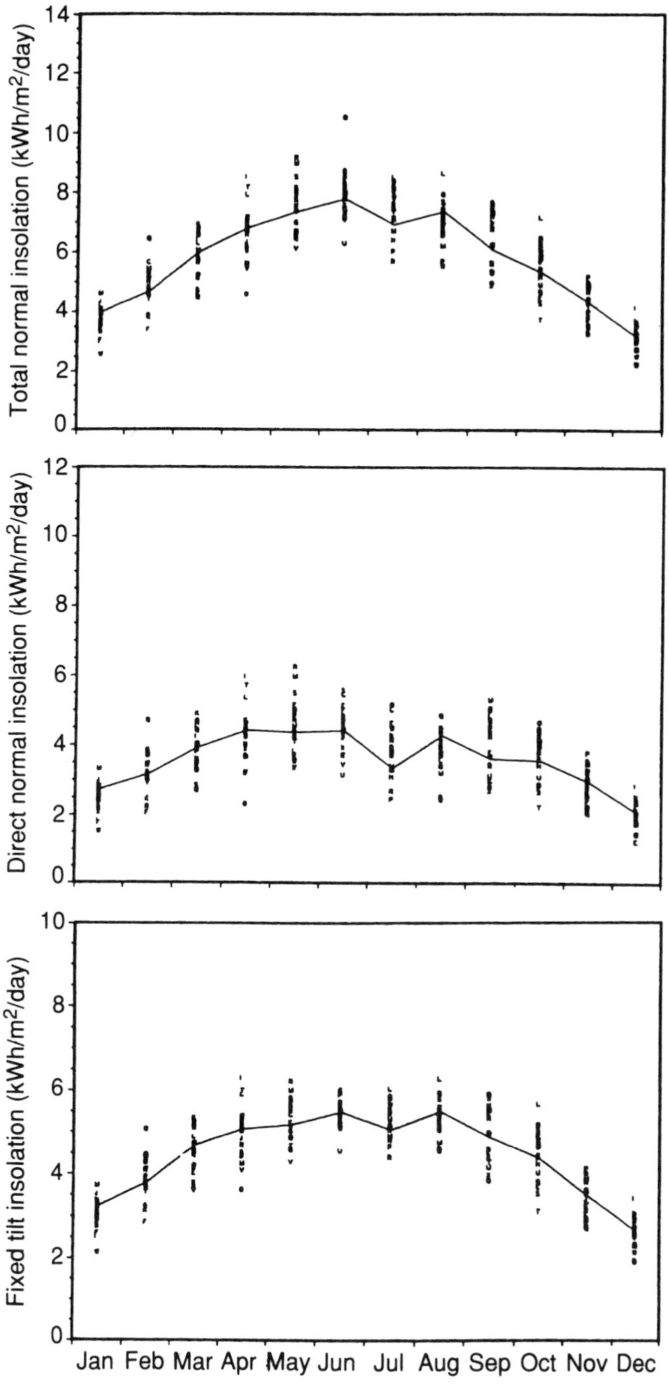

FIGURE 20.4 Typical monthly values of solar radiation for Washington, DC, with annual variation superimposed. SOURCE: Menicucci and Fernandez (1988).

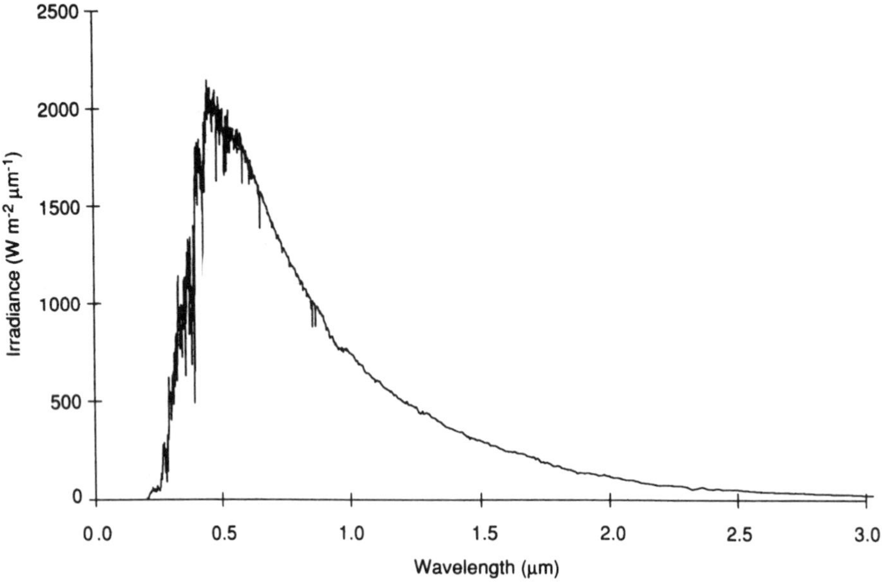

FIGURE 20.5 Extraterrestrial spectrum. SOURCE: Wehrli (1985). (Also given in Riordan and Hulstrom, 1990 and Appendix B.)

differences in the depth of the water vapor absorption bands at a relatively dry site (Denver) compared with a relatively wet site (Cape Canaveral).

Spectral variations are generally considered to affect outdoor PV system performance less than variations of total incident solar radiation and operating temperature affect such performance. Spectral variations are considered in the

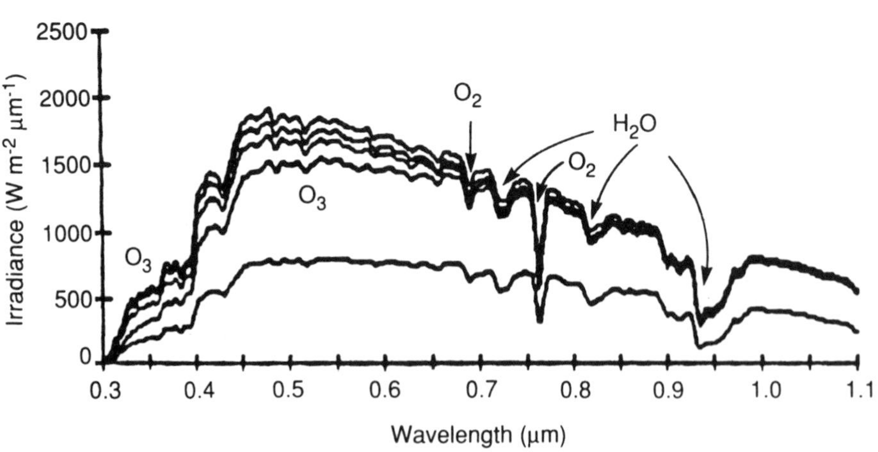

FIGURE 20.6 Measured spectra at Cape Canaveral, FL, with a few of the major atmospheric absorption features identified.

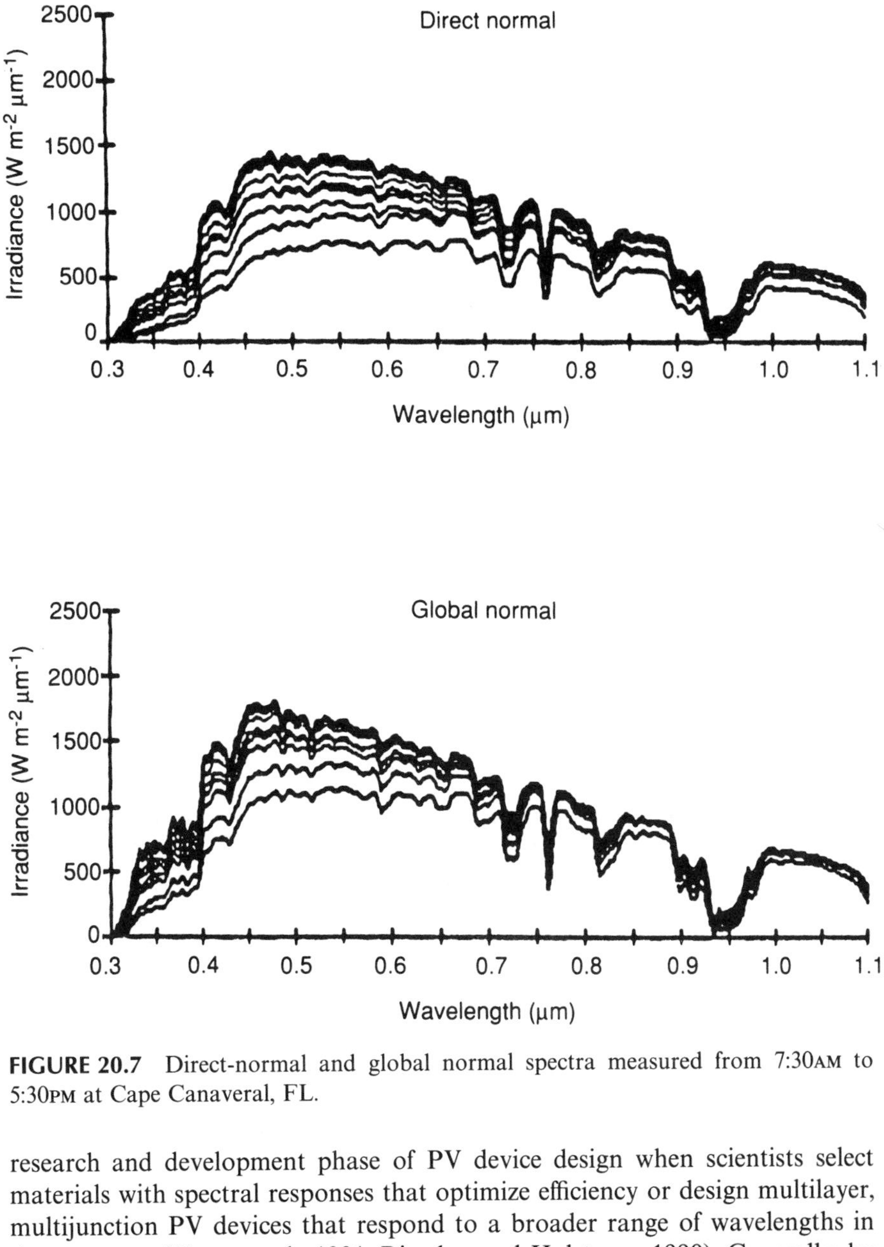

FIGURE 20.7 Direct-normal and global normal spectra measured from 7:30AM to 5:30PM at Cape Canaveral, FL.

research and development phase of PV device design when scientists select materials with spectral responses that optimize efficiency or design multilayer, multijunction PV devices that respond to a broader range of wavelengths in the spectrum (Kurtz et al., 1991; Riordan and Hulstrom, 1990). Generally, by the time PV devices reach the market as modules there has been adequate consideration of spectral variations so that performance is not seriously compromised by natural outdoor variations.

On the other hand, outdoor PV system performance measurements do show some daily and seasonal variations of efficiency (PVUSA Project Team,

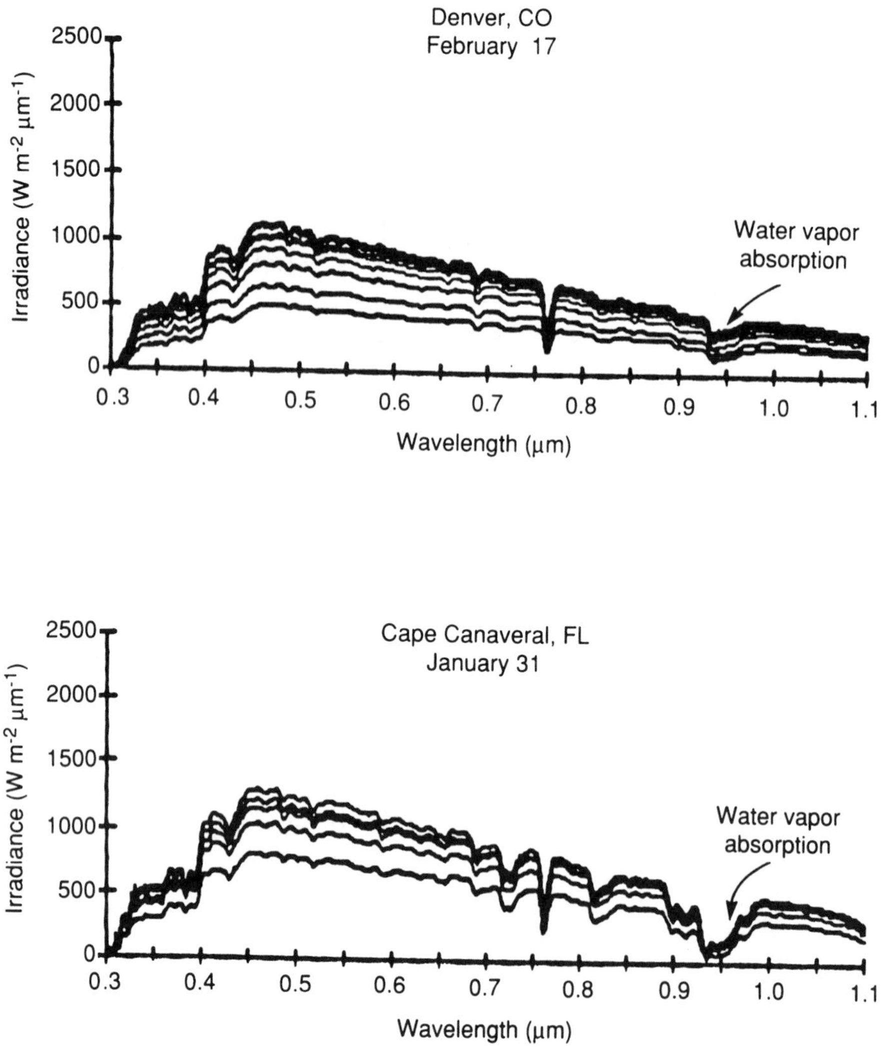

FIGURE 20.8 Spectra measured at Denver and Cape Canaveral showing differences in the depths of the water vapor absorption bands.

1990; Candelario et al., 1991). Analysis of efficiency variations has shown that they can be due to spectral irradiance variations (Riordan and Hulstrom, 1990), but it has proven to be difficult to separate spectral from temperature effects in the outdoor measurements (Whittaker et al., 1991).

Because of the importance of being able to compare PV device efficiency measurements made at different times by various entities, these measurements are often reported for "reference" conditions that include a specific spectrum. The American Society for Testing and Materials (ASTM) has published standards of spectral irradiance, which are E 891 (direct-normal) and E 892 (global for a 37° tilted surface) (see Section 3) (American Society for Testing and Materials, 1991). Procedures exist for translating efficiency measurements to reference or standard conditions (Emery and Osterwald, 1988; Hansen, 1991).

3. EXTRATERRESTRIAL AND TERRESTRIAL SPECTRAL SOLAR RADIATION REFERENCE DATA SETS

The extraterrestrial (or "air-mass zero" or AM0) spectrum is important for space applications of PV systems and as a starting point for spectral irradiance modeling at the earth's surface. ASTM has adopted a standard extraterrestrial spectrum (American Society for Testing and Materials, 1991); however, improved data based on satellite measurements are available. One of these improved extraterrestrial spectra was published by the World Radiation Center in Davos, Switzerland, in 1985 (Wehrli, 1985) and reproduced (with permission from the author) in a report by the Solar Energy Research Institute (SERI, now the National Renewable Energy Laboratory) (see Fig. 20.5 and Appendix B) (Riordan, 1987). Another updated extraterrestrial spectrum was compiled by the Air Force Phillips Laboratory in 1988 (Shettle et al., Hall 1988). The integrated spectral irradiance (called the *solar constant*) for the more recent spectra is 1,367–1,370 Wm^{-2}, depending on the source of the data. These values are higher than the older value of 1,353 Wm^{-2} represented by the ASTM standard.

The ASTM standard direct-normal and global spectra, E 891 and E 892, are based on spectral irradiance data sets produced by SERI (now NREL) (Hulstrom et al., 1985). These data sets are reproduced in Appendix A and are shown in Fig. 20.9. (There are slight differences in the published ASTM standards, which are probably typographical errors.) These spectra represent selected atmospheric conditions (e.g., atmospheric turbidity at 500 mm of 0.27, precipitable water of 1.42 cm) and a solar zenith angle of 48.19° (relative air mass of 1.5), chosen as representative conditions for the United States.

These spectral data sets are used for reporting PV device efficiencies with respect to reference conditions (Emery and Osterwald, 1988). Because they represent selected atmospheric conditions and a particular solar zenith angle, actual outdoor PV device efficiency measurements will vary for different conditions. (This is similar to deviations of PV module performance from name-plate ratings that are derived for a specific set of operating conditions specified in the manufacturer's product literature.)

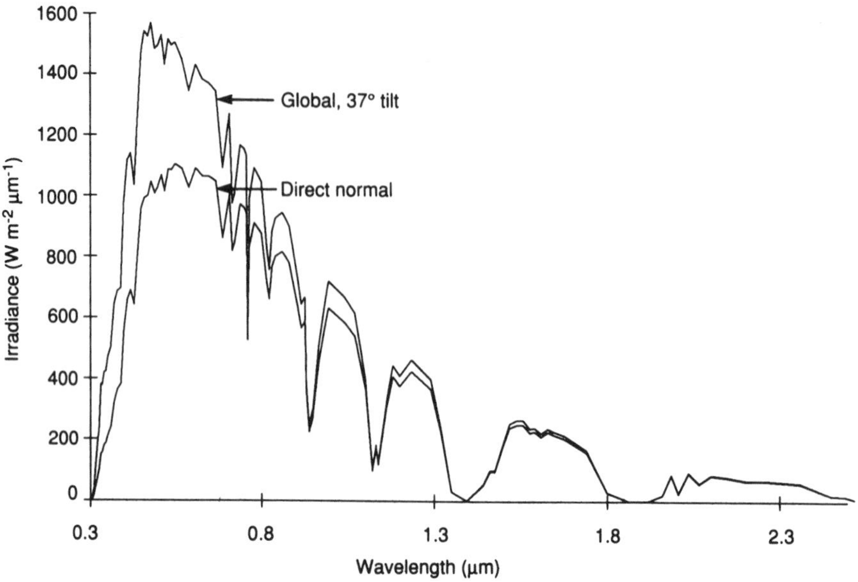

FIGURE 20.9 Air mass 1.5 direct-normal (AM1.5D) and global (AM1.5G) spectra. SOURCE: Hulstrom et al. (1985).

4. SOURCES OF SOLAR RADIATION DATA

Figure 20.10 shows a typical path that might be followed to select an appropriate data set for PV system design and performance simulations. First, one would search for any measured data for the site. Potential sources of these data are the National Weather Service (archived by the National Climatic Data Center, 1981a, b) utility companies, universities, solar industries, and state climatologists. If measured data are located, the reliability of the data should be examined (instrument type and maintenance and calibration history) and the data processed through a quality assessment program to flag questionable data. Short-term data (less than a 10 year record) should be compared with long-term (30 year) records to examine inter-annual variability that can be caused by periods of drought, volcanic eruptions, climate anomalies, and so forth. Long-term records (comprising measured and modeled values) are contained in the 23 year (1952–1975) SOLMET/ERSATZ data base for approximately 250 locations in the United States (National Climatic Data Center, 1978) and in the improved and extended 30 year (1961–1990) National Solar Radiation Data Base (NSRDB) for 239 location in the United States produced by NREL (National Renewable Energy Laboratory, 1992a, b). The NSRDB is the preferred long-term data record because it includes improved measured data collected between 1976 and 1990 and better model estimates for times and data base stations where no measured data exist. Quality flags

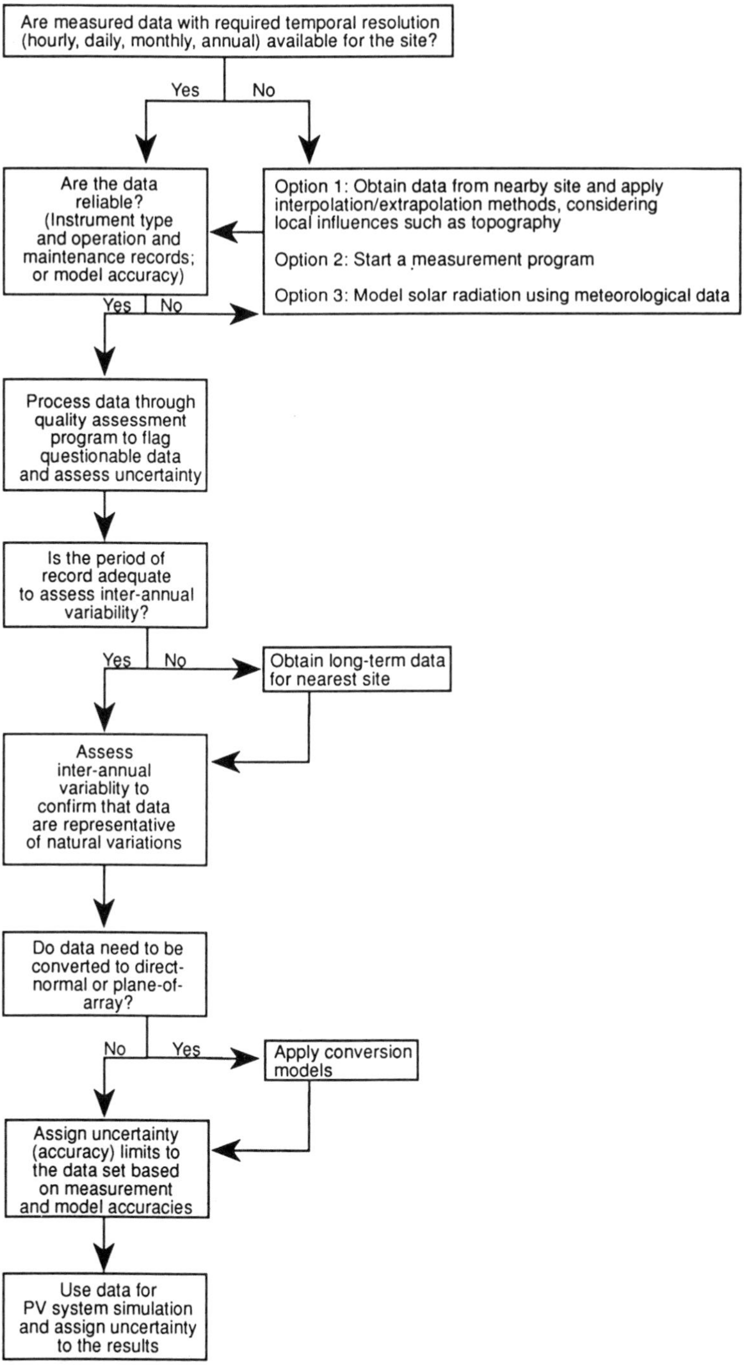

FIGURE 20.10 Typical strategy for selecting appropriate resource data for PV system design.

attached to each hourly value in the NSRDB indicate the source and estimated uncertainty of the data.

If no measured data are available for the site, one can extrapolate or interpolate data from nearby sites, establish a measurement program, or produce model estimates from meteorological data, if available. Extrapolation or interpolation should consider factors such as topography that influence local conditions. Again, the reliability of the data should be examined and the data processed through a quality assessment program. Short-term data should be compared with long-term records to capture information on inter-annual variations.

Instruments routinely used for measuring solar radiation include pyranometers, pyrheliometers, and rotating shadow band radiometers. Pyranometers have a 180° field of view and make measurements of global horizontal radiation and of plane-of-array global radiation for tilted surfaces. Pyrheliometers are mounted on sun trackers and have a narrow field of view for measuring solar radiation coming directly from the sun's disk. Rotating shadow band radiometers measure both global horizontal and diffuse solar radiation. A motor moves the shadow band to shade the sensor from direct-beam solar radiation while diffuse solar radiation is measured. Then the motor moves the shadow band to unshade the sensor while global solar radiation is measured.

Because the most common measurement of solar radiation is global-horizontal, techniques must often be applied to derive direct-normal or plane-of-array global irradiance. The most reliable of the modeling techniques requires hourly data as input. The uncertainty of the modeled values must be evaluated with respect to the uncertainty of the input data and the accuracy of the model.

High-resolution (at least hourly) temporal data are needed for peak load-matching applications. If these data are not available, they must be interpolated or extrapolated from the nearest locations with hourly data. Techniques have been developed to use satellite data to estimate hourly values of solar radiation by extracting meteorological information from the satellite images (e.g., cloudiness) and estimating solar radiation from meteorological conditions. If these satellite techniques can be made operational and routine for solar energy applications, they will be a valuable source of information in combination with selected measurement locations.

There are a variety of sources of solar radiation data for PV system simulations (see References). The sources include the National Climatic Data Center; the American Society of Heating, Refrigerating, and Air-conditioning Engineers (ASHRAE); publications by NREL and Sandia; handbooks prepared by universities and others; entities such as the Commission of European Communities; utilities such as Pacific Gas and Electric Company and Arizona Public Service Company; and the solar energy journals such as *Solar Energy*. Worldwide data are availale in publications such as that by Löf et al. (1966) (see examples in Fig. 20.11a–d). Some commercial software products for PV system design include solar radiation data. In addition, some

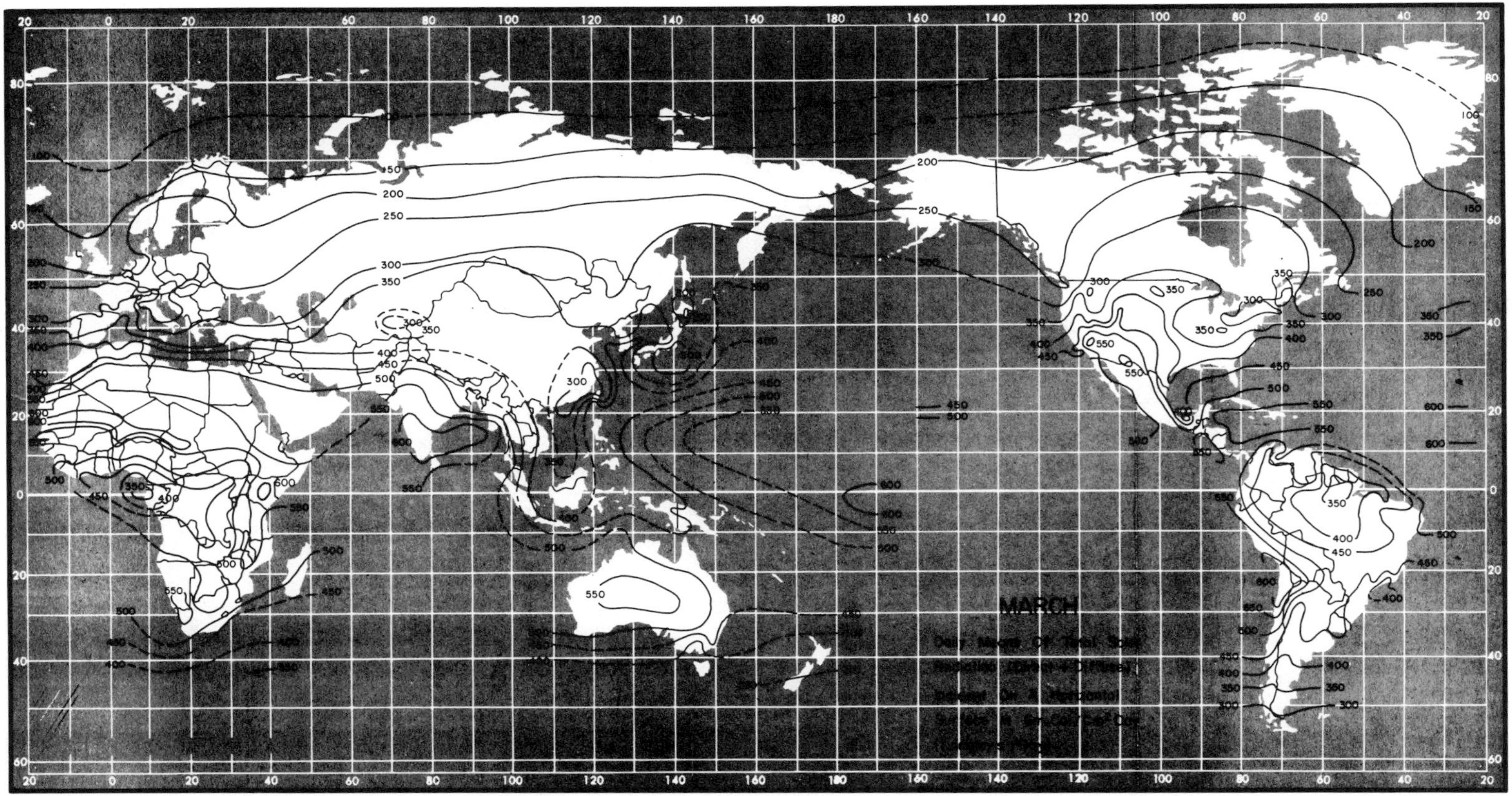

FIGURE 20.11a Example of world solar radiation map for March in Langleys (1 Langley = 0.01163 kW-h/m^2). SOURCE: Löf et al. (1966).

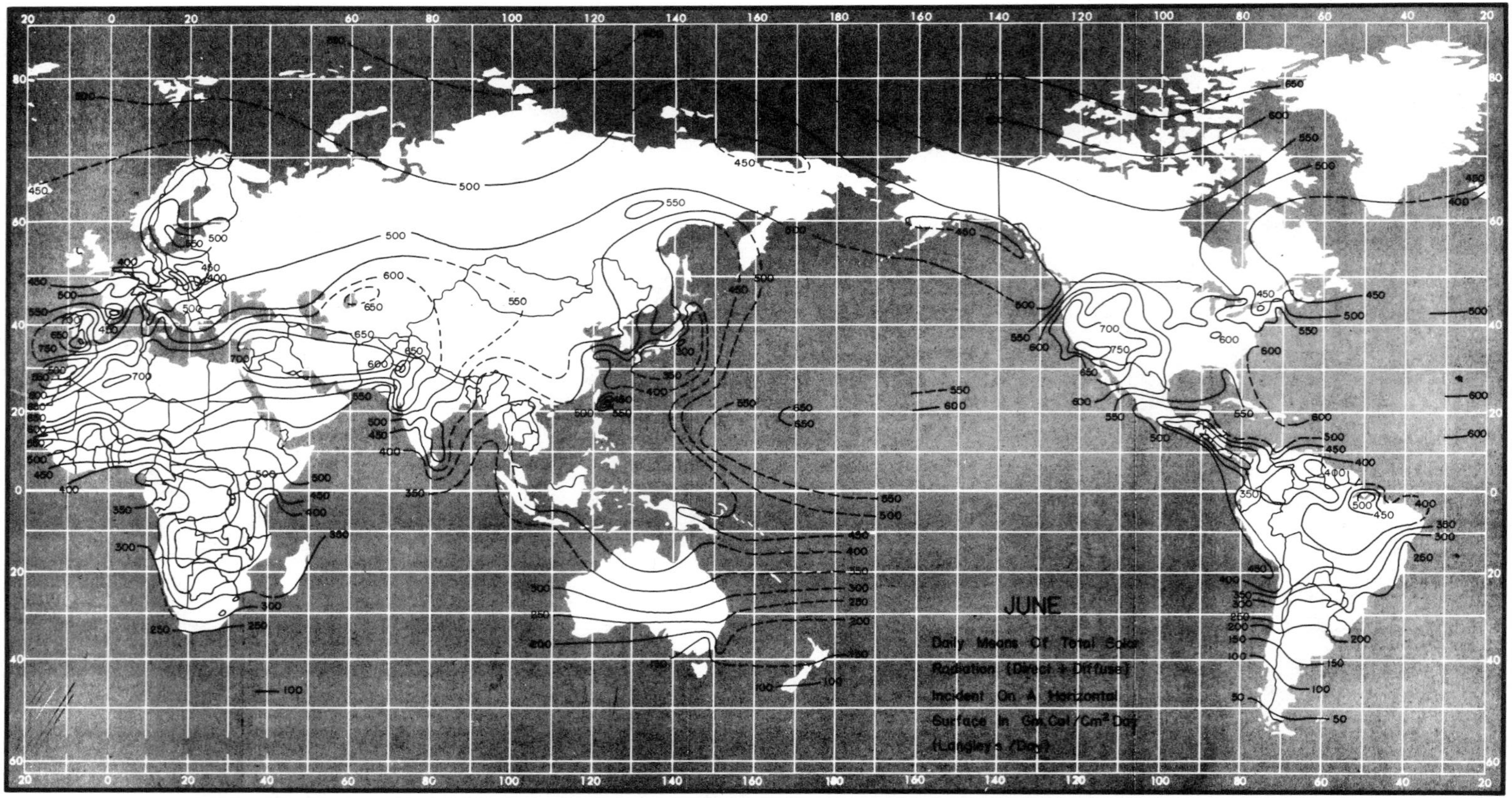

FIGURE 20.11b Example of world solar radiation map for June in Langleys (1 Langley = 0.01163 kW-h/m^2). SOURCE: Löf et al. (1966).

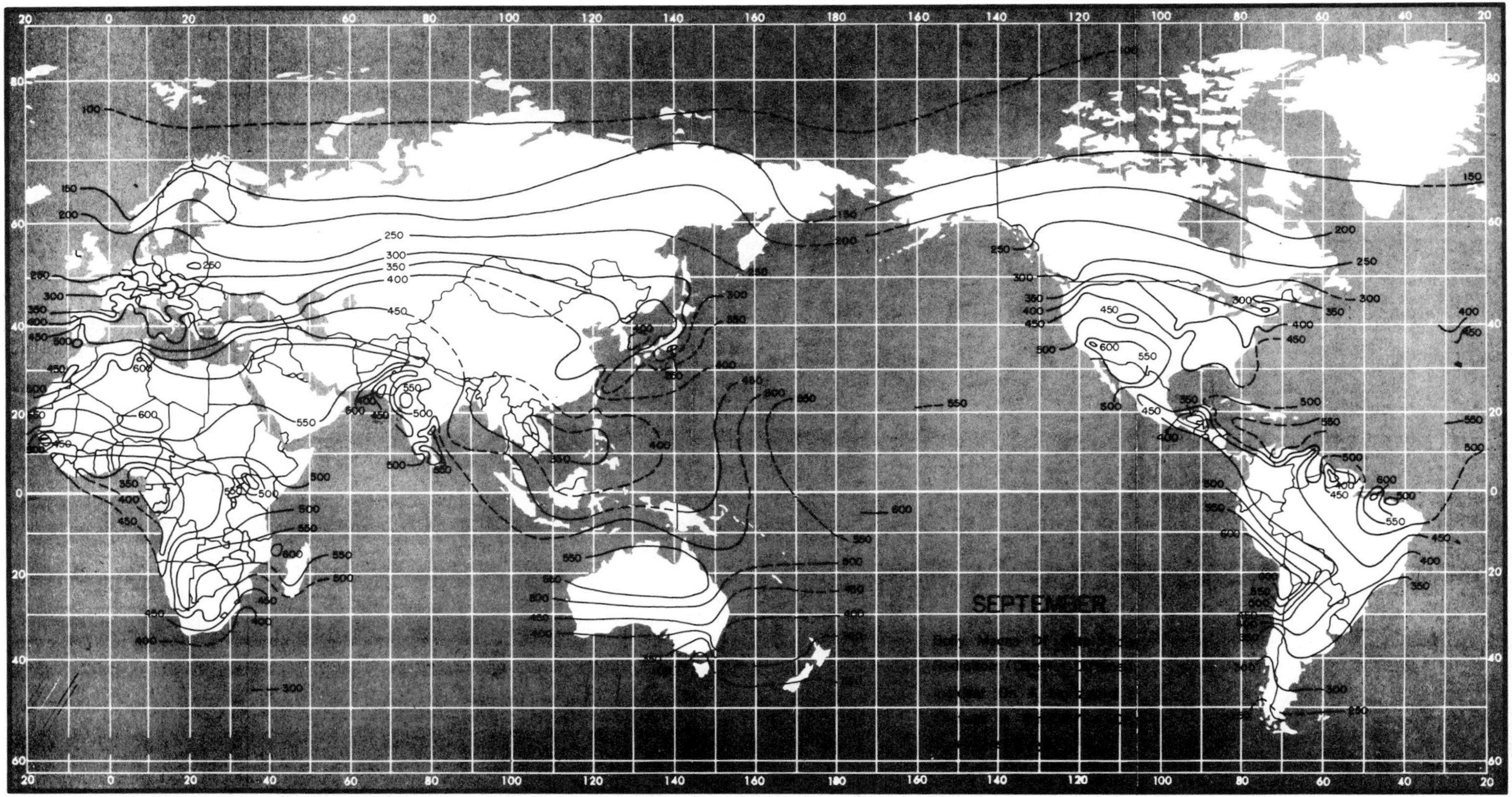

FIGURE 20.11c Example of world solar radiation map for September in Langleys (1 Langley = 0.01163 kW-h/m^2). SOURCE: Löf et al. (1966).

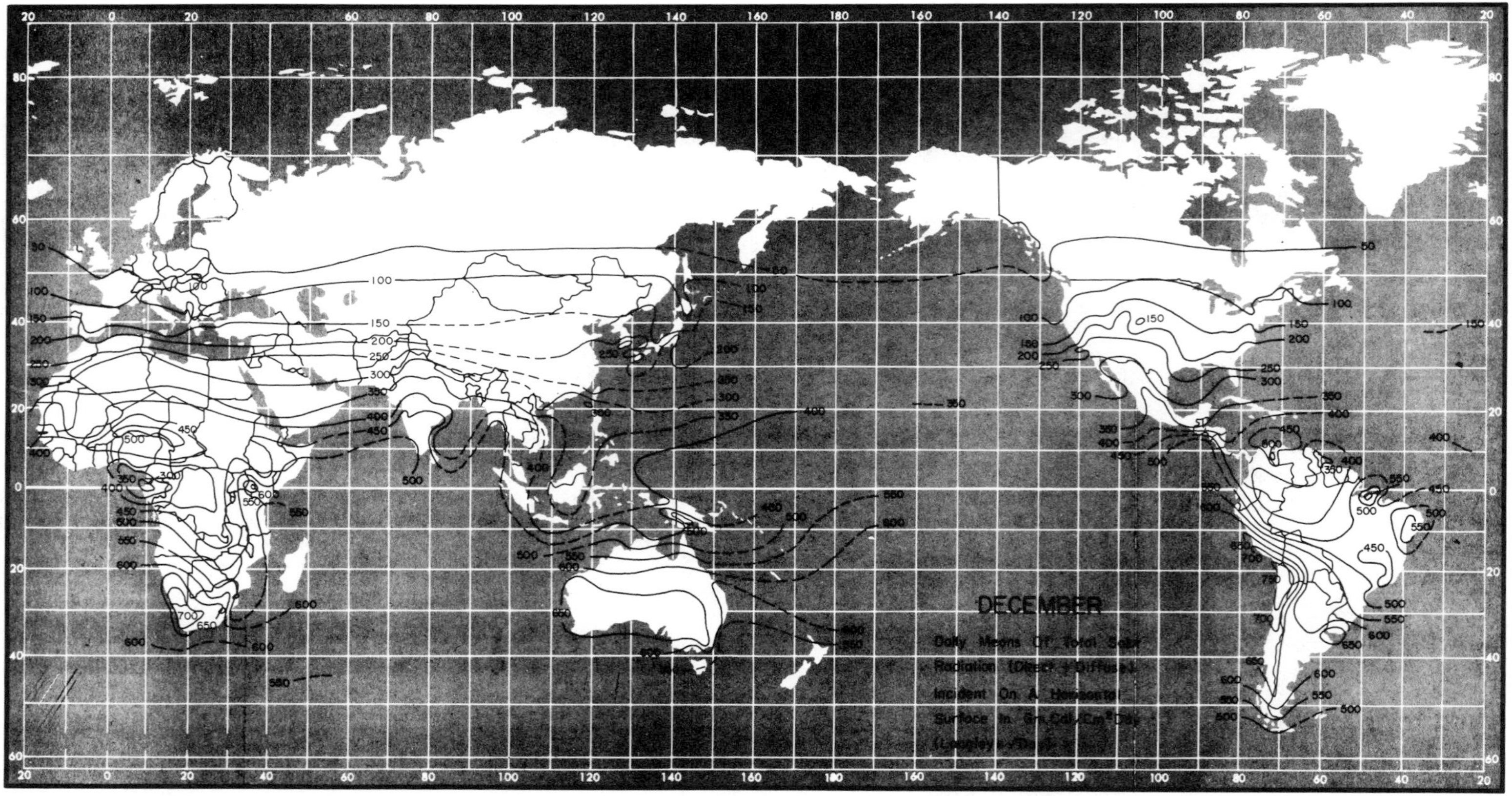

FIGURE 20.11d Example of world solar radiation map for December in Langleys (1 Langley = 0.01163 kW-h/m^2). SOURCE: Löf et al. (1966).

PV industry groups have access to proprietary software and data sets for PV system design. The more recent sources of data are generally preferred because better instrumentation and modeling techniques were available.

The essential questions to ask whenever using any of these data sources are (1) Are the data measured, modeled from meteorological data, or extended from a nearby site? (2) What is the uncertainty of the data, and how does that uncertainty impact the PV system performance simulations? (3) Have the data been processed through a quality assessment procedure to remove questionable data? When the cost of PV array is low compared with the overall system cost, we can account for uncertainties in the solar radiation data and the electric load demand by using "engineering judgment" to increase the size of the PV array. However, as the PV array increases in size and cost, this becomes less acceptable, and we need more accurate solar radiation and electric load demand data to optimize the design and project the cost. Entities that have experience in measuring and modeling solar radiation resources, such as NREL, Sandia, the National Weather Service, utility companies, the PV industry, and universities, can provide advice and assistance on sources of data, measuring and modeling solar radiation, uncertainty estimates, and quality assessment procedures.

5. THE FUTURE OF SOLAR RESOURCE CHARACTERIZATION

The recent generation of the 1961–1990 National Solar Radiation Data Base for 239 locations in the United States (National Renewable Energy Laboratory, 1992a, b) will provide a good long-term baseline record for the United States and will lead to a number of new products, such as a new national atlas and 30 year statistics. The majority of these data are from 30 to 40 measurement sites operated by the National Weather Service over the 30 year period plus modeled estimates of solar radiation based on meteorological data from an additional 200 or so National Weather Service locations. The national solar radiation network, operated by the National Weather Service, is sparse (less than 30 stations in operation); its continued operation is being affected by declining budgets and multiple priorities. Therefore, regional solar radiation monitoring networks, such as Pacific Gas and Electric Company's SIMP (Nelson and Augustyn, 1990) are becoming even more important for obtaining reliable data with the accuracy and temporal and spatial detail to design and reliably predict the performance of PV systems.

The lead center for the U.S. DOE's Solar Radiation Resource Assessment Program is NREL in Golden, CO. This laboratory is working with other national laboratories, electric utility companies, consultants, and instrumentation companies to provide assistance, procedures, and guidelines for the reliable characterization of solar radiation resources in the United States.

ACKNOWLEDGMENTS

I appreciate the review and comments on this text by Roland Hulstrom, Daryl Myers, and Bill Marion at the National Renewable Energy Laboratory. This text was produced by an employee at the National Renewable Energy Laboratory under funding from the U.S. Department of Energy.

REFERENCES

American Society for Testing and Materials (1991), *Annual Book of ASTM Standards*: E 891-87, Standard Tables for Terrestrial Direct Normal Solar Spectral Irradiance for Air Mass 1.5; E 982-87, Standard Tables for Terrestrial Solar Spectral Irradiance at Air Mass 1.5 for a 37° Tilted Surface; E 490-73a (reapproved 1987), Standard Solar Constant and Air Mass Zero Solar Spectral Irradiance Tables, American Society for Testing and Materials, Philadelphia, PA.

Candelario, T.R., Hester, S.L., Townsend, T.U., and Shipman, D.J. (1991), *22nd IEEE Photovoltaic Specialists Conf. Record*, IEEE, New York, pp. 493–500.

Chapman, R.N. (1988), *Sizing Handbook for Stand-Alone Photovoltaic/Storage Systems*, SAND87-1087, Revised, Sandia National Laboratories, Albuquerque, NM.

Chapman, R., and Fernandez, J. (1989), *SIZEPV, A Simulation Program for Stand-Alone Photovoltaic Systems, User's Manual*, SAND89-0616, Photovoltaic Design Assistance Center, Sandia National Laboratories, Albuquerque, NM.

Emery, K.A., and Osterwald, C.R. (1988), *Current Topics in Photovoltaics*, Vol. 3, Academic Press, New York.

Hansen, B.R. (1991), *22nd IEEE Photovoltaic Specialists Conf. Record*, IEEE, New York, pp. 802–806.

Hulstrom, R., Ed. (1989), *Solar Resources*, MIT Press, Cambridge, MA.

Hulstrom, R., Bird, R., and Riordan, C. (1985), *Solar Cells*, **15**, 365–391.

Kurtz, S.R., Olson, J.M., and Faine, P. (1991), *Solar Cells*, **30**, 501–513.

Löf, G.O., Duffie, J.A., and Smith, C.O. (1966), *World Distribution of Solar Radiation*, Engineering Experiment Station, Report No. 21, Solar Energy Laboratory, University of Wisconsin, Madison.

Maxwell, E.L. (1987), *A Quasi-Physical Model for Converting Hourly Global Horizontal to Direct Normal Insolation*, SERI/TR-215-3086, Solar Energy Research Institute, Golden, CO.

Maxwell, E.L. (1991), *Proc. Biennial Cong. Int. Solar Energy Society*, Pergamon, New York, pp. 975–980.

Menicucci, D.F. (1986), *The Determination of Optimum Mounting Configurations for Flat-Plate Photovoltaic Modules Based on a Structured Field Experiment and Simulated Results From PVFORM, A Photovoltaic System Performance Model*, SAND85-1251, Sandia National Laboratories, Albuquerque, NM.

Menicucci, D.F., and Fernandez, J.P. (1986), *Estimates of Available Solar Radiation and Photovoltaic Energy Production for Various Tilted and Tracking Surfaces Throughout the US Based on PVFORM, A Computerized Performance Model*, SAND85-2775, Sandia National Laboratories, Albuquerque, NM.

Menicucci, D.F., and Fernandez, J.P. (1988), *A Comparison of Typical Meteorological Year Solar Radiation Information with the SOLMET Data Base*, SAND87-2379, Sandia National Laboratories, Albuquerque, NM.

Nann, S., and Emery, K. (1992), *Solar Cells*, 27, 189–216.

National Climatic Data Center (1978), *SOLMET, Vol. 1, User's Manual (TD9724) Hourly Solar Radiation—Surface Meteorological Observations*, Environmental Data and Information Service, Asheville, NC.

National Climatic Data Center (1981a), *Typical Meteorological Year User's Manual (TD9734) Hourly Solar Radiation—Surface Meteorological Observations*, Environmental Data and Information Service, Asheville, NC.

National Climatic Data Center (1981b), *SOLMET/ERSATZ* Data Base.

National Renewable Energy Laboratory (1992a), *1961–1990 National Solar Radiation Data Base for the United States*, Golden, CO, archived by the National Climatic Data Center, Asheville, NC.

National Renewable Energy Laboratory (1992b), *Interim Solar Radiation Data Manual*, Golden, CO.

Nelson, R.J., and Augustyn, J.R. (1990), *Pacific Gas & Electric Company's Solar Insolation Monitoring Program, 5.633 Years and Counting*, Pacific Gas & Electric.

Parsons, B.K., and Wan, Y-H. (1992), *Factors Relevant to Utility Integration of Intermittent Renewable Technologies*, National Renewable Energy Laboratory, Golden, CO, draft.

Perez, R., Ineichen, P., Seals, R., and Zelenka, A. (1990a), *Solar Energy*, **45**, 111–114.

Perez, R., Seals, R., Stewart, R., Kapner, M., and Stillman, G. (1991), *Proc. Biennial Cong. Int. Solar Energy Society*, Pergamon, New York, pp. 679–702.

Perez, R., Seals, R., Zelenka, A., and Ineichen, P. (1990b), *Solar Energy*, **44**, 99–108.

PVUSA Project Team (1990), *1989–1990 PVUSA Progress Report*, Strategies Unlimited, Mountain View, CA.

Reindl, D.T., Beckman, W.A., and Duffie, J.A. (1990), *Solar Energy* **45**, 9–17.

Riordan, C. (1987), *Extraterrestrial Special Irradiance Data for Modeling Spectral Solar Irradiance at the Earth*'s Surface, SERI/TR-215-2921, Solar Energy Research Institute, Golden, CO.

Riordan, C., and Hulstrom, R. (1990), in T.J. Coutts and J.D. Meakin, Eds., *Current Topics in Photovoltaics*, Vol. 4, Academic Press, New York, pp. 1–23.

Sandia National Laboratories-Albuquerque (1998), *Stand-Alone Photovoltaic Systems, A Handbook of Recommended Design Practices*, SAND87- 7023, Sandia, Albuquerque, NM.

Shettle, E.P., Anderson, G.P., and Hall, L.A. (1988), *Extraterrestrial Solar Spectrum for Use With LOWTRAN*, Phillips Laboratory, Hanscom Air Force Base, Bedford, Massachusetts.

Solar Energy Research Institute (1981), *Solar Radiation Energy Resource Atlas of the United States*, SERI/SP-642-1037, Solar Energy Institute, Golden, CO (out of print; copies of maps are available from National Renewable Energy Laboratory; however, maps are being updated using the 1961–1990 national solar radiation data base).

Solar Energy Research Institute (1990), *Insolation Data Manual and Direct Normal Solar Radiation Data Manual*, SERI/TP-220-3880, Solar Energy Research Institute, Golden, CO. Updated in the Solar Radiation Data Manual for Flat-Plate and

Concentrating Collectors, National Renewable Energy Laboratory, Golden, CO, NREL/TP-463-5607, April 1994.

Wehrli, C. (1985), *Extraterrestrial Solar Spectrum*, Publication No. 615, Physikalisch-Meteorologisches observatorium and World Radiation Center, Davos, Switzerland.

Whitaker, C.M., Townsend, T.U., Wenger, H.J., Iliceto, A., Chimento, G., and Paletta, F., (1991), *22nd IEEE Photovoltaic Specialists Conf. Record*, IEEE, New York, pp. 608–613.

CHAPTER TWENTY-ONE

Future Space Program Plans, Needs, and Potential

S. RUSTY SAILORS, Power Systems, Spectrum Astro Inc.,
Gilbert, AZ 85234

1. INTRODUCTION

There is an inherent challenge to writing a chapter of this sort as it deals with evaluating the future direction of an ever-changing industry in a tremendously dynamic period of global political change. While many plans and programs may vary with political fluctuations, some fundamental needs should remain relatively constant into the next century, giving us confidence in the potential for the solar power industry in space.

The information presented here is intended to give the reader valuable insight into the current trends within the United States and throughout the rest of the world affected by the solar cell industry. I address questions such as: Why are solar cells important in space programs? What are the current trends in the world's governments and commercial industries concerning space programs, and how do these trends affect requirements for the usage of space-based solar cells? What are the current plans of the major space agencies and companies? How will we be affected by the changes occurring in the way our governments do business? How does the change in the relationship between the United States and the former Soviet Union affect the defense arena? These are important questions that need effective answers. As discussed, there are answers to these and other questions in this chapter that affect the solar cell industry along the lines of funding levels they can expect, the technologies that will be needed to support these requirements, and the

Solar Cells and Their Applications, Edited by Larry D. Partain.
ISBN 0-471-57420-1

potential areas of focus for the future that can lead to competitive advantages and enhanced technical performance.

This chapter is organized into five primary sections that follows this introduction. The first section is a general discussion addressing the importance of solar cells in space, the trends throughout the world in the space arena, and the resultant issues affecting requirements for space programs. The next three sections contain details, concerning the plans, needs, and potentials of the space solar cell industry as a whole. The chapter closes with a summary of the information discussed.

2. GENERAL DISCUSSION

Solar cells provide the primary method of power generation for space missions around or near the earth for all nations currently supporting the development, manufacture, and deployment of space programs. As a result, a large emphasis continues to be placed by the governments of the United States, Russia, Japan, European countries, and Canada as well as by commercial industry on improving this method in order to increase power densities and reduce system and launch costs.

As private and U.S. and foreign government funding levels become more and more of an issue, cost reduction is reaching into a large number of pertinent areas related to system level space-related issues. These pressures are leading to reduced and automated ground station support, and, to realize additional cost reductions through lower replenishment rates, spacecraft components and systems are being designed with greater space environment life durations. With the increased development of long-life components, increased levels of autonomy, and greater values of reliability in other life-limiting spacecraft components, batteries and solar cells remain the life-determining systems. Therefore, a requirements driver exists to increase solar cell performance at end-of-life (EOL). As a result, greater initial power performances must be realized with acceptable degradation rates, or increased radiation hardness must be realized at existing power performance levels.

Improvements continue to be made in silicon cell technology, and Japan and Germany report significant advances in cell efficiencies while maintaining or improving radiation hardness. Silicon will continue to be the mainstay of most programs for the near future just from cost and risk reduction perspectives. Gallium arsenide (GaAs), on its various substrates, continues to challenge the market, and manufacturers continue to improve AM0 average minimum efficiencies. As a result, many new programs propose the use of GaAs in order to realize the system level benefits of increased performance. As is usually the case, future planners continue to seek leaps in technology or magnitudes of improvement in all areas of spacecraft technology in order to reduce mass and cost; this holds true for solar cell technologies as well. In addition to Si and GaAs technologies, there are two main areas of choice for new cell and array development programs; this leads us to the issues concerning the tradeoffs between thin-film cells, which may be significantly greater in radiation hardness and could potentially provide substantial reductions in manufacturing and laydown costs but are lower in efficiencies, and high-efficiency cells, which are more expensive but may result in efficiencies as high as 28% with varying levels of radiation hardness. The primary benefit of high-efficiency cells is the potential reduction in system mass, volume, and cost as a result of the increase in performance in areal density. The choice of future programs to utilize one type or the other will greatly effect the planning and implementation of research and development programs (Benner, 1991). Both areas of cell types require additional development in assembly techniques in order to reduce costs further. As a result, in addition to improved performance, future requirements will continue to be generated concerning assembly issues such as improved welding techniques, simplified interconnects, and improved cell glassing techniques. While these technologies do not presently yield the orders of magnitude improvements desired by planners, a 28% efficient cell could significantly reduce the volume of solar arrays that would be an enabling technology for some proposed programs.

Many investigations of the potential benefits of thin films have yielded valuable comparisons to initiate at least a fundamental level of trades. The history of thin-film development has shown a steady increase in efficiencies since 1978. Many promoters of thin films feel that this trend will continue and that these technologies will compete with the performance of current cell types as demonstrated in Table 21.1. Table 21.2 lists projections of the potential increase in efficiencies of a few thin film examples.

Dr. Dennis Flood of the National Aeronautics and Space Administration (NASA) has also thoroughly investigated the potential of thin films for space applications as a result of NASA's need to reduce power generation costs and mass. Table 21.2 provides some valuable information on thin films he has evaluated for their space programs. In addition to those shown in Table 21.2,

TABLE 21.1 Projections for Future Efficiencies in Space Applications

		Future (%)	
Material	Current (%)	Conservative	Optimistic
Si	18.0	19.5	22.0
GaAs	21.4	22.0	25.0
$CuInSe^2$	11.2	12.0	13.0
Optimal thin film	8.6	12.5	15.0
Thin-film cascade	12.5	18.0	20.0

SOURCE: Landis (1991).

TABLE 21.2 Potential Thin-Film Solar Cells for Space Applications

					Radiation Resistance (P/P_0)	
Cell Type	Cell Structure	Projected Eff. (%)	Lab Eff. (%)	Commercial Eff. (%)	1×10^{15} (1 MeV, e−)	1×10^{13} (1 MeV, p+)
a-Si:H	Single junction, single gap on rigid substrate	10	<9.0	<5.0	0.80	0.65
a-Si:H	Tandem junction, single gap on rigid substrate	12	9.9	<5.0	—	0.75
a-Si:H	Tandem junction, single gap on flexible substrate	10	5.5	—	—	—
a-Si:H	Tandem junction, dual gap on rigid substrate	15	8.6	—	—	—
a-Si:H	Monolithic, multiple band gap on rigid substrate	18	10.9	—	—	—
$CuInSe_2$	3 μm cell, 1 μm window on glass substrate	>13	10.4	—	1.00	0.65
CuIn GaSe	3 μm cell, 1 μm window on glass substrate	>15	8.2	—	—	—
CdTe	Thin film on glass substrate	>18	9.8	—	—	—
a-Si:H/ $CuInSe_2$	Mechanically stacked tandem cell	>20	12.5	—	—	—

SOURCE: Flood (1992a).

there are many different types of materials and manufacturing processes of thin-films solar cells. While some copper-indium-diselenide (CIS) cells are made through the metal organic chemical vapor deposition (MOCVD), others still use a liquid phase epitaxy (LPE) process. Because of these and other differing manufacturing methods, care must be taken when statements are made concerning one category of thin-film cells versus another. This becomes particularly important when radiation testing is performed on different cell materials without comparing different manufacturing processes because the effects of proton and electron insertions are highly position and energy dependent. Perhaps from the results of combined European Space Agency (ESA), NASA, and Department of Defense (DOD) radiation testing, both in space and in ground facilities, *accurate* simulation models could be generated to predict cell performance based on materials, their depths and other mechanical parameters, and synergistic radiation and other environmental effects. Work has begun in this area, but not to the extent necessary to make it a viable alternative to extensive flight and ground testing.

Just as there is a large variety of thin-film cells, there is a significant variety of high-efficiency solar cells. There are mechanically stacked and monolithically grown multi-band gaps under development, (see Table 21.3 for example programs), as well as complex internal structures to create light trapping of certain spectra for increased utilization, while still others reflect light of certain wavelengths to keep the cell cooler and increase performance as a result. For the most part, these are primarily experimental cells in early stages of development, and the variety and extent to which each is pursued will be determined by government and industry funding. Other chapters in this book address the differing types of these cells and their inherent benefits.

The world is changing rapidly, and the space industry is growing as a result. As political walls are reduced to rubble, new relationships are built to replace them. These relationships are being developed through the governments of now friendly nations and sometimes through not so obvious commercial industrial relationships and alliances. In this dynamic economic and political environment, firm definitions of future government programs are not possible. As a

TABLE 21.3 Example United States Air Force (USAF)/BMDO Photovoltaic Technology Programs

Program	Contractor	Technology	Purpose
Multiple-band gap	Spectrolab	GaInP/GaAs + Ge mechanical stack	28% BOL min avg eff
Multiple-band gap	RTI	AlGaAs/Si 2-Term mechanical stack	23%–26% AM0 BOL eff
Si substrate dev	Washington State	GaAs on ZnSe/Si wafers	Use cheaper Si substrates
Improved Single junction GaAs	Astro-Power	GaAs with light-trapping layer	23% BOL min avg eff

result, this part of the chapter attempts to identify factors and current trends that may affect the direction of near- and far-term government programs in addition to changes in the ways and means by which governments may come to do business. As budgets become tighter, governments will seek plausible methods to reduce costs. These methods include major paradigm shifts that will change the way program managers approach and generate contracts. In the United States the USAF procurement procedures are already being modified to reflect this mode change, and the results are more like commercial procurements with inherent streamlines and increased contractor involvement in authorship of contracts. These changes will benefit the contracting world by simplifying the interfaces on government contracts. With these benefits, however, additional responsibilities will fall to the contractors. More up-front design and verification will be required, and more responsibility for nonadherence will fall on the contractor. In addition, to maximize its technology investment dollars, government agencies will focus on contractors that are multiprogram oriented and whose program results may have commercial applications. In the United States a formal declaration has been made in support of this idea in the form of the concepts of Dual Use and Defense Conversion. As the U.S. Congress scrutinizes technology investment funding lines, the more commercial applicability that can be demonstrated, the more secure a program may be. While the government procurement actions trend toward a more commercial nature, industry must respond and streamline costs. Quality is still a major factor, and, as stated above, the contractor will now assume more responsibility for meeting quality and performance criteria. The major changes will require more efficient documentation, smarter testing approaches, and improved performance to contract specific criteria. New Mil-Standards in process will accommodate this! Small aerospace component and spacecraft companies are providing examples of the future trend in the space business and are probably ahead of their time in pursuing this type of commercialized aerospace procurement approach.

The most significant impact on future plans and trends, however, is the change in the relationship between the United States and the former Soviet Union and the evolving world political climate. However, even during the evolution toward the realization of world peace, the verification and maintenance of such peace remains a need. Space systems have played a major role in providing the atmosphere for maintaining a generally stable situation between the major world powers and in providing assistance to the military in resolving specific regional conflicts. To continue to maintain a stable military situation and to provide early detection of threats against remaining world powers, space systems must remain a major player in the game and retain adequate funding for development and sustenance.

As a result of the reduction in budgets, a shift in paradigms is paramount for the separate government agencies to continue to exist and perform effectively. This change in the primary method of doing business involves the cooperation of funding and managing development programs between differing agencies such as the technology development laboratories of the Department of Energy (DOE), DOD, and NASA. Each of these government entities requires advances in solar cell technologies and has suffered significant funding reductions or major scrutiny for applicability of major programs in meeting future missions requirements. It is only sensible that these agencies coordinate their development programs and complement, rather than duplicate, each other's work. This will prove difficult for the private solar cell industry, as duplicate programs are no longer funded and development programs migrate toward such a coordinated and focused government approach.

Just as cooperation is a requirement for efficient functioning of U.S. government agencies, the same is now proving to be true within the cooperative group of nations developing new space programs. It is now feasible and sensible for ESA and Canadian programs, for example, to consider joint ventures with U.S. agencies and American private industry on new space technology development programs, including solar cells, in order to leverage technology investment dollars. As certain political hurdles are overcome and suitable agreements are reached, the same is proving to be true for joint programs between countries such as Japan, Russia, China, and the United States. The point to be made here is that, while the United States has been one of the primary powers in the space industry and the former Soviet Union no longer exists, alliances are forming at tremendous speed between numerous nations. While Russia may not be an independent leader now in the space game, she is forming industrial and political alliances in most of the leading space nations. In addition, she is forming alliances with smaller countries who want to enter the space arena.

As the world's commercial and military markets maintain their current large-sized spacecraft for existing programs, newer commercial and military programs are emerging that propose using large numbers of small spacecraft to perform simplified and dedicated services. Smaller spacecraft of this sort can be launched as multiples from the same booster and can be deployed separately from simple trajectories. The loft of multiple spacecraft in the same launch vehicle creates new challenges for manufacturers wherein cost per pound has a finer resolution as compared with large spacecraft and therefore a greater effect than in today's launch environment where the trades are large step functions. For example, if a manufacturer can use higher efficiency power panels, perhaps he can launch six spacecraft in the same vehicle rather than five; this may justify the increased cost of high-efficiency solar cells. The commercial use of small satellites will largely out-number military small satellites for the near future, and Table 21.4 lists some proposed small satellite constellations and their associated power requirements. This information is only provided to illustrate the potential PV market for such constellations and will change as commercial requirements and baselines fluctuate.

The world order is changing and military requirements will respond to these changes and their resultant perceived threats. All areas of military programs are changing, yet the space surveillance programs should remain relatively

TABLE 21.4 Proposed Small Satellite Commercial Constellations

Satellite	Estimated Mass (lb)	Prime Power (W)	No. of Satellites	Inclination
ARIES	275	107 avg 278 peak	48	90° (polar)
ELLIPSO I	40	22 peak	6	63.5°
ELLIPSO II	385	174 peak	18	63.4°
GLOBALSTAR	510	150 avg		
Conus		875 peak	24	47°
Global			48	47°
GONETS	496	40 avg 48 peak	36	83°
IRIDIUM	851	1429 avg	77	90°
LEOCOM	97	112.4	24	90°
LEOSAT	50–100	23.4 avg	18	40°
ODYSSEY	2,500	1800	12	55°
ORBCOM	331	360 avg		
Inclined orbits			18	40°–60°
Polar orbits			2	90°
STARSYS	110–220	120	24	60°

SOURCE: Hatfield (1991).

stable as the need for such programs remains mostly unchanged. As a result, space-related technologies such as solar power will remain in demand.

The Ballistic Missile Defense Office (BMDO) is a major player in the military industry and has undergone major restructuring in its focus and organization in response to the changing relationships of the world's militaries. There is no longer the perceived threat of an all-out nuclear attack from the former Soviet Union, and, a result, there is a reduced emphasis by government representatives on BMDO funding. In light of this change in funding emphasis, BMDO has begun to focus on smaller satellites with dedicated missions to prove their sensor capabilities. The Miniature Sensor Technology Insertion (MSTI) Program is an example of the emergence of this philosophy and demonstrates the feasibility of the concept of "faster, better, cheaper" with minimal funding, standardized buses, and quick turn arounds of 12 months from concept to launch. This capability will enable the DOD to perform flight experiments on emerging sensors in the time-frame necessary to make the flight data useful and within the funding constraints they are experiencing. If programs of this sort continue to be successful, launches of follow-on experiments are probable. On the larger scale, the demonstration of the technical performance, reduced costs, and quick turn around capabilities of small spacecraft of this sort could support the feasibility of small satellite constellations and "pop up" spacecraft for DOD visibility into theater level incidents.

Therefore, even with declining military budgets, the requirements for solar-powered spacecraft remain intact, and if the proposed small satellite constellations and on-the-shelf "pop ups" are realized, the demand for applicable solar cell technologies will greatly increase.

3. PLANS

The U.S. DOD will also continue to maintain and deploy its current major programs with an emphasis on cost reductions and improved performance. The large-scale spacecraft programs, such as MILSTAR, will be scaled down as the U.S. Congress urges the DOD to utilize more commercial assets for its missions. Follow-on programs to the large-scale ones are also focusing on size reductions and autonomy to realize the resultant cost savings in launch and ground support. Along these lines, U.S. intelligence agencies are being urged to develop standardized buses for their payloads to reduce costs further, yet still meet their specialized mission requirements. It is very difficult to predict the future of U.S. military programs in this dynamic political environment, but it is safe to say that the military space mission will remain intact even with the possible use of smaller spacecraft and commercial communication assets.

The U.S. government also has significant plans and requirements for nonmilitary applications in space through agencies such as NASA and the DOE. NASA's plans include EOS-A (Earth Observation System), SEI (Space Exploration Initiative), and Space Station, just to name a few. The specific requirements of these programs are also extremely dynamic at this point as a result of the change in NASA management and the uncertainty of emphasis by the new Federal administration. Just as the military continues to develop "generic" technologies that can support a variety of applications, NASA has also invested in the development of thin-film and concentrator cells in addition to a focus on two primary array structures, as outlined in Table 21.5.

It is clear, then, that NASA will continue to invest in technologies that will assist them in meeting aggressive goals such as these. In addition, NASA has placed an emphasis on small satellites through its Small Satellite Technology Initiative, (SSTI), which will emphasize the development of specific space technologies for small spacecraft. The major purpose of this initiative is to develop standardized small spacecraft buses that will support the myriad of NASA payloads waiting to be flown. NASA officials have recognized the risk and delays associated with large-scale programs that can result in the loss of decades of funding and preparation by placing a large number of payloads on one large spacecraft rather than distributing that risk and funding across multiple small satellite launches on inexpensive launchers.

The DOE's funding is much more stable than that of the military programs, but the amount of money the DOE intends to invest in space-related PV development programs is still unclear. The National Renewable Energy Laboratory (NREL) is currently coordinating the potential future efforts in this

TABLE 21.5 NASA's Future Photovoltaic Array Programs

Planar arrays	
Goals	300 W/kg; $\geqslant$200 W/m^2
Approach	Flexible lightweight blankets
	Radiation-resistant cells
Availability	1996: InP/Ge (passive substrate), 175 W/kg
(unless developed	1998: InP/Si, 200 W/kg (perhaps a-Si:H or CIS)
by others earlier)	2000: Thin (4 μm) InP, 300 W/kg
Concentrator arrays	
Goals	$\geqslant$330 W/m^2; $\geqslant$100 W/kg (15 years LEO)
	$\geqslant$250 W/m^2; $\geqslant$75 W/kg (15 years GEO)
Approach	Minidome Fresnel lens, rigid panels
	30% Tandem cells in LEO
	InP cells in GEO
Availability	1998: Array engineering model ground demo
(Same conditions)	Either cell available

SOURCE: Flood (1992b).

area, but because of the generally more stable and effective long-range planning conducted by the DOE, they have not yet provided documented direction for this funding line. However, representatives of the DOE feel that the lessons learned from NREL's terrestrial PV programs could well be adapted for space cell programs. NREL has claimed that the cost of space-qualified cells (dollars/watt) could well be reduced by orders of magnitude if terrestrial-based concepts were adopted.

While the stability of government programs continues to fluctuate, the commercial space market continues to climb. Imaging data utilization is on a significant rise in the commercial marketplace, and new data processing techniques are providing new applications for such data. In an interesting twist, some of these data will even become available from classified U.S. programs. This demonstrates another move on the government's part to demonstrate the Dual Use of commercial applications of government resources. Many commercial enterprises are establishing new markets and creating new demands for these emerging imaging processing techniques. From surveying ranges for cattlemen in Australia, to catching subsidized farmers in the European Community cheating on their utilization claims (SPOT & Landsat), image processing is finding a large number of applications. This rising demand has justified a significant increase in satellite imaging programs and will provide a steady market for the satellite-related technologies industry.

Hughes Aircraft Company, Martin Marietta Astro Space (MMAS), TRW, Lockheed, and others provide major telecommunications, imaging, and specific sensor satellites throughout the world. American corporations must continue in their current trend to remain competitive with other manufacturers throughout the world. As we have seen in the launch vehicle battles, there are new rules to the game and American companies and the institutions that govern them must learn these guidelines in order to take advantage of the continually expanding world market. European alliances have set, as their primary goal, to compete heavily with major producers such as Hughes and MMAS; Loral has already become part of such an alliance. This is important because, of the 43 communications satellite contracts let between the end of 1987 and 1991, Space System/Loral, including its European players, won 11. The telecommunications arena is continually expanding and currently provides the greatest share of future market potential for the PV and spacecraft component industries.

The United States and Canada share in the defense of North American airspace and are agreeable to pursuing joint programs to maintain and enhance that defense under the U.S.–Canada Defense Production and Development Sharing Agreements. As a result, there is an opportunity for new programs in such a joint-nation alliance, and power systems for spacecraft will prove to be a large factor. Much of the Canadian government's power work is managed through their Department of National Defense's (DND) Defense Research Establishment of Ottowa (DREO). While Canada develops many of its own technologies, it is heavily dependent on other countries, generally the United States, to provide solar cells for their conventional power systems. Conservative in nature, most of their proposed work includes the use of Si or sometimes even GaAs (Maskell, 1991). Canada is developing several space radar programs, which are power hungry by nature, and will be looking to solar cell manufacturers to take advantage of these opportunities.

New technologies in Great Britain are primarily focused through the Defense Research Agency (DRA) and British Aerospace establishment (BAe), which is currently under bid for an alliance with major European contractors. A takeover such as this may greatly effect BAe's ability to influence the level of Great Britain's involvement in future technology programs. Still, concentrating through the DRA, the pursuit of mutually beneficial joint programs has been dictated by the Information Exchange Program (IEP), which was put in place specifically for this purpose. The Space Technology Research Vehicle (STRV) joint nations program, which is to test emerging solar cell technologies in a harsh space environment, is a good example of international cooperation through the IEP.

The ESA, representing thirteen European countries, greatly influences the direction of technology development through their technology center in Noordwijk, the Netherlands, and by their policy directed through their main offices in Paris, France. However, as a result of the dependence on ESA funding from the varying contributing countries and the percentage each provides, technology sources are in turn greatly influenced. The ESA feels that joint programs will benefit them financially and help them to meet their goals in the long term. However, there is a strong feeling from the Europeans that the United States should learn how to listen in joint programs and not just come to the table with a predefined plan to see who wants to participate. This will take some learning on the part of U.S. participants, but it is essential in creating new and effective joint nation programs.

The ESA has a great deal of planning and investment from its members in future programs. Table 21.6 indicates a fair number of upcoming power-related programs, most of which will require large solar arrays.

Japan's focus on new cells is toward high-efficiency silicon at a target of 18% BOL. This is to meet their requirements of achieving high efficiency at low cost. Table 21.7 summarizes some of the characteristics, including power levels and cell technologies, of a few of the proposed Japanese programs.

The GMS-5 Program will provide Japan with additional capabilities in weather watch by VISSR, in collection of weather data for relay in imaging data, as well as in search and rescue (Kimura, 1992). Programs such as these will keep the demand for PV in Japan on the rise.

Russia has almost completed preparation for its Glonas constellation of 21 active spacecraft with three spares as a complement to the U.S. GPS system of spacecraft.

Korea has launched Kitsat, a Korean student-built small satellite constructed at the University of Surrey in England under a joint technologies agreement. Korea entered another technologies exchange program with MMAS in

TABLE 21.6 ESA Planned Space Programs

Name	Mission	Status	Power Data
ARIANE RECOV. CARRIER	Space Trans.	In definition	2 kW solar array
ISO	Science	Launch 93	580 W solar array
SOHO	Science	Launch 95	1.5 kW solar array
CLUSTER	Science	Launch 95	4 × 420 W solar array
XMM	Science	Under dev.	1 kW solar array
FIRST	Science	Proposed for dev.	1 kW solar array
ROSETTA	Science	Proposed for dev.	RTG approx. 400 W
PRISMA	Science	Phase A study	1 kW solar array
INTEGRAL	Science	Phase A study	1.1 kW solar array
STEP	Science	Phase A study	340 W solar array
MARSNET	Science	Phase A study	N × 6 solar array
ERS-2	Earth observ.	Under dev.	2.1 kW solar array
ARISTOTELES	Earth observ.	Proposed for dev.	1.07 kW solar array
EO MISSIONS ON POLAR PLATFORM	Earth observ.	In definition	7.7 kW solar array
2ND GENERATION METEOSAT	Earth observ.	Phase A study	1.2 kW solar array
EURECA	Large systems	Launch 92	5.5 kW solar array
COLUMBUS FFL	Large systems	Under dev.	19 kW solar array
COLUMBUS PPF	Large systems	Under dev.	7.7 kW solar array
ARTEMIS	Large systems	Under dev.	2.5 kW solar array
DATA RELAY SAT	Large systems	Proposed for dev.	3 kW solar array

TABLE 21.7 Solar Array Specifications for Future Japanese Satellites

Satellite Name[a]	Launch	Power Output (kW)	Solar Cell Type
ETS-VI	Summer 94	4.1	Si BSFR; 50 μm
GMS-5	Winter 94	0.29	Si BSFR; 200 μm
ADEOS	Winter 95	4.5	Si BSFR; 100 μm
COMETS	Summer 97	5.4	GaAs; 200 μm

[a]ETS, *E*ngineering *T*est *S*atellite; GMS, *G*eostationary *M*eteorological *S*atellite; ADEOS, *Ad*vanced *E*arth *O*bserving *S*atellite; COMETS, *Com*munications and broadcasting *E*ngineering *T*est *S*atellite.

the United States in which they purchased two telecommunications satellites but also will receive training, technology exchange, and subcontracts for major subcomponents of the spacecraft. Though Korea lacks many internal natural assets, historically they have achieved large shares of the competitive marketplace through technology exchanges and training of their engineers and manufacturers. The same could occur in the spacecraft manufacturing arena as the demand for inexpensive satellites with simplified functions grows. The United States and Europe have the same opportunity, but Europe is still focusing on large-scale programs and a majority of the U.S. spacecraft manufacturers feel that small satellites are not a priority. As a result, Korea, or perhaps even China, may well take a large jump ahead in this area.

France greatly influences the direction and philosophy of the ESA through its substantial funding and political power. The prime contractors in France, such as Matra Marconni Space (a joint British/French corporation) seem to follow the lead of the ESA, and as a result the PV community is in turn substantially controlled by the French government's policy through the ministry of Space & Research. In 1992, the French were spending at the rate of $2.7 billion U.S. dollars per year on space while the total ESA budget projected for 1993 through 1998 was $24 billion (U.S. $) total. As far as the future market potential in France is concerned, the French government may have long-term plans with the British and other European countries to replace the current Skynet and Syracuse military satellite programs with a military telecommunications space-based system called Eumilsatcom. In the interim, the French government has announced plans to produce the Osiris and Zenon programs, which will provide early warning and eavesdropping capabilities to augment the Helios optical satellite (a joint French, Italian, and Spanish program) and the Syracuse communications system. If U.S. companies desire to, and are allowed to, compete for programs such as this, they must develop a much stronger competitive approach, for they will be competing with growing conglomerates in the European community. For the PV market to date, however, the French have not demonstrated a desire to compete with the primary European players, the Germans.

West Germany has dominated the European space PV market and will invest heavily in future budgets for ESA in order to bring joint programs with Russia into play, even at the cost of funding dollars for their national and internal programs. The cost of modernizing East Germany could become a factor in the equation, but should not affect Germany's primary direction of their major programs and their ESA funding emphasis. Currently, Germany controls a virtual monopoly in the solar cell development and production arena, with small firms in Great Britain and Italy trying to take a minor share through mainly supportive roles. Even these firms, however, are sometimes controlled by German companies such as TST. If the current trend holds, then Germany will be successful at holding off any potential European competitors and will only have to deal with U.S. solar cell manufacturers in bids on European as well as American telecommunications programs.

European alliances between various countries such as Germany, France, Italy, and Great Britain are forming into the competitive giants of the telecommunications and imaging industry. Some of these alliances, such as Space Systems/Loral, include American companies in order to maintain a political lever to involve themselves in the most suitable circumstances in varying political environments. Alliances of this sort are extremely powerful and represent the future mode of the space business.

4. NEEDS

The solar cell industry has basic needs that will allow future users to accomplish their necessary goals. These needs include improved array structures that allow for minimal impacts for cell size and type changes, ease of manufacture, scalability and improved power to mass ratio; improvements in laydown techniques to simplify manufacturability and to reduce costs; leaps in technology for space-based solar cells that will improve efficiencies, increase radiation tolerances, simplify manufacturability to reduce costs, and utilize inexpensive materials; and improved simulation, analysis, and testing that will build confidence in cell performance on orbit.

Just as there are differing cell categories, there are differing array structures in current use and under new development that will also affect the emphasis of cell development. Cell sizing, interconnects, large versus small cells, flexible arrays, and laydown costs are all traded against the projected benefits of each array structure. For example, many agencies focus on flexible array while some ESA programs have extensive experience in this area and are backing away from flexible blankets because of assembly costs outweighing the benefits. This does not mean that a flexible array is not the right choice for future programs. It clearly indicates that if manufacturing and assembly processes are streamlined that this structure may prove quite competitive. These areas are being heavily scrutinized for advancements to meet the overall necessity of reduced costs of manufacturing and laydown. Table 21.8 represents a good example of process improvement planning provided by the ESA.

TABLE 21.8 ESA Assembly Technology Programs

Process	Status	Timeframe	Method
Al interconnect technology	Running	87–92	Development of ultrasonically welded Al interconnects
GaAs module Technology	Running	89–91	Parallel-gap and ultrasonic welding of GaAs cells
Advanced Si-module Technology	Completed	87–91	Improvement of Ag and Mo welding (model and tests)
Solar cell assembly techniques, Si and GaAs	Proposed	92–94	Continue ASTP-3 development for new cell types
Direct glassing	Running	87–91	Direct bonding of coverglass to Si cells (electrostatic)
IR reflecting coating	RFQ	91–92	Reduction of operating temperature by reflector on coverglass
Teflon bonding of glass on cells	RFQ	91–92	Quality Teflon-bonded Si-cell assemblies ("fast process")
Covers for Hi-ETA cells	Proposed	92–94	Optimize glass configuration (optical and geometrical)
Direct bonding of GaAs cells	Proposed	92–94	Apply predeveloped direct glass technology to GaAs cells

Along the same lines, each array structure may require a different cell configuration, and its true benefits may not be applicable until 8 cm × 8 cm versions of the cell of choice is in the manufacturing stages or roll-out versions of thin films are under production. To address this issue, array designers and manufacturers should focus on designing arrays that truly accept differing cell sizes in order to accommodate the emerging cell designs that do survive. The challenge here is to design-in that inherent adaptability, but at a low cost or no cost to change cell types.

Array improvements are hard needs for NASA and the DOD. NASA's Office of Aeronautics and Space Technology briefed at an Interagency Advanced Power Group meeting that the primary objective of their PV energy conversion programs was to provide the technology for PV arrays with improved conversion efficiency, reduced mass, reduced cost, and increased operating life for advanced space missions (Bennett, 1992). NASA went on to state that specific long-range goals are to develop the technology base for PV arrays with specific power densities of 300 W/kg with substantial reductions in size and cost, and increases in EOL power capability. The USAF and BMDO interim goals are to reach specific power densities of 150 W/kg and areal power densities of 100 W/m^2.

A fairly obvious area of needs is that of improvements to efficiencies, radiation hardness, and manufacturability of space solar cells. Improvements of this sort will provide enhancements that will increase the power available to using programs and extend the usable duration of these life-limiting components. Increases in the radiation tolerance of high-efficiency silicon cells is the near-term solution to these concerns and will provide the most acceptable route for conservative programs. An emphasis needs to remain, however, on the research in the area of increasing the efficiency of GaAs-based multiple-band gaps, which should be augmented with a focus on simplified manufacturability and the use of inexpensive materials. For example, Washington State University is developing a technique to grow GaAs cells on cheaper flexible silicon substrates by using a zinc diselenide buffer layer to reduce dislocation densities. Approaches such as this to reduce manufacturing costs are a must for the future of space solar power programs.

There is another need whose solution would assist designers in building confidence in the accuracy and applicability of their designs. In the new space business environment in which smaller and medium-sized programs require quick turn arounds for new designs and large constellation programs need to provide multiple trades, sizing and analysis of cell and array performance through accurate simulations would enable companies to respond to the fluctuating requirements of commercial and experimental military programs. These enhanced simulations and analytical programs would have to be flexible enough to address the performance, radiation hardness, and architecture of emerging cell technologies and generate accurate results using this information as important trade parameters for prescribed orbits and mission modes.

5. POTENTIAL

There is a great deal of potential for growth in the space solar cell arena as well as for new areas of technical improvements. For program growth, Dual Use and Defense Conversion offer new areas of interest and funding for space-related technologies developed under defense programs and can be utilized for commercial applications. This allows for a tremendous augmentation to the commercial space sector of the United States that will be able to utilize the tremendous assets that have been developed under the defense umbrella.

Another area of realistic potential for the future is that of proposed commercial communications constellations that will enhance performance for present high end users of FAX, high quality voice, high rate data, and high resolution graphics, as well as to provide basic service to Third World countries that do not presently have telecommunications services. The specific plans and requirements for these programs fluctuate, as they are still in the proposal stage of the dynamic commercial market, but the potential need for significant numbers of low-cost spacecraft is staggering. One might consider the Calling Network proposed by Calling Communications (Tuck et al., 1993) as an example in which a constellation of 840–924 spacecraft are required. In such an example, high-efficiency yet low-cost solar power is an enabling technology that could drive the program. This program is a good example of the springboard effects of defense conversion in which many of the proposed technologies that make such a system feasible were developed and funded under U.S. government programs.

Still another exciting area of potential is that of bridging between the development, through millions of dollars of investment, of terrestrial solar cells and the development of space solar cells. Mr. John Benner of the NREL and I share the same conviction that there is a bridge to be found between these two worlds that will significantly reduce the costs of space solar power. Consider, for a moment, some modification to terrestrial cells and the use of altered terrestrial array techniques that would be used to construct an array that would probably experience more degradation than space cells, but would survive for 1 or 2 year missions for space-borne experiments. Such an approach would reduce the cost of such an array by almost an order of magnitude and still meet such a mission's requirements. Using these approaches will possibly enable the "Snap Together" spacecraft that is almost a firm necessity for the success of highly populated constellations. The bridging of the terrestrial and space worlds is an exciting possibility for future space missions to simplify cell manufacture and array assembly to meet necessary cost reductions.

Possibly the greatest potential for the future of solar power in space is the opportunity to work joint programs between government agencies and countries that, in the past, would never have been able to work together on such a detailed technical level. Today and into tomorrow, the stage is set for entities throughout the world to share in joint programs and to develop technologies that could significantly benefit all space programs. Already countries are cooperating on joint projects, but the recent changes in world politics have opened the doors for significant increases in this form of cooperation.

6. CONCLUSION

In the midst of unprecedented world political upheavels, government budgets are dynamic across the globe. Even as the world environment is changing, however, solar power remains the baseline for most future military and commercial space missions. The viable choices of solar cells available for use in future programs are increasing as thin-film and high-efficiency cells continue to be developed. These cells can serve as enabling technologies for future high-power applications, for constellations of small satellites, and for standard buses for experimental spacecraft.

As demonstrated in this chapter, many factors affect future requirements and trends: technical factors affected by emerging and enabling technologies; a great variety of cost drivers and cost limiters; improved data processing techniques; and, most influential of all, fluctuating international and national political environments.

These changing times are forcing changes in the way governments must complete their tasks. The most effective way for government agencies to operate is through cooperation and leveraged programs. This requires a change in paradigms as well as in effective training, software, unified data bases, and so forth, for planners and programs managers to utilize joint agency cooperative programs. Our accounting systems and funding strategies must be adapted to accommodate this new approach to doing business.

As many defense programs are being reduced, so many commercial programs are expanding. Commercial markets continue to rise in response to ever-increasing demands for information and imaging data. In response to these changing markets and political pressures, more and more government agencies, in adopting commercial tendencies, will attempt to prove their value by demonstrating the commercial as well as the military value to their programs. In the commercial market, American companies have the opportunity to remain strong and be greatly involved in the manufacture of spacecraft and components for European and Pacific Rim countries. It is important to note that joint programs are aggressively being pursued by other nations and these alliances are challenging the United States' control of the space arena. As a result, the United States must aggressively pursue similar national as well as commercial alliances in spite of industry complaints and grievances in order to remain competitive.

Costs and technological advancement will drive the near-future needs, and these areas will be pursued by both the commercial world and government agencies. The number of technology programs will be reduced, but the remaining programs should be reasonably stable. More and more the commercial programs and government agencies will select their niches in which to focus their technology investment dollars to maximize their returns and enhance the viability of future programs.

Potential for the future includes opportunities for Dual Use and Defense Conversion, allowing the commercial industry to benefit from billions of dollars invested in defense programs. Commercial constellations provide ample opportunities for improved solar cells and array manufacturing techniques. An exciting area for potential efforts is the opportunity to bridge between the terrestrial and space solar industries. This area holds great potential for future investigations and could reap substantial benefits in costs and simplicity of manufacture for space programs as well as further proving the viability of Dual Use of DOE-invested dollars.

Finally, the world is continuing to shrink and international boundaries are eroding. The "New World Order" is actually occurring in many ways, and American cooperation and alliances must extend beyond the boundaries of the

United States and circle the globe. Opportunities for industry and government agencies will continue to exist and to grow if this approach is aggressively pursued.

REFERENCES

Benner, J. (1991), *22nd IEEE Photovoltaic Specialists Conf. Record,* IEEE, New York, pp. 7–11.

Bennett, G.L. (1992), *IAPG Spring Symposium Solar & Systems Working Groups,* March 1992, Vol. I.

Flood, D.J. (1992a), *6th Int. PV Sci. Eng. Conf.*

Flood, D.J. (1992b), *Interagency Advanced Power Group Spring Symposium,* March 1992, Vol. I.

Hatlelid, J.E. (1991), *5th Annu. Conf. Small Satellites.*

Kimura, K. (1992), *Int. Symp. Space Technol. Sci.*

Landis, G.A. (1991), *26th IECEC,* Vol. II, pp. 256–261.

Maskell, C.A. (1991), *26th IECEC,* Vol. I, pp. 340–345.

Tuck, E.F., et al. (1993), *7th Annu. AIAA/USU Conf. Small Satellites.*

CHAPTER TWENTY-TWO

Government Terrestrial Acceleration Programs

WILLIAM WALLACE, National Renewable Energy Laboratory, Golden, CO 80401

1. INTRODUCTION

Beginning in 1971, the U.S. Federal Government initiated a terrestrial photovoltaic (PV) development program. The program was funded first through the National Science Foundation (NSF) and the Energy Research and Development Administration (ERDA) and, beginning in 1977, through the U.S. Department of Energy (DOE). For the past 22 years continuous, but variable, levels of funding have provided support for basic and applied research and development (R&D), manufacturing process research, and system and application development. Recipients of this funding included DOE's national laboratories, universities, the PV industry, and the electric utility industry. Throughout this time period, the primary objective of the Federal terrestrial program has been the development of PV as a significant energy source for the U.S. domestic electric utility market. However, resources have also been used to support remote application development for domestic and international markets and applications for nonutility end-users.

The role of the government today in technology development and commercialization is facing and responding to new challenges. For example, with increased global competition, technology and product development cycles are becoming compressed. Companies are finding it difficult and expensive to pursue technology development in isolation, and cooperative efforts through industrial consortia and government/industry partnerships are increasing in

Solar Cells and Their Applications, Edited by Larry D. Partain.
ISBN 0-471-57420-1

the United States and abroad, especially in Europe and Japan. The actions of other governments directly impact the competitiveness of U.S companies. In addition, competition is accelerating the globalization of major corporations, and it is becoming increasingly difficult to differentiate between domestic and foreign industries. These challenges impose new demands on the relationship between government and industry, but also create new opportunities for enhancing U.S. competitiveness and increasing world competition.

2. U.S. TERRESTRIAL PHOTOVOLTAICS PROGRAM

2.1. Overview

During the past two decades, the U.S. Federal Government through the DOE and its predecessor agencies has provided continuous funding for the development of terrestrial PV technologies. The U.S. funding history for PV, as well as overall funding for solar technologies, is shown in Figure 22.1. The level of PV funding has varied in stages according to priorities set by the U.S. Congress and presidential administrations. During the Carter Administration, the oil crises of the 1970s provided a focus on energy policy and renewable energy development. The Solar Photovoltaic Research, Demonstration, and Development Act of 1978 (McCormack, 1978) set the stage for the accelerated ramp-up in PV funding that occurred during the Carter Administration, reaching a peak of $151 million in Fiscal Year 1980. The Act provided for an

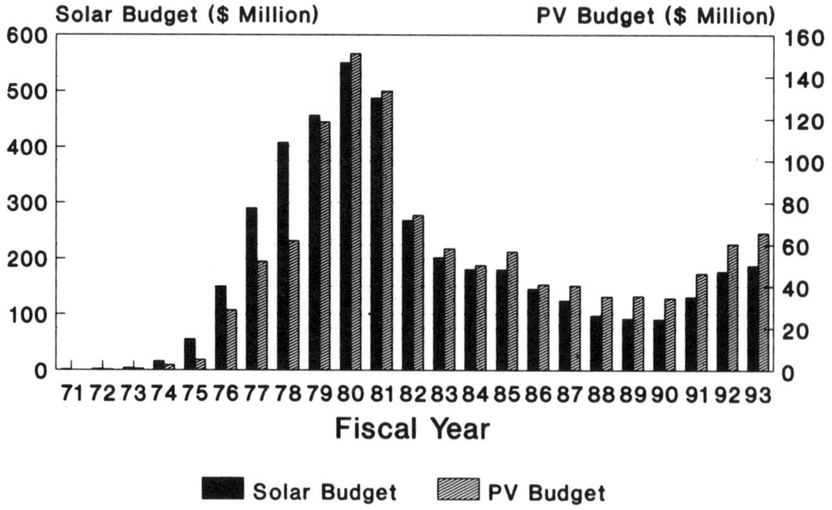

FIGURE 22.1 U.S. Federal Government funding history for PV and Solar (Herwig, 1992).

early photovoltaics demonstration program designed to accelerate the commercialization of PV in domestic markets.

With the election of President Reagan in 1980, a fundamental shift occurred in energy policy that called for an emphasis on government support of basic research for renewable energy technology development. Commercialization was considered the primary responsibility of the private sector. Over the next 10 years, the DOE budget for photovoltaics declined, reaching a low of $35 million in Fiscal Years 1988 through 1990. The impact of lower budgets was the severe curtailment of PV demonstration and commercialization projects, limited mainly to system research and field testing of new technologies and the consolidation of the PV program at two DOE laboratories, the Solar Energy Research Institute (SERI) and the Sandia National Laboratories (SNL). (Note that on September 16, 1992, SERI was designated the National Renewable Energy Laboratory [NREL] by President George Bush.)

Renewed concerns over U.S. energy security and global competitiveness resulted in a new emphasis on the development of national energy and industrial policies during the Bush Administration. In 1991, the DOE prepared a broad-based national energy strategy (U.S. Department of Energy, 1991a) outlining comprehensive policy initiatives for fossil, renewable, and energy efficiency technologies in all major energy-consuming sectors of the U.S. economy. In addition, strong congressional support for renewable energy development resulted in increasing solar and PV budgets during the past 4 years, restoring the PV budget to $65.5 million in Fiscal Year 1993. Congress has also drafted its own national energy strategy (Energy Policy Act of 1992) (House of Representatives, 1992) which includes provisions for government support of private sector joint ventures to accelerate PV commercialization.

Cumulative U.S. Federal Government funding of PV terrestrial technology development to date is approximately $1.2 billion, and approximately another $2 billion (Richard Sellers, Director of Technology Services, Solar Energy Industries Association [SEIA], personal communications, 1992) has been spent by the U.S. domestic PV industry in R&D, manufacturing, and product development. One constant in the Federal program has been the support of basic and applied research for material and device development that has provided the foundation for technology transfer to manufacturing and the constant improvement of module performance. Funding has been variable for system research, demonstration projects, and commercialization activities. For Fiscal Year 1992, the division of funds among business organizations (universities, small and large businesses, and national laboratories) and among technology categories (silicon and high-efficiency devices, thin films, manufacturing support, and module and system development) is shown in Figure 22.2 (Herwig, personal communication). During recent years, DOE policy has consistently been to direct approximately 50% of the PV budget to the private sector in R&D subcontracts and direct cost-sharing of technology projects.

One of the benefits of the U.S. and worldwide investment in terrestrial PV technologies is reflected in the steady market growth of world PV module

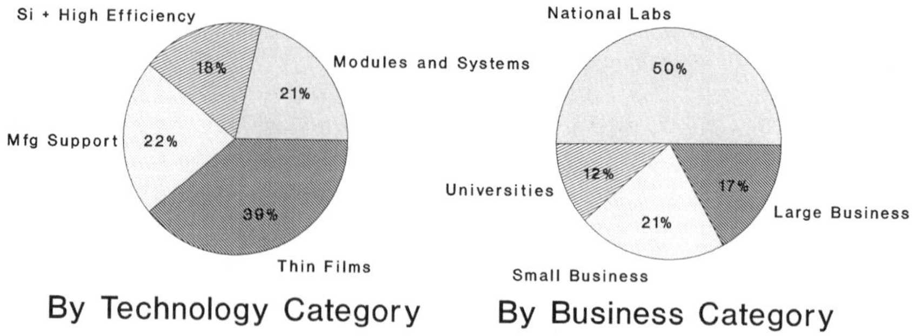

FIGURE 22.2 The funding distribution for the Fiscal Year 1992 U.S. DOE budget among technology and business categories obtained from the annual operating plans of NREL and SNL (1992).

shipments, as shown in Figure 22.3 (Maycock, 1992; Maycock and Stirewalt, 1985). Since 1987, the growth in world module shipments has averaged 17% per year. The U.S. market share of this growing world PV market has been constant at 30%–35% since 1985, with half of sales going to exports. Although thin-film PV technologies have become a significant factor in the marketplace, single crystal and polycrystalline silicon PV modules still represented 73% of module sales in 1991. The foundation of the PV world market has been and continues to be remote applications for industrialized and developing country needs. Although interest in utility-connected PV applications is growing,

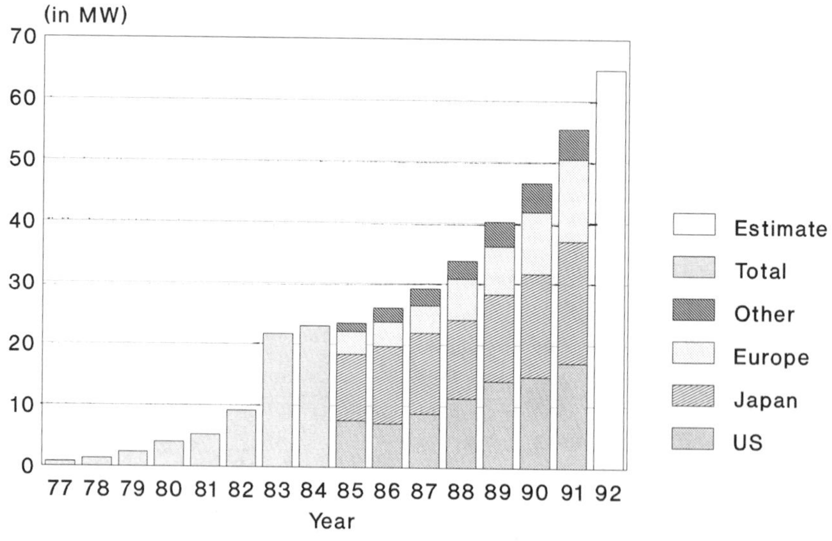

FIGURE 22.3 The PV market as a function of time (Maycock, 1985, 1992).

especially in Europe and the United States, the contribution of such applications to the world market is still small.

Another benefit of the worldwide investment in PV and a major driver for PV market growth has been the decline in PV module and system prices as a function of time. Since 1974, PV module prices have declined by aproximately a factor of 10 in current (unadjusted by inflation) dollars, from $\$70/W_p$ (watt-peak) to today's price of $\$4$–$\$6/W_p$. Module prices currently represent ~50% of total system prices. From inception, a primary objective of the Federal terrestrial PV program has been cost reduction in module and balance-of-system (BOS) components of PV systems. Complementary objectives have been associated with high efficiency, long lifetime (30 years), and high reliability for flat-plate and concentrator cells and modules suitable for utility applications.

Based on a learning curve model (Fig. 22.4), technology cost reduction is not simply a function of time but is related to cumulative production experience impacted by R&D advances, the rate of technology transfer to production, and the rate of investment in advanced manufacturing processes and economy-of-scale production plants. Manufacturing investment is also linked to market development, which in turn is heavily influenced by the price reduction resulting from manufacturing expansion and competition. Any effective cost reduction strategy must address the intrinsic linkage associated with concurrent

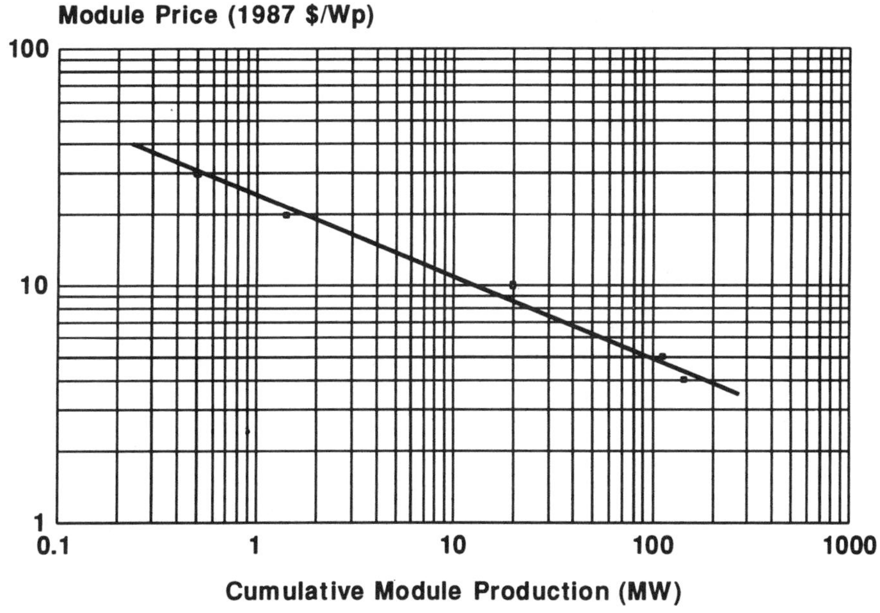

FIGURE 22.4 Early DOE cost goals for PV and PV learning curve for price reduction.

manufacturing investment and market development requirements.

Current DOE PV targets are reflected in the most recent 5 year plan (U.S. Department of Energy, 1991b) and in the recently developed *Solar 2000* strategic plan (Annan, 1992), both prepared by the Office of Solar Energy Conversion (OSEC). The DOE goals are aggressive, calling for PV-generated electricity price reductions from current levels of 25–50 cents/kW-h to 10–20 cents/kW-h in the 1995–2000 time frame (all prices in 1990 dollars). *Solar 2000* also proposes the cumulative installation of 1.5 GW of PV generating capacity by U.S. manufacturers in domestic and international markets, including 300 MW in the domestic grid-connected electric utility market, by the year 2000.

OSEC recognizes that the PV program goals are very aggressive and that the private sector must bear the primary responsibility in achieving the targets. Consequently, a revised DOE strategy is outlined in updated planning documents representing a potential shift in policy for the remainder of the decade. Policy shifts include a renewed emphasis on PV applications in all end-use sectors, including the building, industry, and transportation sectors. The strategy for the electric utility sector also emphasizes the phased introduction of PV in utility markets by taking advantage of current, cost-effective applications and by developing high-value, grid-connected applications that are intermediate between the off-grid and bulk power generation extremes of the utility market. *Solar 2000* also calls for joint multiple stakeholder collaboration and technology development and commercialization through joint ventures.

The U.S. Federal Government's role in PV technology development has involved three major activities: (1) innovation and fundamental research, (2) technology development and validation, and (3) market conditioning. These activities are reviewed in the following sections.

2.2. Innovation and Fundamental Research

The support of basic and applied R&D at the material, cell, and submodule levels has been a permanent component of the U.S. terrestrial photovoltaic program since inception. Fundamental research is necessary to provide the incremental technical advances in new materials and devices that support continuous improvements in PV module and system performance and to provide proof-of-concept demonstrations for innovative ideas. Government supported and cost-shared fundamental research in universities, national laboratories, and industrial laboratories has led to numerous innovations that provide the foundation of today's commercial module technologies. For example, early research in the DOE's Flat-Plate Solar Array (FSA) Project supported the development of bulk polycrystalline and ribbon silicon technologies and improvements in single crystal silicon solar cell processing (Jet Propulsion Laboratory, 1985).

The Federal program has also supported advances in thin-film photovoltaics. This includes early research on hydrogenated amorphous silicon (a-Si : H) materials and devices and the development of multijunction device structures as an approach to achieve stabilized high efficiencies. Government support of industry also assisted the transfer of improvements in a-Si:H materials and device structures to 1 m^2 module production pilot lines (Stafford et al., 1991). Support of polycrystalline thin-film devices based on CdTe, $CuInSe_2$, and thin-film silicon materials has led to current cell efficiencies in the 10%–16% range and concurrent technology transfer to 0.4 m^2 module pilot lines (Ullal et al., 1992). Concentrator devices based on III–V materials have advanced up to 34% in multijunction devices and have high potential for low-cost processing (Benner, 1991). Innovative research includes improvements and engineering advances in balance-of-systems (BOS) components as a corollary strategy to reduce overall PV system costs (Hasti et al., 1990).

Several authors in this book have reviewed the status of R&D for the various PV material technologies mentioned above. The point must be made, however, that government support of fundamental research must continue in order to generate the further advances ultimately needed in successive stages of technology development for cost-effective terrestrial PV in utility and other applications.

2.3. Technology Development and Validation

Technology development at the module and BOS levels and technology validation at the system level have been facilitated by joint government/private sector technology transfer programs throughout the history of Federal government support for terrestrial photovoltaics. A primary measure of success for these programs is the resulting improvements in the technology base and the extent to which research results are transferred to the manufacturing base in industry. Examples of specific partnerships between the government and industry, or end-users, are listed in Table 22.1.

TABLE 22.1 Major DOE PV Government-Industry Partnership Programs

Project	Years	DOE Funding ($ Million)	Industry Funding ($ Million)
Flat-Plate Solar Array Project (FSA)	1975–1986	228	N/A
Amorphous Silicon Research Project (ASRP)	1984–1993	39	33
Polycrystalline Thin-Film Project	1990–1992	12 (est.)	5 (est.)
Photovoltaic Concentrator Initiative	1990–1993	12 (est.)	5 (est.)
Photovoltaic Manufacturing Initiative (PV-MaT)	1990–1995	55 (est.)	55 (est.)
Technical Validation (PV-USA)	1986–1995	30 (est.)	30 (est.)

The Flat-Plate Solar Array (FSA) Project, funded at $228 million over the period 1975–1985, was the first long-term comprehensive technology development program supported by the DOE (Jet Propulsion Laboratory, 1986). FSA was managed by the Jet Propulsion Laboratory and evolved from recommendations presented at the Solar Photovoltaic Energy Conference held at Cherry Hill, New Jersey, in 1973 for the development of a terrestrial PV array technology based on crystalline silicon. The FSA project was an integrated development program including material, cell, and module R&D; manufacturing process development; module qualification testing; and field testing of systems. The project established a strong precedent in the PV Program for joint technology development and technology transfer working with the PV industry, universities, and national laboratories in a cooperative research mode.

The overall 10 year objective of the FSA Project was the development of a viable terrestrial PV module technology based on the following criteria for single crystal and polycrystalline silicon modules: a 10% energy conversion efficiency, 20 year lifetime, and selling price of $\$0.50/W_p$ (1975 dollars). The major path of the cost reduction strategy was based on the development and evaluation of several polycrystalline silicon ingot and ribbon sheet growth processes as an alternative to Czochralski single crystal silicon. During the course of the FSA Project, module efficiencies improved by a factor of three (5%–15%), module warranties of 10 years were introduced by manufacturers, and module prices decreased by a factor of 15 ($\$75/W_p$ to $\$5/W_p$ in 1985 dollars).

One of the most effective components of the FSA Project was the periodic block buys of new modules, implemented in five stages to evaluate new material technologies and manufacturing processes (Jet Propulsion Laboratory, 1985). Module qualification testing and field testing of systems provided direct feedback to manufacturers in process optimization and became a model for technology validation of new technologies in future DOE projects. At the end of the project, several crystalline silicon module technologies were commercialized, and the early system purchases influenced cost reduction in the industry. Experience also showed that an effective technology transfer mechanism is direct support of applied R&D and process development in industry with R&D support from universities and the national laboratories.

In 1984, the Amorphous Silicon Research Project (ASRP) was established at SERI to coordinate research in the a-Si:H thin-film photovoltaics technology to produce technology options for industrial development (Sabisky, 1984; Stafford et al., 1991). The ASRP introduced cost-shared, multiyear subcontracting, implemented in 3 year partnership programs, to pursue applied R&D in industrial laboratories. The ASRP also facilitated information transfer and research support between industrial laboratories, universities, and internal research at SERI/NREL. Direct technology transfer significantly shortened the lead time to get laboratory results to the pilot line and then to prototype manufacturing lines. Multiyear funding also provided stability for planning purposes.

During the 10 year term of the ASRP program from 1984 through 1993, a-Si:H 1 cm^2 cell efficiencies reached 13% and 1 ft^2 submodule efficiencies increased from 4% to 10%. Numerous improvements in transparent conducting metal oxide contacts, doped microcrystalline layers, high and low band gap a-Si:H alloys, and tunnel junctions formed the basis for advanced single junction and multijunction device technologies. These improvements were transferred to large-area (up to 1 m^2) batch and continuous in-line module production lines based on plasma-induced chemical vapor deposition of a-Si:H alloys. The program focused on multijunction devices as an engineering approach to mitigate light-induced degradation and produce high-efficiency modules. Stability and efficiency continue to be issues at the research level and in large-area module production.

Cost-shared, multiyear subcontracts with industry and coordination with internal, national laboratory, and university research have impacted the design of subsequent technology development programs, including the Polycrystalline Thin Film Project at NREL (Zweibel et al., 1991) and the Photovoltaic Concentrator Initiative at SNL (Hasti et al., 1990), shown in Table 22.1. Sandia National Laboratories concentrator research also emphasizes the integration of in-house high-efficiency cell process development into industry production facilities.

The success of the technology transfer strategy in the ASRP and lessons learned in the FSA Project influenced the evolution of the DOE's manufacturing process research initiative established in 1990. The PVMaT (Photovoltaic Manufacturing Technology) Project is based on multiyear, cost-shared subcontracts with industry, selected through competitive solicitations, to improve manufacturing processes for module technologies. Manufacturing process research is conducted on flat-plate module technologies based on crystalline silicon, a-Si:H, and polycrystalline thin films (CdTe and $CuInSe_2$), and on concentrator module technologies (Witt et al., 1991). The Project supports research at individual companies with protected proprietary positions and supports nonproprietary teamed research for generic enabling manufacturing technologies (e.g., module encapsulation). A unique aspect of PVMaT is that direct industry input from the major U.S. PV manufacturers and the Solar Energy Industries Association (SEIA) was factored into project design.

An essential stage of PV technology development is the validation of system performance in the field. Early technology programs in the FSA Project fielded crystalline silicon technologies and demonstrated remote and utility-interconnected applications. In 1986, the Pacific Gas & Electric Company (PG&E) initiated a cooperative program with the DOE and 12 other electric utility and state agency participants to test advanced PV technologies. In the PVUSA (Photovoltaics for Utility Scale Applications) project, new thin-film and advanced crystalline silicon flat-plate and concentrator technologies are being validated in comparative utility-interconnected field tests (Candelaria et al., 1991, 1992). Test systems consist of 20 kW units for emerging technologies and 200–400 kW units for advanced technologies. One major benefit of PVUSA is the direct experience manufacturers obtain in meeting procurement schedules and engineering specifications of the end-user during shake-down of new technologies. Manufacturers have stated that this feedback has enabled them to improve the performance and identify the cost drivers for modules and BOS components.

2.4. Market Conditioning

The U.S. PV industry today is poised for a new phase of manufacturing expansion and cost reduction. Major module manufacturers have plans for economy-of-scale production facilities of 5–25 MW, and ultimately larger, annual production capacity. To achieve cost reductions, new plants will not only take advantage of economy-of-scale operations, but also incorporate advanced manufacturing processes, extensive automation, and advanced PV material/cell/module technologies. BOS improvements and economy-of-scale cost reductions will also be incorporated at the system level, most effectively in turnkey systems designed for replicable end-user applications.

One major barrier to implementing manufacturing expansion plans is the current limited growth of the world PV market ($\sim$17% per year). Expansion plans call for up to an order of magnitude increase in individual manufacturing operations over current levels. The uncertainty associated with the identification of PV applications with sufficient market potential to support planned expansions is an important factor inhibiting manufacturers' abilities to attract investment. In turn, accelerated market expansion is tied to investment in manufacturing operations leading to cost reduction. Manufacturers are seeking proof of market potential in expanded domestic and international remote PV markets and in domestic applications with large market potential, particularly in the U.S. electric utility market, in order to inspire investor confidence and attract investment capital.

In recognition of the role that renewable energy technologies can play in meeting needs, there is a growing interest in PV by utility companies. Over 100 utility companies now have some experience with PV technologies. This experience is mostly in a wide range of current, cost-effective, off-grid applications such as water pumping, cathodic protection, security and commercial lighting, environmental monitoring, telecommunications, and others that have been installed as the lowest cost option to solve a specific problem in the field (Bigger et al., 1991). Increasing recognition is also being given to the benefits of PV as a modular and scalable technology that eliminates environmental and fuel uncertainties. These and additional benefits accrue when PV is used in grid-support applications within a distributed utility system, which is a concept that uses modular electrical generating units placed at the periphery of the grid to displace central power stations (Iannucci and Shugar, 1991). The distributed utility system creates savings by deferring transmission and distribution upgrades and reducing operating and maintenance costs.

An analysis by PG&E of PV associated with a specific distribution substation at Kerman, California, indicates that PV can have a value that is two to three times the traditional avoided capacity and energy credits normally given to the technology by utility companies (Shugar, 1990). Sandia National Laboratories is currently extending the PG&E analysis to four additional utility companies with different system characteristics. An evaluation of the broader benefits of the distributed utility system and the use of modular distributed electric generating and storage technologies is being conducted in a joint project by the DOE, Electric Power Research Institute (EPRI) and Gas Research Institute (GRI) coordinated by PG&E with several utility partners (*Distributed Utility Study*, 1992).

Utility companies are also interested in PV as a demand-side management (DSM) technology. Studies in New York (Perez et al., 1990; Bailey et al., 1991a, b) and other locations (Hoff et al., 1991) show that in many regions of the United States there is a good correlation between the output of PV systems and peak utility loads, especially for summer peaking utilities and for commercial building loads. These studies indicate that PV located at end-user sites with minimal amounts of storage can offset customer peak demand charges and peak time-of-use energy rates, which can be as high as $25/kW and $0.15–$0.27/kW-h, respectively (Casazza, Schultz & Associates, 1990). A preliminary study by PG&E found that customer-owned PV can have a value as high as $5,000–$6,500/kW for residential and commercial customers, respectively (Wenger and Hoff, 1992). NREL is supporting the analysis of DSM value and the use of DSM incentives for PV (Byrne et al., 1992). The development of integrated PV building products is also being pursued in NREL's PV:BONUS Program (Photovoltaics for Building Opportunities in the U.S.), supported by OSEC with participation by the Office of Building Technology at the DOE (U.S. Department of Energy, 1992).

Examples of large-scale utility system studies and several utility-interconnected commercial and residential building studies in the United States are shown in Table 22.2. The large systems have been monitored since 1983/84 (e.g., Wenger et al., 1990); however, both the Carissa Plains and Hesperia Lugo systems have been acquired by Carizzo Solar and are being dismantled for module sales in the remote market. Factors limiting utility market growth include high cost, lack of information and education, inadequate distribution networks and a service infrastructure to support utilities and lack of turnkey products in demonstrated replicable utility applications. These deficiencies will have to be addressed in the course of developing utility markets.

To address issues for PV market development, utilities have formed a nonprofit organization called the Utility PhotoVoltaic Group (UPVG) to accelerate the use of cost-effective and emerging high-value applications of PV for the benefit of electric utilities and their customers (UPVG, 1992). This group is sponsored by the three major utility national trade organizations: the American Public Power Association (APPA), the Edison Electric Institute

TABLE 22.2 Examples of Utility-Interconnected PV Systems in the United States[a]

Utility	Project	Technology	System Size	Operational
Pacific Gas & Electric Company	Carissa Plains	Flat-plate SC-Si (two axis tracking)	5.2 MW AC	January 1984
Southern California Edison	Hesperia Lugo	Flat-plate SC-Si (two axis tracking)	1 MW AC	November 1982
Sacramento Municipal Utility District	SMUD	Flat-plate SC-Si (one axis tracking)	2 MW AC	August 1984
Florida Power Corp, FSEC	Ground mounted DSM	a-Si : H (fixed tilt)	15 kW DC	August 1988
Niagara Mohawk Power Corp.	Roof mounted commercial	SC-Si (one axis tracking)	15 kW DC	July 1990
NEES (Gardner)	Roof mounted multi-residential	Poly-Si (fixed tilt)	60 kW DC	1986
City of Austin 3M Company	Roof mounted commercial	22 × linear Fresnel conc. (two axis tracking)	300 kW DC	1989

[a]SC, single crystal.

(EEI), and the National Rural Electric Cooperative Association (NRECA). The UPVG will pursue an integrated set of five task activities, which include education and application identification, market characterization, development of market acceleration strategies, development of utility system planning tools, and technology transfer through application validation projects. Each task will be directed by a utility-led implementation group.

3. INTERNATIONAL PROGRAMS

Competition with foreign vendors is a major driving force for the development of the U.S. PV industry. As in the United States, PV manufacturers in foreign countries, particularly in Japan and Europe, have benefitted from two decades of support from their respective governments in R&D and market development activities. Highlights of Japanese and European PV activities and U.S. international programs are given in the following sections.

3.1. Japanese Sunshine Project

Terrestrial photovoltaic technology development in Japan has received consistent support from the national government since 1974. Varying exchange rates and differences in accounting make comparison of funding levels between the United States and Japan difficult. However, in terms of research personnel and approximate budget comparisons, Japan's funding of PV R&D has been equal to or greater than U.S. funding over the past decade. The Fiscal Year 1992 budget request in the Japanese program was approximately \$51 million (Goto, 1992). The Japanese are extremely strong competitors with the United States (and Europe) in flat-plate crystalline silicon and thin-film PV technologies, with similar state-of-the-art levels in module production and performance and in advanced research achievements.

Funding for the Japanese terrestrial PV program is conducted in the "Sunshine Project" under the Agency for Industrial Science and Technology (AIST), which is a division of the Ministry of Trade and Industry (MITI). The Sunshine Project is implemented through the New Energy & Industrial Technology Development Organization, formerly called NEDO. In November 1990, a new organization consisting of 28 PV manufacturers and called PVTEC (Photovoltaic Power Technology Research Association) was formed to coordinate joint industrial R&D with the Sunshine Project and the new NEDO organization (Hamakawa, 1991). PVTEC ensures significant industry input into the Sunshine Project and coordination of research among industrial and university contractors and with national laboratories. PVTEC expects to direct Y60 billion ($\sim$\$400 million) in research expenditures between 1990 and 2000.

Sunshine Project goals are similar to U.S. goals and are aggressive. The target for module and BOS costs by 1995 are \$3.30/W_p and \$2.30/W_p, respectively, and by 2000 the targets are \$1.00/W_p and \$1.70/W_p, respectively.

PV production capacity expansion is targeted to increase from 14.5 MW_p/year in 1990 to 500 MW_p/year in 2000. Electricity cost targets are \$0.18/kW-h in 1995 and \$0.07/kW-h in 2000. Cost targets are based on aggressive milestones for efficiency and manufacturing process improvements in single crystal and polycrystalline silicon cells and submodules and in a-Si:H and polycrystalline compound semiconductor thin-film technologies (Hamakawa, 1991).

Japanese Government funding supports an aggressive application and system development program, including: (1) central and distributed utility-interconnected systems (largest system is 1 MW), (2) remote off-grid applications and hybrid systems for island villages, (3) integrated PV building systems, and (4) an aggressive BOS program for development of low-cost power conditioning equipment and battery storage (Morishita et al., 1991). A utility-interconnected test site has been established on Rokko Island in conjunction with the Central Research Institute of the Electric Power Industry (CRIEPI), which involves the utility industry in examining the engineering criteria associated with significant PV penetration in the utility grid (Kobayashi et al., 1991).

The development of terrestrial PV technologies in Japan benefits from several factors. For example, Japanese corporations and the government take a long-term view toward technology development, establish long-term goals, and develop plans based on relatively long-term funding commitments. In addition, PV is integrated into the broad product base of vertically integrated corporations. Businesses performing PV rersearch are able to take advantage of integrated products and technology spin-offs. Examples of integrated product designs include PV building products, such as PV roofing tiles and PV-powered air conditioners (Hasunuma et al., 1992), and automobile sun roofs (Hamakawa, 1991, see Chapter 12).

In terms of the PV market, Japanese module shipments of 20 MW in 1991 were approximately equally divided between consumer products and commercial power modules. Virtually all of the commercial modules were shipped to the United States and Europe. Shipments to the American continent exceeded 2 MW (Maycock, 1992). Government policy also promotes exports. In the Fiscal Year 1992 budget for the Sunshine Project, an additional \$9.7 million was requested to promote the use of PV in public facilities and overseas by the Agency of Natural Resources and Energy (Goto, 1992). Consequently, Japanese investment in expanded production capacity will, to a significant extent, be dependent on the development of utility markets in the United States and Europe and the expansion of developing country markets.

3.2. European Programs

The Commission of the European Communities (CEC) has supported the development of terrestrial PV technologies since 1975 in joint national R&D programs in industrial and research institutions and system demonstration projects (Palz, 1991). During more than a decade of system installation and monitoring, the CEC Demonstration Program supported 104 PV projects with a current installed capacity of 1.3 MW. Most systems were installed in remote residential and village applications (Kaut et al., 1991). These projects, along with others supported by individual countries, provided the early system experience for developing the domestic European market and a significant export market in developing countries. The most recent 3 year CEC program (1989–1992) was supported at a modest level of ~\$9.6 million.

European research has focused on material research and the processing of cells and wafers for single crystal and polycrystalline silicon module technologies, with less, but significant, support going to a-Si:H and $CuInSe_2$ technology development. Crystalline silicon module manufacturers are now expanding production capacity and experienced a 31% growth in module sales in 1992 (Maycock, 1992). Strengths in Europe's terrestrial PV development programs have been a technology focus on flat-plate crystalline silicon, strong support of demonstration projects leading to system and application development, and export assistance from the federal governments to manufacturers. Environmental concerns stimulated by events in Chernobyl and Eastern Europe have strengthened recent interest in PV and renewable energy.

From 1982 through 1991, the Federal Ministry for Research and Technology (BMFT) in Germany has spent ~700 million DM on terrestrial PV. In 1989–1991 80% of the budget went to material, cell, and module development, approximately equally split between crystalline silicon and thin-films. System development and demonstration projects consumed the remainder of the budget (Wollin and Batsch, 1991). Demonstration projects in Germany have been oriented toward utility-interconnected systems and hybrid systems with storage. BMFT supports development of remote applications and village power, including water pumping, lighting, and diesel replacement/augmentation, in developing countries (Posorski, 1991). An accelerated terrestrial program was initiated in 1991 to supply domestic grid-connected power for distributed PV applications, called the *1000 Roofs PV Program* (Rauber and Wollin, 1992; Hotop, 1991).

Since 1987, Switzerland has supported a terrestrial PV acceleration program called *Project Megawatt*. The project supports the development and installation of standardized 3 kW grid-connected residential PV systems with a cumulative installed capacity of 1 MW in over 300 Swiss homes (Real, 1991). A 10 year moratorium on nuclear power plant construction mandated by the Swiss public has focused enhanced interest on PV in Switzerland, and a recent joint venture was announced to expand the project on a multimegawatt scale (Real and Ludi, 1991). Switzerland and Germany have been developing expertise in PV building design and system development, in which PV modules are integrated into the architecture of commercial and residential buildings.

The PV program in France is managed by the Environment and Energy Management Agency (AEME), which has domestic and foreign responsibilies. Although the French PV budget is modest, the AEME supports a broad application development program including rural electrification, residential systems, and remote applications for developing countries. Local domestic and foreign AEME offices and other agencies promote PV through education and training services. Domestic manufacturers benefit from export assistance and local foreign support by the French Government in territories and former colonies, for example, Morocco, French Polynesia, and New Caledonia (Claverie et al., 1992). Italy passed a New Energy Plan in 1990 that calls for the cumulative installation of 25 MW of PV generating capacity by 1995 (Farinelli, 1991). A dependence on imported energy and a tacit ban on nuclear power in Italy has resulted in the support of PV and renewable energy by the country's national utility company, ENEL. ENEL also supports the distributed utility concept and promotes a 1995 objective of installing 1.5 MW of PV each in grid-support and residential applications. France and Italy emphasize crystalline silicon module production.

European terrestrial PV development programs over the past two decades have laid a solid foundation in the technology base for manufacturing, particularly for crystalline silicon modules, and system and application development for domestic and international markets. Success can be measured in an expanding market share and in current increases in manufacturing capacity. The traditional subsidies of governments in Europe for strategic technologies and export assistance provided to manufacturers will result in a strong competitive position in the future for PV. Many companies in Europe are also gaining expertise in product design and development for dispersed utility-interconnected applications, such as integrated PV building systems, that will represent a significant component of the market in the near future.

3.3. U.S. International Activities

The DOE has supported the development of international markets for PV in a number of activities throughout most of the history of the terrestrial PV program. The DOE, through the Office of Solar Energy Conversion, is an active member of the Committee on Renewable Energy Commerce and Trade (CORECT), established by the Renewable Energy Industries Development Act of 1983 (Public Law 98-370). CORECT has provided aid to the U.S. PV industry in identifying federal programs and services that support the export needs of business (Five Year Research Plan, 1991). DOE is also an active member of the FINESSE Project (Financing Energy Services for Small-Scale Energy Users) established in 1988 to conduct market studies and develop business plans for promoting alternative energy products in Southeast Asia.

OSEC also supports energy development initiatives in Latin America and the Caribbean basin. For example, Sandia National Laboratories is providing support from its Design Assistance Center (DAC) to the Mexican National Solidarity Program known as PROSONOL, which is developing and implementing a rural electrification program incorporating PV, wind, small hydro, solar thermal, and biomass technologies (Rannels, 1991). NREL is also providing support to a joint U.S.-Brazil program signed by the DOE in 1992 to provide PV service to 2,000 remote homes in the states of Ceara and Pernambuco (National Renewable Energy Laboratory, 1992).

These international marketing activities, working in concert with financing agencies such as the World Bank and AID, have the potential to expand the PV export market significantly in developing countries for the benefit of U.S. manufacturers in competition with foreign vendors. The activities are also necessary to ensure a level playing field for international competition. Time is required, however, to achieve results due to the education, economic, and other infrastructural barriers associated with foreign markets.

4. THE FUTURE—U.S. COMPETITIVENESS

The Federal Government will continue to play a critical role in the development of terrestrial photovoltaics for the foreseeable future. Continued R&D funding in basic and applied research at the material, cell, and module levels is critical in generating the base of technology advancements needed in manufacturing to support module performance improvements and cost reduction. Complementary support of BOS component development is also necessary to improve system reliability and reduce energy costs. Terrestrial PV systems must meet the rigorous engineering and economic performance criteria imposed by major end-users, particularly in the electric utility industry, in order to become a major contributor to the nation's energy portfolio.

In the 1990s the role of the government is likely to expand beyond the recent emphasis on R&D as issues such as global competitiveness, trade deficits, energy security, and environmental concerns create pressure in the United States to expand energy and industrial policies. The government's precise role in the support of technology development and commercialization will be a function of public debate and current administration policies. However, the use of government–private sector collaborative efforts for technology and market development is likely to increase. Preliminary reports from the Clinton Administration indicate that more emphasis will be put on investing in a strategic twenty-first century technology infrastructure, systematically seeking private sector input for technology policy and competitiveness issues, and enhancing technology transfer to industry to create new jobs and products (*Chemical & Engineering News*, 1992).

A shift in focus at the DOE's national laboratories will also occur in the future. With the end of the Cold War, a partial redirection of resources is being implemented by Congress and the current administration, away from defense spending to the support of civilian commercial activities. The DOE's national laboratory system is being required to play a much stronger role in technology transfer. In the PV program, this transition has been occurring and will accelerate through such mechanisms as cooperative research and development agreements (CRADAs), in which internal laboratory resources are committed to proprietary industrial research. Increasingly, nonproprietary national laboratory research will also be directed toward solving generic industry problems.

Universities are redefining their relationship with government and industry. The United States has the best research infrastructure in the world, the foundation being the U.S. university system. Universities not only participate in creating the research base for technology transfer, but also provide the training centers for industrial personnel. In the PV program, support of university research is vital to produce continuous innovation in the PV industry and supply qualified technical personnel for industry expansion. The DOE dedicated two university research centers of excellence in 1992 at the Georgia Institute of Technology and at the Institute of Energy Conversion in the University of Delaware. A critical funding mass at such centers can enhance basic research, and universities can be more effectively utilized in support of manufacturing process rersearch and technology transfer in manufacturing support centers.

At the federal and state levels regulatory actions can impact PV in the utility industry. Examples of Federal regulatory initiatives proposed in the National Energy Strategy include incorporation of integrated resource planning in the wholesale power market through the Federal Energy Regulatory Commission (FERC) and modifications of PURPA and PUHCA (Public Utility Holding Company Act of 1935) to allow renewable energy technologies to compete more effectively in utility markets. A permanent extension of the 10% federal investment tax credit was recently passed by Congress. State tax incentives have also been important, for example, the state property tax and state income tax credits in California.

Accelerated commercialization through market aggregation programs, market guarantees, and potential government cost-sharing of system purchases at significant levels can stimulate market development. Market aggregation strategies can be controversial, but have been used by the government in the past to accelerate technology development. To be effective, such strategies must be linked to advanced manufacturing activities leading to cost reduction with concurrent technology support for manufacturing and system development. Other government trends include continued export assistance to U.S. industry and international market development, and expanded use of PV in other sectors including buildings and transportation.

ACKNOWLEDGMENTS

This work was prepared under Contract No. DE-AC02-83CH10093 to the U.S. Department of Energy.

REFERENCES

Annan, R. (1992), *Solar 2000—A Collaborative Strategy*, Vols. I and II. Office of Solar Energy Conversion, U.S. Department of Energy.

Bailey, B., Doty, J., Perez, R., Stewart, R., and Huse, W.G. (1991a), *Proc. 26th Annual IECEC Conf.*

Bailey, B., Doty, J., Perez, R., Stewart, R., and Huse, W.G. (1991b), *Proc. 1991 ISES Solar World Cong.*

Benner, J.P. (1991), *22nd IEEE Photovoltaic Specialists Conf. Record*, IEEE, New York, pp. 7–11.

Bigger, J.E., Kern, E.C. Jr., and Russell, M. (1991), *22nd IEEE Photovoltaic Specialists Conf. Record*, IEEE, New York, p. 486.

Byrne, J., Hadjilambrinos, C., and Wang, Y.-D. (1992), *Proc. 11th PV Adv.* R&D Project Rev. Mtg.

Candelario, T.R., Ellyn, W., and Jennings, C. (1992), *1991 PVUSA Progress Report*, Report Number 007.5-92.6, prepared by the PVUSA Project Team.

Candelario, T.R., Hester, S.L., Townsend, T.U., and Shipman, D.J. (1991), *22nd IEEE Photovoltaic Specialists Conf. Record*, IEEE, New York, p. 493.

Casazza, Schultz & Associates, Inc. (1990), *Electric Rate Book*. data for Consolidated Edison Company of New York, Inc. and other northeastern utilities

Chemical & Engineering News (1992), October 12, pp. 21–28.

Claverie, A., Chabot, B., and Anglade, A. (1992), *Proc. 6th Int. PV Sci. Eng. Conf.*, pp. 533–536.

Distributed Utility Study (1992), Joint study by PG&E, DOE, EPRI, and GRI wih several utility partners, results reviewed in a Distributed Utility System Workshop held in Berkeley, California, December 2–3.

Farinelli, U., Ambrosini, G., and Bigotti, R. (1991), *Proc. 10th E.C. PV Solar Energy Conf.*, pp. 1385–1389.

Goto, T. (1992), *Proc. 6th Int. PV Sci. Eng. Conf.*, pp. 521–527.

Hamakawa, Y. (1991), *22nd IEEE Photovoltaic Specialists Conf. Record*, IEEE, New York, pp. 1199–1206.

Hasti, D.E., King, D.L., and McBrayer, J.D. (1990), *21st IEEE Photovoltaic Specialists Conf. Record*, IEEE, New York, p. 217.

Hasunama, M., Takeoka, A., Fukuda, Y., Suzuki, M., Sakoguchi, E., Tokizaki, H., Wakisaka, K., Kishi, Y., and Kuwano, Y. (1992), *Proc. 6th Int. PV Sci. Eng. Conf.*, pp. 489–492.

Hoff, T., Shugar, D.S., and Wenger, H.J. (1991), *22nd IEEE Photovoltaic Specialists Conf. Record*, IEEE, New York, p. 662.

Hotop, R. (1991), *Proc. 10th E.C. PV Solar Energy Conf.*, pp. 1308–1311.

House of Representatives (1992), *Energy Policy Act of 1992*, H.R. 776, Conference Version, October, 7.

Iannucci, J.J., and Shugar, D.S. (1991), *22nd IEEE Photovoltaic Specialists Conf. Record*, IEEE, New York, p. 566.

Jet Propulsion Laboratory (1985), *Flat-Plate Solar Array Project*, JPL 400-279 10/85.

Jet Propulsion Laboratory (1986), *Flat-Plate Solar Array Project Final Report, Vol. VIII, Project Analysis and Integration*, JPL Publication 86-31.

Kaut, W., Riesch, G., Blaesser, G., Gillet, W.B., and Hacker, R.J. (1991), *Proc. 10th E.C. PV Solar Energy Conf.*, pp. 1301–1304.

Kobayashi, H., Takigawa, K., Hashimoto, E., Kitamura, A., and Matsuda, H. (1991), *22nd IEEE Photovoltaic Specialists Conf. Record*, IEEE, New York, pp. 695–700.

Maycock, P. (1992), *Photovoltaic News*, **11** 1–7.

Maycock, P.D., and Stirewalt, E.N. (1985), *A Guide to the Photovoltaic Revolution*, Rodale Press Emaeus, PA, p. 116.

McCormack, M. (1978), *13th IEEE Photovoltaic Specialists Conf. Record*, IEEE, New York, p. 6.

Morishita, H., Agawa, T., Morishige, T., Miyazato, M., Hashimoto, E., Takigawa, K., and Kobayashi, K. (1991), *Proc. 10th E.C. PV Solar Energy Conf.*, pp. 1326–1329.

National Renewable Energy Laboratory (1992), *Solar 2000 Update*, NREL

Paltz, W. (1991), *Proc. 10th E.C. PV Solar Energy Conf.*, pp. 1369–1370.

Perez, R., Seals, R., and Stewart, R. (1990), *Assessing Photovoltaic Interaction With New York Summer-Peaking Utilities*, Report to the New York Power Authority.

Posorski, R. (1991), *Proc. 10th E.C. PV Solar Energy Conf.*, pp. 1353–1355.

Rannels, J.E. (1991), *22nd IEEE Photovoltaic Specialists Conf. Record*, IEEE, New York, p. 479.

Rauber, A., and Wollin, K. (1992), *Proc. 6th Int. PV Sci. Eng. Conf.*, pp. 529–532.

Real, M.G. (1991), *Proc. 10th E.C. PV Solar Energy Conf.*, pp. 1305–1307.

Real, M.G., and Ludi, H. (1991), *22nd IEEE Photovoltaic Specialists Conf. Record*, IEEE, New York, pp. 574–575.

Sabisky, E.S. (1984), *Amorphous Silicon Research Project, Five Year Research Plan 1984–1988*, SERI/SP-211-1350, National Technical Information Services, U.S. Department of Commerce, Springfield, VA.

Shugar, D.S. (1990), *21st IEEE Photovoltaic Specialists Conf. Record*, IEEE, New York, p. 836.

Stafford, B.L., Luft, W., von Roedern, B., Crandall, R., and Wallace, W.L. (1991), *22nd IEEE Photovoltaic Specialists Conv. Record*, IEEE, New York, p. 1329.

Surek, T. (1992), *Overview of NREL's Photovoltaic Advanced R&D Project*, NREL/TP-410-4723.

Ullal, H.S., Stone, J.L., Zweibel, K., Surek, T., and Mitchell, R.L. (1992), *Proc. 6th Int. PV Sci. Eng. Conf.*, pp. 81–87.

UPVG (1992), *Utility PhotoVoltaic Group Program Summary*, October, 1992, UPVG, 202/857-0898.

U.S. Department of Energy (1991a), *National Energy Strategy* (1991a), National Technical Information Service, U.S. Department of Commerce, Springfield, VA.

U.S. Department of Energy (1991b), *Photovoltaics Program Plan FY 1991–FY 1995*, DOE/CH10093-92, National Technical Information Services, U.S. Department of Commerce, Springfield, VA.

U.S. Department of Energy (1992), *PV:BONUS: Building Opportunities in the U.S. for Photovoltaics*, Notice of Program Interest (NOPI), DE-NP02-92-CH10514, National Technical Information Services, U.S. Department of Commerce, Springfield, Va.

Wenger, H., and Hoff, T. (1992), *PV as a DSM Option: Concept Development*, presented at the EPRI Photovoltaics Planning Workshop: Utility- and Customer-Owned Residential and Commercial PV Applications, June 9–11.

Wenger, H.J., Jennings, C., and Iannucci, J.J. (1990), *21st IEEE Photovoltaic Specialists Conf. Record*, IEEE, New York, p. 844.

Witt, E., Herwig, L.O., Mitchell, R., and Mooney, G.D. (1991), *22nd IEEE Photovoltaic Specialists Conf. Record*, IEEE, New York, p. 501.

Wollin, K., and Batch, J. (1991), *Proc. 10th E.C. PV Solar Energy Conf.*, pp. 1381–1384.

Zweibel, K., Ullal, H.S., Mitchell, R.L., and Noufi, R. (1991), *22nd IEEE Photovoltaic Specialists Conf. Record*, IEEE, New York, p. 1057.

PART FIVE

SUMMARY

CHAPTER TWENTY-THREE

Status, Potential, Conclusions

LARRY D. PARTAIN, Edward L. Ginzton Research Center,
Varian Associates, Palo Alto, CA 94304-1025

1. INTRODUCTION

The world market for solar cells grew to 65 MW/year power levels and ~\$0.5 billion/year sales levels by 1992. Crystalline silicon modules were dominant, with ~10% terrestrial efficiency mostly for off-grid industrial and residential applications and with ~15% space efficiencies for use on satellites. However, a significant portion (~20%) was in 2%–4% efficient hydrogenated amorphous silicon (a-Si:H) cells for consumer products, mainly calculators. These contrast with the best laboratory demonstration cells that exceed 30% efficiency levels. The largest power segment was for off-grid terrestrial systems that could grow to the 200–400 MW/year levels by the end of the twentieth century. A strong assist for this would be continuing factor-of-2 decreases in module prices, from the ~\$6.50/W 1992 levels, for each order-of-magnitude growth in cumulative production. The saturated space market was a ~0.1 MW/year niche for radiation hard devices of high-power densities per unit weight and area with \$300–\$1,800/W panel prices in 1992. It could grow modestly (approximately double or quadruple) if the growing personal communications market requires large numbers of new satellites.

The potentially large utility power market (1,000's MW/year and larger) is proving difficult to penetrate likely due to requirements for minimum efficiencies of at least 10% and for ~\$1–\$2/W module prices. The dominant crystalline silicon technology of the early 1990s cannot meet these requirements, with

Solar Cells and Their Applications, Edited by Larry D. Partain.
ISBN 0-471-57420-1

screen-printed grids and high materials costs per unit watt. Single junction solar cells can provide module efficiencies up to 30% using advanced approaches that include single crystal silicon and GaAs. Alternate materials to crystalline silicon need nonsolar and niche applications to help them develop their underlying technology base and to continue far enough down their learning curves to be competitive. Module efficiencies in the 30%–50% range will need multiple junction cells of differing band gaps. This high performance range will likely use III–V materials and possibly not include crystalline silicon. An orderly acceleration program is recommended to develop cost competitive terrestrial modules. This should start with 1992's 10% efficiency and \$6.50/W price levels, continue in roughly 5% efficiency goal steps, with constantly decreasing \$/W prices until limiting performance around 50% efficiency and prices below \$2/W are reached (in 1992 dollars). Unless the industry establishes better profit margins, such advances will only occur in a reasonable time frame (less than 50 years) if well-directed government acceleration programs are implemented. Even with higher profit margins, the geographical regions that wisely accelerate their technology with government support will be among the market leaders.

2. STATUS

The worldwide shipment of terrestrial solar cell modules grew by at least a 17% compounded rate from the late 1970s to the 65 MW/year level in 1992 (see Chapter 22). Off-grid systems, for industrial and residential use, dominated (>50%) these module applications (Chapter 15). Most contained single and multicrystalline silicon (Chapter 2), but a substantial fraction (at least 4 MW/year, Chapter 19) used a-Si:H in consumer products, mostly calculators. Japanese and other Pacific Rim companies dominated a-Si:H manufacturing while the single and multicrystalline silicon market was split about equally among Japan, the United States, and Europe in 1991 (Chapter 22). Crystalline silicon was preferred for power applications, while a-Si:H was the choice when its large-scale integration more effectively gave higher voltages for relatively low current applications.

The terrestrial module manufacturing costs were ~\$5/W in 1992 for single and multicrystalline silicon (Chapter 9). This translates into an ~\$0.4 billion/year terrestrial market in 1992 assuming 30% gross margins with a \$6.50/W sales price. The manufacturing costs of a-Si:H consumer products were substantially higher at ~\$20–\$50/W (Chapter 19), but were still cost effective compared with their storage battery competition (at ~\$1,000/kW-h). The a-Si:H alone contributed on the order of ~\$0.1 billion/year in sales. Terrestrial utility demonstration projects of 20–100 kW system sizes provided 8%–12% array efficiencies with crystalline silicon cells and 2%–4% array efficiencies with a-Si:H cells through 1993 (Chapters 1 and 16).

The space market worldwide in 1992 was on the order of ~0.1 MW/year split about equally between commercial and military applications (Chapters 19 and 21). Space array costs were on the order of \$2,000–\$3,000/W for a total space market of ~\$0.2–\$0.3 billion/year. From 15% to 60% of these costs were for the space solar cell panels themselves, depending on mission design and specific costs attributed to the array. At a minimum, a space panel consists of solar cells, coverglass, interconnect wiring, and bonding and wiring to a substrate with labor for fabrication and test. Additional items that make up the total system include hinges (to connect panels to each other and the yoke), the yoke (to connect to the space craft), hold-down and deployment mechanisms, batteries, and power conditioning electronics. Thus space panel costs (~\$300–\$1,800/W) were approximately two orders of magnitude higher (or more) than terrestrial module costs, and their total dollar market size (~\$30–\$180 million/year) was ~6%–36% of the terrestrial dollar volume in 1992. However, space solar cells are a highly leveraged enabling technology. They allow the space launch and communications satellite market to exist in its present form at the billions-of-dollars-per-year total market size. The space array market was relatively stable in 1993, with declining military demand offset by slowly growing commercial demand.

GaAs space cells individually sold for five to six times the price of silicon space cells in 1992 according to P. Iles, R. Neff, and D. Flood (authors of Chapters 4, 13, and 17, respectively). Their typical 18% efficiencies exceed the 14%–15% values typical for silicon. The most important parameters that can make GaAs cost competitive with silicon in space, at the total launch systems level, are specific power (W/kg and W/m^2), particularly when GaAs is grown on Ge substrates that are lighter than GaAs substrates.

The terrestrial crystalline silicon module market essentially used the configuration developed in the early 1980s under the Jet Propulsion Laboratory's \$228 million Flat-Plate Solar Array (FSA) Project. This Project was for manufacturing development with competitive block purchases that stretched the technology (Chapters 9 and 22). The FSA program generated an incremental jump in module shipments in 1983 and fostered the steady growth of the following 10 years. The 18 MW of modules shipped by U.S. companies in 1991 were mostly crystalline silicon, and they gave yearly sales equal to about half of the total FSA program expenditure assuming an ~\$6.50 module sales price. The balances of the 1991 module shipments were ~20 MW by Japan (crystalline and amorphous silicon), 13 MW by Europe, and ~4 MW by the rest of the world (Chapter 22). The a-Si:H consumer products market for calculators grew exponentially in the 1982–85 time frame before saturating around 4 MW/year in the 1987–1989 time period (Chapter 19).

Less than 10% of the terrestrial module efficiencies field measured in the United States in 1988 and 1989 were greater than 90% of the manufacturers' rated values. None equalled or exceeded their rated values even when corrected for temperature (Chapter 16). Some were only 60% of rated value typically determined from flash simulator measurements. These discrepancies should be eliminated.

The simulator measured efficiencies of the "best-ever" laboratory cells greatly exceed those measured on large terrestrial systems in the field under actual outdoor sunlight conditions. The best terrestrial single junction laboratory cell efficiencies were 23% for flat-plate and 28% for concentrator systems by 1993. The corresponding large field demonstrated efficiencies were only 40%–50% of these laboratory values. The best terrestrial laboratory a-Si:H, tandem junction solar cell was 13.7% by 1993 (see Chapter 7), but the corresponding large field test efficiencies were only 20%–30% of this.

The highest laboratory cell efficiencies come with light concentration and with multiple band gap junctions that more effectively convert sunlight to electricity. This combination provided 33%–35% terrestrial efficiencies for GaAs/GaSb two junction, 100× concentrator cells (after correction for the 1991 Sandia recalibration) (Fraas et al., 1990; Benner, 1991) (see Chapters 1 and 6). The latter was a major milestone giving the first solar cell efficiency to exceed 30%. Later nonconcentrating (flat-plate) cells approached this milestone. Bertness et al. (1993) achieved 29.5% terrestrial efficiency with a laboratory $GaInP_2$/GaAs two junction, flat-plate cell.

3. POTENTIAL

Terrestrial module manufacturing was on a simple learning curve from 1987 through 1992. Then module prices dropped about a factor of two for every order-of-magnitude increase in cumulative module production in MW (Chapter 22). If this continues, module prices could drop below $2/W when cumulative production increases between one and two orders of magnitude beyond 1992 levels. When the standard crystalline silicon plant capacities grow from their 1992 1 MW/year production levels to 10 MW/year levels, the module manufacturing costs should fall from $5/W to the $3/W level based on the old Jet Propulsion Laboratory FSA block purchase technology (Chapter 9). The latter would correspond to $4–$5/W sales prices. By 1993 this FSA-based technology was 14% individual cell efficiencies with 400 μm thick wafers sawn from silicon ingots. At this point materials costs account for over half the total manufacturing costs and begin to fundamentally limit further cost reductions (Chapter 9).

The major uses of terrestrial solar cells in 1992, in off-grid systems, are probably in their initial upward swing of a classic "S-shaped" market development curve (Chapter 19). When they might enter exponential growth phases and at what level they will saturate is difficult to predict, since this market segment was so diffuse and poorly characterized by 1993. Nevertheless projected growth to 200–400 MW/year in module shipments by the end of the twentieth century appears reasonable (Chapter 19), particularly if prices keep falling a factor of two for every order of magnitude growth in cumulative production. However, continued price reductions will require technology beyond the old FSA block purchase module configurations dominant in 1993.

The 1,000's of MW cumulative production for this segment, likely needed to drive prices below \$2/W (in 1992 dollars), may not occur before the first or second decade of the twenty-first century. There is a key need here for a comprehensive market study of the off-grid segment that enlists wide cooperation like the Solar Energy Industries Association and its members and key individuals who have tracked these trends over many years.

The installed U.S. electric utility grid capacity is on the order of a few hundred thousand MW (Armstrong-Russel et al., 1983). Even a solar cell market penetration of a 0.1%/year would exceed 1993's total worldwide terrestrial market volume. With 10% penetration and 30 year module lifetimes, the average U.S. replacement market would be on the order of 700 MW/year. The latter is 10 times the 1993 terrestrial world market size.

Solar cell penetration of non-U.S. utility companies offers additional market opportunities. Entry may be easier in the developing countries where traditional utilities are not as well established. The problem is meeting utility power needs where cost is a major deterrent. Perhaps the first cost-effective, utility solar cell system was installed in Kerman, CA, in 1993 based on a unique (and rare) combination of circumstances (Chapter 16). Most prior utility demonstrations have not been cost competitive with ~\$0.10/kW-h power bought from traditional sources. The energy equivalent costs of the best large solar cell systems through 1993 were more in the ~\$0.30 to ~\$0.60/kW-h range, assuming 20–30 year system lifetimes. Greater than 10% utility penetration in the twenty-first century may require coordination with cost-effective energy storage like natural hydro, another solar renewable resource.

The space market was saturated by 1993, with slow overall growth at best (Chapter 21). Deployed space panels approached within 80%–90% of "best-ever" laboratory cell performance for contemporary devices of the same general design (Chapter 4). There are continuing space needs for increased power per unit weight and area, particularly at end-of-life. The latter encourages radiation hardness. Alternative power sources are unlikely to supplant space solar cells for the foreseeable future. Potential market growth depends on how new technologies, such as in personal communications, develop. The Iridium program for personal communications was financed in 1993, and it created a demand for 0.1 MW of space solar array power (Chapter 21). This is on the order of the total space solar array production in 1993.

There are two approaches to improving the price per watt of solar cell modules so that market expansion can continue and accelerate, particularly for terrestrial applications. The first is improvement in performance, with efficiency improving more than related materials and fabrication costs. The second is to decrease costs more rapidly than related efficiency losses. The ideal is efficiency increases with simultaneous decreases in cost per unit area. At high production volumes, material costs fundamentally limit price reductions. A good benchmark for comparison is the FSA block purchase silicon modules with 400 μm thick wafers and ~\$6.50/W 1992 selling prices for large-scale systems of ~10% efficiencies.

Efficiency is fundamentally limited by the maximum possible values of short circuit current density, open circuit voltage, and fill factor. Short circuit current is limited by 100% quantum efficiency for photons with energies greater than the band gap. By 1993 the best single crystal silicon and GaAs cells had short circuit current densities within 80%–90% of these maxima (Chapter 1). Open circuit voltages are fundamentally limited to the splitting of the quasi-Fermi levels. The majority- and minority-carrier concentrations and the solar cell band gap primarily determine this splitting (Chapter 1).

Doping determines the majority carrier concentration that can at most be in the $10^{18}\,cm^{-3}$ range for standard designs. Beyond this, reduced minority-carrier lifetime and mobility decrease the short circuit current density, according to the standard Shockley p/n junction model (Chapters 1 and 3). The low injection assumption of the Shockley model limits minority-carrier concentrations to $\sim 10^{17}\,cm^{-3}$ in standard configuration cells. Absorption of sunlight photons generates these minority-carrier concentrations. The observed open circuit voltage (V_{oc}) of crystalline silicon and GaAs cells indicates that minority-carrier concentrations, on the order of $10^{14}\,cm^{-3}$ and less, were being reached by 1993 (Chapter 1). These best V_{oc}s correspond to 70%–85% of the V_{oc}s that would be achieved in silicon at $10^{17}\,cm^{-3}$ minority-carrier concentrations and to 85%–95% of the V_{oc}s that would similarly be reached in GaAs (Chapter 1).

The open circuit voltage and the exponential dependence of current on voltage, normalized by kT/q, determine the fill factor according to the standard Shockley p/n junction model (Chapter 1). For the best devices through 1993, the measured fill factors were 90%–95% of these Shockley model limits in crystalline silicon and 92%–99% of these values in GaAs. Combined, the efficiencies of the best crystalline silicon cells were 55%–75% of the Shockley model maximum. In GaAs they were 65%–75% of this limit, all by 1993.

At the $10^{14}\,cm^{-3}$ minority-carrier concentration levels indicated by the V_{oc}s of the best 1993 technology cells, the Shockley model maximum efficiencies for single junction devices occur around a 1.4 eV band gap. This is for peak efficiency values $\sim$30% for AM1.5 terrestrial conditions and $\sim$27% for AM0 space conditions. These maxima are independent of whether or not one uses one sun or concentrated sunlight (Chapter 1). At the $10^{17}\,cm^{-3}$ minority-carrier concentration limits for low Shockley model injection, the band gap for maximum single junction performance shifts down to $\sim$1.2 eV with peak efficiencies $\sim$37% for AM1.5 terrestrial conditions and $\sim$33% for AM0 space conditions. At the $10^{8}\,cm^{-3}$ minority-carrier concentrations (corresponding to the best single junction a-Si:H V_{oc}s reported by 1993), the band gap for best Shockley model performance is $\sim$1.5 eV, with maximum efficiencies around 19%–20% for AM1.5 conditions and $\sim$18% for AM0 space conditions.

Considering two junctions of differing band gaps, the maximum Shockley model efficiencies for AM1.5 terrestrial devices are $\sim$40% at $10^{14}\,cm^{-3}$ minority carrier concentrations and $\sim$50% at $10^{17}\,cm^{-3}$ minority-carrier concentrations, but only ~26% for $10^8\,cm^{-3}$ minority-carrier concentrations (Chapter 1). Space AM0 efficiencies are typically a relative ~10% below terrestrial values. At open circuit voltages corresponding to minority-carrier concentrations of $10^8\,cm^{-3}$ or less (such as a-Si:H and its related alloys), the resulting low voltage from the lower band gap component cell limits the efficiency gain from the second junction to a relative 25% or less.

4. CONCLUSIONS

As stated above, the dominant terrestrial solar cell module technology in 1993 was based on crystalline silicon largely developed in the early 1980s under the Jet Propulsion Laboratories (JPL) Flat-Plate Solar Array Project. This specific technology's progress along its cumulative learning curve is a significant barrier to the entry of competing technologies. Its height was ~300 MW and over $2 billion in cumulative experience through 1993. Single crystal silicon was the dominant space array technology, with a cumulative effort exceeding 0.5 MW and $0.4 billion through 1993. Both of these leverage the hundreds of billions of dollars of underlying crystalline silicon technology developed for computers and other electronic applications.

Important underlying technology is being developed in III–V devices for optoelectronics and high-speed, low-power electronics and in a-Si:H for flat-panel displays. The uniformity and yields needed to make flat-panel displays as large as projection televisions are closely related to what is needed to make similarly sized, a-Si:H, solar cell panels with efficiencies over 10%. Other alternate solar cell technologies, like copper indium diselenide and cadmium telluride, are having a difficult time progressing far enough down their learning curves to compete with crystalline silicon. Such alternates need nonsolar applications to leverage growth of their underlying technology bases.

A new technology often begins with overall price and performance characteristics that are inferior to the entrenched one. The advantages of the new approach have to be so significant and its learning curve so steep that it overtakes the prior technology with only a small fraction of the resources already invested in the prior technology. This is for direct competition. An alternate strategy is to start in a protected niche market not served by the entrenched one so the new approach moves along its own learning curve without direct competition until it can outperform the old one.

Typically rapid adoption of a new technology requires orders-of-magnitude level advantages in relevant characteristics compared with the competition's, provided the results are cost effective. Factor-of-two level improvements do not appear to be sufficient particularly when they cost more. Batteries powered the first Soviet Sputnik satellite for 21 days. Solar cells powered the closely following U.S. Vanguard I for 6 years (Chapter 17). The solar cell panel hardly affected the overall costs. This cost-effective, two-orders-of-magnitude advantage immediately led to adoption of solar cell power for satellites. The a-Si:H solar cell power for portable calculators eliminated the inconvenient and expensive (~\$1,000/kW-h, Chapter 19) battery replacement and provided a more reliable product. The cost of a portable calculator with solar cell power equals the price of three or four battery replacements. Such cost-effective enabling applications go beyond order of magnitude to make battery-free portable calculators possible at all. The relative low cost of integrated a-Si:H cells (compared with hand-assembled small crystalline silicon ones) allowed the selling price with solar cells to be no more than traditional calculators with batteries. In this calculator power market, a-Si:H never really had to compete with crystalline silicon.

Future penetration into the portable computer market remains to be seen, with a limitation being the limited power available. GaAs cells could provide 5–10 times the efficiency and power density of a-Si:H in an area where competing prices are \$20–\$50/W. One can grow GaAs on semi-insulating substrates that allow large-scale integration like a-Si:H. Laser beaming of power through optical fibers for electricity (see Section 5.2.2 in Chapter 4) already uses this GaAs integration. For the portable computer power niche, the potential market is much larger than the 1992 level of ~\$0.1 billion/year for solar powered calculators, and it is likely protected from direct competition with crystalline silicon.

There are numerous examples of rapid adoption of an order-of-magnitude improved technology once it becomes cost effective. These range from transistors, to instant photography, to xerography, to FAX machines, to personal computers, to VCRs and so on. For solar cells themselves, the enabling characteristic is fuel-free energy conversion. The key remaining challenge is also to be cost effective in systems that adequately match energy supply with demand. When an alternate solar cell technology challenges crystalline silicon, a major problem is to come up with even factor-of-two advantages, particularly at competitive prices. A further complication is that crystalline silicon is a moving target that constantly improves.

A rough rule-of-thumb threshold, used to assess potential entry into very large terrestrial energy conversion markets, is a minimum 10% field efficiency and a module sales price in the \$1–\$2/W range (in 1992 dollars). The actual base line values in Chapter 16 for \$0.11/kW-h utility electricity are 11% field efficiencies and \$1.21/W modules in sunny locations that, over the year, average 1 kW/m^2 sunlight intensities for 6.8 h/day (i.e., 2,486 kW-h/m^2-yr). Chapter 16 did not consider efficiencies below 8% because the advantages of increased efficiency are so large and the penalties of low values are so harsh. Even with zero solar cell costs, efficiencies significantly below 10% cannot provide utility competitive power due to area-related costs. One can calculate specific values for the latter zero-cost, efficiency minimum from expressions like those in Chapter 16 based on a number of related assumptions. Materials costs that exceed \$1.50/W fundamentally limit the price of the standard "'JPL-like" crystalline silicon modules of Chapter 9 even with 10% efficiencies. A very positive factor for increased market growth is substantially higher efficiency, but with little or no increase in costs per unit area. Figure 16.3 in Chapter 16 illustrates this.

The terrestrial solar cell module market is highly competitive, and it was just becoming profitable in the early 1990s. Low gross margins in the 30% range typically characterize low profitability that slightly exceeds costs. Consequently the industry itself can only support minimal R&D in the range of 5% or less of sales. This level is characteristic of mature industries that evolve slowly with slow innovation and technology advances. In contrast, fast growing profitable markets can have gross margins of 60% and higher that support R&D at 20%–30% of sales levels that rapidly drive technology development. Government programs can play a key role in accelerating technology development.

Solar cell module prices of \$5–\$6/W (and below!) may have been charged for JPL-style crystalline silicon according to back calculations from total system installed costs for some utility demonstration projects arranged in 1993 and 1994, according to F. Goodman (coauthor of Chapters 11 and 16). Chapter 9 indicates that such prices are below manufacturing costs. If real, such losses could be underwritten for promotional purposes, but they cannot be sustained on a large scale over time. Large losses do not contribute to the health and growth of the industry. Solar cell companies would be well advised to focus on market niches that allow reasonable profits. Government acceleration programs should realize and accommodate the need for profitability. Applications that require \$1–\$2/W modules will evolve naturally, once enough progress is made on the learning curve to produce such products profitably.

The efficiencies of the best laboratory solar cells rose dramatically in the 1980s largely from government-supported research. However, commercial products of the early 1990s incorporated little of this because no one developed ways to make them cost effective.

There are two notable examples of rapid incorporation of advanced technology into commercial solar cell products in a cost-effective manner. One was the JPL FSA development program for crystalline silicon modules. The other was the Japanese development of a-Si:H for consumer products dominated by solar-powered calculators. The JPL program forced a reasonable compromise between high performance and manufacturability, stability, and costs through large block purchases of products that had to pass increasingly stiff qualification tests closely tied to the needs of potential terrestrial solar cell buyers. This effort met all of its program goals except module costs. Nevertheless, it nurtured the most cost-effective product available technically at the time. The Japanese creatively identified consumer wants that they cost effectively coupled with the unique interconnect advantages of a-Si:H on an insulating substrate.

Large-scale follow-up "JPL-like" programs should probably be pursued. They would target terrestrial module efficiency performance in the field starting at 15%. Then they would move progressively through 5% increments to 20%, 25%, 30%, and so forth until cost effective 50% efficiency levels are reached near current theoretical maximum values. Initially at each efficiency step there should be no limits on concentration. However, concentration limit goals should follow at each increment, dropping first to $\leqslant 2000\times$ and then to $\leqslant 20\times$ and finally to no concentration at all. High-concentration systems may only become cost effective at the higher efficiency levels if they continue far enough down their learning curves to be competitive.

Improved geographical and temporal distribution data are being obtained on solar insolation for the United States (Chapter 20). If concentrator use becomes widespread, more detailed information on "direct normal" insolation will also be required. If developing country markets become a large market share, better information on worldwide insolation will also be needed. Such information will decrease the margins for over design, decrease systems costs, and improve the projections of systems power output.

Demonstration field efficiencies should be determined by dividing array power output by a single solar power input value like that falling on an optimally tilted fixed surface. Then a user accurately knows the relative differences in electrical output he can expect from a given aperture area of installed modules in a given location (Hester and Hoff, 1985). An end user has no interest in details like a concentrator rejecting the diffuse component of sunlight. At least 15% of the demonstration sites should be in high mountain desert regions where skies are clear, temperatures are low, and the insolation strong. This is where the best environment for energy conversion on a utility scale may exist and where concentrators have their strongest advantages.

Perhaps the same couple of hundred-million-dollar-level effort as the original JPL program can reach the 15% efficiency mark for modules on a large scale because of the experience already gained. However, subsequent program costs will grow since the market will be larger and the new advances will have to compete on a learning curve with ever larger dominant technologies. It is unlikely that any one political entity will have the resources, willingness, leadership, and judgement to go alone successfully to 50% levels. Nevertheless such government efforts will likely accelerate 15% products by several years, the mid-level improvements by a decade or more, and 50% performance by 30–50 years, compared with normal market cycles, if profitability stays low.

Even with higher profits, government acceleration can provide early market dominance in new advances for its domestic manufacturers. Such dominance should persist, at least at a major share level, for many years. The United States became the majority supplier of crystalline silicon modules after the JPL program, and it still retained a nearly equal market share in the early 1990s with comparable-size suppliers in Japan and Europe (Chapter 19). Japan was the majority supplier of a-Si:H for calculators initially. It retained a major share in 1992 in competition with lower cost suppliers from other Pacific Rim countries (Chapter 19).

The improvement goals mentioned above for efficiency and costs include orders-of-magnitude reductions in sunlight concentration at a given efficiency level. Small, logical, and accomplishable steps should be used as goals. For comparison, an orderly progression of factor-of-two increases in memory chip capacity and constantly falling cost per bit of storage fostered the hundred-billion-dollar-per-year market for single crystal silicon based computers. This ultimately gave exponential growth in both market size and product performance. Here the early U.S. lead and market dominance were lost to Pacific Rim competitors in the mid 1980s. U.S. market share recovered in the early 1990s as U.S. chip manufacturers began to cooperate and the U.S. government increased its R&D funding focused on near-term customer needs (e.g., SEMATECH). Rapid solar cell industry growth should become more self-sustaining and profitable for market leaders as it grows to 1, 10, and possibly to 100 billion dollars per year levels. Then, even a small percentage of sales supports a sizable R&D program. At such high production levels there are challenging safety problems to be solved for solar array manufacturing and disposal (Chapter 18).

All the elements necessary for success of Japanese-style development programs are not immediately obvious. However, an essential one is "Quality Functional Deployment," described in part by Hauser and Clausing (1988). Their paper is an extreme version that contains key conceptual elements adaptable to wider, more generalized applications. Prime importance should be placed on listening to the "voice of the customer" in his or her own words without interpretation, including how much they would be willing to pay for improvements, starting at the initial stages of any new developments.

Government-sponsored, basic R&D on solar cells should probably continue at levels comparable to those in the 1980s, which led to dramatic increases in "best laboratory cell" performance that approach ideal Shockley diode behavior. This will blaze the trail and build the foundation for future advances. However, the scope and effort should be expanded also to address applied problems impeding near-term progress. The largest near-term improvements can come from decreasing the factor of two to four differences between terrestrial field demonstrations and "best laboratory cell" efficiencies, drawing particularly from space panel experience (Chapter 4). For crystalline silicon, the exact reasons can be separately identified by carefully measuring individual cells in a simulator before connecting them into a "typical" test array with enough voltage, current, and temperature monitoring points to pinpoint exact losses as the array is exposed to outdoor sunlight. Then one can specify how much of the difference is due to "average" efficiencies being less than the "best," to mismatch losses from differences (standard deviations) among cell properties and their interconnect patterns, to temperature rises and to surface fill factors being less than one.

The next largest improvements can come from gaining enough control that "typical" cells approach as closely to the "ideal" Shockley p/n junction limiting performance as the "best" cells (see Table 1.6, Chapter 1). The existing differences are particularly due to fill factors being less than the Shockley ideal for the measured open circuit voltages (as indicated in Fig. 1.3 in Chapter 1). This fill factor deficit is caused by "leakage currents" not being negligible at the voltages of maximum power and above. Control will require improved basic understanding of the "leakage current" parameters. Perimeter contributions to "leakage currents" are substantially reduced in crystalline silicon by "coin stacking" and plasma etching cell edges (Chapter 2) and in GaAs and other III–V materials by trench etching the perimeter (Partain et al., 1990) using a wet chemical etch (Fang and Look, 1993). Even with reduced perimeter currents, significant AlGaAs (and GaAs and likely crystalline silicon) "leakage currents" remain. A second "generation-recombination" diode current mechanism (Hovel, 1975) does not give a good description of this problem particularly when the temperature dependence is considered (Partain et al., 1989). Separation of perimeter from bulk "leakage currents" is a challenge (DeMoulin et al., 1988), but this separation can be achieved with "guard rings" (Barbe and Westgate, 1970).

Single junction efficiency over 30% can theoretically be achieved in materials with a band gap between ~0.8 and ~1.7 eV in devices that operate like ideal Shockley p/n junctions, which most current high-efficiency devices do (Chapter 1). A key to approaching this limiting performance is optical generation of high minority-carrier concentration levels. Light trapping in thinner devices enhances the latter. Thinness also reduces price-limiting materials cost. Light concentration can immediately give high minority-carrier concentrations, but this has trade offs. For example, more complex two axis sun tracking is essential, the diffuse component of sunlight usually cannot be used, 10%–25% of the available light is lost in the concentration optics, and low series resistance becomes a limiting factor. One sun production of high minority-carrier levels is preferred, although there is no known way to achieve $10^{17}\,cm^{-3}$ levels in GaAs or crystalline silicon without concentration at the present time (Chapters 1–3). Very thin cells may need some external light trapping structures (Landis, 1990).

The underlying assumptions of the Shockley model break down at minority-carrier concentrations on the order of the majority carrier ones. There is no fundamental reason to limit minority carriers below such "high injection" levels, but one needs a consensus on new theories that assess if this offers even higher performance potential. The references to Chapter 3 summarize the current state of "high-injection" theories.

Efficiencies in the 30%–50% range will require multiple junctions of differing band gaps if Shockley model devices are used. There are no obvious ways to light trap efficiently in thin, monolithic, multiple junction devices at the present time. The first 30%–50% efficient modules will probably use concentrated light. Early demonstration of laboratory cell efficiencies over 40% will likely use better developed junction pairs like 1.42 eV GaAs and 0.73 eV GaSb or 1.9 eV $GaInP_2$ and 1.1 eV Si with significant differences in their band gaps (see Figs. 1.15 and 1.16, Chapter 1). In such demonstrations, physically splitting concentrated light into separate long and short wavelength beams with a multilayer dielectric mirror (Moon et al., 1978) will allow each cell to operate at different concentration ratios for maximum performance by spacing each a different distance from the concentrating lens. Cell separation in space will also allow the most flexible four terminal interconnection and permit straightforward use of standard light-trapping techniques.

There are devices from a large variety of materials that do not obey Shockley p/n junction theory and give low-efficiency solar cells (Chopra and Das, 1983). Indeed, a defect-controlled, space-charge-limited-current model predicts that solar cell behavior can be seen from any material that exhibits band gap optical absorption once asymmetrical electrical contacts are attached (Partain, 1987). In particular, the ideal Shockley p/n junction equations do not describe a-Si:H cells well. One a-Si:H model entirely describes their low fill factors by the voltage dependence of their short circuit current (Faughn and Crandall, 1984) and in another solely by the voltage variation of their "diode" current (Partain, 1987). In reality it must be a combination of both (Yeung et al., 1991). First principles modeling with few simplifying assumptions provide explanations of a-Si:H behavior (Hack and Shur, 1985; Kuwano et al., 1982), but there are so many fitting parameters that predictive ability is limited, and the numerical integrations are so involved that ease of use is compromised. Until a-Si:H devices are better understood theoretically, it will be difficult to assess how or whether their performance can approach the levels indicated by the Shockley model for a material of the same band gap.

Field demonstration of 100 kW and greater power levels, with module efficiencies up to 30%, should be possible with single junction cells of either single crystal silicon or GaAs (Chapter 1). Carefully fabricated single proof-of-concept modules have demonstrated over 20% efficiencies using a small number of hand-selected and -matched, high-performance devices, first in GaAs at 1,000 × concentration (Chapter 5) and later in single crystal silicon with no concentration (Zhao et al., 1993). The largest impediment to such performance in commercial single crystal silicon cell modules are the latter's use of screen-printed grids. This prevents full exploitation of the long carrier lifetime potential of higher quality silicon substrates like float zone refined material (Chapter 2).

The next goals for 50–100 W proof-of-concept single modules should be 25% efficiency, then incremented in 5% steps up to 50%, starting with no concentration limits and proceding through restrictions to $\leqslant 200\times$, $\leqslant 20\times$, and finally no concentration. The high-concentration versions should be developed at least through 20 kW level demonstrations. Above that, concentrator systems need to show cost effectiveness.

The wide discrepancies between manufacturer-predicted efficiencies and actual field-measured performance should be reduced. A major contributor is inaccuracy in setting the intensity of simulators used to measure devices and modules. Following the lead of the space community, certified reference cells can handle this if their relative spectral response is sufficiently close to that of the test devices and modules being characterized. National centers like the National Renewable Energy Laboratory in the United States, the Fraunhofer-Institut for Solare Energiesysteme in Europe, and the JMI Institute in Japan should calibrate the reference cells. These centers should be certified by an international board of recognized experts who monitor center consistency in continuing round-robin tests with new device types added as needed. The correct values of open circuit voltage and fill factor can be measured in ideal Shockley p/n solar cells when the simulator intensity is adjusted to give the proper short circuit current. In cells not well described by the Shockley model, the same light-generated current can give different values of open circuit voltage and fill factor if the spectral content of the simulator light is varied (Rothwarf et al., 1987; Partain and Grounner, 1989). A warning signal is when measured fill factors are below Schockley ideal values for the observed open circuit voltages (Chapter 1) in a way not justified by a series resistance that is independently determined. There are enough fitting parameters to match the linear current-voltage characteristics of almost any cell in the light using a double diode model with series and shunt resistances (Hovel, 1975). However, the resulting fit parameter values may have no physical basis. Reality checks should be made for all cell types by periodic measurements of efficiency in actual sunlight.

Space solar cell panels have a potential for at least an order-of-magnitude reduction in price since their 1992 costs were two orders-of-magnitude more than terrestrial modules. Such reductions will not drastically reduce space solar array costs unless similar reductions occur in power conditioning and battery storage. Power conditioning costs could be reduced by an order of magnitude if a limited number (three or four) of standard system configurations were defined to cover the vast majority of space missions. A key element in cost reduction is elimination of most "quality control" inspections, including individual measurements of all component cells. The latter is an essential tenet of Edward Deming's philosophy (1986) of achieving higher quality at lower cost exploited so well by the Japanese over the past two decades. The latter requires more robust design like "parallel first" cell connection in arrays that desensitize the system to individual variations in cell parameters, when techniques like statistical process control reduce their spread. Screen-printing and plating the wiring interconnects onto the insulating substrates, before cells are attached, can offset the increased cost of "parallel first" interconnection in either space or terrestrial arrays. Space systems are a niche that can provide a sheltered market to develop radiation hard devices like GaAs and InP since the highest performance single crystal silicon cells do not survive well under space radiation exposure (Chapters 2, 4, 13, and 17).

The space solar cell market size will likely not expand greatly from space panel price decreases. This market segment is near saturation and its costs are only a fraction of the total systems costs for space communications, observation, and launch activities. The economic incentive is to develop more radiation hard devices with higher end-of-life efficiencies and lower weight. Then higher solar cell panel prices could be charged and still achieve lower life cycle costs for the total system. Lower space panel prices will likely occur as a "spin-off" of the much lower priced terrestrial modules.

By 1992 the solar cell market had grown steadily to 65 MW/year sales volumes. Crystalline silicon dominated sales volume for terrestrial off-grid modules of ~10% efficiency selling for ~\$6.50/W and for space satellite panels of ~15% efficiency selling for ~\$300–\$1,800/W. However, there was a significant fraction in a-Si:H for consumer calculators of 2%–3% efficiency selling for ~\$20–\$50/W. The largest near-term growth potential is for terrestrial off-grid systems, since calculator and space satellite applications appear to be near saturation. The long-term potential is for increased efficiencies up to 50%, decreased terrestrial prices in the \$1–\$2/W range (in 1992 dollars), and expanded market growth in installed power by two to three orders of magnitude or more, over the next 30–50 years. The latter depends on how much penetration the technology can make into electric utility markets.

REFERENCES

Armstrong-Russell, M.K., Freedman, W., and Spitler, E.E. (1983), *Proc. Soc. Photooptic Instrum. Eng.* **407**, 132–145.

Barbe, D.F., and Westgate, C.R. (1970), *J. Phys. Chem. Solids* **31**, 2679–2687.

Benner, J.P. (1991), *22nd IEEE Photovoltaic Specialists Conf. Record*, IEEE, New York, pp. 7–11.

Bertness, K.A., Friedman, D.J., Kibbler, A.E., Kramer, C., Kurta, S.R., and Olson, J.M. (1993), *AIP Conf. Proc.*, 12th National Renewable Energy Laboratory Photovoltaic Program Review, Denver, CO.

Chopra, K.L., and Das, S.R. (1983), *Thin Film Solar Cells*, Plenum, New York, p. 422.

Deming, E. (1986), *Out of Crisis*, Center for Advanced Engineering Study, MIT, Cambridge, MA.

DeMoulin, P.D., Tobin, S.P., Lundstrom, M.S., Carpenter, M.S., and Melloch, M.R. (1988), *IEEE Electronic Devices Lett.* **9**, 368–370.

Fang, Z.-Q., and Look, D.C. (1993), *J. Electr. Mtls.* **22**, 1361–1363.

Fraas, L.M., Avery, J.E., Martin, J., Sundaram, V.S., Girard, G., Dinh, V.T., Davenport, T.M., Yerkes, J.W. and O'Neill, M.J. (1990), *IEEE Trans. Electron Devices* **37**, 443–449.

Faughnan, B.W., and Crandall, R.S. (1984), *Appl. Phys. Lett.* **44**, 537–539.

Hack, M., and Shur, M. (1985), *J. Appl. Phys.* **58**, 997–1020.

Hauser, J.J., and Clausing, D. (1988), *Harvard Business Review*, May–June, pp. 63–73.

Hester, S., and Hoff, T. (1985), *22nd IEEE Photovoltaic Specialists Conf. Record*, IEEE, New York, pp. 777–779.

Hovel, J.J. (1975), *Solar Cells, Semiconductors And Semimetals*, Vol. 11, R.K. Willardson and A.C. Beer, Eds., Academic Press, New York, pp. 51–62.

Kuwano, Y., Tsuda, S., and Ohnishi, M. (1982), *Jpn. J. Appl. Phys.* **21**, 235–241.

Landis, G.A. (1990), *21st IEEE Photovoltaic Specialists Conf. Record*, IEEE, New York, pp. 1304–1307.

Moon, R.L., James, L.W., Vander Plas, H.A., Yep, T.O., Antypas, G.A., and Chai (1978), *13th IEEE Photovoltaic Specialists Conf. Record*, IEEE, New York, pp. 859–867.

Partain, L.D. (1987), *J. Appl. Phys.* **61**, 5458–5466.

Partain, L.D., Chung, B.-C., Schultz, J.C., and Virshup, G.F. (1989), *Technology Demonstration of a High-Efficiency, Monolithic, Three-Junction Cascade Solar Cell for Space Applications*, Contract #F33615-88-C-2845 Final Report, Wright Patterson Air Force Base, Dayton, OH, pp. 59–63.

Partain, L., and Grounner, M. (1989), *Mater. Res. Soc. Symp.* **148**, 403–408.

Partain, L.D., Schultz, J.C., Virshup, G.F., and Ristow, M.L. (1990), *Appl. Phys. Lett.* **57**, 1840–1842.

Rothwarf, A., Wyeth, N.C., and Phillips, J. (1978), *13th IEEE Photovoltaic Specialists Conf. Record*, IEEE, New York, pp. 399–405.

Yeung, P., Shapiro, F.R., and Rothwarf, A. (1991), *22nd IEEE Photovoltaic Specialists Conf. Record*, IEEE, New York, pp. 1369–1373.

Zhao, J., Wang, A., Taouk, M., Wenham, S.R., Green, M.A., and King, D.L. (1993), *IEEE Electronic Devices Lett.* **14**, 539–541.

The AM1.5G and AM1.5D Spectral Irradiances and the Corresponding Maximum One Sun Short Circuit Current Densities Versus Band Gap

The standard direct-normal (AM1.5D) and global-tilt (AM1.5G) spectral irradiances for the terrestrial conditions are those of Hulstrom et al. (1985) from the National Renewable Energy Laboratory in Golden, Colorado. The global-tilt spectrum is for a sun-facing surface tilted 37° from the horizontal. The total irradiances are 767.2 W/m^2 for AM1.5D and 962.5 W/m^2 for AM1.5G, obtained with trapezoidal rule, numerical integration. The AM1.5G irradiance values can be normalized to a convenient total irradiance of 1,000 W/m^2 by multiplying the irradiance at each wavelength and the short circuit current density at each band gap energy by the constant 1,000/962.5.

The (unnormalized) spectral irradiance $\phi(\lambda)$ at each wavelength λ was converted into photon flux density $F(\lambda)$ using $F(\lambda) = \lambda\phi(\lambda)/hc$, where h is Planck's constant and c is the speed of light. This in turn was converted into maximum one sun short circuit current densities by trapezoidal rule numerical integration of Eq. (7) of Chapter 1 assuming zero front surface reflectivity, unity quantum efficiency for $\lambda \leqslant \lambda_G$, and zero quantum efficiency for $\lambda > \lambda_G$. The λ_G is the wavelength of the photon whose energy equals the band gap energy.

REFERENCE

Hulstrom, R., Bird, R., and Riordan, C. (1985), *Solar Cells* **15**, 365–391.

Wavelength (μm)	Spectral Irradiance (W/[m^2 μm])		Band Gap (eV)	Maximum One Sun J_{sc} (mA/cm^2)	
	Direct-Normal	Global-Tilt		Direct-Normal	Global-Tilt
0.305	3.4	9.2	4.066	0.0012	0.0031
0.310	15.6	40.8	4.000	0.0047	0.0122
0.315	41.1	103.9	3.937	0.0119	0.0300
0.320	71.2	174.4	3.875	0.0230	0.0567
0.325	100.2	237.9	3.815	0.0397	0.0975
0.330	152.4	381.0	3.758	0.0603	0.148
0.335	155.6	376.0	3.701	0.0831	0.202
0.340	179.4	419.5	3.647	0.108	0.260
0.345	186.7	423.0	3.594	0.136	0.322
0.350	212.0	466.2	3.543	0.201	0.461
0.360	240.5	501.4	3.444	0.284	0.629
0.370	324.0	642.1	3.351	0.387	0.829
0.380	362.4	686.7	3.263	0.503	1.04
0.390	381.7	694.6	3.179	0.652	1.31
0.400	556.0	976.4	3.100	0.849	1.65
0.410	656.3	1116.2	3.024	1.07	2.03
0.420	690.8	1141.1	2.952	1.30	2.40
0.430	641.9	1033.0	2.884	1.55	2.80
0.440	798.5	1254.8	2.818	1.87	3.29
0.450	956.6	1470.7	2.756	2.23	3.84
0.460	990.8	1541.6	2.696	2.60	4.41
0.470	998.0	1523.7	2.638	2.99	5.00
0.480	1046.1	1569.3	2.583	3.39	5.60
0.490	1005.1	1483.4	2.531	3.79	6.19
0.500	1026.7	1492.6	2.480	4.22	6.81
0.510	1066.7	1529.0	2.431	4.65	7.42
0.520	1011.5	1431.1	2.385	5.09	8.04
0.530	1084.9	1515.4	2.340	5.56	8.69
0.540	1082.4	1494.5	2.296	6.04	9.35
0.550	1102.2	1504.9	2.255	7.03	10.68
0.570	1087.4	1447.1	2.175	8.01	11.98
0.590	1024.3	1344.9	2.102	9.03	13.33
0.610	1088.8	1431.5	2.033	10.11	14.73
0.630	1062.1	1382.1	1.968	11.20	16.15
0.650	1061.7	1368.4	1.908	12.32	17.59
0.670	1046.2	1341.8	1.851	13.36	18.92
0.690	859.2	1089.0	1.797	14.41	20.25
0.710	1002.4	1269.0	1.746	14.83	20.76
0.718	816.9	973.7	1.727	15.14	21.13
0.724	842.8	1005.4	1.712	15.97	22.13

Wavelength (μm)	Spectral Irradiance (W/[m^2 μm])		Band Gap (eV)	Maximum One Sun J_{sc} (mA/cm^2)	
	Direct-Normal	Global-Tilt		Direct-Normal	Global-Tilt
0.740	971.0	1167.3	1.676	16.70	23.00
0.753	956.3	1150.6	1.648	16.98	23.35
0.758	942.2	1132.9	1.637	17.21	23.61
0.763	524.8	619.8	1.626	17.42	23.86
0.768	830.7	993.3	1.616	18.10	24.67
0.780	908.9	1090.1	1.590	19.23	26.03
0.800	873.4	1042.4	1.550	20.05	27.00
0.816	712.0	818.4	1.520	20.40	27.40
0.824	660.2	765.5	1.505	20.77	27.83
0.832	765.5	883.2	1.491	21.22	28.35
0.840	799.8	925.1	1.476	22.33	29.62
0.860	815.2	943.4	1.442	23.44	30.91
0.880	778.3	899.4	1.409	24.71	32.37
0.905	630.4	721.4	1.370	25.14	32.87
0.915	565.2	643.3	1.355	25.57	33.35
0.925	586.4	665.3	1.341	25.75	33.55
0.930	348.1	389.0	1.333	25.90	33.72
0.937	224.2	248.9	1.323	26.10	33.95
0.948	271.4	302.2	1.308	26.58	34.48
0.965	451.2	507.7	1.285	27.16	35.14
0.980	549.7	623.0	1.265	27.80	35.86
0.994	630.1	719.7	1.248	30.10	38.50
1.040	582.9	665.5	1.192	31.53	40.13
1.070	539.7	614.4	1.159	32.72	41.45
1.100	366.2	397.6	1.127	33.14	41.90
1.120	98.1	105.0	1.107	33.26	42.03
1.130	169.5	182.2	1.097	33.35	42.13
1.137	118.7	127.4	1.091	33.82	42.64
1.161	301.9	326.7	1.068	34.45	43.33
1.180	406.8	443.3	1.051	35.20	44.14
1.200	375.2	408.2	1.033	36.57	45.64
1.235	423.6	463.1	1.004	38.78	48.04
1.290	365.7	398.1	0.961	39.70	49.05
1.320	223.4	241.1	0.939	40.11	49.49
1.350	30.1	31.3	0.919	40.19	49.57
1.395	1.4	1.5	0.889	40.33	49.72
1.443	51.6	53.7	0.860	40.51	49.90
1.463	97.0	101.3	0.848	40.67	50.08
1.477	97.3	101.7	0.840	40.99	50.41
1.497	167.1	175.5	0.828	41.56	51.01
1.520	239.3	253.1	0.816	42.13	51.61
1.539	248.8	264.3	0.806	42.72	52.24
1.558	249.3	265.0	0.796	43.32	52.87

Wavelength (μm)	Spectral Irradiance (W/[m^2 μm])		Band Gap (eV)	Maximum One Sun J_{sc} (mA/cm^2)	
	Direct-Normal	Global-Tilt		Direct-Normal	Global-Tilt
1.578	222.3	235.7	0.786	43.72	53.29
1.592	227.3	238.4	0.779	44.22	53.83
1.610	210.5	220.4	0.770	44.79	54.42
1.630	224.7	235.6	0.761	45.26	54.91
1.646	215.9	226.3	0.753	46.15	55.85
1.678	202.8	212.5	0.739	47.69	57.46
1.740	158.2	165.3	0.713	48.49	58.29
1.800	28.6	29.2	0.689	48.63	58.43
1.860	1.8	1.9	0.667	48.64	58.44
1.920	1.1	1.2	0.646	48.70	58.51
1.960	19.7	20.4	0.633	48.91	58.72
1.985	84.9	87.8	0.625	49.09	58.91
2.005	25.0	25.8	0.618	49.37	59.20
2.035	92.5	95.9	0.609	49.74	59.58
2.065	56.3	58.2	0.600	50.15	60.01
2.100	82.7	85.9	0.590	50.80	60.68
2.148	76.2	79.2	0.577	51.43	61.33
2.198	66.4	68.9	0.564	52.28	62.22
2.270	65.0	67.7	0.546	53.31	63.29
2.360	57.6	59.8	0.525	53.98	63.98
2.450	19.8	20.4	0.506	54.14	64.15
2.494	17.0	17.8	0.497	54.23	64.24
2.537	3.0	3.1	0.489	54.54	64.57
2.941	4.0	4.2	0.422	54.58	64.61
2.973	7.0	7.3	0.417	54.63	64.66
3.005	6.0	6.3	0.413	54.69	64.72
3.056	3.0	3.1	0.406	54.76	64.80
3.132	5.0	5.2	0.396	54.83	64.87
3.156	18.0	18.7	0.393	54.95	65.00
3.204	1.2	1.3	0.387	54.97	65.02
3.245	3.0	3.1	0.382	55.12	65.17
3.317	12.0	12.6	0.374	55.17	65.23
3.344	3.0	3.1	0.371	55.39	65.46
3.450	12.2	12.8	0.359	55.79	65.88
3.573	11.0	11.5	0.347	56.36	66.47
3.765	9.0	9.4	0.329	56.06	67.20
4.045	6.9	7.2	0.307		
Totals (W/m^2)	767.2	962.5			

APPENDIX B

The AM0 Spectral Irradiance and the Maximum One Sun Short Circuit Current Density Versus Band Gap

The revised and updated AM0 spectral irradiance values of Wehrli (1985) are given here. They are based on the publications of Brasseur and Simon (1981) for 0.200–0.310 μm, Arvesen et al. (1969) for 0.310–0.330 μm, Neckel and Labs (1984) for 0.330–0.869 μm, and Smith and Gottlieb (1974) for 0.870 μm through the infrared wavelengths as described by Riordan (1987). The irradiance values and the first derivatives were forced to be equal at the wavelength limits between the various spectra, and the total spectrum was then scaled to yield a total (integrated) irradiance value, or solar constant, of 1,367 W/m^2. The spectral irradiance was converted into photon flux density using the expressions in Appendix A. This was converted into maximum short circuit current density by trapezoidal rule numerical integration of Eq. (7) of Chapter 1 assuming zero front surface reflectivity, unity quantum efficiency for $\lambda \leqslant \lambda_G$, and zero quantum efficiency for $\lambda > \lambda_G$. The λ_G is the wavelength of the photon whose energy equals the band gap energy.

The tabulated irradiances are for the sun–earth distance at its mean value of 1.496×10^8 km that occurs on approximately April 4th and October 5th every year (Iqbal, 1983). These irradiances are increased by a maximum factor of 1.0351 when the distance is at its lowest value (perihelion) of the elliptical orbit on approximately January 3rd each year. They are decreased by a minimum factor of 0.9666 when the distance is at its highest value (aphelion) on approximately July 4th each year.

Calculations of terrestrial irradiances from mean-distance AM0 values are too high for systems in the northern hemisphere during the summer when most solar flux strikes the earth's surface and the earth–sun spacing is greatest. The opposite is true in the southern hemisphere. The uncertainty in the value of the solar constant is on-the-order-of 1% if several different satellite data sets are considered (Riordan, 1987). The solar constant value obtained by trapezoidal rule, numerical integration of the column two spectral irradiances of this appendix is 1,365 W/m^2, or 0.15% lower than the 1,367 W/m^2 Wehrli (1985) value.

REFERENCES

Arvesen, J. C., Griffin, R. N., and Pearson, B. D. (1969), *Appl. Optics* **8**, 2215–2232.

Brasseur G., and Simon, P. C. (1981), *J. Geophys. Res.* **86**, 7343–7362.

Iqbal, M. (1983), *An Introduction to Solar Radiation*, Academic Press, New York, Fig. 1.2.1, Table 1.2.1, pp. 1–3.

Neckel, H., and Labs, D. (1984), *Solar Physics* **90**, 205–258.

Riordan, C. (1987) *Extraterrestrial Spectral Solar Irradiance Data for Modeling Spectral Solar Irradiance at the Earth's Surface*, Publication No. SERI/TR-215-2921, Solar Energy Research Institute, Golden, CO.

Smith, E. V. P., and Gottlieb (1974), *Space Sci. Rev.* **16**, 771–802.

Wehrli, C. (1985), *Extraterrestrial Solar Spectrum*, Publication No. 615, Physikalisch-Meteorologisches Observatorium and World Radiation Center, Davos, Switzerland.

Wavelength (μm)	Spectral Irradiance (W/[m^2 μm])	Band Gap (eV)	Maximum One Sun J_{sc} (mA/cm^2)
0.1995	5	6.2155	0.000097
0.2005	7	6.1845	0.00021
0.2015	7	6.1538	0.00033
0.2025	8	6.1235	0.00047
0.2035	9	6.0934	0.00062
0.2045	9	6.0636	0.00078
0.2055	10	6.0341	0.00094
0.2065	10	6.0048	0.0011
0.2075	11	5.9759	0.0013
0.2085	15	5.9472	0.0017
0.2095	24	5.9189	0.0021
0.2105	28	5.8907	0.0026
0.2115	34	5.8629	0.0032
0.2125	30	5.8353	0.0037
0.2135	32	5.8080	0.0043
0.2145	41	5.7809	0.0050
0.2155	37	5.7541	0.0056
0.2165	34	5.7275	0.0062
0.2175	36	5.7011	0.0069

Wavelength (μm)	Spectral Irradiance (W/[m^2 μm])	Band Gap (eV)	Maximum One Sun J_{sc} (mA/cm^2)
0.2185	45	5.6751	0.0078
0.2195	48	5.6492	0.0086
0.2205	48	5.6236	0.0094
0.2215	39	5.5982	0.010
0.2225	51	5.5730	0.011
0.2235	66	5.5481	0.012
0.2245	58	5.5234	0.013
0.2255	54	5.4989	0.014
0.2265	41	5.4746	0.015
0.2275	41	5.4505	0.016
0.2285	54	5.4267	0.017
0.2295	48	5.4031	0.018
0.2305	56	5.3796	0.019
0.2315	50	5.3564	0.020
0.2325	55	5.3333	0.021
0.2335	46	5.3105	0.021
0.2345	39	5.2878	0.022
0.2355	57	5.2654	0.023
0.2365	49	5.2431	0.024
0.2375	53	5.2211	0.025
0.2385	42	5.1992	0.026
0.2395	46	5.1775	0.027
0.2405	43	5.1559	0.028
0.2415	52	5.1346	0.029
0.2425	72	5.1134	0.030
0.2435	65	5.0924	0.032
0.2445	62	5.0716	0.033
0.2455	51	5.0509	0.034
0.2465	51	5.0304	0.035
0.2475	57	5.0101	0.036
0.2485	45	4.9899	0.037
0.2495	58	4.9699	0.038
0.2505	59	4.9501	0.039
0.2515	47	4.9304	0.040
0.2525	44	4.9109	0.041
0.2535	55	4.8915	0.042
0.2545	61	4.8723	0.044
0.2555	89	4.8532	0.046
0.2565	107	4.8343	0.048
0.2575	129	4.8155	0.051
0.2585	134	4.7969	0.054
0.2595	108	4.7784	0.056
0.2605	102	4.7601	0.058
0.2615	103	4.7419	0.060
0.2625	121	4.7238	0.063

Wavelength (μm)	Spectral Irradiance (W/[m^2 μm])	Band Gap (eV)	Maximum One Sun J_{sc} (mA/cm^2)
0.2635	175	4.7059	0.068
0.2645	274	4.6881	0.074
0.2655	280	4.6704	0.080
0.2665	260	4.6529	0.086
0.2675	270	4.6355	0.091
0.2685	260	4.6182	0.10
0.2695	252	4.6011	0.10
0.2705	293	4.5841	0.11
0.2715	232	4.5672	0.11
0.2725	215	4.5505	0.12
0.2735	204	4.5338	0.12
0.2745	137	4.5173	0.13
0.2755	200	4.5009	0.13
0.2765	258	4.4846	0.14
0.2775	240	4.4685	0.14
0.2785	166	4.4524	0.14
0.2795	89	4.4365	0.15
0.2805	112	4.4207	0.15
0.2815	231	4.4050	0.16
0.2825	307	4.3894	0.16
0.2835	330	4.3739	0.17
0.2845	244	4.3585	0.17
0.2855	141	4.3433	0.18
0.2865	320	4.3281	0.19
0.2875	371	4.3130	0.20
0.2885	307	4.2981	0.20
0.2895	456	4.2832	0.22
0.2905	623	4.2685	0.23
0.2915	600	4.2539	0.24
0.2925	545	4.2393	0.26
0.2935	545	4.2249	0.27
0.2945	509	4.2105	0.28
0.2955	548	4.1963	0.29
0.2965	492	4.1821	0.31
0.2975	531	4.1681	0.32
0.2985	413	4.1541	0.33
0.2995	485	4.1402	0.34
0.3005	403	4.1265	0.35
0.3015	445	4.1128	0.36
0.3025	484	4.0992	0.37
0.3035	631	4.0857	0.39
0.3045	610	4.0722	0.40
0.3055	580	4.0589	0.42
0.3065	575	4.0457	0.43
0.3075	645	4.0325	0.45

Wavelength (μm)	Spectral Irradiance (W/[m^2 μm])	Band Gap (eV)	Maximum One Sun J_{sc} (mA/cm^2)
0.3085	613	4.0194	0.46
0.3095	484	4.0065	0.47
0.3100	495	4.0000	0.47
0.3104	507	3.9948	0.48
0.3108	588	3.9897	0.49
0.3112	707	3.9846	0.49
0.3116	747	3.9795	0.50
0.312	707	3.9744	0.51
0.3124	644	3.9693	0.51
0.3128	663	3.9642	0.52
0.3132	710	3.9591	0.53
0.3136	691	3.9541	0.54
0.314	689	3.9490	0.54
0.3144	722	3.9440	0.55
0.3148	673	3.9390	0.56
0.3152	695	3.9340	0.56
0.3156	765	3.9290	0.57
0.316	675	3.9241	0.58
0.3164	569	3.9191	0.58
0.3168	623	3.9141	0.59
0.3172	749	3.9092	0.60
0.3176	830	3.9043	0.61
0.318	813	3.8994	0.61
0.3184	673	3.8945	0.62
0.3188	642	3.8896	0.63
0.3192	768	3.8847	0.64
0.3196	759	3.8798	0.64
0.32	712	3.8750	0.65
0.3204	778	3.8702	0.66
0.3208	844	3.8653	0.67
0.3212	847	3.8605	0.68
0.3216	736	3.8557	0.68
0.322	695	3.8509	0.69
0.3224	773	3.8462	0.70
0.3228	758	3.8414	0.71
0.3232	646	3.8366	0.71
0.3236	603	3.8319	0.72
0.3240	604	3.8272	0.73
0.3244	618	3.8224	0.73
0.3248	654	3.8177	0.74
0.3252	646	3.8130	0.75
0.3256	682	3.8084	0.75
0.326	852	3.8037	0.76
0.3264	1049	3.7990	0.78
0.3268	1111	3.7944	0.79

Wavelength (μm)	Spectral Irradiance (W/[m^2 μm])	Band Gap (eV)	Maximum One Sun J_{sc} (mA/cm^2)
0.3272	1108	3.7897	0.80
0.3276	1050	3.7851	0.81
0.328	965	3.7805	0.82
0.3284	914	3.7759	0.83
0.3288	913	3.7713	0.84
0.3292	952	3.7667	0.85
0.3296	1043	3.7621	0.86
0.33	1144	3.7576	0.87
0.3304	1137	3.7530	0.88
0.3305	1006	3.7519	0.90
0.3315	968	3.7406	0.93
0.3325	921	3.7293	0.95
0.3335	905	3.7181	0.98
0.3345	940	3.7070	1.00
0.3355	982	3.6960	1.03
0.3365	765	3.6850	1.05
0.3375	866	3.6741	1.07
0.3385	916	3.6632	1.10
0.3395	937	3.6524	1.13
0.3405	992	3.6417	1.15
0.3415	936	3.6310	1.18
0.3425	995	3.6204	1.21
0.3435	985	3.6099	1.23
0.3445	719	3.5994	1.25
0.3455	967	3.5890	1.28
0.3465	919	3.5786	1.30
0.3475	902	3.5683	1.33
0.3485	948	3.5581	1.36
0.3495	865	3.5479	1.38
0.3505	1119	3.5378	1.41
0.3515	993	3.5277	1.44
0.3525	871	3.5177	1.47
0.3535	1115	3.5078	1.50
0.3545	1133	3.4979	1.53
0.3555	1058	3.4880	1.56
0.3565	938	3.4783	1.59
0.3575	891	3.4685	1.61
0.3585	627	3.4589	1.63
0.3595	1136	3.4492	1.66
0.3605	979	3.4397	1.69
0.3615	894	3.4302	1.72
0.3625	1175	3.4207	1.75
0.3635	958	3.4113	1.78
0.3645	1015	3.4019	1.82
0.3655	1263	3.3926	1.85

Wavelength (μm)	Spectral Irradiance (W/[m^2 μm])	Band Gap (eV)	Maximum One Sun J_{sc} (mA/cm^2)
0.3665	1249	3.3834	1.89
0.3675	1214	3.3741	1.92
0.3685	1088	3.3650	1.96
0.3695	1331	3.3559	1.99
0.3705	1075	3.3468	2.03
0.3715	1307	3.3378	2.07
0.3725	1065	3.3289	2.09
0.3735	838	3.3199	2.12
0.3745	878	3.3111	2.15
0.3755	1141	3.3023	2.18
0.3765	1101	3.2935	2.22
0.3775	1291	3.2848	2.26
0.3785	1341	3.2761	2.30
0.3795	1000	3.2675	2.33
0.3805	1289	3.2589	2.37
0.3815	1096	3.2503	2.40
0.3825	733	3.2418	2.42
0.3835	684	3.2334	2.44
0.3845	1027	3.2250	2.47
0.3855	954	3.2166	2.51
0.3865	1071	3.2083	2.54
0.3875	966	3.2000	2.57
0.3885	912	3.1918	2.60
0.3895	1227	3.1836	2.64
0.3905	1223	3.1754	2.68
0.3915	1398	3.1673	2.72
0.3925	955	3.1592	2.74
0.3935	489	3.1512	2.77
0.3945	1101	3.1432	2.81
0.3955	1378	3.1353	2.84
0.3965	650	3.1274	2.86
0.3975	1040	3.1195	2.91
0.3985	1538	3.1117	2.96
0.3995	1655	3.1039	3.01
0.4005	1649	3.0961	3.07
0.4015	1796	3.0884	3.12
0.4025	1803	3.0807	3.18
0.4035	1658	3.0731	3.23
0.4045	1602	3.0655	3.29
0.4055	1672	3.0580	3.34
0.4065	1624	3.0504	3.39
0.4075	1545	3.0429	3.45
0.4085	1824	3.0355	3.51
0.4095	1706	3.0281	3.56
0.4105	1502	3.0207	3.61

Wavelength (μm)	Spectral Irradiance (W/[m^2 μm])	Band Gap (eV)	Maximum One Sun J_{sc} (mA/cm^2)
0.4115	1819	3.0134	3.67
0.4125	1791	3.0061	3.73
0.4135	1758	2.9988	3.79
0.4145	1739	2.9916	3.85
0.4155	1736	2.9844	3.91
0.4165	1844	2.9772	3.97
0.4175	1667	2.9701	4.02
0.4185	1686	2.9630	4.08
0.4195	1703	2.9559	4.14
0.4205	1760	2.9489	4.20
0.4215	1799	2.9419	4.26
0.4225	1584	2.9349	4.31
0.4235	1713	2.9280	4.37
0.4245	1770	2.9211	4.43
0.4255	1697	2.9142	4.49
0.4265	1700	2.9074	4.55
0.4275	1571	2.9006	4.60
0.4285	1589	2.8938	4.65
0.4295	1477	2.8871	4.70
0.4305	1136	2.8804	4.75
0.4315	1688	2.8737	4.81
0.4325	1648	2.8671	4.87
0.4335	1733	2.8604	4.93
0.4345	1672	2.8539	4.98
0.4355	1725	2.8473	5.05
0.4365	1931	2.8408	5.11
0.4375	1808	2.8343	5.17
0.4385	1569	2.8278	5.23
0.4395	1827	2.8214	5.30
0.4405	1715	2.8150	5.36
0.4415	1933	2.8086	5.43
0.4425	1982	2.8023	5.50
0.4435	1911	2.7959	5.57
0.4445	1975	2.7897	5.64
0.4455	1823	2.7834	5.70
0.4465	1893	2.7772	5.78
0.4475	2079	2.7709	5.85
0.4485	1975	2.7648	5.92
0.4495	2029	2.7586	6.00
0.4505	2146	2.7525	6.07
0.4515	2111	2.7464	6.15
0.4525	1943	2.7403	6.22
0.4535	1972	2.7343	6.29
0.4545	1981	2.7283	6.37
0.4555	2036	2.7223	6.44

Wavelength (μm)	Spectral Irradiance (W/[m^2 μm])	Band Gap (eV)	Maximum One Sun J_{sc} (mA/cm^2)
0.4565	2079	2.7163	6.52
0.4575	2102	2.7104	6.59
0.4585	1973	2.7045	6.67
0.4595	2011	2.6986	6.74
0.4605	2042	2.6927	6.82
0.4615	2057	2.6869	6.90
0.4625	2106	2.6811	6.97
0.4635	2042	2.6753	7.05
0.4645	1978	2.6695	7.12
0.4655	2044	2.6638	7.20
0.4665	1923	2.6581	7.27
0.4675	2017	2.6524	7.35
0.4685	1996	2.6467	7.42
0.4695	1992	2.6411	7.50
0.4705	1879	2.6355	7.57
0.4715	2020	2.6299	7.65
0.4725	2043	2.6243	7.72
0.4735	1993	2.6188	7.80
0.4745	2053	2.6133	7.88
0.4755	2018	2.6078	7.95
0.4765	1958	2.6023	8.03
0.4775	2077	2.5969	8.11
0.4785	2011	2.5914	8.19
0.4795	2078	2.5860	8.27
0.4805	2037	2.5806	8.35
0.4815	2092	2.5753	8.43
0.4825	2025	2.5699	8.51
0.4835	2021	2.5646	8.59
0.4845	1971	2.5593	8.66
0.4855	1832	2.5541	8.73
0.4865	1627	2.5488	8.79
0.4875	1832	2.5436	8.87
0.4885	1916	2.5384	8.94
0.4895	1962	2.5332	9.02
0.4905	2009	2.5280	9.10
0.4915	1898	2.5229	9.18
0.4925	1898	2.5178	9.25
0.4935	1890	2.5127	9.33
0.4945	2060	2.5076	9.41
0.4955	1928	2.5025	9.49
0.4965	2019	2.4975	9.57
0.4975	2020	2.4925	9.65
0.4985	1868	2.4875	9.72
0.4995	1972	2.4825	9.80
0.5005	1859	2.4775	9.87

Wavelength (μm)	Spectral Irradiance (W/[m^2 μm])	Band Gap (eV)	Maximum One Sun J_{sc} (mA/cm^2)
0.5015	1814	2.4726	9.95
0.5025	1896	2.4677	10.03
0.5035	1936	2.4628	10.10
0.5045	1871	2.4579	10.18
0.5055	1995	2.4530	10.26
0.5065	1963	2.4482	10.34
0.5075	1908	2.4433	10.42
0.5085	1921	2.4385	10.50
0.5095	1918	2.4338	10.58
0.5105	1949	2.4290	10.66
0.5115	1999	2.4242	10.74
0.5125	1869	2.4195	10.82
0.5135	1863	2.4148	10.89
0.5145	1876	2.4101	10.97
0.5155	1902	2.4054	11.05
0.5165	1671	2.4008	11.12
0.5175	1728	2.3961	11.19
0.5185	1656	2.3915	11.26
0.5195	1830	2.3869	11.34
0.5205	1833	2.3823	11.42
0.5215	1908	2.3778	11.49
0.5225	1825	2.3732	11.57
0.5235	1896	2.3687	11.65
0.5245	1960	2.3642	11.74
0.5255	1932	2.3597	11.81
0.5265	1676	2.3552	11.89
0.5275	1830	2.3507	11.97
0.5285	1899	2.3463	12.05
0.5295	1920	2.3418	12.13
0.5305	1954	2.3374	12.21
0.5315	1965	2.3330	12.29
0.5325	1773	2.3286	12.37
0.5335	1925	2.3243	12.45
0.5345	1860	2.3199	12.54
0.5355	1992	2.3156	12.62
0.5365	1873	2.3113	12.70
0.5375	1884	2.3070	12.78
0.5385	1906	2.3027	12.87
0.5395	1834	2.2984	12.94
0.5405	1772	2.2942	13.02
0.5415	1883	2.2899	13.10
0.5425	1827	2.2857	13.19
0.5435	1881	2.2815	13.27
0.5445	1881	2.2773	13.35
0.5455	1903	2.2731	13.43

Wavelength (μm)	Spectral Irradiance (W/[m^2 μm])	Band Gap (eV)	Maximum One Sun J_{sc} (mA/cm^2)
0.5465	1881	2.2690	13.52
0.5475	1835	2.2648	13.60
0.5485	1865	2.2607	13.68
0.5495	1897	2.2566	13.76
0.5505	1864	2.2525	13.85
0.5515	1873	2.2484	13.93
0.5525	1848	2.2443	14.01
0.5535	1884	2.2403	14.10
0.5545	1900	2.2362	14.18
0.5555	1899	2.2322	14.26
0.5565	1823	2.2282	14.35
0.5575	1848	2.2242	14.43
0.5585	1789	2.2202	14.51
0.5595	1810	2.2163	14.59
0.5605	1845	2.2123	14.68
0.5615	1826	2.2084	14.76
0.5625	1852	2.2044	14.84
0.5635	1863	2.2005	14.93
0.5645	1856	2.1966	15.01
0.5655	1800	2.1927	15.09
0.5665	1831	2.1889	15.18
0.5675	1889	2.1850	15.26
0.5685	1812	2.1812	15.35
0.5695	1862	2.1773	15.43
0.5705	1772	2.1735	15.51
0.5715	1825	2.1697	15.60
0.5725	1894	2.1659	15.69
0.5735	1878	2.1622	15.77
0.5745	1869	2.1584	15.86
0.5755	1832	2.1546	15.94
0.5765	1848	2.1509	16.03
0.5775	1859	2.1472	16.11
0.5785	1786	2.1435	16.20
0.5795	1830	2.1398	16.28
0.5805	1840	2.1361	16.37
0.5815	1855	2.1324	16.46
0.5825	1875	2.1288	16.54
0.5835	1859	2.1251	16.63
0.5845	1862	2.1215	16.72
0.5855	1786	2.1178	16.80
0.5865	1832	2.1142	16.89
0.5875	1850	2.1106	16.98
0.5885	1752	2.1071	17.06
0.5895	1614	2.1035	17.14
0.5905	1815	2.0999	17.22

Wavelength (μm)	Spectral Irradiance (W/[m^2 μm])	Band Gap (eV)	Maximum One Sun J_{sc} (mA/cm^2)
0.5915	1789	2.0964	17.31
0.5925	1810	2.0928	17.39
0.5935	1798	2.0893	17.48
0.5945	1776	2.0858	17.57
0.5955	1785	2.0823	17.65
0.5965	1807	2.0788	17.74
0.5975	1783	2.0753	17.82
0.5985	1760	2.0718	17.91
0.5995	1777	2.0684	17.99
0.6005	1748	2.0649	18.08
0.6015	1753	2.0615	18.16
0.6025	1721	2.0581	18.25
0.6035	1789	2.0547	18.33
0.6045	1779	2.0513	18.42
0.6055	1766	2.0479	18.51
0.6065	1762	2.0445	18.59
0.6075	1760	2.0412	18.68
0.6085	1745	2.0378	18.76
0.6095	1746	2.0345	18.85
0.6105	1705	2.0311	18.93
0.6115	1748	2.0278	19.02
0.6125	1707	2.0245	19.10
0.6135	1685	2.0212	19.19
0.6145	1715	2.0179	19.27
0.6155	1715	2.0146	19.35
0.6165	1611	2.0114	19.44
0.6175	1709	2.0081	19.52
0.6185	1726	2.0049	19.61
0.6195	1709	2.0016	19.69
0.6205	1736	1.9984	19.78
0.6215	1692	1.9952	19.86
0.6225	1715	1.9920	19.95
0.6235	1668	1.9888	20.03
0.6245	1658	1.9856	20.12
0.6255	1634	1.9824	20.20
0.6265	1699	1.9792	20.29
0.6275	1699	1.9761	20.37
0.6285	1699	1.9730	20.46
0.6295	1679	1.9698	20.58
0.631	1641	1.9651	20.75
0.633	1653	1.9589	20.92
0.635	1658	1.9528	21.09
0.637	1656	1.9466	21.26
0.639	1653	1.9405	21.43
0.641	1616	1.9345	21.60

Wavelength (μm)	Spectral Irradiance (W/[m^2 μm])	Band Gap (eV)	Maximum One Sun J_{sc} (mA/cm^2)
0.643	1623	1.9285	21.76
0.645	1629	1.9225	21.93
0.647	1605	1.9165	22.10
0.649	1560	1.9106	22.26
0.651	1608	1.9048	22.43
0.653	1601	1.8989	22.60
0.655	1534	1.8931	22.75
0.657	1386	1.8874	22.91
0.659	1551	1.8816	23.07
0.661	1573	1.8759	23.24
0.663	1557	1.8703	23.41
0.665	1562	1.8647	23.57
0.667	1537	1.8591	23.74
0.669	1548	1.8535	23.90
0.671	1518	1.8480	24.07
0.673	1523	1.8425	24.23
0.675	1512	1.8370	24.40
0.677	1510	1.8316	24.56
0.679	1500	1.8262	24.73
0.681	1494	1.8209	24.89
0.683	1481	1.8155	25.05
0.685	1457	1.8102	25.21
0.687	1469	1.8049	25.37
0.689	1463	1.7997	25.54
0.691	1450	1.7945	25.70
0.693	1450	1.7893	25.86
0.695	1438	1.7842	26.02
0.697	1418	1.7791	26.18
0.699	1427	1.7740	26.34
0.701	1388	1.7689	26.49
0.703	1390	1.7639	26.65
0.705	1417	1.7589	26.81
0.707	1402	1.7539	26.97
0.709	1386	1.7489	27.13
0.711	1387	1.7440	27.29
0.713	1375	1.7391	27.45
0.715	1368	1.7343	27.60
0.717	1355	1.7294	27.76
0.719	1329	1.7246	27.91
0.721	1332	1.7198	28.07
0.723	1349	1.7151	28.23
0.725	1351	1.7103	28.38
0.727	1347	1.7056	28.54
0.729	1320	1.7010	28.70
0.731	1327	1.6963	28.85

Wavelength (μm)	Spectral Irradiance (W/[m^2 μm])	Band Gap (eV)	Maximum One Sun J_{sc} (mA/cm^2)
0.733	1319	1.6917	29.01
0.735	1310	1.6871	29.16
0.737	1308	1.6825	29.32
0.739	1279	1.6779	29.47
0.741	1259	1.6734	29.62
0.743	1287	1.6689	29.77
0.745	1280	1.6644	29.93
0.747	1284	1.6600	30.08
0.749	1271	1.6555	30.23
0.751	1263	1.6511	30.39
0.753	1260	1.6467	30.54
0.755	1256	1.6424	30.69
0.757	1249	1.6380	30.84
0.759	1241	1.6337	31.00
0.761	1238	1.6294	31.15
0.763	1242	1.6252	31.30
0.765	1222	1.6209	31.45
0.767	1186	1.6167	31.60
0.769	1204	1.6125	31.74
0.771	1205	1.6083	31.89
0.773	1209	1.6041	32.04
0.775	1189	1.6000	32.19
0.777	1197	1.5959	32.34
0.779	1188	1.5918	32.49
0.781	1188	1.5877	32.64
0.783	1177	1.5837	32.79
0.785	1181	1.5796	32.94
0.787	1178	1.5756	33.09
0.789	1175	1.5716	33.24
0.791	1159	1.5676	33.38
0.793	1144	1.5637	33.53
0.795	1135	1.5597	33.68
0.797	1153	1.5558	33.82
0.799	1136	1.5519	33.97
0.801	1143	1.5481	34.12
0.803	1130	1.5442	34.26
0.805	1116	1.5404	34.41
0.807	1121	1.5366	34.55
0.809	1096	1.5328	34.69
0.811	1115	1.5290	34.84
0.813	1116	1.5252	34.99
0.815	1108	1.5215	35.13
0.817	1105	1.5177	35.27
0.819	1065	1.5140	35.42
0.821	1081	1.5104	35.56

Wavelength (μm)	Spectral Irradiance (W/[m^2 μm])	Band Gap (eV)	Maximum One Sun J_{sc} (mA/cm^2)
0.823	1074	1.5067	35.70
0.825	1076	1.5030	35.84
0.827	1077	1.4994	35.99
0.829	1073	1.4958	36.13
0.831	1069	1.4922	36.27
0.833	1034	1.4886	36.41
0.835	1053	1.4850	36.55
0.837	1052	1.4815	36.70
0.839	1042	1.4779	36.84
0.841	1045	1.4744	36.98
0.843	1028	1.4709	37.12
0.845	1033	1.4675	37.26
0.847	1025	1.4640	37.39
0.849	971	1.4605	37.53
0.851	1003	1.4571	37.66
0.853	973	1.4537	37.79
0.855	877	1.4503	37.92
0.857	1011	1.4469	38.06
0.859	997	1.4435	38.20
0.861	997	1.4402	38.34
0.863	999	1.4368	38.47
0.865	970	1.4335	38.60
0.867	880	1.4302	38.73
0.869	967	1.4269	38.87
0.871	986	1.4237	39.01
0.873	978	1.4204	39.14
0.875	981	1.4171	39.28
0.877	984	1.4139	39.42
0.879	959	1.4107	39.56
0.881	960	1.4075	39.69
0.883	948	1.4043	39.83
0.885	963	1.4011	39.96
0.887	947	1.3980	40.10
0.889	949	1.3948	40.23
0.891	944	1.3917	40.37
0.893	934	1.3886	40.50
0.895	936	1.3855	40.64
0.897	939	1.3824	40.77
0.899	912	1.3793	40.90
0.901	905	1.3762	41.04
0.903	905	1.3732	41.17
0.905	893	1.3702	41.30
0.907	891	1.3671	41.43
0.909	861	1.3641	41.55
0.911	870	1.3611	41.68

Wavelength (μm)	Spectral Irradiance (W/[m^2 μm])	Band Gap (eV)	Maximum One Sun J_{sc} (mA/cm^2)
0.913	876	1.3582	41.81
0.915	866	1.3552	41.93
0.917	859	1.3522	42.06
0.919	858	1.3493	42.19
0.921	830	1.3464	42.31
0.923	821	1.3434	42.43
0.925	825	1.3405	42.55
0.927	828	1.3376	42.68
0.929	833	1.3348	42.80
0.931	826	1.3319	42.93
0.933	832	1.3290	43.05
0.935	818	1.3262	43.17
0.937	802	1.3234	43.30
0.939	808	1.3206	43.42
0.941	800	1.3177	43.54
0.943	784	1.3150	43.66
0.945	799	1.3122	43.78
0.947	793	1.3094	43.90
0.949	777	1.3066	44.02
0.951	778	1.3039	44.14
0.953	771	1.3012	44.25
0.955	760	1.2984	44.37
0.957	774	1.2957	44.49
0.959	771	1.2930	44.61
0.961	767	1.2903	44.73
0.963	767	1.2876	44.85
0.965	764	1.2850	44.97
0.967	757	1.2823	45.09
0.969	776	1.2797	45.21
0.971	763	1.2770	45.33
0.973	764	1.2744	45.44
0.975	750	1.2718	45.56
0.977	768	1.2692	45.68
0.979	768	1.2666	45.80
0.981	762	1.2640	45.93
0.983	766	1.2614	46.05
0.985	771	1.2589	46.17
0.987	756	1.2563	46.29
0.989	767	1.2538	46.41
0.991	764	1.2513	46.53
0.993	755	1.2487	46.65
0.995	756	1.2462	46.77
0.997	743	1.2437	46.89
0.999	743	1.2412	47.10
1.0025	745	1.2369	47.40

Wavelength (μm)	Spectral Irradiance (W/[m^2 μm])	Band Gap (eV)	Maximum One Sun J_{sc} (mA/cm^2)
1.0075	737	1.2308	47.70
1.0125	734	1.2247	48.00
1.0175	721	1.2187	48.29
1.0225	704	1.2127	48.58
1.0275	708	1.2068	48.87
1.0325	688	1.2010	49.16
1.0375	692	1.1952	49.45
1.0425	681	1.1894	49.73
1.0475	685	1.1838	50.02
1.0525	661	1.1781	50.30
1.0575	650	1.1726	50.57
1.0625	642	1.1671	50.85
1.0675	643	1.1616	51.12
1.0725	638	1.1562	51.40
1.0775	630	1.1508	51.67
1.0825	620	1.1455	51.94
1.0875	614	1.1402	52.21
1.0925	612	1.1350	52.48
1.0975	599	1.1298	52.74
1.1025	608	1.1247	53.01
1.1075	601	1.1196	53.28
1.1125	603	1.1146	53.55
1.1175	589	1.1096	53.28
1.1225	579	1.1047	54.07
1.1275	569	1.0998	54.33
1.1325	566	1.0949	54.59
1.1375	563	1.0901	54.84
1.1425	557	1.0853	55.10
1.1475	556	1.0806	55.36
1.1525	545	1.0759	55.61
1.1575	554	1.0713	55.87
1.1625	540	1.0667	56.12
1.1675	530	1.0621	56.37
1.1725	533	1.0576	56.62
1.1775	525	1.0531	56.86
1.1825	514	1.0486	57.11
1.1875	512	1.0442	57.35
1.1925	511	1.0398	57.60
1.1975	502	1.0355	57.84
1.2025	496	1.0312	58.08
1.2075	494	1.0269	58.32
1.2125	489	1.0227	58.56
1.2175	500	1.0185	58.80
1.2225	481	1.0143	59.04
1.2275	481	1.0102	59.28

Wavelength (μm)	Spectral Irradiance (W/[m^2 μm])	Band Gap (eV)	Maximum One Sun J_{sc} (mA/cm^2)
1.2325	484	1.0061	59.52
1.2375	477	1.0020	59.75
1.2425	477	0.9980	59.99
1.2475	466	0.9940	60.23
1.2525	474	0.9900	60.46
1.2575	463	0.9861	60.69
1.2625	444	0.9822	60.92
1.2675	438	0.9783	61.14
1.2725	439	0.9745	61.37
1.2775	453	0.9706	61.60
1.2825	435	0.9669	61.82
1.2875	437	0.9631	62.05
1.2925	442	0.9594	62.28
1.2975	438	0.9557	62.51
1.3025	438	0.9520	62.74
1.3075	429	0.9484	62.96
1.3125	419	0.9448	63.18
1.3175	416	0.9412	63.40
1.3225	416	0.9376	63.63
1.3275	411	0.9341	63.84
1.3325	405	0.9306	64.06
1.3375	400	0.9271	64.28
1.3425	398	0.9236	64.49
1.3475	394	0.9202	64.70
1.3525	387	0.9168	64.91
1.3575	382	0.9134	65.12
1.3625	378	0.9101	65.33
1.3675	370	0.9068	65.53
1.3725	369	0.9035	65.73
1.3775	368	0.9002	65.94
1.3825	364	0.8969	66.14
1.3875	364	0.8937	66.34
1.3925	358	0.8905	66.54
1.3975	357	0.8873	66.74
1.4025	353	0.8841	66.94
1.4075	350	0.8810	67.14
1.4125	346	0.8779	67.33
1.4175	344	0.8748	67.53
1.4225	343	0.8717	67.73
1.4275	348	0.8687	67.93
1.4325	337	0.8656	68.12
1.4375	331	0.8626	68.31
1.4425	327	0.8596	68.50
1.4475	318	0.8566	68.68
1.4525	323	0.8537	68.87

Wavelength (μm)	Spectral Irradiance (W/[m^2 μm])	Band Gap (eV)	Maximum One Sun J_{sc} (mA/cm^2)
1.4575	307	0.8508	69.05
1.4625	317	0.8479	69.24
1.4675	311	0.8450	69.42
1.4725	311	0.8421	69.60
1.4775	307	0.8393	69.79
1.4825	303	0.8364	69.97
1.4875	298	0.8336	70.15
1.4925	303	0.8308	70.33
1.4975	300	0.8280	70.51
1.5025	296	0.8253	70.69
1.5075	295	0.8226	70.86
1.5125	290	0.8198	71.04
1.5175	290	0.8171	71.22
1.5225	286	0.8144	71.39
1.5275	290	0.8118	71.57
1.5325	282	0.8091	71.74
1.5375	274	0.8065	71.91
1.5425	275	0.8039	72.08
1.5475	274	0.8013	72.25
1.5525	273	0.7987	72.42
1.5575	272	0.7961	72.59
1.5625	269	0.7936	72.76
1.5675	263	0.7911	72.93
1.5725	260	0.7886	73.09
1.5775	259	0.7861	73.25
1.5825	255	0.7836	73.42
1.5875	252	0.7811	73.58
1.5925	246	0.7786	73.73
1.5975	246	0.7762	73.89
1.6025	247	0.7738	74.05
1.6075	242	0.7714	74.21
1.6125	244	0.7690	74.37
1.6175	243	0.7666	74.52
1.6225	240	0.7643	74.68
1.6275	244	0.7619	74.84
1.6325	241	0.7596	75.00
1.6375	237	0.7573	75.15
1.6425	234	0.7549	75.31
1.6475	235	0.7527	75.46
1.6525	234	0.7504	75.62
1.6575	234	0.7481	75.78
1.6625	233	0.7459	75.93
1.6675	229	0.7436	76.09
1.6725	228	0.7414	76.24
1.6775	220	0.7392	76.39

Wavelength (μm)	Spectral Irradiance (W/[m^2 μm])	Band Gap (eV)	Maximum One Sun J_{sc} (mA/cm^2)
1.6825	221	0.7370	76.53
1.6875	219	0.7348	76.68
1.6925	219	0.7326	76.83
1.6975	214	0.7305	76.98
1.7025	217	0.7283	77.13
1.7075	212	0.7262	77.27
1.7125	203	0.7241	77.41
1.7175	212	0.7220	77.56
1.7225	205	0.7199	77.70
1.7275	196	0.7178	77.83
1.7325	190	0.7157	77.96
1.7375	189	0.7137	78.10
1.7425	191	0.7116	78.23
1.7475	185	0.7096	78.36
1.7525	187	0.7076	78.49
1.7575	189	0.7055	78.62
1.7625	184	0.7035	78.75
1.7675	182	0.7016	78.88
1.7725	177	0.6996	79.01
1.7775	173	0.6976	79.13
1.7825	171	0.6957	79.25
1.7875	170	0.6937	79.37
1.7925	169	0.6918	79.50
1.7975	173	0.6898	79.62
1.8025	169	0.6879	79.74
1.8075	168	0.6860	79.86
1.8125	160	0.6841	79.98
1.8175	160	0.6823	80.10
1.8225	159	0.6804	80.21
1.8275	156	0.6785	80.33
1.8325	156	0.6767	80.44
1.8375	150	0.6748	80.55
1.8425	153	0.6730	80.67
1.8475	151	0.6712	80.78
1.8525	148	0.6694	80.89
1.8575	145	0.6676	80.99
1.8625	143	0.6658	81.10
1.8675	143	0.6640	81.21
1.8725	135	0.6622	81.31
1.8775	135	0.6605	81.41
1.8825	140	0.6587	81.52
1.8875	138	0.6570	81.62
1.8925	137	0.6552	81.73
1.8975	138	0.6535	81.83
1.9025	133	0.6518	81.93
1.9075	136	0.6501	82.04

Wavelength (μm)	Spectral Irradiance (W/[m^2 μm])	Band Gap (eV)	Maximum One Sun J_{sc} (mA/cm^2)
1.9125	138	0.6484	82.14
1.9175	136	0.6467	82.25
1.9225	134	0.6450	82.35
1.9275	132	0.6433	82.45
1.9325	132	0.6417	82.56
1.9375	131	0.6400	82.66
1.9425	129	0.6384	82.76
1.9475	127	0.6367	82.86
1.9525	126	0.6351	82.96
1.9575	122	0.6335	83.05
1.9625	126	0.6318	83.15
1.9675	125	0.6302	83.25
1.9725	125	0.6286	83.35
1.9775	129	0.6271	83.45
1.9825	125	0.6255	83.55
1.9875	123	0.6239	83.65
1.9925	121	0.6223	83.75
1.9975	123	0.6208	83.84
2.0025	116	0.6192	84.03
2.0125	114	0.6161	84.22
2.0225	113	0.6131	84.40
2.0325	110	0.6101	84.57
2.0425	107	0.6071	84.75
2.0525	104	0.6041	84.92
2.0625	100	0.6012	85.17
2.0775	101	0.5969	85.42
2.0925	98	0.5926	85.66
2.1075	93	0.5884	85.89
2.1225	87	0.5842	86.11
2.1375	85	0.5801	86.33
2.1525	81	0.5761	86.54
2.1675	80	0.5721	86.74
2.1825	75	0.5682	86.94
2.1975	73	0.5643	87.13
2.2125	75	0.5605	87.33
2.2275	75	0.5567	87.60
2.2475	72	0.5517	87.79
2.2625	71	0.5481	88.05
2.2825	69	0.5433	88.30
2.3025	66	0.5385	88.52
2.3225	53	0.5339	88.73
2.3425	58	0.5293	88.96
2.3625	65	0.5249	89.19
2.3825	55	0.5205	89.40
2.4025	54	0.5161	89.62
2.4225	57	0.5119	89.83

Wavelength (μm)	Spectral Irradiance (W/[m^2 μm])	Band Gap (eV)	Maximum One Sun J_{sc} (mA/cm^2)
2.4425	51	0.5077	90.09
2.4675	53	0.5025	90.35
2.4925	54	0.4975	90.61
2.5175	47	0.4926	90.84
2.5425	46	0.4877	91.08
2.5675	44	0.4830	91.30
2.5925	42	0.4783	91.52
2.6175	41	0.4737	91.73
2.6425	39	0.4693	91.97
2.6725	38	0.4640	92.22
2.7025	36	0.4588	92.45
2.7325	35	0.4538	92.68
2.7625	34	0.4489	92.94
2.7975	32	0.4433	93.19
2.8325	31	0.4378	93.43
2.8675	29	0.4324	93.69
2.9075	28	0.4265	93.95
2.9475	26	0.4207	94.19
2.9875	25	0.4151	94.41
3.025	24	0.4099	94.70
3.075	23	0.4033	94.97
3.125	21	0.3968	95.23
3.175	20	0.3906	95.54
3.235	19	0.3833	95.83
3.295	18	0.3763	96.10
3.355	16	0.3696	96.40
3.425	15	0.3620	96.68
3.495	14	0.3548	96.99
3.575	13	0.3469	97.32
3.665	12	0.3383	97.62
3.755	11	0.3302	97.95
3.855	10	0.3217	98.27
3.965	9	0.3127	98.61
4.085	8	0.3035	98.96
4.225	7	0.2935	99.32
4.385	6	0.2828	99.69
4.575	5	0.2710	100.08
4.805	4	0.2581	100.47
5.085	3	0.2439	100.86
5.445	2	0.2277	101.30
5.925	2	0.2093	101.82
6.615	1	0.1875	102.50
7.785	1	0.1593	103.32
10.075	0	0.1231	
	Total 1,367 W/m^2		

Index